INTERACTION OF ATOMIC PARTICLES WITH A SOLID SURFACE

VZAIMODEISTVIE ATOMNYKH CHASTITS S POVERKHNOST'YU TVERDOGO TELA

ВЗАИМОДЕЙСТВИЕ АТОМНЫХ ЧАСТИЦ С ПОВЕРХНОСТЬЮ ТВЕРДОГО ТЕЛА

INTERACTION OF ATOMIC PARTICLES WITH A SOLID SURFACE

Ubai A. Arifov

Director, Institute of Nuclear Physics
Academy of Sciences of the Uzbek SSR
Tashkent, USSR

Translated from Russian by
G. D. Archard

 Springer Science+Business Media, LLC 1969

Ubai Arifovich Arifov was born in 1909 in the city of Kikande in the Uzbek SSR. In 1931 he received his first degree from the Physico-Mathematical Faculty of the Pedagogical Academy in Samarkand. He was for several years a member of the Physics Faculty of the Cotton Institute of Kikande, and later of the Physics Department of the Physico-Mathematical Faculty of the Mid-Asian State University in Tashkent. After serving as an officer in the Red Army during the Second World War, he defended his candidate's dissertation in 1945 and was made Director of the Physico-Technical Institute of the Academy of Sciences of the Uzbek SSR. Arifov defended his doctoral dissertation in 1954 and was made a professor in 1955. In 1956 he was elected to the Academy of Sciences of the Uzbek SSR, became a member of the Presidium, and Chairman of the Department of Physico-Mathematical Sciences. In the same year he was made Director of the Institute of Nuclear Physics of the Academy of Sciences of the Uzbek SSR, and from 1962 to 1966 he served as President of the Academy of Sciences of the Uzbek SSR. From 1962 to 1966 he was also a Deputy to the Supreme Soviet of the USSR.

Since 1945 Arifov has headed the Department of Electronics in the Physical Institutes of the Academy of Sciences of the Uzbek SSR, and since 1967 he has worked as Director of the Institute of Electronics. Since 1963 he has headed the Department of Physical Electronics of the Physical-Engineering Faculty of Tashkent Polytechnic Institute. Since 1965 he has been Editor-in-Chief of the journal "Geliotekhnika." An English translation of Arifov's earlier work, "Interaction of Atomic Particles with a Metal Surface," was published in 1963.

Library of Congress Catalog Card Number 79-76223

The original Russian text, published by Nauka Press in Moscow in 1968, has been corrected by the author for this edition.

Арифов Убай Арифович

ВЗАИМОДЕЙСТВИЕ АТОМНЫХ ЧАСТИЦ
С ПОВЕРХНОСТЬЮ ТВЕРДОГО ТЕЛА

© 1969 Springer Science+Business Media New York
Originally published by Consultants Bureau, New York in 1969.
Softcover reprint of the hardcover 1st edition 1969

ISBN 978-1-4899-4811-3   ISBN 978-1-4899-4809-0 (eBook)
DOI 10.1007/978-1-4899-4809-0

FOREWORD TO THE AMERICAN EDITION

Some four years ago I was extremely pleased to learn from my American colleagues, at a conference in Belgrade, that my book "Interaction of Atomic Particles with a Metal Surface," published in 1961, had been translated into English.

In the book now under consideration, which constitutes a continuation and an extension of the former monograph, I have included the results of many Soviet and American colleagues as well as those of scientists in other countries working in this field of physics. However, my intention was not to provide a detailed review of all the literature but to set out the results of work carried out by my own colleagues and myself over the last twenty years in the Academy of Sciences of the Uzbek SSR. At the end of this edition I have appended a brief review of our research in 1967-1968, i.e., work carried out since the text of the Russian edition was prepared.

I am happy to note that the activities of Uzbek scientists in this field are attracting the attention of researchers in the USA and other countries of the world.

I shall be very pleased if English-speaking readers find some useful ideas and information in this book. I hope that this edition will promote a fuller exchange of scientific information between Soviet and American physicists and the ultimate benefit of the field of physical science to which I and many of my colleagues have devoted our lives.

CONTENTS

FOREWORD TO THE 1968 RUSSIAN EDITION

In the last five or ten years, interest in the phenomena accompanying the bombardment of solid surfaces by atomic particles (ions, electrons, atoms, and molecules) has been increasing continuously. Physical processes earlier only studied in connection with the operation of ion and electron devices came under the scope of such important branches of science and technology as plasma physics and magnetohydrodynamic motors, the physics of thermionic converters, photo- and thermoelectric converters and gas lasers, the supersonic motion of solids in the ionosphere and the erosion of the moon's surface, the ionic doping of semiconductors, and radiation physics. Investigations into the effect of ion beams on single crystals are developing vigorously. Secondary ion—electron and electron—electron emission are now widely used for measuring very small currents in mass spectrometry; there have been attempts to use these processes for obtaining images of surfaces. Ion bombardment is used for ionic etching, for obtaining thin films by cathodic sputtering, etc.

In recent years the flow of publications has increased continuously and newer and newer spheres in which the bombardment of atomic particles plays a dominant part have been discovered. It may well be asserted that the study of interactions between atomic particles and solid surfaces has taken on a leading significance in solid-state physics, nuclear physics, electronic technology, ionic—electronic technology, cosmonautics, and the power of the future. This leads us to hope that our many years of research in this field may be of interest to readers specializing in a wide range of subjects.

In 1961 we published our monograph "The Interaction of Atomic Particles with a Metal Surface" (Tashkent, published by the Academy of Sciences of the Uzbek SSR). The book now laid before the reader is based on the former monograph, the material of which has been largely reworked and expanded, chiefly by way of a review of our own research and that of other authors between 1961 and 1967. In addition to this, the book includes questions on the study of nonmetallic surfaces, as reflected by the title of the present treatment.

We thought it desirable to reproduce the Foreword to the 1961 edition so as to emphasize the continuity of the two books.

As a result of investigations carried out during the last six years, a great deal of new information has been secured, and this has made it necessary to redistribute the material, to extend some chapters, and to add others on the interaction of slow ions with a solid surface, the angular distribution of secondary ions, secondary processes in semiconducting and dielectric films, cathodic sputtering, and electron—electron emission. The chapter devoted to theoretical questions has also been rewritten. The whole book has been considerably revised from the point of view of the very latest ideas regarding the nature of the phenomena treated.

As in the earlier publication, the author was not concerned with producing a teaching manual or an exhaustive handbook; his aim was to acquaint the reader with the work carried out during the past 20 years in Tashkent in the Academy of Sciences of the Uzbek SSR. The

1

new version of the well-known book of Professor L. N. Dobretsov [233] and the recently published book by Kaminsky [730] may serve as handbooks for the reader studying the subject. The author has tried to pay special attention to questions not treated in these books.

The author finds it his pleasant duty once again to proffer his sincere thanks to colleagues who took part in preparing both the earlier book and this one, and also to A. A. Aliev, D. D. Gruich, A. Kh. Kasymov, A. I. Kulagin, and V. A. Shustrov for help in the work on the present edition.

The author wishes to salute the memory of his friend and mentor, Academician S. V. Starodubtsev, one of the originators of investigations into the interactions of atomic particles with solid surfaces, in cooperation with whom the earliest of the investigations described in this book were carried out.

FOREWORD TO THE 1961 RUSSIAN EDITION

Although the collisions of atoms and ions with metal surfaces have been studied for a long time, their investigation has not only still not lost its current importance, but is also arousing great interest in connection with the development of new fields of physics and technology. The study of these phenomena, which has been conducted in recent years, has revealed a whole series of complex and strictly interrelated processes where hitherto only isolated facts had been known, creating a simplified and partly inaccurate picture. Despite the large number of works published in recent years, there is still no consistent analysis of the results obtained, nor any systematic exposition in a single book. Individual sections and chapters in various monographs devoted to this question fail to provide a complete presentation of the processes associated with the interaction of atomic particles and solid surfaces.

This book has been written with the aim of at least partly eliminating this lack and giving a short chronological review of investigations in this field. It does not pretend to give an exhaustive description of all such processes. The book includes materials relating to the interaction of atomic particles with metallic surfaces only (pure or coated with films of adsorbed atoms). We consider the phenomena of ion—electron, ion—ion, neutral atom—ion, and neutral atom—electron emission, and also the neutralization of ions at a metallic surface. The investigations embrace an energy range extending from a few eV to thousands of electron volts.

Questions associated with cathodic sputtering, electromagnetic radiation, and the adsorption and desorption of atoms are not considered in this book, as they require treatment all of their own. Certain other processes, for example surface ionization, are only described as an illustration of the application of the double-modulation method.

Special attention is devoted to expounding the double-modulation oscillographic method developed for the rapid, inertia-free study of various secondary and subsidiary processes accompanying the collision of an atomic particle with the surface of a metal. The hope that this method may prove useful in other fields of research has inspired us to collect, systematize, and generalize the results of theoretical and experimental investigations carried out by the staff of the Electronics Department with the participation and under the direction of the author over the last 15 years in Tashkent, in the Physicotechnical Institute and in the Institute of Nuclear Physics of the Academy of Sciences of the Uzbek SSR.

The author considers it his pleasant duty to offer sincere thanks to Academician S. V. Starodubtsev, who took upon himself the labor of editing the book, and also Directors of Laboratories in the Electronics Department of the Institute of Nuclear Physics of the Academy of Sciences of the Uzbek SSR A. Kh. Ayukhanov and R. R. Rakhimov, who read the manuscript and made a number of valuable comments. The author is also indebted to the Director of the Theoretical Section of the Department, E. S. Parilis, and Senior Research Fellow L. G. Gurvich, who wrote the theoretical tenth chapter, read the manuscript, and commented on the theory of the matter.

The author also expresses his thanks to his colleagues Kh. Kh. Khadzhimukhamedov, L. M. Kishinevskii, and P. U. Arifov for help in preparing the manuscript for publication, and also to the whole staff of the Electronics Department for the many years work which have preceded the publication of this book.

NOTATION

a (T) — Probability of evaporation as an atom or ion

A^* — Ratio of the statistical weights of an atom and ion on the surface of a metal

d_p — Depth of penetration of ions into a metal

d_{ee} — Depth of the effective generation of secondary electrons in a metal on electron bombardment

d_{ie} — Depth of the effective generation of secondary electrons in a metal on ion bombardment

E_0 — Energy of incident (primary) particles (atoms, ions, or electrons)

E — Energy of a once elastically scattered particle (atom, ion, or electron)

E_{max} — Maximum energy of a multiply scattered particle (neutral atom or ion)

E_p — Energy transferred in the inelastic collision of atoms

E_{rec} — Energy transferred in the elastic collision of atoms (rec = recoil)

E_e — Effective energy of an atom

E_{min} — Minimum energy of elastically scattered particles (atom or ion)

E_g — Minimum energy of an incident atom or ion for which kinetic emission of electrons takes place

$\mathscr{E}$ — Energy of the relative motion of colliding atoms

I_0 — Current of incident particles (atoms, ions, or electrons)

I — Current of secondary particles (ions or electrons)

I_d — Current of diffusion ions

I_e — Current of evaporated ions

I_s — Current of scattered ions

I_t — Current of thermionic electrons

I_Σ — Total current of secondary positive ions

j_0 — Density of the flow of incident particles

K — Coefficient of secondary ion—ion emission

K_d — Emission coefficient of diffusion ions

K_e — Emission coefficient of evaporated ions

K_{sl} — Coefficient of slow secondary ions

K_s — Coefficient of elastically scattered ions

K_s^0 — Coefficient of neutral-ion emission

m_1 — Mass of a target atom

m_2 — Mass of incident atoms or ions

m — Reduced mass of colliding atoms

m_{eff} — Effective mass of a target atom

m — Mass of an electron

N — Coefficient of cathodic sputtering

r — Coefficient of elastically scattered electrons

t — Time

T –   Temperature

U_0 –   Accelerating voltage

v_0 –   Velocity of incident particle (atom, ion, or electron)

v –   Velocity of scattered particles (atom, ion, or electron)

v_g –   Minimum velocity if an incident atom or ion for which kinetic emission of electrons takes place

v_i –   Ionization potential of an atom

Z_1 –   Charge on the nucleus of a target atom

Z_2 –   Charge on the nucleus of an incident particle (ion or atom)

α –   Degree of ionization

α –   Sum of the angle of incidence of the primary particle and the escape angle of the secondary particle

β –   Coefficient of a positive ionization surface

β –   Scattering angle of incident particles

γ –   Coefficient of secondary ion—electron emission

γ_k –   Coefficient of kinetic electron emission

γ_p –   Coefficient of potential electron emission

γ_m –   Coefficient of potential electron emission under the influence of metastable atoms

δ –   Coefficient of true secondary electron—electron emission

η –   Ratio of the energy of single elastic scattering of an ion to the energy of the incident ion

θ –   Escape angle of secondary particles

θ –   Degree of covering of a metal surface by adsorbed atoms

ϑ –   Scattering angle in the center-of-mass system

$\varkappa$ –   Ion-beam accommodation coefficient

λ –   Range of electrons in a metal

λ_c –   Particle condensation energy

λ_+ –   Heat of evaporation of an ion

λ_0 –   Heat of evaporation of an atom

σ –   Coefficient of secondary electron—electron emission

τ_a –   Life of an adatom (more exactly an atom connected to a surface)

τ_d –   Life of a diffusion ion

τ_e –   Life of an evaporated ion

φ° –   Incident angle of incident particles

φ –   Electron work function

INTRODUCTION

When a flow of atomic particles interacts with a solid surface a number of complex phenomena take place simultaneously. In the early stage of the investigation it was considered that the bombarding atomic particle caused the emission of secondary ions, neutral atoms (cathodic sputtering), and electrons. This simplified picture of the phenomena is illustrated in Fig. 1.

In actual fact, as later work showed, the picture of the secondary processes taking place when atomic particles interact with a solid surface is far more complex both in respect of the variety of phenomena and in respect of the type of secondary particles. The secondary processes and types of particles are indicated in Fig. 2.

A bombarding particle (a positive or negative ion, an electron, or a neutral atom 1), falling on the surface of the target 2, may experience elastic or inelastic scattering in the form of a positive 3 or negative 5 ion or a neutral particle 4. Scattering may thus involve a change in the charge on the particle.

Depending on the conditions of its first collision with the target atoms, the bombarding particle may not only be back-scattered but also penetrate into the depths of the target 6 or be adsorbed on the surface 7. A penetrating atom, as a result of many collisions, gradually loses its energy and starts participating in the thermal motion of the target atoms. Some of these atoms 8 may come out to the surface 7 as a result of diffusion. These form a layer of adsorbed atoms 7 on the surface. Later, as a result of thermal motion, the atoms migrate (7 → 9) and at high temperatures evaporate from the surface of the metal, leaving it as positive ions 10, negative ions 12, and neutral atoms 11.

At the same time as the bombarding particles interact with the target atoms, atoms of target material are ejected (sputtered) in the form of positive 14 (including multiply charged) and negative 16 ions as well as neutron atoms (including metastable atoms 15); electrons 20 may also be ejected. If there are any adsorbed atoms 7 on the surface, these may also be ejected as a result of the impact of the primary beam 2.

In addition to this, at high temperatures there may be a thermal emission of target particles in the form of positive 10 and negative 12 ions and neutral atoms 11, and also the emission of thermoelectrons (thermionic emission) 17. We must also add the emission of electromagnetic radiation (2 → 18) and secondary electrons (2 → 19) at the moment of impact, both from the bombarding atoms 2 and the target atoms 13 and adsorbed atoms ("adatoms") 7.

Quantitatively, each phenomenon is characterized by a corresponding coefficient of secondary emission: the number of secondary particles associated with one incident primary particle. For example, the coefficient of secondary ion—electron emission is

$$\gamma = \frac{I^-}{I_0}.$$

$$\tag{1}$$

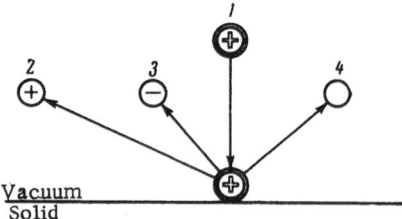

Fig. 1. Simplified diagram of secondary processes taking place on bombarding a solid with positive ions. 1) Bombarding particle; 2) secondary reflected ions; 3) secondary ejected electrons; 4) sputtered particles.

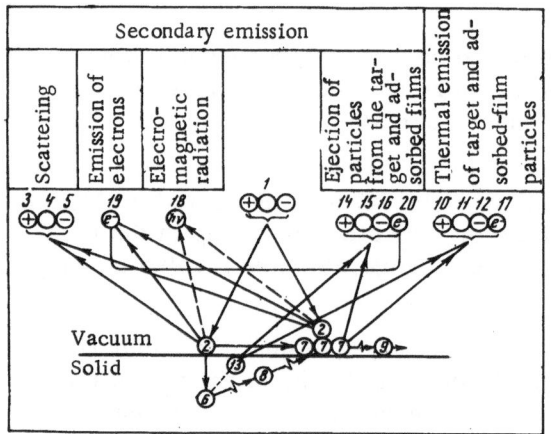

Fig. 2. Modern picture of secondary processes taking place on the interaction of atomic particles with a metal surface.

The coefficient of secondary ion—ion emission is

$$K = \frac{I}{I_0}, \qquad (2)$$

where I^- is the current of secondary electrons, I_0 is the current of positive primary ions, and I is the current of secondary ions.

The coefficients of the other forms of emission are defined analogously.

Usually one determines the values of the coefficients γ, K, and their dependence on mass, energy, ionization potential, primary particle incident angle, the nature and state of the surface, and also the target temperature. In addition to this, one studies the composition, charged state, and angular and energy distributions of the flux of secondary particles. One also studies the energy thresholds of the secondary phenomena.

In order to understand the problems which arise when studying these processes and the manner in which this field of physics is now being developed, let us briefly consider the history of the development and perfection of experimental methods and of the deepening and broadening of our views on the nature of the phenomena in question.

A large number of papers have been devoted to the study of these complicated phenomena, which for the sake of brevity we shall subsequently call secondary processes. All experimental investigations of secondary processes use apparatus consisting of a target, bombarded by atomic particles, and surrounding the target a collector which, depending on the magnitude and sign of the voltage imposed between target and collector, collects secondary particles of one sign or the other; in order to protect it from the direct action of the primary particles, the collector is placed in a guard cylinder. Sometimes apparatus for analyzing the composition and energy of the secondary particles is also included.

Another component of the apparatus is the source of atomic particles. In the majority of investigations research workers have used ions obtained in a glow or arc discharge taking place in an atmosphere of inert or ordinary gases, and also ions of the alkali metals obtained by heating compounds of these metals or by surface ionization.

Fast neutral gas atoms are obtained by resonance charge exchange or the recharging of ions on solid surfaces.

Secondary processes began to be studied at the end of the 19th century. Despite the fact that the conditions of these experiments were far from ideal, the results obtained are of great interest even in modern times.

Let us briefly consider certain experiments relating to the earliest period in the study of secondary processes, when the experimental conditions were still inadequately controlled, but when the investigations were nevertheless able to give a fair idea of the principal laws governing secondary processes.

The secondary electron emission of metals bombarded by positive ions was first noted in 1899 by Villard [4], who explained the formation of cathode rays in a discharge tube as being due to the ejection of electrons by positive ions from the metal serving as cathode. Similar emission which appeared on bombarding metals with α-particles was observed in 1905 by Rutherford [8], Thomson [9], and in 1907 by Logeman [10]. These authors regarded the secondary ion—electron emission of metals as a factor interfering with the study of the fundamental phenomenon. For example, in determining the charge on the particles emitted by Ra and Po, Rutherford found that the secondary particles emitted by the bombarded target, which prevented him from accurately determining the charge on the bombarding particles, were electrons.

The true study of the secondary processes was started by Campbell [13], who in 1915 was the first to study the γ and K factors for a nondegassed target on bombarding the latter with Na^+ and Al^+ ions having an initial energy E_0 of up to 50 keV. Campbell found that γ increased with rising E_0 up to 38 keV and fell for higher energies ($E_0 > 40$ keV); he explained the fall in γ with increasing primary ion energy as being due to the penetration of the bombarding ions into the lower regions of the target. The factor K was found to be independent of the energy of the primary ions.

In 1917, Cheney [15] studied the effect of the nature of the bombarding ion and the target material on the value of γ. In these experiments Pt and Al targets were bombarded with Li^+, K^+, and Rb^+ ions obtained by roasting the sulfides of these elements. Cheney established the effect of the nature of the ion and target on the value of γ; thus, according to his results, $\gamma_{Al} > \gamma_{Pt}$. For the same primary ion energies, Li^+ was most and Rb^+ least efficient in ejecting electrons from both Pt and Al targets. Cheney also showed that γ rose with increasing contamination of the target surface.

In 1925, Klein [32] directed his attention to these phenomena and studied the γ and K factors and the energy distribution of the secondary particles from a Ni target on bombarding with Na^+ and K^+ ions obtained by heating aluminum phosphate. The volt—ampere characteristics of the secondary currents were obtained for target temperatures of 430-440°K. It was found that γ increased with rising E_0 and saturated for $E_0 > 380$ eV. The maximum energy of the secondary electrons was never more than 2 eV. On studying the energies of the secondary ions, two groups of these were found: one with an energy of up to 2 eV and the other with an energy close to that of the primary ions.

Klein first indicated the existence of a group of fast secondary ions with energies close to those of the primary ions. The existence of fast ions in secondary emission failed for a long time to find confirmation in the results of other authors, and the very existence of these ions was regarded as doubtful.

In 1926 and 1927, Jackson [36, 44] studied the effect of the nature of the colliding particles and the energy of the bombarding ions on the values of γ and K. Jackson bombarded Mo, Al, and Ni targets (degassed or otherwise) with K^+, Na^+, Rb^+, and Cs^+ ions from a Kunsmann [35] mixture consisting of compounds of the alkali metals in iron oxide crystallites. The energy of the ions could be varied over a range up to 1000 eV. Jackson found a considerable fall in γ on heating the target, or a rise in the same factor on increasing the numbers of adsorbed atoms and the energy of the primary ions.

Studying the effect of the target material on the value of γ, Jackson obtained a relation in the form $\gamma_{Al} > \gamma_{Ni} > \gamma_{Mo}$, explaining this as being due to an inverse dependence of γ on the work function of these surfaces, i.e., $\varphi_{Al} < \varphi_{Ni} < \varphi_{Mo}$. Studying the effect of the nature of bombarding ions on γ and K, Jackson found that the most efficient emission was due to the K^+ ions, while Na^+, Rb^+, and Cs^+ had a much weaker effect. According to Jackson's data, the maximum energy of the secondary electrons for all the targets studied was no greater than fractions of a volt. The coefficient K depended on the nature of the ions and target material, but in every case it was under 4% and independent of the energy of the primary ions.

It should be mentioned that in all these investigations the vacuum was rather low, the targets were not freed from contaminations, nor the ion beams from the presence of natural particles. Hence the quantitative results of the foregoing authors must be regarded with some caution, all the more so in view of the fact that their results were partly contradictory, even in fundamental conclusions. However, these investigations, although carried out under inadequately pure conditions, nevertheless led to the important qualitative conclusion that the secondary phenomena were extremely sensitive to a large number of different factors: the energy and nature of the primary ions, and the nature and state of the bombarded target surface.

In subsequent investigations great attention has been paid to the creation of clean conditions on the target surface and to studying the effect of various contaminants. At the present time the cleanness of the surface under examination is given preeminent attention, since the secondary processes taking place on bombarding solid surfaces depend very greatly on the state of the surface.

CHAPTER 1

STATIC (GALVANOMETRIC) METHOD OF STUDYING PROCESSES OCCURRING DURING THE BOMBARDMENT OF A SOLID BY ATOMIC PARTICLES

§1. Brief Review of Investigations into Secondary Emission under the Influence of Positive Alkali Ions

At the beginning of the thirties, research workers started working with heated, i.e., cleaned targets. Thus, Oliphant [47] studied the effect of gas films adsorbed by the target on γ and K for various primary ion energies. The target was baked by electron bombardment. The adsorbed films on the target surface were created by leaking the corresponding gases into the apparatus. Targets of Mo and Ni were bombarded with K^+ ions from a Kunsmann source [35].

Oliphant found a rise in γ and K as the contaminant films developed on the target surface. The coefficient K for a clean target was about 1%. After the target had been held in air (atmospheric) this value increased to 3%. For targets covered with an adsorbed film there was a rise in γ and a fall in K as the energy of the bombarding particles increased. Studying the dependence of γ on the film thickness, Oliphant found that the emission of electrons reached saturation after a quarter of a molecular layer had been deposited. The maximum energy of the secondary electrons, according to these results, reached 10 eV. It should be noted that the technical aspect of this investigation was inadequate to allow the accurate measurement of film thickness. Hence the author's assertion that γ and K saturated on reaching a quarter of a molecular layer would appear none too well based.

Moon [70] first studied γ at a high target temperature, between 920 and 1270°K. The Ni target was bombarded with K^+ ions having energies up to 5000 eV. The factor γ fell with increasing target temperature and rose with increasing E_0. The author also noted a rise in γ with increasing contamination. This investigation is interesting because it represents the first attempt at studying secondary processes with a high target temperature and makes mention of a large number of scattered ions, although in this most interesting part of the work no serious quantitative results were given.

Moon and Oliphant [72] bombarded heated W with K^+ ions and found that the secondary ion current continued to exist several seconds after switching the primary current off. The authors explained this phenomenon as being due to the penetration of the primary ions into the depths of the target and their subsequent thermal evaporation.

Using a special apparatus for multiple acceleration, Linford [95] studied the secondary emission of electrons for various metal surfaces bombarded with mercury ions having energies up to 1.3 MeV. The total emission from Ni, Mo, Al, Cu, W, Sn, Hg, Ag, Au, and C was between 8 and 35 electrons per ion, depending on the conditions on the metal surface. Linford came to the conclusion that for a primary ion energy of over 0.3 MeV the secondary emission was almost independent of the nature of the target and the energy of the bombarding ions. The energy of the secondary electrons in this case lay between 0 and 15 eV.

Pavlov and Dobrolyubskii [93] determined γ and K for Cu and Pt targets bombarded with Li$^+$ ions having an energy of up to 100 eV emitted by the mineral spodumene [115] in the heated state. The Cu and Pt targets were studied at red heat and afterward in the cold state.

The authors noted that the energy of the secondary electrons was small compared with those of the primary ions. In addition to this there was an indication in this paper that the absolute values of γ were always greater than the absolute values of K. The value of γ fell with increasing target temperature and both γ and K rose with increasing primary ion energy. The maximum energy of the secondary particles of both signs, according to these authors, was no greater than 10 eV.

Thus Pavlov and Dobrolyubskii's data [93] do not confirm the results of other investigators as to the existence of fast secondary ions, but they do confirm the fall in γ with increasing target temperature. The work was carried out under conditions which permitted the target to become contaminated with a beam of molecules from the heated spodumene.

Koch [105] studied the dependence of γ and K on the degree of coverage θ after bombarding a cold W target with Cs$^+$ ions, using an ion beam of energy 200–1000 eV free from neutral atoms, obtained by separation in a cylindrical condenser. The target was first degassed at 1550°K. However, the measurements themselves were made at room temperature. The author obtained curves giving γ and K as functions of the thickness of the coating. On increasing the thickness of the coating γ gradually rose to a maximum value of 5%, while K fell to 12%; thus, an equilibrium value was achieved for γ and K, and this underwent no further change, despite the continued application of the ion beam. According to the author's calculations, saturation set in for $\theta = 0.1$ of a monatomic coating. Koch also noted that the value of K for a hot target was greater than for a cold one.

Pavlov and Starodubtsev [115] carefully studied a number of problems associated with secondary processes for a large number of different materials. These research workers studied the influence on γ and K of the target material, its properties (particularly electrical and thermal conductivity), the state of the surface, the work function, the temperature, the atomic volume, and the energy of the incident ions; they studied the energy distribution and maximum energy of the secondary particles of both signs. Using Li$^+$ ions obtained by heating spodumene, the authors bombarded targets of Pt, Cu, Te, Sn, Al, and Zn, and also the oxides CuO, Cu$_2$O, and ZnO and the salt NaCl. The energies of the primary ions varied between 100 and 1000 eV.

The authors found that the value of γ could not be correlated with any of the physical parameters of the target under consideration. The absolute values of the coefficients γ obtained from these experiments also failed to agree with those of other authors. The values were much (ten times) greater than those of Jackson [36, 44], although the character of the curves was similar to that of the analogous curves obtained by Campbell [13], Jackson [36], and Moon [70]. For this reason, Pavlov and Starodubtsev came to the conclusion that all the deviations between the results of different research workers and between different observations of the same author were due to varying conditions on the surface of the targets studied, i.e., to the impossibility of physically determining and creating specified conditions on the surface. These authors therefore considered that the only data which could be considered as absolutely definite and final were those relating to the coefficient γ for a specific target, the work function being measured at the same time; they also indicated that as an independent factor the target temperature had no marked effect on γ.

A failing of this paper [115] is the narrow range of target temperatures (800–1200°K). These temperatures were insufficient to remove adsorbed atoms completely. In addition to this, the evaporation of neutral particles from the spodumene was not adequately taken into account.

In another paper [116], Pavlov and Starodubtsev studied the same phenomena, but with smaller energies (10–70 eV) of the primary ions, using W, Mo, and Pt targets and also Pt targets covered with oxide and W targets with various thorium coatings. The targets were bombarded with Na^+ and K^+ ions obtained by heating a W wire coated with pure NaCl and KCl. The target surface was cleaned by brief baking at 1600–1800°K. However, the majority of the measurements were carried out with a W target temperature of about 1200°K.

The authors studied $\gamma(E_0)$ for W, Mo, and Pt targets at about 1200°K and obtained curves identical with their previous curves for E_0 values between 100 and 1000 eV. The emission properties of the individual targets were given by the relation $\gamma_W > \gamma_{Mo} > \gamma_{Pt}$. With increasing target temperature there was a slight but systematic rise in γ. The paper in question also indicated a rise in γ as the adsorbed layer on the target surface became thicker.

On studying the value of γ for W—Th targets the authors observed a connection between γ and the work function of the target surface φ, the value of φ being varied by varying the coverage θ. This relation had the form $\gamma = Be^{-c\varphi}$, where φ is the work function of the surface of the target bombarded and C and B are constant coefficients.

Veith [118] studied the energy distribution of secondary electrons. Targets of Cu, Al, and Pt were bombarded with K^+ ions having energies between 70 and 1140 eV. The author observed the existence of two groups of secondary electrons: the first with energies up to 3 eV, constituting the main bulk of the secondary electrons, and the second with a much higher energy, extending as far as 50 eV for primary ion energies of 1100 eV.

Paetow and Walcher [124] studied the effect of various films, deposited on the pure metal, on γ and K and on the energy distribution of the secondary particles. The method of investigation and the construction of the apparatus were exactly the same as those of Koch [105]. The target was baked for several hours at 2000°K. A pure W target and also a W target coated with adsorbed K, Cs, H_2, and O_2 atoms were bombarded with Cs^+ and K^+ ions freed from neutral particles by separation in a cylindrical condenser. The energy of the ions varied between 500 and 1000 eV. The authors found that, on continuously bombarding the surface of the cold target, i.e., on thickening the coating, the value of γ gradually increased with time and reached saturation; K fell and also approached a saturation value. According to the calculations of these authors, the saturation of γ and K set in at a coverage of $\theta = 1$. After reaching saturation, further bombardment had no effect on the values of γ and K. According to these authors, the energy of the secondary electrons extended from 0 to 20 eV and that of the ions from 0 to 40 eV for a primary ion energy of 500 eV.

In the experiments of Koch, and of Paetow and Walcher, the secondary electron current was measured for an ion-retarding field of about 60 V; it is now known that for primary K^+ ions with energies of 200–1000 eV this only gives incomplete retardation of the secondary ions. Under these conditions the authors were naturally unable to determine the true value of γ for pure tungsten.

Nemenov and Fedyurko [131] studied the variation in the value of γ and the energy distribution of the secondary electrons as functions of the E_0 of the primary ions; they bombarded a Pt target with ions of the Li^{7+} isotope having energies between 3 and 30 keV, separated by a magnetic mass analyzer. The authors observed a rise in γ and the energy of the secondary electrons with increasing E_0, and also a fall in γ for high energies. The fall in γ with increasing E_0 confirms the idea that high-energy primary ions penetrate into the lower regions of the target.

Gurtovoi [144] studied the dependence of γ on the actual surface of a W target covered with thorium and on the value of E_0 on bombardment with K^+ and Cs^+ ions having energies of 200-1000 eV; he found that γ increased in proportion to the area of the coating for constant E_0 and obtained a nonlinear dependence of γ on the work function.

Dukel'skii and Zandberg [197] made some comparative measurements of the coefficient of secondary electron emission for a cold platinum target under the influence of I^- and I^+ ions with energies of 300-1040 eV and Na^- and Na^+ with energies of 290-2200 eV in a vacuum of $(2-3) \cdot 10^{-6}$ mm Hg. In both cases the coefficient γ was much greater for the negative ions than for positive ions of the same energy. In order to explain these results, the authors suggested that in the case of the negative ions some of the observed secondary electrons were formed by the decomposition of the actual negative ions on the target surface.

Eremeev and colleagues [220, 237] studied γ (for temperatures of 300-1200°K) and K (300-2500°K) for Ta, W, and Sn bombarded with K^+ and Li^+ ions having energies of 2-14 keV. The K^+ and Li^+ ions were obtained by heating a tungsten spiral impregnated with a mixture of $K_2O \cdot Al_2O_3 \cdot 2SiO_2$ and $Li_2O \cdot Al_2O_3 \cdot 2SiO_2$. In the opinion of the authors, this mixture, on heating to the temperature of the surface ionization of the corresponding alkali metals, emitted only K^+ or Li^+ ions, there being no neutral particles in the composition of the ion beam. These authors give curves of (γE_0) for cold targets (degassed or otherwise) and also W and Ta targets heated to 1200°K, using an ion-retarding field of 50 V; they obtained a linear dependence of γ on E_0 and noted a fall in emission with increasing target temperature.

Eremeev and his colleagues found no dependence of K on the energy of the primary (Li^+) ions between 2 and 14 keV, nor any on the target temperature between 300 and 2500°K. The values of K for nondegassed cold Ta and W targets were lower than for the cold degassed targets. The values of this coefficient were almost the same for both sorts of ions and targets, being approximately 20%.

Measurements of K at various target temperatures above 1000°K showed that the scattering coefficients of the K^+ ions on Ta were of the order of 50% and those of the K^+ ions on W, 70%. The authors noted a sharp jump in this coefficient between 800 and 1000°K.

Eremeev and colleagues [220, 237] indicated the appearance of scattered ions with very high energies at a high target temperature. The authors, although noting the sharp rise in scattering coefficient associated with fast K^+ ions between 800 and 1000°K, took no account of the appearance of surface-ionization currents of K^+ on W and Ta.

In the papers just mentioned, a mass analyzer was also used to study the velocity distribution and maximum energies of secondary K^+ and Li^+ ions scattered from a Ta target as functions of the energy of the primary ions. In both cases the authors found that the maxima on the distribution curves corresponded to the energy values calculated for the elastic collision of the ions with target atoms. The maximum energy of the secondary ions for a specified scattering angle (90°) agreed with the value

$$E = E_0 \frac{m_1 - m_2}{m_1 + m_2}, \tag{1.1}$$

where E and E_0 are, respectively, the maximum energy of the scattered ions and the initial energy of the bombarding ions, while m_1 and m_2 are the masses of the target atoms and bombarding ions.

Eremeev and Matskevich [241] studied γ and K for a cold, uncleaned Sn surface and for a liquid Sn surface cleaned to various degrees. The authors asserted that both the absolute values of γ and K and the energy distribution of the secondary ions from the cold Sn surface differed in no way from those of cold Ta and W targets.

It was found that, as the surface of the Sn target became cleaner (for a constant energy of the primary beam) the number of slow scattered ions fell, while in the case of a really clean target surface no slow ions were found at all. In addition to this, scattering only took place in the form of fast ions from the clean target, and there was no secondary electron emission. The authors see the reason for the formation of the fast ions in the elastic collision of incident ions and target atoms.

Ploch [221], considering the dependence of γ on a large number of parameters, studied γ as a function of the mass of the bombarding ion, keeping all the other parameters of the colliding particles constant, including the state of the target surface and identity of the bombarding ions. As bombarding ions the following isotopes, separated out by a mass spectrometer, were used: Li^6 and Li^7, K^{30} and K^{41}, Ne^{20} and Ne^{22}. Targets of degassed and nondegassed Mo, Pt, and Be, and nondegassed Cu were studied. The mass spectrometer made it possible, by a single change of magnetic field, to compare the γ of one particular surface for bombardment by different isotopes of the same element. The experiments in all cases showed that for various energies the γ corresponding to the light isotope was greater than that corresponding to the heavy one. On this basis the author considered that in the empirical relationship $\gamma \sim E_0 = m_2 v_0^2/2$ the decisive role in releasing electrons from the target was played by the velocity of the ion. Recalculating the values of γ for an equal velocity, Ploch showed that the $\gamma (v_0)$ curves were the same for both ions. However, the simple velocity law was not followed for all the ions in the alkali series. The author therefore suggested that for ions having the same electron configurations γ depended on the ionic radius r_2 as well; he found support for this view in that, for the alkali ions γ varied analogously to r_2^3/m_2.

As already mentioned, the experimental conditions were not too good. The vacuum was about 10^{-5} mm Hg, and no special measures were taken to clean the surface. The surfaces were thus naturally covered with adsorbed layers of indefinite composition, with a wide spread of masses, and this may very well have led to the apparent absence of any influence of the mass of the bombarding ions.

Analogous experiments aimed at studying the influence of the multiplicity of the charge and the mass of the primary ions on γ were carried out by Dunaev and Flaks [267, 286]. The value of γ for Pt was studied with K^+, Zn^+, Hg^+, Tl^+, and Pb^+ ions. For the K^+, K^{2+}, K^{3+}, and Tl^+, Tl^{2+}, Tl^{3+}, and Tl^{4+} ions the $\gamma (E_0)$ relationship was studied between 1 and 40 keV; for the remaining ions, γ was measured only for the same values of energy and different charge multiplicities. It was found that a considerable rise in γ occurred when the transition from an ion with a charge $(Z - 1)$ to one with a charge Z was associated with a considerable change in ionization potential. The relation between γ and the energy of the primary ions K^+, K^{2+}, and K^{3+} was of a parabolic nature, and this form depended very little on the charge multiplicity. For Tl^+ ions the $\gamma (E_0)$ relationship was almost linear up to between 12 and 15 keV, after which it became parabolic. For ions having a similar electron shell, the character of the $\gamma (E_0)$ relationship was clearly determined by their mass. For heavier ions such as Rb^+ and Cs^+ or Br^+ and I^+, the $\gamma (E_0)$ curves had linear sections, while for light ions such as Li^+, Na^+, K^+ or F^+ and Cl^+, they were very nearly parabolas. Over the whole range of ion energies the difference in the ordinates of the $\gamma (E_0)$ curves for negative and positive ions of the same element never exceeded 100%. It is also interesting to note that for the positive ions Li^6 and Li^7 these authors' results regarding the independence of γ on the mass of the Li isotope for the same ion velocity supported the conclusions of Ploch [221].

These interesting results, however, relate to a target covered with an adsorbed layer of unknown composition. In addition to the original experimental conditions, the results of the measurements bear witness to this. For example, the parabolic relationship $K(E_0)$ is well known [14, 280] to be characteristic for contaminated targets. Hence, the correlation between γ and K observed in the experiments is also apparent, only characterizing certain states of the surfaces bombarded.

In another paper by Flaks [314] the secondary electron emission from a thermally untreated platinum target in a vacuum of $5 \cdot 10^{-6}$ mm Hg is compared for Zn^+, Zn^{2+}, Hg^+, Hg^{2+}, Hg^{3+}, Tl^+, Tl^{2+}, Tl^{3+}, Tl^{4+}, Pb^+, Pb^{2+}, Pb^{3+}, and Pb^{4+} ions of the same kinetic energy (between 1 and 20 keV). The secondary electron current was measured in the target circuit with a potential of 80 V on the collector relative to the target; in the author's opinion this was sufficient to ensure total retardation of all the secondary ions. It was found that multiply charged Zn^+, Hg^+, Tl^+, and Pb^+ ions and also K^+ produced an increased emission of electrons. The data for the secondary electron emission agreed qualitatively with the idea that the kinetic extraction of the electrons was accompanied by a potential extraction effect. Using Tl as an example, the author showed that the potential extraction of the electrons depended very slightly on the kinetic energy of the ion.

Veksler and Shuppe [261], studying the composition of the secondary emission of ions on bombarding two metallic targets of different types with positive Hg^+ ions, asserted that, since no negative Hg^- ions were found in the secondary emission under the experimental conditions, the probability of there being a conversion of the mercury ions was under 10^{-5} if indeed it existed at all.

Zandberg [315] studied the secondary emission of nondegassed and degassed but not atomically clean Pt and W targets bombarded with positive and negative ions of alkali metals and halogens as well as ions of antimony, bismuth, and magnesium. The ions passed from the source through a magnetic mass-monochromator and fell onto the receiving apparatus. The energy of the ions varied between 300 and 1250 eV. The resolving power of the apparatus enabled the isotopes of bromine to be separated out. For the nondegassed targets γ was higher for bombardment with negative ions than positive. For degassed targets γ was about the same for positive and negative ions, but smaller than for nondegassed targets.

Scattering of the primary ion beam of the degassed target, with the retention of a considerable proportion of the original energy of the ions, was observed. Here the negative ions of the alkali metals and the positive halogen ions reversed their charges. These results on the scattering and recharging of the ions may be explained as being due to the elastic interaction of the primary ions with free atoms and groups of atoms in the target. The sign change in some of the scattered ions may be related to exchange interactions taking place in the course of surface ionization.

Brunnee [343, 383] studied ion—electron emission and the reflection of ions of all the alkali metals with an energy of 0.4–2.0 keV from a clean Mo surface in a vacuum of $5 \cdot 10^{-8}$ mm Hg. The ions were obtained by thermionic emission from a tungsten plate and after acceleration passed through a 60° mass spectrometer. The target, degassed at 1700°K, was bombarded with specific isotopes. The reflected ions were studied by the retarding field method. The author, studying $K(E_0)$, observed a sharp fall in K between 0 and 1.5 keV; as the energy of the ions rose further K remained almost constant and fell on passing from Li^+ to Cs^+. The dependence of K on the mass of the ion and the structure of the electron shell was examined. For this purpose K was measured with different isotopes and the slope $\Delta K/\Delta m_2$ was compared with the slope of the curve relating K to the mass of the alkali atoms. The author attributed the agreement between these values to the fact that K was mainly determined by the mass of the ions, while its electron configuration played a minor part. According to Brunnee's results, the energy distribution of secondary ions has roughly the same character for all the ions studied, namely, a fall at low energies, the existence of a region of approximately constant values, and a sharp drop to zero at a certain limiting energy, depending on the mass of the ions. The constancy of the energy distribution of the ions and its independence of energy between 1.5 and 4.0 keV indicates, in the author's opinion, that the scattering of the ions is isotropic in the center-of-mass system over this range of energies. Brunnee explains the substantial rise in K at low energies as being due to an increase in the influence of the bonds between the target atoms; he considers that the classical theory of single collisions gives a fairly accurate picture of the dependence of the reflection coefficient on the mass of the ions, explains the energy distribution, and predicts the limiting energies. The fact that the reflection coefficient for ions with masses greater than that of the target atoms fails to vanish, as well as the formation of ions with low energies, may be explained, according to Brunnee, on the basis of multiple collisions.

As regards ion—electron emission, the factor γ rises linearly with increasing energy from a certain threshold ion energy E_{min}. For constant ion energy, the yield of secondary electrons falls with increasing mass of the ion and increasing ionization potential of the alkali atoms. The energy distribution of the emitted electrons has a sharp maximum near zero, and for an energy of 10 eV falls to 1% of the energy at the maximum. The mean energy of the electrons ejected by the alkali ions falls from Li^+ to Cs^+. In the author's opinion, these results indicate that the secondary electrons arise in the course of ionizing the metal atoms.

Mitropan and Gumenyuk [386] studied the dependence of the extraction coefficient of negative ions, K^-, on E_0 when bombarding Al, Cu, and stainless steel targets with protons and deuterons having E_0 = 200-1000 keV. It was found that K^- fell monotonically on raising the energy of the ions, varying between 10^{-3} and 10^{-4}; it depended on the nature of the target and the velocity of the ion, but not on the mass of the bombarding particle. Degassing the target led to a fall in K^-, but γ remained constant.

Mironov and Nemenov [382] used a cyclotron to study γ for Al and Ni films in a vacuum of 10^{-5} mm Hg under the action of 2-7.3 MeV protons and found that over this range of energies the γ of Al and Ni fell on average from 1.8 to 0.5.

Veksler and Ben'yaminovich [384] carried out a mass-spectrometric study of the products of ionic emission from Mo under the impact of Cs^+ ions after depositing K and Na films with various pressures of oxygen in the apparatus. In the absence of alkali films the current of secondary Cs^+ ions increased with increasing O_2 pressure, the rate of growth being greater for low oxygen pressures. After switching off the heating of the Mo target, the current of secondary Cs^+ ions reached saturation after a time during which all parts of the target surface were (on average) subjected to the action of the Cs^+ ions once only. The saturation current of secondary Na^+ and Cs^+ ions fell rapidly with increasing rate of deposition up to a rate of one monolayer per second, and then fell more slowly. The analogous picture obtained after depositing Na films showed that for low rates of deposition oxygen had the same effect on the emission of Na^+ and Cs^+ ions, the current of Cs^+ ions greatly exceeding that of Na^+. However, for high rates of deposition there was a redistribution of the composition of the emitted particles and the Na^+ current considerably exceeded that of the Cs^+.

Goutte and Guillaud [385, 426] examined the dependence of γ on the tensile stress of a metal sample; they studied the secondary electron emission of metallic samples of Au, Ag, and Ni under the influence of Na^+ ions (1500 eV) and Ar^+ ions (5 keV). The samples, in the form of metal strip, were subjected to mechanical tension in a vacuum of $2 \cdot 10^{-5}$ mm Hg until rupture ensued. The targest were cleaned by ion bombardment.

With increasing tension γ rose by several percent. At the breaking stress the relative rise in γ reached between 3 and 6% for all the metals studied.

Watters [424], employing high-vacuum conditions [(2-4) $\cdot$ 10^{10} mm Hg], measured the γ and the energy distribution of the electrons on bombarding atomically clean and gas-covered W surfaces with 150-1500 eV Li$^+$ and Cs$^+$ ions. The results showed that the probability of secondary electron emission was vanishingly small (10^{-6}) when the energy of the ions was of the order of 150 eV and rose to $6.2 \cdot 10^{-2}$ and $1.2 \cdot 10^{-2}$, respectively, for Li$^+$ and Cs$^+$ ions for an energy of 1500 eV and an atomically clean W surface. The adsorption of a monatomic layer of O_2 or N_2 increased the efficiency of this process by at least an order for ions with energies of under 350 eV and by 2 or 3 times for ions with energies of over 500 eV. In addition to this, it was found that the adsorption of gases by the W surface made no appreciable difference to the energy distribution of the secondary electrons.

Akishin and Vasil'ev [461] used a 72-cm cyclotron to study the emission of electrons from an activated copper—beryllium alloy, widely used for preparing the cathodes and emitters of electron multipliers used in the recording of fast ions. Singly, doubly, and triply charged Li, B, and N ions with energies up to 10 MeV were extracted from the cyclotron and focused on the target. Before measuring γ the target was kept in air at atmospheric pressure for several days. It was found that, over the whole energy range studied, for equal energies the greater γ corresponded to the ions of larger mass. In addition to this, for each form of ion the $\gamma(E_0)$ curve had a maximum.

Veksler [498] studied the energy spectra of the scattered and sputtered Cs$^+$, Mo$^+$, and Ta$^+$ ions on bombarding Mo and Ta targets with positive Cs$^+$ ions having an energy of 900-2150 eV. The experiments were carried out in a vacuum of $(5-8) \cdot 10^{-8}$ mm Hg. The targets were degassed by prolonged baking at 2300-2500°K. From the resultant total spectra of secondary ion emission the scattered and sputtered components were separated with a mass spectrometer. This greatly reduced the observed limiting energy in the spectrum of scattered Cs$^+$ ions for Mo targets. The width of the energy spectrum of sputtered ions was 30-35 eV for Mo$^+$ and 35-50 eV for Ta$^+$ ions, which considerably exceeded the value of 5 eV obtained from the literature, and indicated that the probability of ionization of a sputtered atom leaving the surface rose as its energy increased.

Belyakov and Mittsev [398] showed that the depth of penetration of Li$^+$ ions into a thin layer of aluminum (for bombarding ion energies between 1 and 2 KeV) reached a value equal to tens of atomic layers.

The discrepancies arising between the results of the earlier investigations into the secondary emission properties of metals described in this section naturally suggest the presence of certain uncontrollable factors, manifested to different extents in the results of the earlier authors. We may suppose that such factors include the inadequate cleanness of the surface studied, and the different conditions affecting the latter. The secondary emission properties of the surfaces under consideration may have been considerably affected by the prehistory and the method of initially treating (cleaning) the target, and also by the contaminations which may have arisen during the experiments as a result of the adsorption of residual gases and vapors from the vacuum and the adsorption of the primary beam ions themselves on bombarding a target of frequently unknown composition.

In the nature of things, a considerable proportion of the early investigations concerned with the secondary emission of "pure" targets were carried out not with pure targets at all but with targets covered by adsorbed films of various natures, compositions, and thicknesses, and hence different properties. Hence, in order to study the secondary emission properties of the materials under consideration it is very important to understand the effect of films on the secondary emission properties of pure metals and the conditions under which this influence may be neglected. In the following sections we shall set out the results of the early investigations of Arifov and Ayukhanov [219, 238, 282, 342] carried out by the static method, in which attempts were made to study secondary processes under pure conditions with due allowance for the state of the target surface.

§2. Use of the Galvanomagnetic Method for Studying Secondary Processes Under Relatively Clean Surface Conditions

Experimental Method. Apparatus

All investigations into secondary processes were carried out in apparatus differing little in principle from that shown schematically in Fig. 1.1. The main components are an ion source 4 and a receiving section for primary and secondary particles. The receiving part of the apparatus consists of a screening cylinder 1 with a double diaphragm for introducing the beam and giving it the desired shape and size, a collector 3 for receiving the secondary particles, and a target 2. On the side of the primary-beam entrance, the collector 3 has the form of a hemisphere, which then transforms into a cylinder.

The target constitutes a flat piece of the metal under consideration, usually $30 \times 5 \times 0.05$ mm in size, heated by ac or dc. In order to check the state of the surface by reference to its work function, whisker-type targets were employed in a number of experiments, with cylindrical collectors. The thick leads of the target enabled the latter to be heated to its melting point.

Preparation of the Apparatus for Measurements

Special attention was devoted to the cleanliness of the target surface. The apparatus was evacuated with a two-stage rotary backing pump and two-stage mercury or oil-vapor diffusion pumps. After reaching a vacuum of 10^{-6}-10^{-7} mm Hg, with the necessary heating of the glass parts of the apparatus and degassing of the metal parts, we proceeded to clean the target.

The target was purified from adsorbed and absorbed gases by prolonged heat treatment at high temperatures, liquid nitrogen necessarily being present in the traps of the vacuum equipment. In purifying each new target the latter was held at a temperature close to the melting point (2500°K for Ta and W) for a considerable time (tens of hours).

The target was degassed until the surface reached a state in which the coefficients of secondary electron and ion emission γ and K became reproducible. In the experiments in question, for example, if a tantalum target had been insufficiently degassed the coefficient obtained on bombardment of 400-eV ions of any of the alkali metals was K < 20%; after careful degassing, however, it reached a constant value of about 30%. The value of γ for ions with energies up to 1 keV was very low (under 1%). As the target surface became contaminated with adsorbed molecules of residual gas, there was a continuous rise in γ and a fall in K.

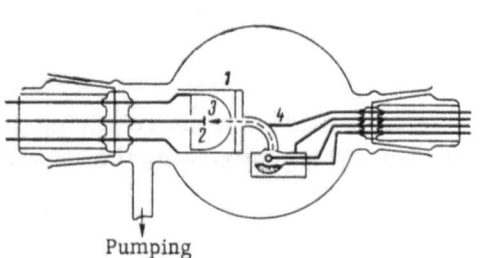

Pumping

Fig. 1.1. Arrangement of vacuum apparatus.

Source of Alkali and Alkaline Earth Ions

Analysis of the operation of ion sources used by the authors of [145, 146, 177, 182, 183, 219, 220] shows that in addition to undesirable foreign impurities the ion beams always contained neutral atoms and molecules of the salt from which the ions were obtained. In addition to singly charged ions there were also multiply charged ions, and in the majority of cases the sources operated in an unstable manner. Yet, in order to study secondary processes taking place as a result of the interaction of the ions with the surface of a pure metal, it would be essential to have an ion source in which the ions would be, as far as possible, identical in energies, charges, and masses; there ought never to be any neutral particles in the ion beams. In addition to this, such a source should operate stably for a long period, as regards both the strength and the direction of the ion beam, and should give the required intensities in a relatively high vacuum.

These requirements are satisfied by a source in which the ions are formed by the surface ionization of alkali and alkaline earth halide molecules, with subsequent separation from neutral atoms. In the experiments described in the following pages, separation of the ions with respect to velocities and with respect to neutral atoms was effected by deflecting the flow of charged particles in the electric field of a cylindrical condenser.

A 127° cylindrical condenser of this type has found wide application in studying the energy distribution of thermoelectrons [229], the characteristic energy losses suffered by electrons striking a solid surface [54, 58, 234, 257, 367, 453, 608, 616, 712], and also in high-resolution mass spectrography [712]. At the present time the electron—optical properties of a cylindrical condenser are generally well known; the theory of these may be found in many textbooks [234, 257, 529]. We shall only consider certain details associated with the experimental application.

It was shown theoretically and experimentally in [54, 58] that a radial field may be used for the energy analysis of charged particles. Such a field is created by applying a potential difference between the plates of a cylindrical condenser.

If a slightly diverging monoenergetic beam of charged particles with an aperture angle of $2\alpha_0$ passes into the condenser, to a first approximation we may consider that the particles will move along circular paths; according to [54], these will intersect for a condenser aperture angle of $\varphi = 127°17'$.

The condition of motion in a circle will be

$$-e\mathscr{E} = \frac{mv^2}{R}. \tag{1.2}$$

Expressing the kinetic energy of the charged particle in terms of the accelerating potential ($mv^2 = 2eU_0$) and the electric field $\mathscr{E}$ in terms of dU/dR, and separating the variables, we obtain the equation

$$\frac{dU}{U_0} = \frac{2dR}{R}, \tag{1.3}$$

determining the resolving power of the analyzer. It follows from (1.3) that the resolving power of the analyzer will be greater for slow particles than for fast ones. Integrating this equation, we obtain the relation

$$U_\kappa = 2U_0 \ln \frac{R_2}{R_1} \tag{1.4}$$

(R_1 and R_2 are the radii of the inner and outer plates of the cylindrical condenser) relating the accelerating voltage U_0 to the critical voltage U_K between the condenser plates for which the charged particles moves along a circle of radius $R_0 = (R_1 + R_2)/2$.

More detailed study [54] shows that only particles passing along an arc with radius R_0 move on the circle, and that for a diverging beam of charged particles with the same energy the focus at the analyzer exit will be not linear (rectangular cross section of the beam) but a strip of width

$$S = \frac{4\alpha_0 R_0}{3},\tag{1.5}$$

where α_0 is expressed in radians. For small angles α_0 correction (1.5) may be neglected.

The ion source consists of an ionizing section together with separating and focusing systems (Fig. 1.2).

The ionizing part of the source consists of a nickel box 2 and a tungsten spiral 1. Inside the source, in a suitable dish, is the alkali or alkaline earth halide 5 (depending on the particular type of ions required).

The separating and focusing system of the source consists of a cylindrical condenser 4 and nickel plates with inner and outer radii of curvature respectively equal to 20 and 24 (or 16 and 20) mm. The outer plate of the condenser is sealed through suitable insulation and the inner plate is sealed directly to the plate 6, which has a slit 3 of size 1 x 12 mm. A funnel 7 is welded to the top of the source box 2 from underneath (the funnel passes down into the box), while the lid in turn is securely welded to the plate 6; slits 3 and 8, 1 x 12 mm in size, are drawn together by a loose tungsten grid not shown in Fig. 1.2.

The focusing part of the source consists of a system of diaphragms ABCD. Of these, A, C, and D are made of nickel plate 0.5 mm thick with slits of 1 x 12 mm; the diaphragm B is cut from red copper 4 mm thick with a slit of 5 x 16 mm.

Diaphragms A and C are rigidly connected to each other and are welded to the first diaphragm D, while diaphragm B is fixed through corresponding insulators to diaphragms A and C. A positive potential relative to A, C, and D is usually applied to diaphragm B. The distance between diaphragms A, B, and C is about 5 mm. This construction of the ion source enabled us to obtain a fine, intense beam of primary ions from the source in accordance with the geometry of the electrodes and the potential applied between the diaphragms. The adjustment and operating principles of the electrical circuit of the source are described below.

On heating the tungsten spiral 1 with ac or dc, vapor is formed from the salts; falling on the surface of the spiral, this partly dissociates into metal and halogen adatoms. Then some of these adatoms (depending on the ionization potential V_i of the adatom and the work function of the filament surface φ) are ionized by surface ionization and fly off from the surface of the spiral with thermal velocities. Subject to the sign of the potential applied between the spiral 1 and the body 2 of the source, metal ions are accelerated in the space 1—3, and some of these monokinetic ions, passing through slits 8 and 3, fall into the cylindrical condenser.

The electron or ion beam, diverging as it enters the condenser, is focused at the exit, the focal distance depending on the condenser aperture.

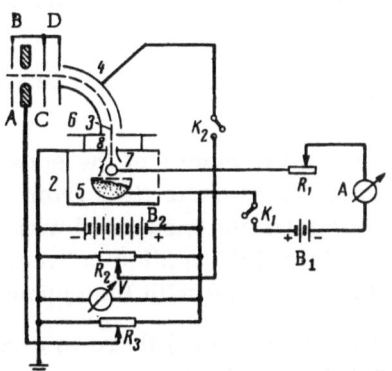

Fig. 1.2. Ion source based on the principle of surface ionization.

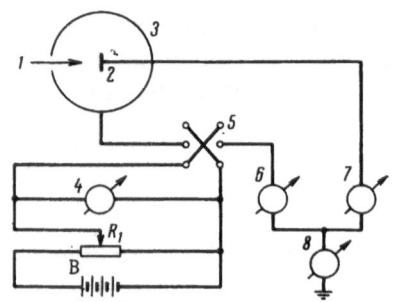

Fig. 1.3. Principal circuit of the static method of studying secondary processes.

Thus a source of alkali ions based on surface ionization, combined with a cylindrical condenser, gives fairly monokinetic ion beams not containing any neutral atoms or molecules. A special experiment showed that no neutral particles in fact reached the target surface (sensitivity of the method of detection 10^{10} atoms/sec). In addition to this, the ions could be focused on a small part of the surface, and the shape of the focal spot could also be prescribed.

In working with this source, its stability and long service without salt recharging proved extremely valuable. With an appropriate choice of heating for the spiral and with a steady temperature of the body of the source, the latter may operate for a long time with an unvarying ion current. It should be noted that by improving the geometry and selecting optimum extraction potentials, a very strong ion beam may be obtained.

Measurements

Before the measurements, the target was subjected to prolonged heat treatment. In order to reduce the contamination of the target surface resulting from the adsorption of residual-gas molecules, in all our galvanometric measurements the target was held at temperatures close to the threshold at which thermionic (electron) emission occurred.

After preparing the target for the measurements, the ion beam was only applied to the target surface during measurements taking a time not greater than a few periods of the measuring galvanometer. Before each measurement, the target was first heated to a high temperature (to 2500°K) for a longer period (3-4 min), and then immediately before the measurement it was subjected to a brief "flash." Then, after reducing the target temperature to the necessary value, the ion beam was applied by feeding an appropriate potential to the outer plate of the cylindrical condenser and the measurement was made.

The shape and size of the ion beam was constantly checked on the surface of the target by reference to the fluorescence of the point irradiated, which was viewed through a special "window" in the screens and collector cylinders. Through the same aperture the temperature of the target was measured with an optical pyrometer.

The measurements were made with a mirror galvanometer having a current sensitivity of about 10^{-10} A · mm/m with a period of up to 5 sec, and an electrometer of sensitivity 10^{-12}–10^{-13} A · mm/m.

The principal circuit of the static method of measuring secondary processes is shown in Fig. 1.3.

The primary ion beam 1 is applied to the target 2 placed in the center of the spherical or hemispherical collector 3. Between the collector and the target an electric field from the battery B is applied; the magnitude and direction of this may be varied and measured with the help of the switch 5, the rheostat R_1, and the voltmeter 4. The electric field applied between the target and the collector retards particles of one sign and accelerates those with the opposite sign.

On changing the potential difference between electrodes 2 and 3 ($V_{2,3}$) there is a change in the currents measured by galvanometers 7 in the target circuit and 6 in the collector circuit, while the readings of galvanometer 8, measuring the total current of primary ions I_1, remain constant. For suitable potential differences between electrodes 2 and 3 the secondary electron currents may be separated from the secondary ion currents and the coefficients of secondary ion—ion emission K and secondary ion—electron emission γ may be estimated individually. For brevity we shall call K the coefficient of secondary ion emission and γ the coefficient of secondary electron emission. Measurement of the currents I_6 or I_7 with due allowance for the potential difference between electrodes 2 and 3 enables us to obtain curves representing the energy distribution of the secondary charged particles.

The arrangement here discussed enables us to measure secondary currents both in the collector circuit and in the ion circuit for positive and negative potentials on the target, but not the primary current I_0. The currents in the target circuit I_7 and in the collector circuit I_6, and the total current I_0, after completely separating the positive charged particles from the negative, are related by the obvious expressions

$$I_0 = [I_7]_- - [I_6]_+, \tag{1.6}$$

$$I_0 = [I_7]_+ + [I_6]_-. \tag{1.7}$$

The sign on the square brackets shows the polarity of the potentials applied to the target or collector.

It was found in the course of experiments that Eq. (1.6) was only true for a pure target when a fairly high positive potential (differing for different types of ions) was applied to the collector. This may be explained by the presence of rather fast positive ions among the secondary particles; these were not stopped by the application of a positive potential to the collector.

In order to establish this fact exactly we made some control experiments. Instead of the target we placed a Faraday cylinder, which played the part of a "black body" with respect to the ions, in the apparatus. We found that the total current was equal to the current passing to the Faraday cylinder, i.e., to the current passing to the target, in the absence of any direct incidence of the primary current on the collector, for positive, negative, or zero potentials of the collector. The direct current to the collector was always zero. Secondary ion and electron currents were completely absent from the collector circuit, i.e., the following conditions held:

$$I = [I_7]_{0, -, +}, \quad [I_6]_{0, -, +} = 0. \tag{1.8}$$

This experiment proved that the ions in the primary beam never bypassed the target, and the nonfulfillment of condition (1.6) for low collector potentials was due to the presence of fast secondary positive ions.

In order to secure a correct interpretation of the resultant data we must allow for the following factors, which may introduce certain errors into the results of the measurements: 1) induced currents associated with the presence of variable electric and magnetic fields near the experimental apparatus; 2) parasitic conductivities between the circuit elements; 3) errors introduced by the measuring apparatus; 4) differences in the parasitic capacities during the measurements in the collector circuit and the measurements of total current.

These factors, however, were controllable, and they were therefore reduced to a minimum in preparing the experiments. For example, induced currents were eliminated by screening the receiving part of the apparatus and the leads. Errors introduced by the measuring apparatus were greatly reduced by measuring the total current of the primary ions and secondary currents with the same measuring device.

We must particularly mention the following sources of error: 1) secondary emission from the collector; 2) the "end effect"; 3) the loss of secondary particles through the opening in the collector; 4) the inadequate determinacy of the state of the bombarded surface. These factors are not directly controlled by the experiment and errors thus introduced have to be judged indirectly. Let us consider some of these individually.

1. Secondary Emission from the Collector. A certain uncontrollable error in measuring the coefficient of secondary ion—electron emission may arise as a result of secondary emission from the collector surface due to fast ions, atoms, and secondary electrons reflected from the target.

Secondary ion—electron emission from the collector, for sufficiently positive values of the collector potential, will not introduce any serious error into the coefficient of ion—electron emission from the target being determined. In view of their low energy, these electrons can only leave the surface of the collector for negative values of the collector potential, and thus only introduce a possible error when determining the coefficient of secondary ion emission.

In finding the coefficient of secondary ion—electron emission, a direct error arises from the elastic reflection of the electrons at the collector surface. It is well known, however, that the proportion of elastically reflected electrons in the overall balance of electron—electron emission is quite small, and falls with increasing electron energy. Hence the value of this error will also be negligible for reasonably positive values of the collector potential.

2. Edge Effect. In determining the absolute values of the coefficient of secondary ion—electron emission a certain error may come from the so-called edge effect of the diaphragms placed in the path of the primary ion beam as the latter enters the receiving part of the apparatus. The presence of the diaphragms has the effect that, in addition to the narrow primary ion beam cut out by the diaphragm, a certain number of secondary products, resulting from interaction between the primary ions and the edge of the diaphragm, may fall into the receiving section (target—collector system). These secondary particles will also fall into the collector (for appropriate potentials of the latter) in addition to the secondary currents from the target which it is desired to measure. The edge effect may be reduced by using a system of diaphragms with a specific ratio of transverse slit dimensions.

The diaphragm system used in our experiments with pure metals enabled us to reduce the value of γ to 1%; this may at the same time serve as an estimate of the error introduced by the diaphragms.

3. Loss of Secondary Particles through the Aperture in the Collector. This loss was calculated from the ratio of the area of the apertures to the area of the whole collector, allowing for the law governing the angular distribution of secondary particles. This correction is in general negligible.

4. Inadequate Determinacy of the State of the Bombarded Surface. The desired determinacy of the conditions on the test surface was created by carefully cleaning and degassing the target. However, the statistics of the experimental data show that the results of individual experiments with the same target may vary considerably. This is the most important contribution to the total error.

If special measures were not taken to clean the target, the errors in determining the coefficients γ and K were so large that the experiments lost reproducibility. By taking the special measures listed in the foregoing, we were able to secure reproducibility of the experiments within the limits of measuring error. Control experiments showed that the experimental errors (determined by comparing the spread of the points on the curves and the reproducibility of the curves plotted in the forward and backward directions) obtained under pure conditions never exceeded 2-3% in different experiments.

§3. Influence of Adsorbed Films on Secondary Processes

In measuring the secondary emission from a pure cold target with E_0 of the order of 300-400 eV we found that the application of a small retarding field (30-50 V) between the target and

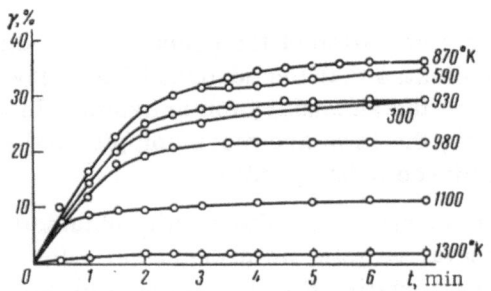

Fig. 1.4. Time dependence of secondary electron emission from adsorbed films.

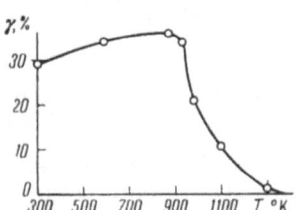

Fig. 1.5. Temperature dependence of the secondary electron-emission saturation current.

the collector at the very beginning of the experiments, when the target was clean, was insufficient to stop the secondary ions completely. The galvanometer in the collector circuit always first shows a current of positive ions. The value of this ion current falls with increasing retarding field and may equal zero for fairly large values. However, as time passes, the current in the galvanometer gradually passes through zero and shows a predominant current of negative ions in the collector circuit. This is because, as time passes, an adsorbed film is formed on the surface of the originally clean target (in the future we shall call this "contamination" for short), leading to a gradual fall in the energy of the secondary ions and at the same time to the development and multiplication of secondary electrons.

In order to study this phenomenon we obtained the time curves of secondary ion—electron emission for various target temperatures. A beam of primary ions was directed onto a well-cleaned target. Between the target and the collector we established an ion-retarding field such as to prevent the secondary ions from the target from reaching the collector. In all cases (for our particular targets and ions), the electron emission increases as the target cooled.

By way of example, Fig. 1.4 shows the time curves of secondary electron emission for a tantalum target on bombarding with Cs^+ ions at various temperatures. We see from the figure that all the curves show clear saturation, the value of the saturation currents falling with increasing temperature. However, there is an optimum temperature (thickness of contaminant film), in the present case 870°K, at which there is maximum emission, and also a maximum temperature for which (within the limits of experimental error) no secondary electron emission occurs.

Figure 1.5 indicates the γ saturation values as a function of target temperature. The curve shows that, at high temperatures (about 1350°K), for ion energies of up to 1 keV, the secondary electron emission from tantalum is negligibly small ($\gamma < 1\%$, i.e., less than the experimental error). In individual experiments we may achieve an accuracy of 1% in measuring γ, and nevertheless, no secondary electron emission is detected.

The idea that this fall in secondary electron emission at high temperatures might be the result of the appearance of a large number of fast ions was checked by applying fields retarding ions with energies of up to $0.9E_0$ between the target and the collector. It was found that for all target temperatures studied the saturation secondary electron-emission currents were independent of the value of the ion-retarding potential after the latter had reached a certain specific value.

Thus, the presence of an adsorbed film on the surface of the test target is the main cause for the appearance of ion—electron emission. The fact that the ion—electron emission is small ($\gamma < 1\%$) for high target temperatures (up to the threshold of thermionic emission) leads to the conclusion that the emission of electrons by metals of the tantalum and tungsten type on bombardment by low-energy positive alkali ions is not characteristic of the metals themselves, but is mainly due to the formation of adsorbed films on their surfaces.

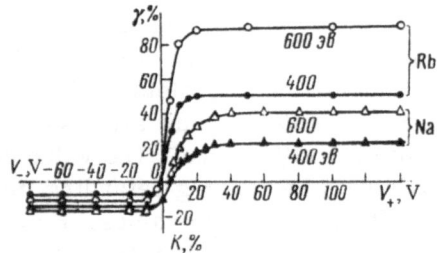

Fig. 1.6. Volt—ampere characteristics of secondary ion emission from tantalum.

The time variation of the secondary ion currents also indicates that the adsorbed films have an influence on the secondary ion emission. In this case, however, the laws are more complicated and will be considered subsequently.

For comparison with the results obtained from pure metal targets, we present the volt—ampere characteristic (Fig. 1.6) for a cold tantalum target with equilibrium coatings consisting of Na and Rb atoms, formed by the adsorption of the bombarding ions. The figure gives the ion and electron components of the secondary currents for two values of the primary ion energy E_0 (400 and 600 eV) for both types of films and ions. The curves show the direct dependence of the electron component of the secondary currents on the primary ion energy and also its dependence on the nature of the bombarding ions.

The secondary ion currents in this case also show a complicated dependence on the nature and energy of the primary ions. It is important to note that, in the presence of the films, the secondary emission (ion) coefficient K rises with increasing primary ion energy. We may clearly consider that the secondary emission ions from the equilibrium layer consist principally of ions from the film of adsorbed atoms ejected by cathode sputtering. The same volt—ampere characteristics give an idea of the continuity of the energy distribution of the secondary electrons and ions from the adsorbed films.

The resultant experimental data show that, for studying the secondary emission properties of the surfaces of pure metals, we must absolutely eliminate the possibility of adsorbed films forming on the test target. The most favorable conditions for eliminating the formation of adsorbed films are created at a high target temperature, since the "lifetime" of the adatoms on the surface is very short and, therefore, equilibrium coatings on the surface of the test targets will be insignificant. In view of this, subsequent galvanometric studies of secondary processes and all corresponding measurements were carried out at a target temperature close to the threshold of thermionic (electron) emission.

§4. Migration of Adsorbed Particles on a Metal Surface

Atoms and ions adsorbed on the surface of a metal may move along the surface. Migration is the first stage in freeing the surface from its covering film (the effect of this on secondary processes at various target temperatures was considered in the preceding section).

In order to observe the surface migration of easily dissociating and evaporating molecules, it is extremely important to monitor the minute movements of the particles. In order to create a suitable apparatus, Arifov and Lovtsov [213] used the principle of the cylindrical electron projector, in which the point of emission of electrons from a filament stretched along the axis of a cylinder is projected on a cylindrical screen. In a similar way we may record the point of emission of an atomic particle which has migrated to this point along the surface of the filament. In contrast to the electron projector, a system of collectors arranged along the surface of a cylinder was used instead of a cylindrical screen.

The arrangement of the apparatus employed is shown in Fig. 1.7. The filament 9, on which a narrow molecular beam of KCl falls through the diaphragm 15 with the open magnetic shutter 16 from the furnace 14, is surrounded by a cylinder 8 with a slit 10, at a distance of 1 mm from the surface of the cylinder. Inside the latter in the lower left-hand quadrant we placed six measuring plates, each of which covered an angle of about 15°. A similar measuring

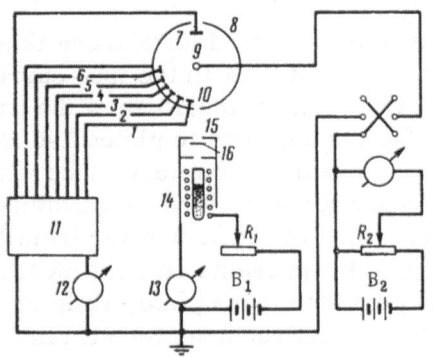

Fig. 1.7. Electrical circuit for studying the migration of atoms from a metal surface.

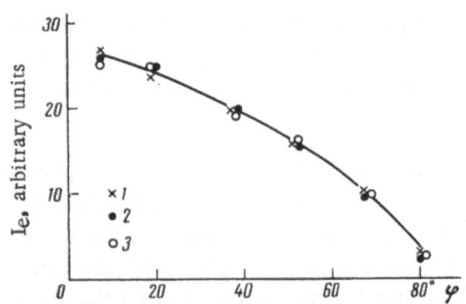

Fig. 1.8. Distribution of the ion-current density with respect to escape angles. 1) Calculated values; 2) experimental values for T = 1420°K; 3) experimental values for T = 1000°K.

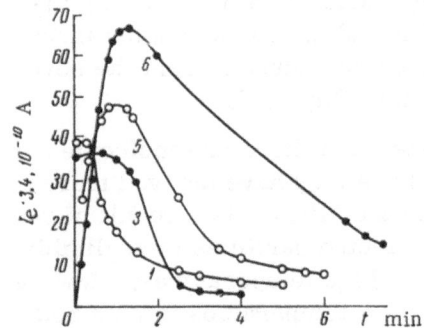

Fig. 1.9. Time dependence of the ion currents.

plate 7 was placed at the surface of the cylinder on the prolongation of the line from slit 10 to the target 9. A commutator 11 enabled us to measure the current 12 from each plate by itself, from all the plates together, and from the cylinder, and also the total current 13 from the plates and the cylinder. The incandescent filament 9 had a diameter of 125 μ; its temperature was determined with an optical pyrometer. Between the filament and collector was a voltage of about 100 V.

The number of molecules falling from the beam onto unit surface of the filament equals I cos φ, where I is the intensity of the molecular beam, and φ is the angle between the direction of the beam and the radius to the surface element in question. If the adsorption properties of the filament surface are homogeneous and the beam intensities not too large, then the number of adsorbed molecules is proportional to the number of incident molecules, and these are distributed over the part of the filament surface opposite to the slit in accordance with a cosine law. If the surface of the filament is uniform in relation to surface ionization, then the number of ions forming is proportional to the number of adsorbed and dissociated molecules on each surface element.

On applying an extracting electric field between the filament and the collector, we obtain an ion current with an intensity distribution obeying a cosine law if there is no migration of particles on the surface of the filament. If, however, there is appreciable migration on the surface, the ion-intensity distribution will not follow a cosine law. With the illustrated arrangement of plates 1-6 in the apparatus, the greatest intensity should occur on plate 1 if there is no migration and if the cosine distribution is obeyed, while the smallest should occur on plate 6. No ions should fall on plate 7. Since the area of the plates is known, we may, by applying the cosine law, calculate the relative value of the currents passing to the plates and compare the results with the ion currents measured at various temperatures.

Figure 1.8 shows the results of measurements for temperatures of 1000°K and above. For such temperatures the distribution of the ion currents in the plates obeys the cosine law within the limits of measuring error. This indicates that there is very little particle migration along the surface. Hence, at temperatures above 1000°K, no marked migration occurs, and the molecules falling on the filament fly away from it after moving for a short distance along the surface, dissociating, and evaporating in the form of ions.

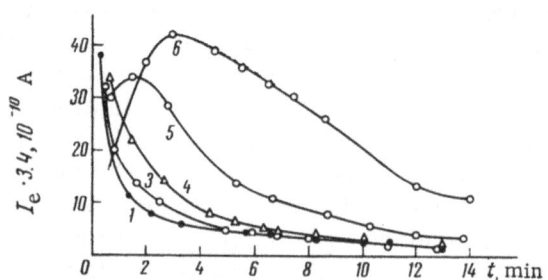

Fig. 1.10. Variation in the currents passing to different plates with time at 810°K.

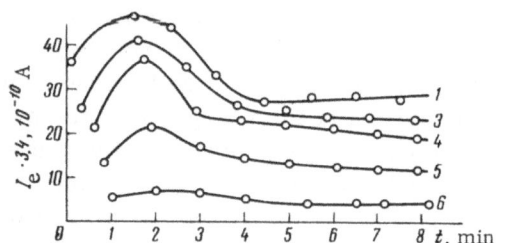

Fig. 1.11. Variation in the currents passing to different plates with time at 970°K.

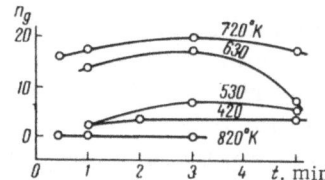

Fig. 1.12. Time dependence of the current to the seventh plate.

At temperatures below 1000°K there is a deviation from the cosine law in the ion—current distribution. Figures 1.9 and 1.10 show the time dependence of the ion currents to plates 1-6 at T = 810°K. The measurements were made with a filament previously heated to high temperatures. The results presented in Fig. 1.9 differ from Fig. 1.10 in that Fig. 1.9 represents current measurements taken in turn from each plate, while Fig. 1.10 represents current readings for all six plates in one series of measurements, the density of the molecular beam being larger in the first case. The resultant pictures are qualitatively the same. From these we see quite clearly that the ion distribution differs considerably from a cosine distribution; the current rises from the first to the sixth plate, this difference becoming greater with time. All this indisputably suggests the migration of molecules along the surface before they dissociate and their ionized atoms leave the surface.

With increasing temperature the migration diminishes. For T ~ 860°K, the ion current to plates 1-4 is greater than to plates 5 and 6. For T = 970°K, the distribution is close to the cosine law, although there is still a change in the distribution with time (Fig. 1.11).

For temperatures of under 800°K, the surface-ionization currents fall sharply. In order to prove migration at low temperatures, a beam of molecules was applied to the filament for a definite period (1 min). Then in a specified time interval the target temperature was rapidly increased to 2200°K and the galvanometer throw measuring the electron current from the seventh plate was noted. The results of these measurements are shown in Fig. 1.12.

The existence of ion currents to the seventh plate indicates that there are molecules and atoms on the side of the filament opposite to the slit and that these may have arrived there either by migration from the side of the filament turned toward the slit, or as a result of reflection from various parts of the apparatus. The adsorption of reflected particles is negligibly small, since this would produce a current to the seventh plate at high temperatures. However, since this never occurs, the ion current to the seventh plate at low temperatues is obtained simply as a result of the migration of particles on the filament.

From the results presented in Fig. 1.12 we may conclude that the velocity of migration depends on the temperature. As we should expect, the migration velocity rises as temperature increases. However, with increasing temperature the evaporation rate as well as the migration velocity becomes greater. For temperatures of the order of 1000°K and over, the rate of evaporation is so high, i.e., the lifetime of an adatom on the filament surface is so short, that the ions fly away from the filament without moving any appreciable distance along it.

§ 5.   Energy Distribution of Secondary Electrons
and Ions and Limiting Secondary-Ion Energies

As we showed earlier, starting from a certain value of the voltage retarding the ions, the secondary electron-emission saturation currents are independent of the value of this voltage. A study of this value, which determines the limiting energy of the secondary ions, is of independent interest, since by knowing it we may discover the mechanism underlying the interaction of the primary ion beam with the surface of the target under examination. We therefore plotted the secondary emission volt—ampere curves, i.e., the relation between the currents to the collector and the magnitude and sign of the voltage applied between the target and the collector for various primary ion energies.

The volt—ampere characteristics obtained on bombarding pure tantalum heated to 1300°K with positive sodium ions (Na^+) having energies of 200, 300, and 400 eV are shown in Fig. 1.13. Analysis of these curves shows the following: 1) the secondary electron emission for a target temperature of 1300°K and the primary ion energies in question are small ($\gamma < 1\%$); 2) the limiting energies of the secondary ions are large and depend linearly on the energy of the primary ions; 3) the coefficient of secondary ion emission is very large (over 30%) and increases with falling energy of the primary ion beam.

The small value of the ion—electron emission from a pure target ($\gamma < 1\%$) was at first glance a strange factor. In view of this we considered possible methodical reasons which might lead to an apparent reduction in the electron component of the secondary current. The greatest danger came from the magnetic field of the current directly heating the target. This field might rotate slow electrons and prevent them from reaching the collector. We found by control experiments, however, that this effect was comparatively slight.

In addition to this we found that at temperatures corresponding to the appearance of thermionic emission from the target the electron saturation current was established at a relatively low target-collector potential difference. We later found that, on disconnecting the heating, the secondary-emission electron current did not appear immediately, but only after the accumulation of adsorbed gas layers on the cold target. On subsequent oscillograph observation we discovered that the secondary electron emission was negligibly small both from a pure hot tantalum target and from a pure cold target.

In the case of bombarding a pure tantalum surface with 400-eV Na^+, K^+, Rb^+, and Cs^+ atoms at a temperature of 1300°K there was also a linear dependence of the limiting energy on the energy of the primary ions. However, the coefficients of proportionality for each ion were different and showed a certain dependence on the mass of the bombarding ions. The results of the measurements are shown in Fig. 1.14 in the form of composite volt—ampere secondary emission curves for these ions. We see from the figure that the coefficient of secondary electron emission is practically zero ($\gamma < 1\%$); the coefficients of secondary ion scattering for all the 400-eV alkali ions studied are almost the same, and equal to about 30%. However, we notice from the volt—ampere characteristics that the secondary emission contains a group of slow ions as well as the fast ions. The current of these ions is different for different alkali elements and only appears at high target temperatures under the condition $V_i < \varphi$. These currents are currents of surface ionization accompanying secondary emission at high target temperatures.

The relation between the limiting energies of the secondary scattered ions and the energy and mass of the primary ions, for the case of perpendicular incidence of the beam on the target surface and scattering at a large angle β, coincided with the well-known expression for elastic collisions:

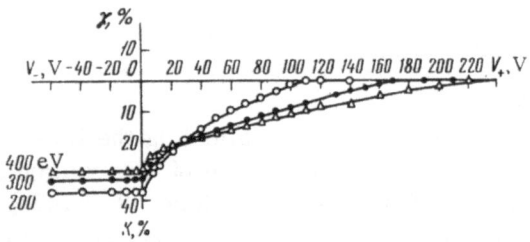

Fig. 1.13. Family of volt—ampere characteristics of secondary emission from pure tantalum.

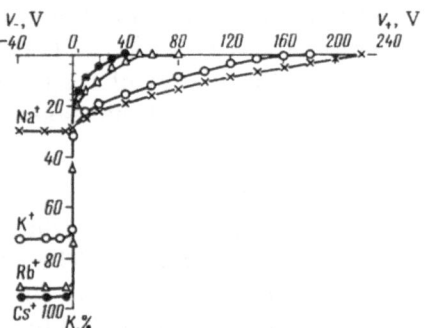

Fig. 1.14. Group of volt—ampere curves for secondary emission from pure tantalum bombarded with 400-eV Na$^+$, K$^+$, Rb$^+$, and Cs$^+$ ions at a target temperature of 1200°K.

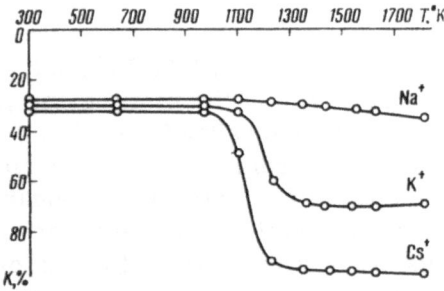

Fig. 1.15. Temperature dependence of the coefficient of secondary emission from pure tantalum bombarded by Na$^+$, K$^+$, and Cs$^+$ ions with energies of 400 eV.

$$E = \frac{E_0\left(\frac{m_1}{m_2} - 1\right)^2}{\left[\cos\beta \pm \sqrt{\left(\frac{m_1}{m_2}\right)^2 - \sin^2\beta}\,\right]^2}. \qquad (1.9)$$

The calculated values of $\eta_T = \frac{E}{E_0} = \left(\frac{m_1 - m_2}{m_1 + m_2}\right)^2$ for $\beta = 180°$ and the experimental values of $\eta_e = E/E_0$ for Na$^+$, K$^+$, Rb$^+$, and Cs$^+$ ions agree quite closely.

The agreement between the limiting energies of the secondary ions obtained in the present experiment and the values calculated from formula (1.9) for elastic collisions (particularly in the case of Na$^+$ and K$^+$ ions) is clearly explained by the small number of secondary ions leaving the surface of the target at large angles to the normal,* and also possibly by the fact that the target—collector system failed to satisfy the Lukirskii conditions, and the measured values of limiting energies corresponded more to the normal component.

The volt—ampere characteristics obtained also give an idea of the energy distribution of secondary ions. We see from Figs. 1.13 and 1.14 that the secondary ions have a continuous energy spectrum, from thermal to extremely large values, determined by the law of elastic collisions between the colliding particles. This result is hard to explain if we suppose that the secondary ions are only scattered by multiple collisions with the surface atoms of the target. This point of view, expressed by certain authors, requires certain assumptions regarding the mechanism of energy transfer. It is natural to suppose that the primary ions mainly suffer multiple collisions with target atoms lying at a depth of a few atomic layers. Then the continuous distribution obtained experimentally is a good confirmation of this assumption.

§6. Composition of Secondary Ion Emission

We see from the volt—ampere characteristic of secondary ion emission from pure

_____

*Special experiments on the intensity distribution of the secondary ions with respect to escape angle confirmed this view (see Chapter 6, §2).

tantalum bombarded with positive Na^+, K^+, Rb^+, and Cs^+ ions with energies of 400 eV at a target temperature of about 1300°K (Fig. 1.14) that the scattering coefficients of fast alkali ions from tantalum are almost identical and equal to about 30%. It also follows from these curves that, for target temperatures of around 1300°K, the coefficients of, as we shall provisionally call them, "total ion emission," increase sharply for all the alkali ions except Na^+, being different for different types of ion. For example, for $K^+ \sim 72\%$, $Rb^+ \sim 91\%$, while for Cs^+ we have 98%. The energy distribution of these "secondary" ions shows that they only have thermal velocities. On reducing the target to room temperature, the "secondary" ions with thermal velocities vanish completely. In the case of Na^+, slow ions are not found either at high or low temperatures; this may be understood if we consider the very low coefficient of surface ionization for Na on Ta. Considering that ionization occurs in accordance with the Saha—Langmuir formula

$$\alpha = \frac{n_+}{n_0} = \frac{1}{2} \exp\left(-\frac{V_i - \varphi}{kT}\right), \tag{1.10}$$

and putting $V_i = 5.12$ eV and $\varphi = 4.12$ eV, T = 1350°K, we obtain $\alpha \approx 0.1\%$, which is imperceptible for our measuring accuracy.

Figure 1.15 gives the temperature dependence of the total ion-emission coefficient for bombarding K^+, Na^+, and Cs^+ ions on pure tantalum. The ordinates represent the coefficients of secondary emission and the abscissas the target temperature. It follows from these curves that up to about 1000°K the secondary ions consist mainly of scattered primary ions (about 30%), while on increasing the target temperature above 1000°K the total ion-emission coefficient of K^+ and Cs^+ rises, and that of Na^+ remains unaltered.

We may conclude from the curves of Figs. 1.14 and 1.15 that for high target temperatures the total ion emission corresponding to K, Cs, and Rb contains a group of "secondary" ions with thermal velocities, the number of which is determined by the temperature of the target surface and the difference between the work function of the surface φ and the ionization potential of the bombarding ion V_i.

It is natural to suppose that a proportion of the alkali ions are scattered from the target still retaining some of their initial energy, while the rest lose their initial energy completely on the surface or deeper down into the metal and are converted into adsorbed or absorbed atoms, respectively. At high temperatures the adsorbed atoms will partly evaporate in the form of ions and partly in the form of neutral atoms.

The evaporating ions will have, first, thermal velocities and, secondly, a certain time inertia with respect to the primary ion beam, since the statistical lifetime of the adatoms determined by the heat of adsorption, the temperature, and other target parameters will vary between fractions of a microsecond and many hours.

These types of ions have no relation to the true secondary ions. Hence, in this case, the use of the general terms "coefficient of secondary ion emission," "scattering coefficient," and particularly "coefficient of reflection" is quite unjustifiable, although some authors do this.

However, the inertial properties of this type of ion cannot be studied by the galvanometric method in which apparatus with a large intrinsic period, often exceeding the period of the actual phenomenon by hundreds of times, is used. A special method was accordingly developed for studying these phenomena in time.

It follows from the galvanometric experiments in question that the processes of secondary emission from a pure metal target are very complicated. A very great part is played by the films of adsorbed atoms rapidly forming on the surface; the time taken to form these is much shorter than the period of one measurement by the static galvanometric method. It was clear that for a complete study of the mechanism of such complicated and rapidly occurring phenomena the existing method was quite inadequate. It was essential to develop new methods and procedures enabling us to study the phenomena in time with as far as possible pure surface conditions, so that the processes due to various factors could be examined individually and their mutual relation established.

CHAPTER 2

DYNAMIC (OSCILLOGRAPHIC) METHOD OF INVESTIGATION

§1. Fundamental Principles of the Inertia-Free Method of Studying Secondary Ion and Electron Processes

As already indicated, one of the difficulties of studying secondary emission by the static method is the fact that, in the course of measurement, the target surface is required to remain clean, yet at the same time measuring the energy distribution of the secondary particles by the static method, i.e., taking one volt—ampere characteristic, is a long process, requiring several hours or more, and it is almost impossible to keep the surface clean for this period.

First, the surface becomes contaminated with gases in a bad vacuum. The target surface is covered with a monomolecular layer of adsorbed residual-gas molecules in a time determined by the formula

$$t = \frac{\sqrt{2\pi m_g kT}}{d_g^2 p q_g},$$
(2.1)

where p is the pressure of the residual gas, q_g is the adhesion coefficient of the gas molecule to the surface, d_g is the Goldschmidt diameter of the molecule, and m_g is its mass. Table 2.1 shows the values of t for various temperatures and pressures of the residual gases (O_2 and N_2) for $q_g = 1$.

Secondly, the primary ion beam itself is a source of target-surface contamination. An ion beam of density j_0 creates a monomolecular coating on the target surface in a time equal to

$$t = \frac{e}{q_g j_0 d_g^2}.$$
(2.2)

It follows from Table 2.2 that this time lies between a few seconds and a few minutes, depending on the density of the beam and the kind of ions.

The majority of experiments were carried out with a comparatively low vacuum. Hence, essentially it was not pure metals which were being studied, but films of unknown composition adsorbed on the targets under examination. By substantially shortening the time between the cleaning of the target and the measurement, and also reducing the measuring period, the working conditions on the target surface may be considerably improved. In this respect a reduction of two or three orders in the measuring time is equivalent to the same reduction in the residual pressure (improvement to the vacuum), as indicated by Tables 2.1 and 2.2.

A rapid recording method enabling us to take the whole volt—ampere curve in a fraction of a second after cleaning the target, even at 10^{-7} mm Hg, would enable us to work in target-surface conditions such as could only be obtained at a vacuum of 10^{-9}-10^{-10} mm Hg when using

Table 2.1. Time for the Formation of a Monatomic Layer of
Residual Gases for Various Temperatures and Pressures

Temp., °K	Pressure, p, mm Hg	Gas	
		O_2 ($m_g = 32 \cdot 1.66 \cdot 10^{-24}$ g)	N_2 ($m_g = 28 \cdot 1.66 \cdot 10^{-24}$ g)
300	10^{-6}	28.4 sec	26.6 sec
	10^{-7}	4.7 min	4.4 min
	10^{-8}	47.4 min	44.4 min
	10^{-9}	7.9 h	7.4 h
1100	10^{-6}	54.5 sec	51.0 sec
	10^{-7}	9.1 min	8.5 min
	10^{-8}	1.5 h	1.4 h
	10^{-9}	15.1 h	14.2 h
1500	10^{-6}	63.6 sec	59.5 sec
	10^{-7}	10.6 min	9.9 min
	10^{-8}	1.7 h	1.6 h
	10^{-9}	17.6 h	16.5 h

Table 2.2. Time for the Formation of a Monatomic Layer by
Adsorption from the Beam at Room Temperature and Various
Current Densities

Atom	$j_0 = 10^{-4}$ A/cm^2	$j_0 = 10^{-5}$ A/cm^2	$j_0 = 10^{-6}$ A/cm^2
Li	13.2 sec	2.2 min	22 min
Na	8.3	1.4	13.8
K	4.5	45.3 sec	7.5
Rb	3.6	36.0	6.0
Cs	2.9	29.4	4.9
Ne	3.1	31.2	5.2
Ar	2.2	21.7	3.6
Kr	2.0	20.4	3.4
Xe	1.7	16.8	2.8

the galvanometric method, which requires some 5 sec per point, or about an hour to take the whole volt—ampere characteristic.

It would appear that a simple solution to the problem would be to hold the target at a temperature sufficient for the desorption of the adsorbed films. However, this method has usually been ignored by research workers, apparently on the grounds of the following considerations. If the heated target is bombarded with alkali ions, then the adsorbed atoms from the primary beam may evaporate in the form of ions as a result of surface ionization. The evaporation of positive ions from metal surfaces starts at about 900–1200°K. These ions, not being in essence secondary, will nevertheless be recorded by the galvanometer circuit as secondary particles, and this will lead to serious errors in determining the coefficient of secondary ion emission. For a target temperature of over 1400–1500°K, there is usually a great deal of thermionic electron and ion emission, so that the measurement of secondary processes becomes almost impossible. This difficulty, which is associated with the inertia of the galvanometric method, is the main obstacle to the study of rapid processes taking place when a solid surface is bombarded with atomic particles.

The foregoing considerations indicate the necessity of developing an experimental method which would, first, effect the measurements in a very short time so as to eliminate the effects of adsorption and, secondly, separate the truly "secondary" ions (those escaping from the target as a result of the impact of a primary ion) from ions ejected from the target surface by the thermal motion of the atoms (evaporation).

Even in a very high vacuum and with complete purification of the beam from neutral particles, the results are still distorted by the adsorption of ions from the primary beam. This difficulty is of a fundamental character and cannot be entirely eliminated. An exception occurs for experiments in which bombardment is carried out by means of ions belonging to the same element as the target [240]. Even in this case, however, one has to remember the lattice distortions created by the primary ions. Nevertheless, the effect of adsorption from the beam may be almost eliminated by reducing the value of the primary current, increasing the target temperature, and shortening the time during which the ion beam acts on the target.

The first method seemed particularly promising, since in measuring the primary and secondary currents sensitive electrometers may be used and the value of the currents measured may be reduced by four or five orders. However, a practical trial of this method revealed serious difficulties which almost cancelled its advantages. It was found that, on sharply reducing the working currents, "parasitic" conduction currents, the intrinsic emission of thermoelectrons and thermoions from the heated target, induced currents arising from the capacitive couplings, photoeffects, etc., acquired particular significance. Experiments showed that it was undesirable to work with primary currents smaller than 10^{-10} A.

The second method, involving a reduction in the adsorption of adatoms on the surface by raising the target temperature, is also not irreproachable from the point of view of principle, and is in general undesirable. The point is that, in order to establish the nature of the phenomenon, it is often necessary to determine the temperature dependence of the secondary processes as a primary factor, whereas usually the influence of temperature arises by virtue of the varying density of the adsorbed film. As an example of this incorrect interpretation of the role of temperature we may cite one of the papers in [220], in which it was asserted, without considering the influence of adsorption, that the secondary emission coefficient of electrons associated with ionic bombardment fell with increasing target temperature.

The third method, in which the effect of adsorbed atoms is reduced by substantially shortening the target-bombardment period, proved not only convenient in the experimental respect, but also excellent from the fundamental point of view.

The author of this monograph estimated the possibilities of measuring secondary processes by methods using ballistic equipment, amplitude indicators, and oscillographic recording [284].

As already indicated, one volt—ampere characteristic in practice took several hours to obtain by the galvanometric method, since the target had to be cleaned afresh by heating before taking each point on the curve.

The process of recording secondary and primary currents may be accelerated and automated by using an oscillographic circuit without fundamentally altering the measuring method. Let us suppose that, instead of the galvanometers 6 and 7 in Fig. 1.13 we have the input of a current amplifier connected to the vertical plates of an oscillograph. Then the vertical deflection of the electron beam in the oscillograph will be proportional to the current I_6 or I_7, respectively. If we vary the voltage between the target and collector $V_{2,3}$, the variation in current I_6 or I_7 will correspond to the relationship $I_6 = I_6(V_{2,3})$. For periodic changes $V_{2,3} = V_{2,3}(t)$, the current $I_6(t)$ will also be a periodic function. This enables us to synchronize $I_6(t)$ to the oscillograph sweep and obtain a motionless picture on the screen.

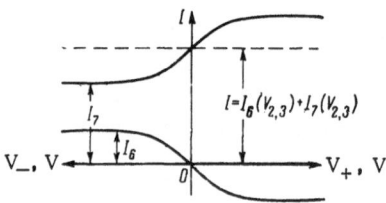

Fig. 2.1. Example of the volt—ampere characteristics of secondary emission.

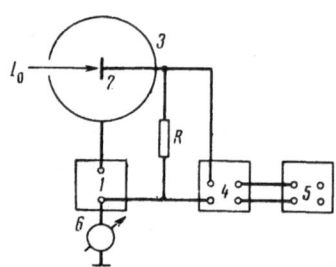

Fig. 2.2. Fundamental electrical circuit of the oscillograph method of measuring secondary currents.

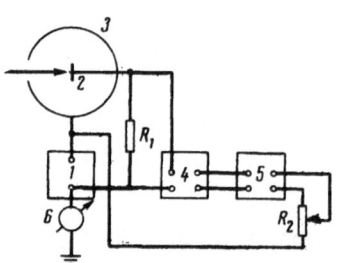

Fig. 2.3. Modification of the circuit in the oscillograph method of measuring secondary currents.

Analysis of this picture is in general complicated, and can only be easily carried out in the case in which the voltage $V_{2,3}$ varies on the same law as the voltage on the x plates of the oscillograph. Since, in ordinary oscillographs, the sweep is linear, we find that if the voltage $V_{2,3}$ also varies linearly, a picture similar to that shown in Fig. 2.1 will appear on the oscillograph screen (with a change in the scales along the x and y axis, of course).

In practice, the oscillographic recording of characteristics by the method in question may be carried out with the help of the circuit shown in Fig. 2.2, where 1 is a saw-tooth voltage generator, 2 is the target, 3 is the collector, 4 is a linear amplifier, 5 is an oscillograph with a linear sweep, 6 is a galvanometer, I_0 is the primary beam, and R is a resistance shunting the capacitive input of the accelerator. In order to keep the operation of the sawtooth voltage generator and the generator of the sweep voltage in the oscillograph synchronized, it is better to replace these by a single source of periodically varying voltage. Our own circuit for operating the measuring part in this case is shown in Fig. 2.3. The intensity of the primary beam is checked by the galvanometer 6. The generator of the periodically varying voltage 1 creates a variable potential difference $V_{2,3}$ between the target 2 and the collector 3, and at the same time, by way of the potential divider R_2, specifies the sweep voltage on the x plates of the oscillograph 5.

This arrangement is convenient because, independently of the shape of the periodic pulses from the generator 1, the picture on the oscillograph always reproduces (on a certain scale) the curve relating the secondary current to the retarding potential difference. If, in fact, the linear amplifier 4 and the oscillograph circuit give a sensitivity K_y along the y axis, while the potentiometer R_2 moves the voltage $V_{2,3}$ along the x axis, we have

$$y = K_y I(t), \quad x = K_x V_{2,3}(t_1).$$

For synchronization, however, $t = t_1$. Since I and V are related to one another by the distribution curve

$$I = f(V), \tag{2.3}$$

x and y will be related to each other by the expression

$$\frac{y}{K_y} = f\left(\frac{x}{K_x}\right). \tag{2.4}$$

Thus the curve (see Fig. 2.1) on the oscillograph screen will reproduce the relation (2.3) between the secondary current and the potential difference $V_{2,3}$ on a certain scale.

Since, on using the arrangement of Fig. 2.3, the shape of the voltage—time relationship is not particularly important, it is permissible to use a source of sinusoidal voltage of specific frequency, a sawtooth generator, or any other source as generator. It must be remembered,

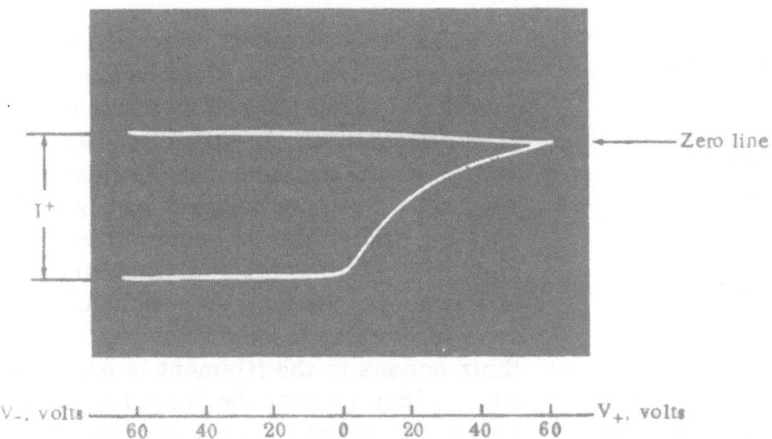

Fig. 2.4. Oscillogram of the volt—ampere characteristic of
secondary emission from pure cold tantalum.

however, that only for a voltage varying linearly with time (sawtooth voltage) will the motion of
the electron beam of the oscillograph along the x axis be uniform. For example, a sinusoidal
voltage will be inconvenient because the part of the oscillogram close to zero, which corresponds to the transition from a retarding to an accelerating potential, will be traversed with a
maximum velocity, and the line on the oscillograph corresponding to the transition from ion to
electron currents will be poorly visible on the photograph.

The arrangement of Fig. 2.3 was checked experimentally. For this we used the apparatus
described in Chapter 1 (see Fig. 1.1). As target we employed pure cold tantalum bombarded
with positive Rb$^+$ ions after passing the beam through a cylindrical condenser to separate the
ions from the neutral atoms.

The oscillogram relating the secondary ion current I$^+$ (ordinate) to the potential difference V$_\pm$ (abscissa) is shown in Fig. 2.4. This was obtained in the following way. In a fraction
of a second the horizontal line, which we shall subsequently call the zero line of the oscillograph (abscissa), is photographed. Then a beam of positive ions is directed at a just-cleaned
and cooled target. Instead of the zero straight line, a line containing a break (lower line in
Fig. 2.4) appears. This is photographed on the same film as the zero line. If we disconnect the
ion current, the zero line will again appear on the screen. Thus, by taking two photographs, the
volt—ampere characteristic of the secondary processes may be secured.

§2. Oscillographic Method of Studying Phenomena

of Adsorption on a Metal Surface

The use of the oscillographic method is most effective when studying such rapidly occurring processes as adsorption, evaporation, and the ionization of atoms on the surface of a metal.

Starodubtsev [190] studied the phenomenon of adsorption and surface ionization on bombarding a heated surface with a flow of particles unable to penetrate into the depths of the target because of their low energy. Alkali ions or atoms served as particles of this kind. The investigations were carried out by the modulated molecular-beam method. This method made it
possible to estimate two important characteristics of the adsorption of electropositive atoms:
the heat of adsorption and the absolute coefficient of surface ionization.

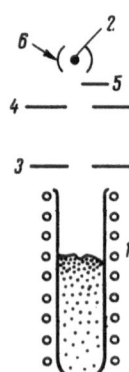

Fig. 2.5. General arrangement of the electrodes in the apparatus.

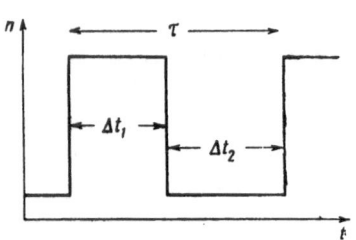

Fig. 2.6. Modulation shape of the molecular beam.

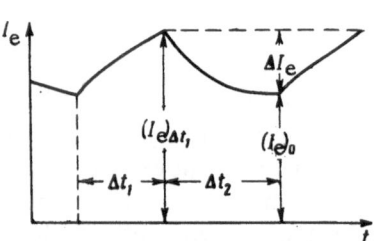

Fig. 2.7. Oscillogram of the time variation of I_e.

Let the furnace 1 (Fig. 2.5) heat a certain substance, creating a certain vapor pressure of atoms or molecules capable of being ionized on the incandescent surface of the metal filament 2. Let the diaphragms 3 and 4 cut out a narrow beam capable of being interrupted by a mechanical device 5. The interruption may be effected either by a rotating disc with slots cut in it or by a vibrating elastic plate, or by some other suitable means. Thus, for a certain time Δt_1 the atoms or molecules in the beam have access to the incandescent filament 2, while for a time Δt_2 their access to the filament is prevented. The sum $\Delta t_1 + \Delta t_2 = \tau$ determines the period of modulation of the atomic beam, the modulation shape of which is shown in Fig. 2.6; the current of atoms or molecules experiencing surface ionization is collected by the collector 6 when there is an adequate potential difference between this and the filament 2.

Naturally the ion current will be modulated at the same frequency as the molecular beam. However, the shape of the modulated ion current and the depth of modulation will not be the same as in the molecular beam. Clearly, an oscillographic study of the time dependence of the ion current will enable us to determine the "life" of the adsorbed atoms and the heat of adsorption.

The ion current I_e may be calculated by means of the formula (derivation given later)

$$I_e(t) = (I_e)_\infty + [en_0 W_+ - (I_e)_\infty] \exp[-a(T)t], \qquad (2.5)$$

where

$$a(T) = A \exp\left(-\frac{\lambda_0}{kT}\right) + B \exp\left(-\frac{\lambda_+}{kT}\right),$$

in which λ_0 and λ_+ are related by the Schottky relation [16]

$$\lambda_- - \lambda_0 = V_i - \varphi, \qquad (2.6)$$

while A and B vary only very slightly with temperature and are related to the probabilities of the evaporation of the atom W_0 and the ion W_+ in the following way:

$$W_0 = A \exp\left(-\frac{\lambda_0}{kT}\right), \quad W_+ = B \exp\left(-\frac{\lambda_+}{kT}\right). \qquad (2.7)$$

At the instant at which the beam of molecules ceases to be applied to the filament, i.e., after an interval of time Δt_1, the quantity I_e becomes equal to

$$(I_e)_{\Delta t_1} = (I_e)_\infty + [en_0 W_+ - (I_e)_\infty] \exp[-a(T)\Delta t_1].$$

On the section Δt_2 (see Fig. 2.6), the ion-emission current will fall exponentially:

$$(I_e)_{\Delta t_2} = en_0 W_+ = (I_e)_{\Delta t_1} \exp[-a(T)(t - \Delta t_1)]. \qquad (2.8)$$

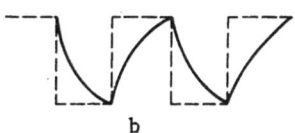

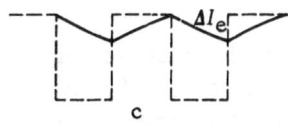

Fig. 2.8. Different forms
of oscillograms.

At the moment of time $\tau = \Delta t_1 + \Delta t_2$,

$$(I_e)_{\Delta t_2} = (I_e)_{\Delta t_1} \exp\left[-a(T)\Delta t_2\right]. \qquad (2.9)$$

If a quasi-stationary condition is reached, the final state of one period, i.e., $(I_e)_{\Delta t_1}$, should give the density of the initial current $en_0 W_+$ for the next period

$$\{(I_e)_\infty + [(I_e)_0 - (I_e)_\infty] \exp\left[-a(T)\Delta t_1\right]\} \exp\left[-a(T)\Delta t_2\right] = (I_e)_0,$$

whence

$$(I_e)_0 = (I_e)_\infty \left[1 - e^{-a(T)\Delta t_1}\right] e^{-a(T)\Delta t_2} (1 - e^{-a(T)\tau})^{-1}.$$

Thus the ion current will not follow the variation of the molecular beam; it will vary between the limiting values $(I_e)\Delta t_1$ and $(I_e)_0$ in accordance with Eqs. (2.8) and (2.9) along sections of exponential curves (Fig. 2.7).

Calculation of Heats of Adsorption from the Oscillograms of the Ion Currents

If the voltage drop in a certain resistance due to the ion current is fed to the vertical plates of an oscillograph, and the horizontal sweep is synchronized with the frequency $\nu = 1/\tau$, we may obtain a picture of the varying component of ion current on the screen (Fig. 2.7). An oscillogram of this kind enables us to find the value of $a(T)$. In this respect there are three possible cases, differing methodically from one another (Fig. 2.8).

1. $a(T)\Delta t_1 \gg 1$ (Fig. 2.8a). The current oscillogram is very close in shape to the curve representing the variation in the molecular beam. In this case it is usually impossible to plot or photograph the exponential part of the current rise-and-fall curve; however, by using the fact that the "depth of modulation" is 100%, we may measure the deviation (on the time scale) of the curve from the vertical for an ordinate corresponding to half the amplitude of the curve; this will be the time for the current to be reduced by a factor of two, and hence,

$$a(T) = \ln\left(\frac{2}{\Delta t_1}\right).$$

2. $a(T)\Delta t_1 \approx 1$ (Fig. 2.8b). Here we have a well-expressed oscillogram of the exponentially rising and falling parts of the curves. We may photograph this and then, after plotting on a semilogarithmic scale, find $a(T)$ to a high degree of accuracy.

3. $a(T)\Delta t_1 \ll 1$ (Fig. 2.8c). In this case, the shape of the oscillogram will not allow us to determine $a(T)$ accurately, since only a small part of the exponential curve falls on the graph. However, even in this case we may determine $a(T)$ if we note the maximum change in the current and also measure the average current I_e with a galvanometer.

It follows from (2.9) that

$$\frac{\Delta I_e}{(I_e)_{\Delta t_1}} = \frac{[(I_e)_{\Delta t_1} - (I_e)_0]}{(I_e)_{\Delta t_1}}.$$

Using the fact that $a(T)\Delta t_2$ is small, and expanding the exponential term in series, we obtain

$$\frac{\Delta I_e}{(I_e)_{\Delta t_2}} = a(T)\Delta t_2 - \frac{1}{2}[a(T)\Delta t_2]^2 + \ldots,$$

and, analogously,

$$\frac{\Delta I_e}{(I_e)_0} = a\,(T)\,\Delta t_2 + \frac{1}{2}\,[a\,(T)\,\Delta t_2]^2 + \ldots$$

However, the average value of the current measured by the galvanometer in series with the collector lies between $(I_e)_0$ and $(I_e)_{\Delta t_2}$. Hence,

$$\frac{\Delta I_e}{(I_e)_0} > \frac{\Delta I}{I_e} > \frac{\Delta I_e}{(I_e)_{\Delta t_2}}.$$

Thus if we replace $(I_e)_0$ or $(I_e)_{\Delta t_2}$ by the mean value of the steady current I_e, we make an error not exceeding $[a(T)\Delta t_2]^2$

The measurement of $a(T)$ in all three forms enables us to cover a wide range of temperatures, even without changing the frequency of modulation ν, and to obtain a series of values of $a(T)$. However, it follows from (2.5) that, if we plot a graph of $\ln a(T)$ as a function of the reciprocal temperature $1/T$, then in general we should not obtain a straight line. Using relation (2.6), let us transform $a(T)$ from (2.5):

$$a\,(T) = A e^{-\lambda_0/kT}\left[1 + \frac{B}{A}\,e^{-(V_i - \varphi)/kT}\right] = B e^{-\lambda_+/kT}\left[1 + \frac{A}{B}\,e^{-(\varphi - V_i)/kT}\right]. \qquad (2.10)$$

We see from (2.10) that the value of $a(T)$ obtained experimentally may serve to calculate the heats of evaporation of the ions and neutral atoms. However, in order to calculate these quantities we must not set up $\ln a(T)$ and $f(1/T)$ in the manner of Moon, Oliphant, and Evans [72, 77, 78].

From formula (2.10) we may deduce a linear dependence on the reciprocal temperature of the following expressions:

$$Y_0 = \ln \frac{a\,(T)}{1 + (B/A)\,e^{-(V_i - \varphi)/kT}} = \ln A - \lambda_0/kT, \qquad (2.11)$$

$$Y_+ = \ln \frac{a\,(T)}{1 + (B/A)e^{-(\varphi - V_i)/kT}} = \ln B - \lambda_+/kT. \qquad (2.12)$$

The slope of the straight lines (2.11) and (2.12) determines the heat of vaporization of the neutral atoms and ions.

Experimental Application of the Method

In order to verify the applicability of the method experimentally, we used the apparatus shown in Fig. 2.9 [190]. On the ground-glass joint 1 we mounted a furnace with a Nichrome winding 10 in order to produce the K or Na vapor. The molecular beam was formed by means of three diaphragms: in the furnace 2, at the screen 3, and in the receiving cylinder 4. A tungsten filament 5 of diameter $80\,\mu$ and 6 cm long lay in the beam of atoms passing through the diaphragm. The ions formed by surface ionization on the filament were collected by the field into the measuring cylinder 6. In the path of the atomic beam was an iron slide 7 fixed by means of an elastic tungsten spring 12 in a rigid base. By means of the iron core of the coil 9 and the magnetic conductor 8, the slide 7 could be set into intensive oscillatory motion.

In the simplest case oscillatory motion was communicated to the plate 7 at a frequency of 100 cps. For this the coil 9 was supplied with a 50-cps current. By rotating a ground-glass joint set perpendicular to the plane of the sketch, the filament could be moved relative to the atomic beam and the slide 7, thus regulating the relative time intervals Δt_1 and Δt_2. The beam of atoms passing through all the diaphragms and not falling on the filament was caught in the upper part of the apparatus by the liquid-air trap 11. The rest of the operation of the system is clear from the sketch.

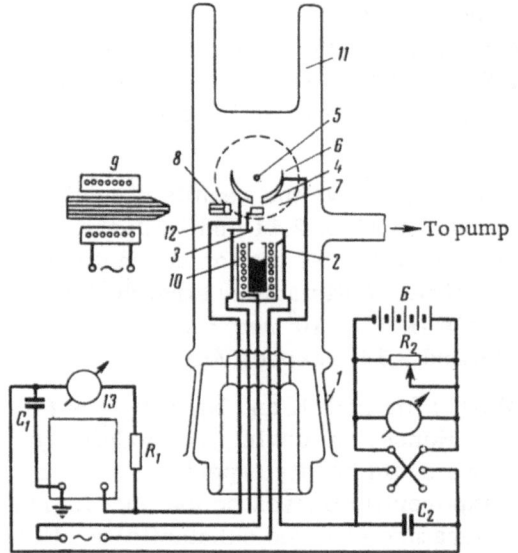

Fig. 2.9. Apparatus and electric circuit for the measurements.

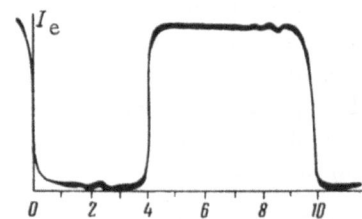

Fig. 2.10. Typical oscillograms corresponding to a high temperature of the filament.

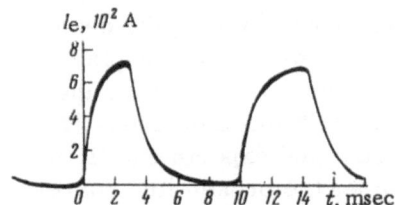

Fig. 2.11. Oscillogram for K at a filament temperature of 1250°K.

After ordinary preparation of the apparatus for the measurements, the work function was determined in order to check the state of the surface. Then the furnace was started, the current was fed to the coil of the electromagnet, and the slide was set in oscillatory motion. By rotating the ground-glass joint with the filament and simultaneously observing the form of the oscillogram at T = 2400°K, the necessary relation between Δt_1 and Δt_2 was established. Then by regulating the heating of the furnace we selected a beam density giving a reasonably clear oscillogram of the ion current. We measured the current $\bar{I}_i$ with the galvanometer 13 and also the filament-heating current, and then photographed (or measured with a transparent coordinate grid) the oscillogram of the varying component of the ion current.

One advantage of using periodically repeating processes is the fact that the oscillograms average out the random errors and at the same time enable the scatter of the curves to be estimated. For photographing the oscillograms an exposure of the order of 15–20 sec is required. A similar time is required for visual observations. In this time the process repeats itself about 2000 times, and if in this period all the curves merge into one then this gives confidence in the absence of random errors, which are inevitable when plotting curves point by point. For recording the currents we used an oscillograph connected in accordance with Fig. 2.9. This was earlier calibrated for the purpose and showed a maximum current sensitivity of $2.8 \cdot 10^{-9}$ A/mm.

The leads and all the measuring apparatus were screened. The galvanometer and batteries were decoupled with condensers. Even so, at the highest sensitivity we obtained interference with a maximum scale value of 5 mm. Ordinary measurements were therefore made with an oscillograph current sensitivity of $4 \cdot 10^{-9}$ A/mm or under.

In all cases, the work function of the tungsten filament lay between 4.53 and 4.59 eV.

Figure 2.10 shows some typical oscillograms corresponding to a high filament temperature. For sodium, curves of this kind were obtained with T > 1550°K and for potassium with T > 1400°K. On reducing the temperature, the oscillograms gradually changed their form, the exponential parts of the curves appearing more distinctly. The picture appears most clearly for $a(T) \approx 10^3$, when the exponential damping half period is close to a thousandth of a second. Here almost all the current change occurs during the time Δt_2. Figure 2.11 shows an oscillogram for potassium at a filament temperature of 1250°K.

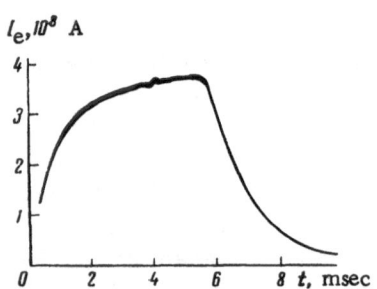

Fig. 2.12. Oscillogram for Na at a
W-filament temperature of 1400°K.

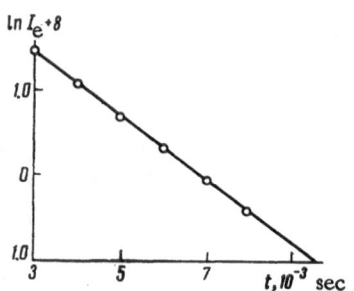

Fig. 2.13. Analysis of the oscillogram of Fig. 2.11.

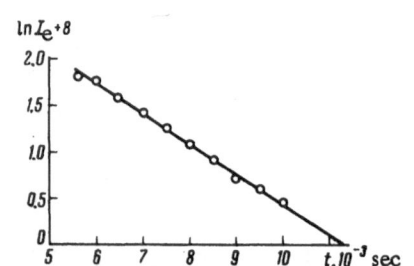

Fig. 2.14. Analysis of the oscillogram of Fig. 2.12.

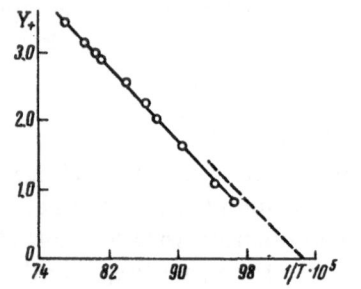

Fig. 2.15. Variation of Y_+ with
$1/T$ for K^+.

Figure 2.12 shows a similar oscillogram for sodium, but at a temperature of 1400°K. The ion current (Figs. 2.11 and 2.12) is modulated almost completely. This enables us to verify the exponential character of the curves.

Figures 2.13 and 2.14 give the results of analyzing the oscillograms shown in Figs. 2.11 and 2.12; the ordinates represent the logarithms of the ion current and the abscissas represent time. Satisfactory straight lines are obtained; the slope is $8.6 \cdot 10^2$ sec^{-1} and $7.3 \cdot 10^2$ sec^{-1} for Figs. 2.13 and 2.14, respectively.

The value of $a(T)$ for both K and Na was very temperature dependent. Hence, on leaving the modulation frequency constant we could hardly expect to observe large parts of the exponential curves on the oscillograms if the filament temperature changed substantially. Even the oscillogram shown in Fig. 2.12 with $a(T) = 7.3 \cdot 10^2$ sec^{-1} is not situated completely on the graph. The axis of abscissas is drawn below and its position determined by equating the average current through the galvanometer to the average current determined from the oscillograms. In order to extend the range of temperatures, measurements of $a(T)$ were carried out under all three conditions. The results of these measurements are shown in the graphs. Figure 2.15 gives the relation between Y_+ (Fig. 2.12) and $1/T$ for K^+. For potassium, which has an ionization potential of $V_i = 4.32$ eV, and a tungsten filament with a work function of 4.54 eV,

$$Y_+ = \ln [a(T)/(1 + 2e^{-2550/T})],$$

where the factor 2 in front of the exponential function corresponds to the ratio of the statistical weights of the atom and the ion.

In the temperature range between 1050 and 1450°K in which the measurements were made, the denominator is close to unity and $Y_+ = \ln a(T)$ to three-place accuracy. This enabled Starodubtsev to compare the results of his own measurements with those of Evans [77, 78]. The curve in Fig. 2.15 obtained experimentally is represented by a straight line; the broken line corresponds to Evans' data. Starodubtsev's straight line leads to a heat of evaporation of the ion equal to 2.55 eV; the slope of Evans' straight line gives a value of 2.43 eV for K^+ ions. The agreement may be considered satisfactory, particularly in view of the fact that the data relate to different temperature ranges.

For sodium we cannot calculate the heat of evaporation of the ions by means of the simplified formula, re-

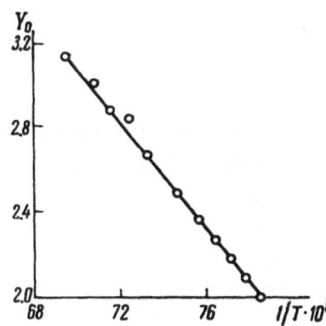

Fig. 2.16. Variation of Y_0 with $1/T$ for Na.

garding $Y = \ln a(T)$. For this element the ionization potential $V_i = 5.12$ eV. Hence, in this case instead of (2.12) we have

$$Y_+ = \ln [a(T)/(1 + 2e^{6500/T})].$$

Thus, the change in the denominator with temperature is quite commensurable with the change in the numerator, and hence, if we take the simplified formula $Y_+ = \ln a(T)$ in order to calculate the evaporation of the Na$^+$ ion from tungsten, we shall obtain an incorrect result.

It is easy to show that for sodium the rate of change of $a(T)$ with temperature is determined by the heat of evaporation of the neutral atom. It follows from (2.11), in fact, that

$$Y_0 = \ln [a(T)]/(1 + 1/2e^{-6500/T}) \approx \ln a(T).$$

In Fig. 2.16 the ordinates represent the values of Y_0 and the abscissas represent $1/T$. The graph constitutes a straight line from which the heat of evaporation may be calculated as 2.73 eV. The heat of evaporation of the Na$^+$ ion is found from the relation $\lambda_+ = \lambda_0 + V_i - \varphi = 3.3$ eV. Calculation by formula (2.11) of course leads to the same result.

The value of the heat of evaporation of the Na$^+$ ion from a tungsten surface cannot be compared with any reliable measurements of other authors or even with any calculations, since none have been carried out for the heat of evaporation of Na$^+$ ions from tungsten.

We might be able to use the method proposed by Evans for cesium, rubidium, and potassium. Evans compared his own data with the results obtained by calculating the heat of adsorption of the ion, supposing this to be entirely due to image forces, i.e., $\lambda_+ = e^2/4r_2$, where r_2 is the radius of the ion taken from x-ray measurements of the lattice constant of the alkali halides. Although Evans obtained comparatively slight differences between the experimental and calculated values, this method of comparison is unconvincing, since it suggests that the heat of evaporation of the ion depends only on the structure of the ion itself and it is entirely independent of the nature of the adsorbing metal.

In view of this we should note that Starodubtsev's value of $(\lambda_+)_{Na} = 3.3$ eV differs little from that calculated by means of the formula for the image forces $(\lambda_+)_{Na} = 3.0$ eV; however, he did not regard this as a confirmation of his own data. Nevertheless, the oscillograms presented and the regular arrangement of the points on the straight lines $\ln I_e = f(t)$ and $Y_+ = f_1(1/T)$ confirm the validity of the method and give confidence in the reliability of the results.

The foregoing discussions and experiments relate to a slight covering of the surface with alkali atoms. It is thus very important to estimate the surface densities of the adsorbed atoms. In all cases,

$$\frac{I_e}{a(T)} = en \frac{Be^{-\lambda_+/kT}}{a(T)} = en \left\{ \frac{e^{-(V_i - \varphi)/kT}}{2 + e^{-(V_i - \varphi)/kT}} \right\},$$

whence

$$n = [I_e/a(T)e][2 + e^{-(V_i - \varphi)/kT}]/e^{-(V_i - \varphi)/kT}.$$

For the lowest temperature in the experiments with sodium (1300°K), $I_i = 2 \cdot 10^{-8}$ A, $a(T) = 116$ sec^{-1}. These data give $n_{max} = 4 \cdot 10^{13}$ atoms/cm^2, which corresponds to $\theta < 0.1$. For the highest temperature (1450°K), θ falls to about 10^{-3}. In experiments with K$^+$, the value of θ varies from 10^{-2} to 10^{-4}.

§3. Oscillographic Method of Determining the Heat of Evaporation of Atoms and Ions

Starodubtsev's molecular-beam modulation method not only made it possible to expand the temperature range of the investigation in which adsorption was being studied, but also enabled the experiments to be extended to atoms of types difficult to ionize, having an ionization potential exceeding the work function of the adsorbent. However, this method was rather complex. It was essential to find a suitable method for studying the adsorption of electropositive atoms in a simpler setting, permitting the use of a sealed-off apparatus, and this was done by Arifov, Lovtsov, and Ayukhanov [192].

Let us consider an incandescent metal filament in an atmosphere of alkali vapor, periodically changing the sign of its potential. For a negative potential on the filament, adsorbed electropositive atoms can of course only evaporate in the form of neutral particles, because the number of negative ions is very small. For a positive potential the adatoms evaporate more intensively, since neutral and charged particles evaporate at the same time. Hence, for a negatively charged filament there is an accumulation of adatoms on the surface, while for a positive charge the number of adsorbed particles diminishes, asymptotically approaching a certain equilibrium value.

Let us consider only slight degrees of filament coverage. Then the probability W_0 of the evaporation of an adatom in 1 sec may be regarded as a constant value for the particular temperature, while the probability W_+ of the evaporation of a positive ion may be regarded as independent of the coverage. It is clear that the equilibrium concentrations of the adatoms at a potential retarding the ions is

$$n_+ = \frac{N_0}{W_0},$$

and at a potential attracting the ions

$$n_- = \frac{N_0}{W_0 + W_+},$$

where N_0 is the number of neutral atoms falling in 1 sec on the incandescent filament from the gas phase.

There is no doubt that we shall always have $n_+ > n_-$. On changing the sign of the filament charge there is a transitional process in which the concentration of the adatoms changes from n_+ to n_-. A study of the process of equilibrium establishment offers the possibility of determining the heat of adsorption. By developing a discussion analogous to that of the preceding section, we may consider the processes involved in the accumulation and evaporation of adatoms when the potential of the ionizing surface changes.

<u>Accumulation Process.</u> A negative potential is applied to the filament. Let N_0 atoms per sec fall on the filament. Then

$$\frac{dn}{dt} = -W_0 n + N_0,$$

whence

$$n = \left(n_\rangle - \frac{N_0}{W_0}\right) \exp\left(-W_0 t\right) + \frac{N_0}{W_0}.$$

Thus for a negative potential on the filament the rate of accumulation of adatoms is determined exclusively by the probability of the evaporation of neutral atoms W_0. If the initial number of adsorbed particles $n_0 > N_0 / W_0$, there is a falling curve; if, however, $n_0 < N_0 / W_0$, then the number of adsorbed atoms rises exponentially.

Evaporation Process. If a positive potential is applied to the filament, then the differential equation of the process of equilibrium establishment takes the form

$$\frac{dn}{dt} = -(W_0 + W_+) n + N_0,$$

whence

$$n = \left(n_0 - \frac{N_0}{W_0 + W_+}\right) \exp \left[-(W_0 + W_+) t\right] + \frac{N_0}{W_0 + W_+}, \qquad (2.13)$$

and hence the process of establishing equilibrium is in this case determined by the total probability of the evaporation of an adatom in the neutral or ionized form.

It would only be possible to observe the accumulation of adatoms by measuring the change in the electron currents. However, for small coverages this is not practicable. For this reason the equilibrium-establishment process was studied in a different way, using the phenomenon of surface ionization.

According to (2.13), the current of ions from the filament equals

$$I = en B \exp \left(-\frac{\lambda_+}{kT}\right) = eB \exp \left(-\frac{\lambda_+}{kT}\right) \left\{\frac{N_0}{W_0 + W_+} + \left(n_0 - \frac{N_0}{W_0 + W_+}\right) \exp \left[-(W_0 + W_+) t\right]\right\}.$$

The steady value of the current $(I_e)_\infty$ equals

$$(I_e)_\infty = eB \exp \left(-\frac{\lambda_+}{kT}\right) \frac{N_0}{a(T)}.$$

For T = const we obtain

$$\Delta I_e = I_e - (I_e)_\infty = eB \exp \left(-\frac{\lambda_+}{kT}\right) \left(n_0 - \frac{N_0}{W_0 + W_+}\right) \exp \left[-a(T) t\right],$$

but, since

$$n_0 = \frac{N_0}{W_0},$$

we have

$$\Delta I_e = \frac{N_0}{W_0} eB \exp \left(-\frac{\lambda_+}{kT}\right) \frac{W_+}{W_0 + W_+} \exp \left[-a(T) t\right].$$

Hence the difference between the total current (at the instant of time t) and the equilibrium ion current varies on an exponential law.

The value of $a(T)$ is easy to find by subjecting the oscillogram to a logarithmic transformation

$$\ln \Delta I_e = \text{const} - a(T) t, \qquad (2.14)$$

as a result of which we obtain a straight line in the coordinate system of $\ln \Delta I_e$ and t, the slope determining the value of $a(T)$.

As shown by Starodubtsev, the heat of evaporation of atoms and ions may be determined by plotting the straight lines (2.11) and (2.12). The slopes of these straight lines lead to the heats of evaporation λ_0 and λ_+.

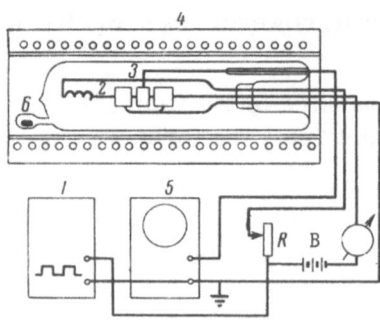

Fig. 2.17. Principal electrical circuit of the apparatus for the oscillographic determination of the surface-ionization coefficient of atoms and the heats of evaporation of atoms and ions.

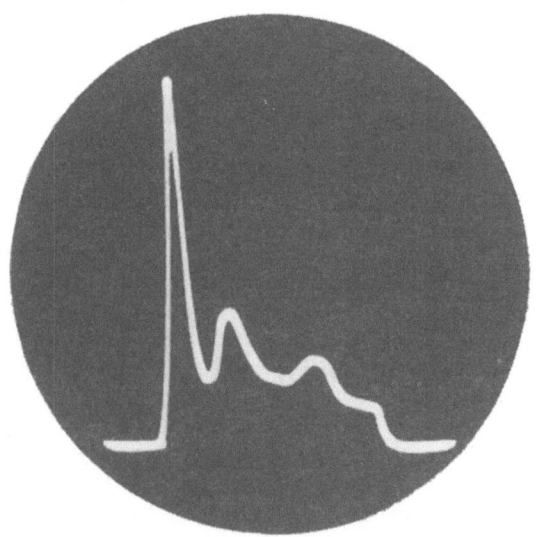

Fig. 2.18. Typical oscillogram of the ion currents of the transitional process, distorted as a result of parasitic oscillatory processes in the circuit.

Measurements and Discussion of Results

The measuring part of the apparatus consisted as usual of a tungsten filament 2 stretched along the axis of three coaxial cylinders 3, the end members of which constituted a guard ring, while the middle one (Fig. 2.17) was the collector. In Fig. 2.17, we have the following notation: 1) "top-hat" pulse generator; 4) furnace for controlling the saturation vapor pressure; 5) measuring oscillograph; 6) substance being studied; the heating circuit contains a heating battery, an ammeter, the filament, and a rheostat. The temperature of the filament is measured by reference to the filament heating current.

The order of the experiments was as follows. The apparatus was heated in the furnace to a temperature at which an appreciable ion current from the filament appeared on the middle measuring cylinder of the collector. The current from the guard cylinders bypassed the measuring devices, the arrangement of which was the same as in the previous experiments. After reaching current stability, the periodic, pulsed "top-hat" voltage was applied between the filament and the collector, and the curves representing the establishment of equilibrium were presented on the oscillograph.

Special attention was given to producing photographs of the oscillograms with a really sharp central part, representing the transitional process, and also to the elimination of all parasitic oscillatory and aperiodic processes which might distort the shape of the oscillogram (an example of such an oscillogram of ion current is shown in Fig. 2.18).

A change in the filament temperature leads to a sharp change in the shape of the oscillogram, since the index of the exponential $a(T)$ changes. Raising the temperature shortens the "life" of the particles on the surface, and the slope of the curves becomes sharper. Calculations showed that the fall in the ion current obeyed Eq. (2.14), i.e., the relation between $\ln \Delta I_e$ and t was a straight line (Fig. 2.19). The slope of this straight line gives the value of $a(T)$.

On the basis of the values of $a(T)$ determined for various temperatures of the filament and Eq. (2.12), the value of λ_+ was calculated. In Fig. 2.20, the ordinates represent Y_+ and the abscissas the reciprocal temperature. We see that the points fall on a straight line. The slope of the straight line gives the value of $\lambda_{K+} = 2.2$ eV, where λ_{K+} is the heat of adsorption of a positive K^+ ion.

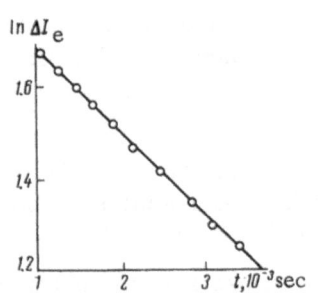

Fig. 2.19. Relation between $\ln \Delta I_e$ and the relaxation time t.

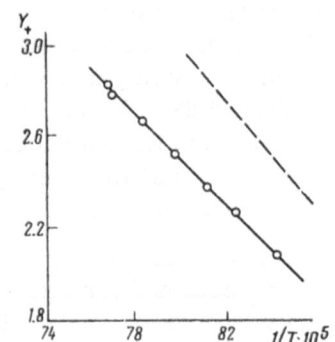

Fig. 2.20. Relation between Y_+ and 1/T.

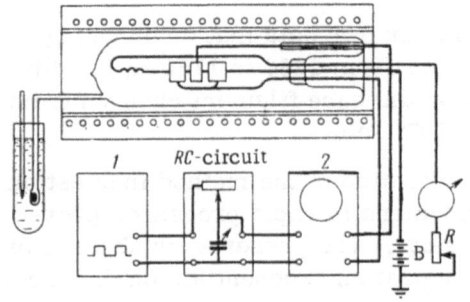

Fig. 2.21. Principal arrangement of the apparatus for determining the heat of adsorption of ions and atoms, using a functional sweep of the horizontal beam of the oscillograph.

The value obtained for λ_{K^+} may be compared with the results of other authors. The broken line in Fig. 2.20 is taken from the work of Starodubtsev [190], who used modulated molecular beams to determine the heat of evaporation. According to Starodubtsev, the slope of the straight line corresponds to λ_{K^+} = 2.52 eV. Still earlier, Evans obtained λ_{K^+} = 2.43 eV. Clearly, the deviation from the data of other authors should not be associated with inaccuracy in the method itself.

The experimental method in question cannot yet give satisfactory results for elements with high ionization potentials, since, in this case, it is difficult to accumulate a substantial excess of the atoms in a field retarding the ions. In addition to this, the photocurrents appearing for high filament temperatures restrict the possibilities of investigation.

For a more precise determination of λ_+ the experimental method may be improved by using periodic fields at high frequencies. This enables us to move the exponential along the horizontal and thus carry out a more precise determination of $a(T)$. It is probable that our value of $a(T)$ was rather lower than that of the other authors for the very reason that we used a frequency of 50 cps, for which the exponentials in the temperature range studied are extremely steep.

The method described for determining the heat of adsorption requires substantial computing operations in order to determine $a(T)$. If for this purpose we use the method proposed by Tolstoi and Feofilov [191, 207] for studying relaxation processes, we may determine the value of $a(T)$ in 1-2 min without photographing the oscillograms, plotting graphs, and making a cumbersome analysis of the curves. This method is also more accurate.

The method was verified by using an apparatus with a tungsten filament in potassium vapor, with commutation of the collector potential. The arrangement of the apparatus is shown in Fig. 2.21. The nonlinear (exponential) sweep of the oscillograph beam was achieved by means of an RC circuit (taumeter). "Top-hat" voltage pulses were applied to the input of the taumeter and between the filament and the collector from a special tube generator 1, and the voltage taken from the RC circuit (of exponential shape at the output) was applied to the horizontal amplifier of the oscillograph 2. The rest of the apparatus is clear from Fig. 2.21.

In order to determine the parameter $\tau_a = 1/a(T) = RC$ it is sufficient to achieve rectification of the curve on the oscillograph by an appropriate choice of R and C (R in MΩ, C in μF). The

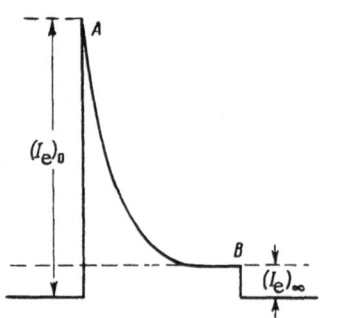

Fig. 2.22. Oscillogram of the time variation of the ion current obtained with a linear sweep of the horizontal oscillograph beam.

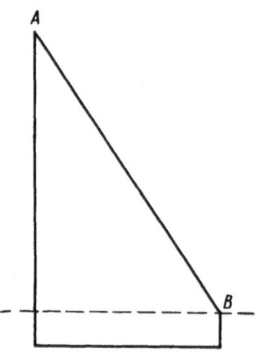

Fig. 2.23. Oscillograph representing the time variation of the ion current obtained with a nonlinear sweep of the horizontal oscillograph beam.

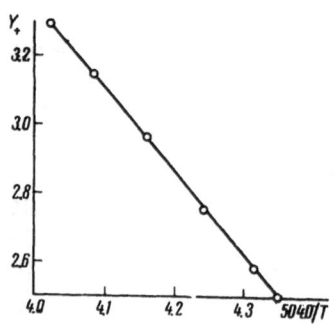

Fig. 2.24. Values of Y_+ as a function of $1/T$.

resistance R consisted of two decade resistances and one smoothly varying resistance r of 1000 Ω:

$$R = n_1 \cdot 10{,}000 + n_2 \cdot 1000 + r,$$

where n_1 and n_2 could be varied from 0 to 10. Thus, from the position of the taumeter scales we could immediately calculate the value of $\tau_a = RC$.

Figure 2.22 shows an oscillogram obtained with a linear sweep of the oscillograph beam and Fig. 2.23 a nonlinear version achieved by rectification. The law governing the fall in the ion current (see Figs. 2.25 and 2.26) is expressed by formula (2.14). If we develop the fall-off curve on the same law, the section representing the fall in ion current (section AB in Fig. 2.22) will be expressed by a straight line between the corresponding points (Fig. 2.23). In the next half period there is no ion current, and the oscillograph beam will return along the horizontal to its original position; then the process will be repeated.

The graph of Fig. 2.24 was plotted from the resultant values of $a(T) = 1/\tau_a$ obtained for different filament temperatures, using Eq. (2.12). The slope of the straight line gives the heat of evaporation of K$^+$ ions from tungsten as $\lambda_{K^+} = 2.41$ eV. This agrees closely with Starodubtsev's experiments [190] and also those of Evans [77, 78], which gave the heat of evaporation of K$^+$ ions as 2.52 and 2.43 eV, respectively.

We see from our own data that on changing the filament temperature from 1250 to 1150°K, the "life" of the K adatoms on the pure filament changes from $5.2 \cdot 10^{-4}$ to $4.1 \cdot 10^{-3}$ sec.

Thus the application of the method in question to adsorption phenomena in rapid processes proved completely successful. The presence or absence of characteristic evaporation exponentials on the secondary ion—emission oscillograms of incandescent surfaces may constitute an illustration of the part played by surface ionization in secondary ion—ion emission processes.

§4. Oscillographic Method of Determining the Coefficient of Ionization on the Surface of an Incandescent Metal

The coefficient of surface ionization is defined as the ratio of the number of ions leaving a heated metal surface to the number of neutral atoms falling on the same surface under equilibrium conditions.

The idea of the method of determining the absolute coefficient by means of an oscillograph employed by Arifov, Lovtsov, and Ayukhanov [192, 193] is as follows. Suppose that N_0 atoms per second fall on the filament for an alkali vapor pressure of p. If the temperature of the filament is high enough, the number of incident and evaporating atoms will be the same. We may write two relations, depending on the direction of the filament—collector field.

For (+) on the collector:

$$N_0 = n_+ \left[A \exp\left(-\frac{\lambda_0}{kT}\right) \right],$$ (2.15)

in which case the adatoms may only evaporate unrestrictedly in the form of neutral atoms.

For (–) on the collector:

$$N_0 = n_- \left\{ \left[A \exp\left(-\frac{\lambda_0}{kT}\right) \right] + \left[B \exp\left(-\frac{\lambda_+}{kT}\right) \right] \right\},$$ (2.16)

in which case the atoms may evaporate either as neutral atoms or in the form of ions.

If we suddenly change the direction of the filament—collector field, putting (–) on the collector, the ion current will have a maximum value at the initial instant and then fall off to a certain constant value (Fig. 2.25).

For the initial and final ion currents we may write the following expressions:

$$(I_e)_0 = e n_+ B \exp\left(-\lambda_+/kT\right),$$
$$(I_e)_\infty = e n_- B \exp\left(-\lambda_+/kT\right).$$

From these we obtain

$$\frac{(I_e)_0}{(I_e)_\infty} = \frac{n_+}{n_-}.$$

Equating the right-hand sides of Eqs. (2.15) and (2.16), we have

$$n_+ A \exp\left(-\lambda_0/kT\right) = n_- [A \exp\left(-\lambda_0/kT\right) + B \exp\left(-\lambda_+/kT\right)],$$

whence

$$\frac{n_+}{n_-} = \frac{A \exp\left(-\lambda_0/kT\right) + B \exp\left(-\lambda_+/kT\right)}{A \exp\left(-\lambda_0/kT\right)} = \frac{W_0 + W_+}{W_0} = 1 + \frac{W_+}{W_0}.$$

The expression for the ionization coefficient takes the form

$$\beta = \frac{W_+}{W_0 + W_+} = \frac{(I_e)_0 - (I_e)_\infty}{(I_e)_0}.$$ (2.17)

For sufficiently high filament temperatures (T > 1200°K), the excess of ions retarded by the field evaporates in a very short time; hence, in order to make the process periodic and observe a steady picture on the oscillograph, we must apply pulses of strictly rectangular shape to the collector.

The repetition frequency of the periodic pulses should be chosen in such a way that in a half period $\tau/2$ the excess of adsorbed atoms may evaporate completely. In our own experiments, $\tau/2$ was constant and equal to $1 \cdot 10^{-2}$ sec with a duty factor of 50%.

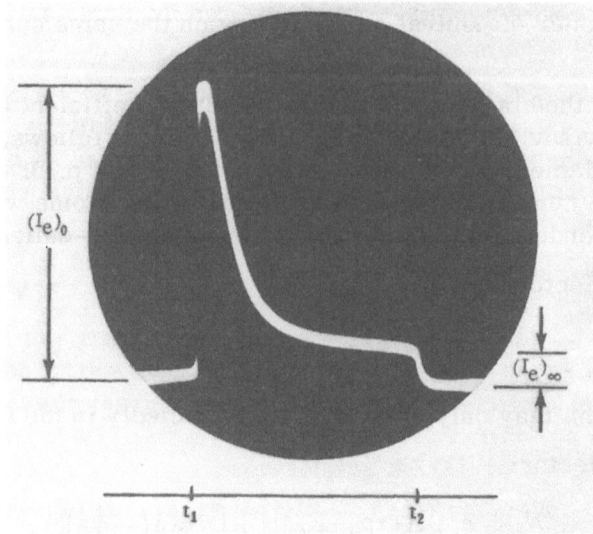

Fig. 2.25. Variation in the positive ion current from
a tungsten filament heated to 1300°K in an atmosphere
of potassium vapor as a result of the application of a
"top-hat" modulated field to the filament—collector
system ($t_1 = t_2 = \frac{1}{100}$ sec).

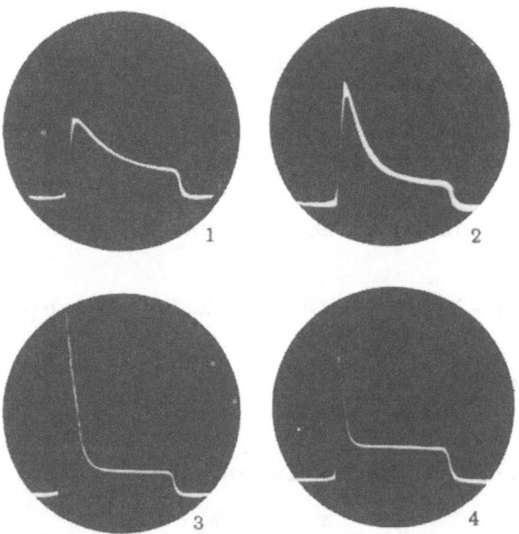

Fig. 2.26. Examples of oscillograms showing the
variation in the shape of the curves representing the
positive K^+ ion currents with the temperature of the
tungsten filament.

    Figure 2.25 shows a typical oscillogram of the variation in ion current. The analysis of
this oscillogram reduces to the following. Up to the instant t_1 a field retarding the ions is applied to the filament. At the instant t_1 the field changes sign (making the collector negative)
and the ions are able to leave the filament. At this moment a current $(I_e)_0$ appears. The cur-

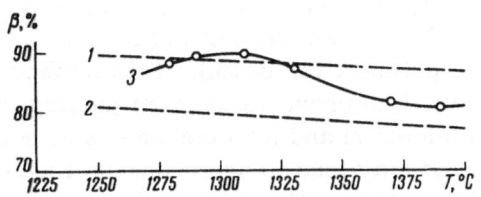

Fig. 2.27. Relation between the ionization coefficient of potassium and the
temperature of the tungsten filament.

rent starts falling exponentially; at the instant t_2 the
ion current reaches $(I_e)_\infty$, the field changes sign
again (making the collector positive), and the current
suddenly falls to zero.

Figure 2.26 shows examples of an oscillogram
illustrating the change in the shape of the ion-current
curves as the filament temperature increases. Curves
1, 2, 3, and 4 correspond to filament temperatures
1190, 1215, 1310, and 1390°K. Curve 3 corresponds
to the filament temperature at which maximum ionization occurs.

Calculation of the absolute ionization coefficients is carried out by measuring the ordinates
of the break points on the oscillogram, $(I_i)_0$ and $(I_i)_\infty$, and substituting the results in the formula
for determining β.

The results of such calculations are presented in Fig. 2.27, which represents data corresponding to temperatures of the saturated potassium vapor (curve 3) such that the coefficients
are independent of the saturated vapor pressure, the coverage of the filament by the potassium
being sufficient for the observation of the phenomenon and at the same time small enough to
justify classifying the surface as clean.

The same graph shows the values of the ionization coefficient of potassium calculated
from the Saha—Langmuir formula:

$$\beta = \frac{\exp\left(-\dfrac{V_i - \varphi}{kT}\right)}{A^* + \exp\left(-\dfrac{V_i - \varphi}{kT}\right)}$$

(A^* is a constant equal to the ratio of the statistical weights of the ions and atoms of the test
substance), the ionization potential of potassium V_i being taken as 4.32 eV. The work function
for the tungsten of our filament was close to 4.55 eV. Curve 1 is calculated for a value of $A^* = 1$,
as taken by Copley and Phipps [102], and curve 2 for $A^* = 2$.

We see from Fig. 2.27 that the experimental curve 3 agrees with curve 1 at the ionization
maximum, then falls more sharply and approaches curve 2. This indicates that the phenomenon
is a complex matter. There is little doubt that the crystallographic inhomogeneity of the tungsten surface plays a major part, this not having been taken into account in the investigation. It
is also possible that even for a slight coverage of the filament by potassium (of the order of a
few hundredths of a monolayer) there will be a certain change in the ionization coefficient.

The close agreement between the results obtained and the theoretical calculations in-
spires confidence in the fact that the method in question correctly estimates the coefficients of
surface ionization for coverages changing rapidly in time (10^{-3}-10^{-5} sec); it may thus be used
for calculating the surface ionization of adsorbed atoms when studying the secondary emission
under ion bombardment.

§ 5. Method of Double Modulation

The oscillographic method developed by Arifov, Ayukhanov, and Starodubtsev [260, 281], of
taking volt—ampere characteristics outlined in §1 of this chapter, enables us to accelerate the process of obtaining the energy distribution of secondary particles by hundreds of times; it introduces an element of objectivity into the measurements and enables us to obtain results for constant conditions on the target surface. Nevertheless, the method cannot be regarded as perfect

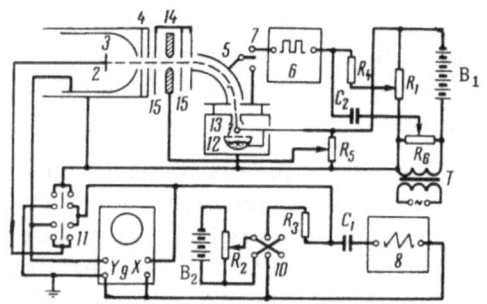

Fig. 2.28. Principal electric circuit of the apparatus for studying secondary processes by the double-modulation method.

in this form. A great disadvantage is the necessity of obtaining a zero line by an independent approach (elimination of the primary ion beam). This method of obtaining the zero line prevents us from judging the true values of the electron and ion components of secondary emission and may lead to an erroneous idea of the processes taking place. The point is that the use of an ac amplifier at the oscillograph input leaves the steady component of the secondary current, and hence the vertical displacement of the whole picture on the oscillograph screen, indeterminate.

In order to establish the position of the zero line on the oscillograms exactly, it is sufficient to switch the primary current being measured on and off rapidly. In this case the apparatus will record the peak value of the secondary current for the particular time interval (corresponding to the true secondary current) each time the system is switched on. In other words, we must intensity modulate the primary current with "top-hat" shaped pulses. If we choose an ion-beam modulation frequency half that of the sweep, then the volt—ampere characteristic will appear on the oscillograph screen in one period and the zero line in the other. The best result, however, is obtained for an ion-beam modulation frequency tens of times greater than that of the sweep (see Fig. 2.29). In this case the zero line is represented automatically, and the volt—ampere characteristic is the envelope of the "top-hat" pulses of secondary currents.

However, the automatic recording of the zero line is not the main aim of double modulation. As will be seen from the sequel, the use of ion beams modulated in the form of "top-hat" pulses not only simplifies the method of measuring, but also enables us to make a deeper study of phenomena taking place on a solid surface, and in particular to separate inertial secondary processes from inertia-free processes, and modulated from unmodulated. The inertia-free components of the secondary current, in fact, entirely repeat the modulation of the primary, whereas the inertial components are either not modulated at all or the form of their modulation is distorted. The character of this distortion enables us to separate different components of the secondary currents and to study their nature.

The modulation of the primary beam has yet another extremely important advantage. As indicated earlier, in order to reduce the adsorption of ions from the beam in the course of measurement, it is essential to reduce either the beam intensity or the time of bombardment. The double-modulation method enables both of these to be achieved.

In fact, by reducing the duration of the primary current pulses and increasing their duty factor, we sharply reduce the time of target bombardment without thereby reducing the total time of taking the curve. The short, rare pulses, as it were, "probe" the position of the volt—ampere characteristic, which constitutes their envelope. In the same way the mean beam intensity during the measurements is sharply reduced, without spoiling the accuracy of the measurements, since the pulsed current measured is far greater than any parasitic currents.

Thus we employ two modulations in the measuring system, and the secondary current recorded on the oscillogram is twice modulated: at the frequency of measuring the collector potential, and at the frequency of interrupting the primary ion current. This method has become known as the double-modulation method [281].

Let us use the arrangement illustrated in Fig. 2.28 to illustrate the principle of measuring the secondary emission by means of the double-modulation method.

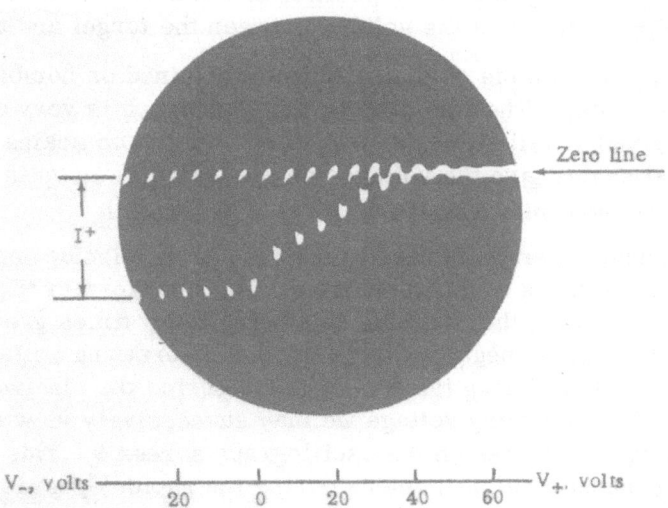

Fig. 2.29. Oscillogram of a volt—ampere characteristic of secondary ion emission from a clean cold
tantalum target.

A beam of ions from alkali halides 12, obtained by surface ionization on the spiral 1,
passes through the funnel 13, is accelerated by the voltage of the battery B_1, separated from
neutral particles inside the cylindrical condenser 5, and focused by means of the electrostatic
lenses 14 and 15. The secondary particles arising from the bombardment of the target 2 are
collected by the collector 3, surrounded by a guard cylinder 4. Modulation of the primary ion-
beam intensity is achieved by connecting the generator 6 (which generates rectangular pulses
with an amplitude of several tens of volts) in series with a steady voltage taken from battery B_1
by means of the potentiometer R_1 to the plates of the cylindrical condenser 5. The switch 7
may be used to prevent the ordinary beam from reaching the target completely. The voltage
between the target 2 and the collector 3 is varied periodically around a certain constant value
specified by the battery B_2 through the potentiometer R_2. The periodic variation in the voltage
is achieved by using a sawtooth generator 8. The secondary emission current is fed to the input of the vertical amplifier of the oscillograph 9, the horizontal plates of which receive a sawtooth voltage from the generator 8, which is at the same time used to create a voltage between
the target and the collector.

The frequency and duty factor of the pulses from generator 6 are chosen in accordance
with experimental conditions, while the frequency of the sawtooth generator is kept at around
25 cps. In order to obtain a stable image on the oscillograph screen, the frequencies of generators 6 and 8 must be tuned so that of generator 6 will be a multiple of that of generator 8.
Switching is carried out by means of switches 10 and 11.

Let us briefly consider the operating principles of the system for the case in which the
frequency of generator 6 is several tens of times greater than that of generator 8. We suppose
first that the target surface is cold but reasonably clean, that there is no evaporation of adsorbed atoms, and that the process of releasing the secondary ions and electrons is inertia-free.
Then the character of the oscillogram will be as follows. The principal line will repeat an integral distribution curve of the type shown in Fig. 2.4, as in the arrangement without modulation of the primary current; however, each time the current is interrupted by the voltage of the
"top-hat" pulse from generator 6, the current will fall to zero, showing the position of the zero
line at that particular instant of time. Thus, in this case, we show an advantage over the ar-

rangements of Figs. 2.2 and 2.3, in that the position of the zero line will be recorded automatically in a single cycle of variation of the voltage between the target and the collector.

Figure 2.29 shows an example of an oscillogram obtained on bombarding a clean cold tantalum target with Rb^+ ions. The zero line on the photograph is very clearly visible. On this oscillogram and others which will appear subsequently, we attach scales for convenience in reading: the horizontal scales give the voltages on the collector in volts ($V_\pm$) and the vertical scales give the emission currents, positive (I^+) and negative (I^-).

The use of the transformer T (without battery B_1) in the circuit enables us to study bombardment with ions and electrons at the same time. The transformer T receives a sinusoidal voltage from the mains network, the period of this being many times greater than the length of the "top-hat" pulses. During the negative half period of alternating voltage from the source the positive ions will be attracted, during the positive half-period the electrons will be attracted. Thus in a single period of alternating voltage we may successively view the ion—electron and electron—electron emission pictures on the oscillograph screen 9. This method is particularly valuable when studying transitional phenomena during the rapid deposition of a film of foreign atoms on a substrate.

§6.   Use of the Double-Modulation Method

for Studying Secondary Processes in Pure Metals

It follows from our consideration of the principles underlying the oscillographic double-modulation method that the latter is well adapted to make high-speed measurements. The values of the secondary emission coefficients and information regarding the energy distribution of the secondary particles may be obtained in fractions of a second, immediately after cleaning the target by heating it to a high temperature. The obtaining of data relating to secondary emission from a clean surface is only limited by the cooling time required in order to eliminate secondary ionization currents. It was found, however, that the double-modulation method could also be used for studying secondary processes in pure metals at high target temperatures. In this case the indeterminacy of the true secondary ion currents, due to the development of surface-ionization currents and thermionic emission (which occur in the case of static measuring methods), is entirely eliminated, and a new possibility of separating the surface-ionization currents and thermionic emission from the "true" secondary currents and studying each of these forms of emission independently emerges.

Let us consider that the bombarded target is heated to a certain temperature, such that the alkali atoms adsorbed on the target surface may evaporate. It is natural to suppose that a proportion of the alkali ions (if these are used to bombard the target) will bounce off the target without inertia in the form of neutral or charged particles, while the others will be adsorbed. The adsorbed atoms will evaporate partly in the form of ions and partly in the form of neutral atoms. The statistical life of an adatom is determined by the heat of adsorption and the target temperature. This time may vary with temperature from a fraction of a microsecond to many hours. In any case, the time of evaporation of the ions from an adsorbed layer is many times greater than the time of a single collision between a primary ion and a target atom.

If for the moment we leave aside the question of the possible deep penetration of the primary ions into the target and consider that a bombarding ion either bounces off, retaining considerable energy, or is adsorbed on the surface, then theoretically we may predict the change in the shape of the oscillograms due to surface ionization. Here the theory of transitional processes developed by Starodubtsev [190] for modulated atomic beams is completely applicable.

Let a current of ions I_0 fall on the target at the instant at which the voltage of generator 6 is removed (see Fig. 2.28). The probability that the ion will transform into an adsorbed atom we call q; then the rate of incidence of the alkali atoms into the adsorbed layer equals

$$U_0 = q \, \frac{I_0}{e} \,, \tag{2.18}$$

where e is the charge on the electron. The adsorption of the ions leads to the creation of a certain coverage of the target with adatoms and to the development of a process in which ions and neutral atoms evaporate from the surface. The probability of the evaporation of atoms and ions may be expressed in the form

$$W_0 = A \exp(-\lambda_0/kT), \; W_+ = B \exp(-\lambda_+/kT),$$

where A and B may be considered to a first approximation as independent of temperature, while λ_0 and λ_+ are the heats of evaporation of the atoms and ions, respectively.

For the nonstationary process we have

$$\frac{dn}{dt} = q \frac{I}{e} - n(W_0 + W_+). \tag{2.19}$$

This relation determines the rate of change of the number of adatoms on the surface.

The initial conditions are in general determined by the fact that there is already a number of adatoms n_0 on the target surface at the instant of the arrival of the ion beam, these not having been able to evaporate during the previous period of the generator 6.

The solution of Eq. (2.19) with the initial conditions $n = n_0$ at $t = 0$ has the form

$$n = n_0 \exp[-a(T)t] + \frac{qI}{e}\{1 - \exp[-a(T)t]\}, \tag{2.20}$$

where

$$a(T) = W_0 + W_+.$$

The secondary-emission ion current which owes its origin to surface ionization during that part of the period of the voltage from generator 6 in which the primary beam is able to reach the target, is expressed by the relation

$$I_e = neB \exp[-\lambda_+/kT] = neW_+, \tag{2.21}$$

or

$$I_e = n_0 eB \exp[-a(T)t] \exp(-\lambda_+/kT) + \frac{qBI}{a(T)}\{1 - \exp[-a(T)t]\} \exp(-\lambda_+/kT). \tag{2.22}$$

If we introduce the ionization potential of the atoms V_i and the work function of the surface φ, we may eliminate the heat of evaporation from (2.22). In fact,

$$\frac{BI}{a(T)} \exp(-\lambda_+/kT) = \frac{qBI}{A} \frac{\exp\frac{1}{kT}(V_i - \varphi)}{1 + \frac{B}{A} \exp\frac{1}{kT}(V_i - \varphi)} = (I_e)_\infty. \tag{2.23}$$

The quantity $(I_e)_\infty$ constitutes that part of the primary ion current which during the continuous action of the primary ion beam is realized in the form of slow surface-ionization ions flying off from the target.

By using relation (2.23) we may rewrite expression (2.22) in the form

$$I_e = (I_e)_\infty + [en_0 W - (I_e)_\infty] \exp[-a(T)t]. \tag{2.24}$$

This component of secondary current increases during that part of the period of the voltage from generator 6 in which the primary beam reaches the target, and at the instant of time t = $t_0/2$ (t_0 being the period of the voltage from generator 6) takes the value

$$I_e\left(\frac{t_0}{2}\right) = (I_e)_\infty + [en_0 W_+ - (I_e)_\infty] \exp[-a(T)t_0/2]. \tag{2.25}$$

In the quasistationary condition, the value indicated in (2.25) is the initial one for the half period in which the primary current is absent. During this time interval there will be an exponential fall in the current, given by the law

$$I_e(t) = I_e\left(\frac{t_0}{2}\right) \exp\left[-a(T)\left(t - \frac{t_0}{2}\right)\right], \tag{2.26}$$

and at the end of the part of the period under consideration, the secondary current component will take the value

$$I_e(t_0) = I_e\left(\frac{t_0}{2}\right) \exp\left[-a(T)\frac{t_0}{2}\right], \tag{2.27}$$

which is the initial value for the following period, i.e., it should be equal to (2.24) if we put t = 0 in this. Hence, we may determine the value of $en_0 W_+$. From (2.24)-(2.26) we have

$$en_0 W_+ = \frac{(I_e)_\infty \left\{\exp\left[-a(T)\frac{t_0}{2}\right] - \exp[-a(T)t_0]\right\}}{1 - \exp[-a(T)t_0]} \equiv (I_e)_{min}. \tag{2.28}$$

The quantity $(I_e)_{min}$ constitutes the minimum value of that part of the secondary emission current due to surface ionization.

In order to obtain the total ion emission, we must take the surface-ionization current I_e and add the true secondary ion-emission current, which is distinguished by a lack of inertia and considerable energies of the ions:

$$I^+ = I_e + K_s I, \tag{2.29}$$

where K_s is a coefficient indicating what proportion of the primary ions are scattered on collision with the target.

Hence, the proportion of ions escaping with considerable energies from the heated target (constituting an inertia-free component) will synchronously reproduce the shape of the primary beam modulation, i.e., they will give rectangular teeth on the oscillogram. The ions adsorbed on the target and transformed into adatoms will also give ion currents when the primary beam is absent. These surface-ionization currents will fall off exponentially, distorting the rectangular shape of the oscillogram teeth. The shape of the pulses will be altered in different ways for different target temperatures, and this will offer the possibility of separating the inertia-free escaping secondary particles from the "sticking" ions, which evaporate independently of the ion bombardment of the target.

The existence of two groups of ions in the secondary emission is confirmed by the oscillograms in Figs. 2.30, 2.31, and 2.32, taken at temperatures of 300, 1100, and 1350°K for a tungsten target bombarded with Rb^+ ions, when the frequency of the generator 6 only exceeds that of generator 8 by a few times.

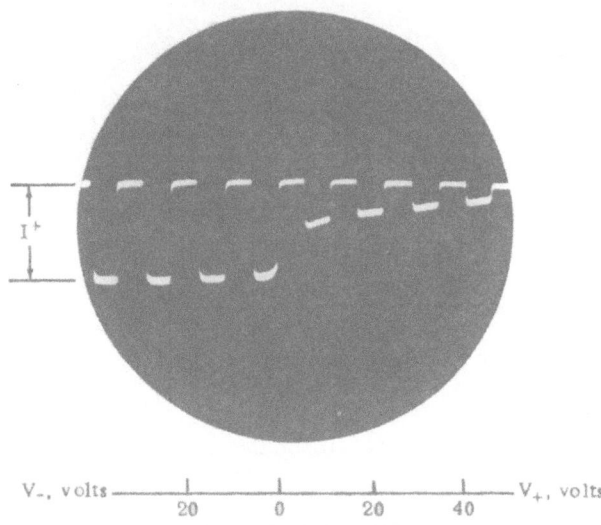

Fig. 2.30. Oscillogram of the volt—ampere characteristic of secondary ion emission from clean cold
tungsten.

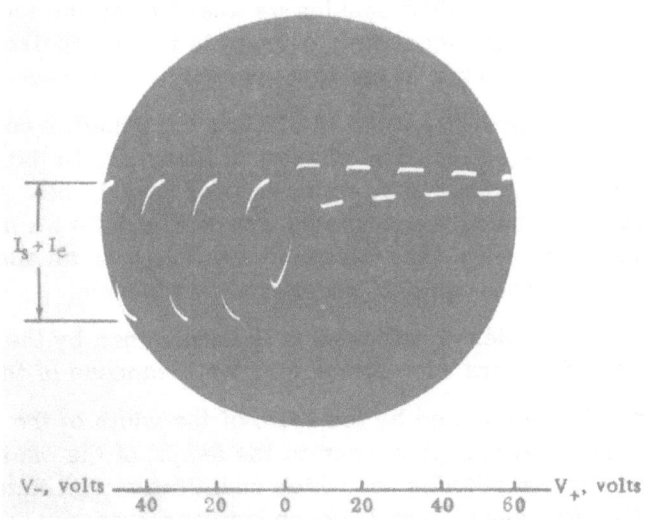

Fig. 2.31. Oscillogram of the volt—ampere characteristic of secondary ion emission from pure tungsten at 1100°K.

We see from the oscillograms (Fig. 2.30) that at low temperature only inertia—free
formation of secondary ions occurs. On raising the temperature the total secondary ion current increases and at the same time a peculiar delay develops (Fig. 2.31); ion currents exist
even after disconnecting the primary beam, and these vary exponentially. On raising the temperature to 1350°K (Fig. 2.32) these exponentials fall so sharply with time that the curves once
more become rectangular. However, the height of the ordinates of the curves is greater at this
target temperature; this clearly reveals the existence of a considerable group of secondary ions
formed by surface ionization.

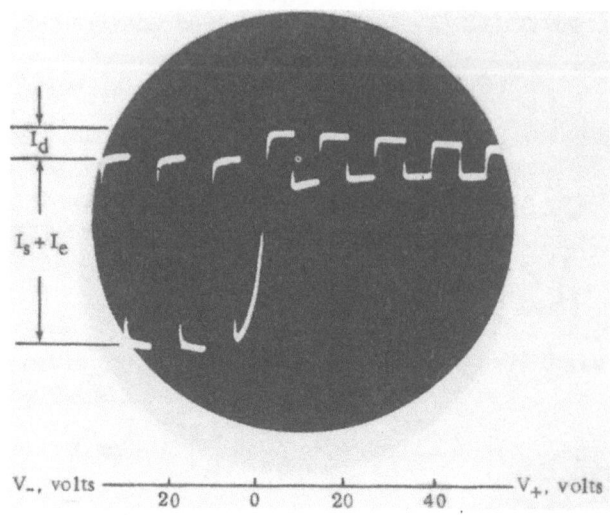

Fig. 2.32. Oscillogram of the volt—ampere characteristic of secondary ion emission from pure tungsten bombarded with Rb$^+$ ions with an energy of 400
eV at a target temperature of about 1350°K.

Thus the double-modulation method enables us to establish the fact that the influence of
temperature on the increase in total secondary current is mainly indirect; it is associated with
the evaporation of some of the adatoms in the form of ions.

As already indicated, the time required to plot one distribution curve and the corresponding zero line is about $\frac{1}{25}$ of a second instead of several hours, as in the galvanometric measuring method. This opens another possibility, which is qualitatively new when compared with galvanometric measurements. By using the arrangement of Fig. 2.28 we may follow the changes
in the distribution curves of the secondary particles $I = f(V_{2,3})$ as an adsorbed layer gradually
builds up on a clean target surface, carefully degassed in advance.

The error in the method under consideration is determined by the graphical accuracy of
locating the oscillograms and errors associated with the cleanness of the target.

The graphical error is determined by the ratio of the width of the horizontal line drawn
by the electron beam on the oscillograph screen to the height of the oscillograph picture (the
pulse). For the large current amplitudes for which quantitative estimates were made, this error
was no greater than 2-3%. As regards small secondary currents, we confined ourselves to draw-
ing qualitative conclusions as to the presence or absence of the particular component of secondary current in question.

Errors depending on the cleanness of the target may be extremely large. In each particular case in our experiments we took measures to secure the best possible cleanness of the
surface, as may be judged from the reproducibility of the curves and from the fact that the coefficients γ and K_s reached their limiting values. Oscillograms taken under identical conditions agreed to within the line thickness. Hence the error due to target contamination was no
greater than that committed in measuring the oscillograms.

In this connection we should mention that the oscillograms to be presented here do not
represent the record of a single process, but the superposition of 10-15 repetitions of the oscil-

lograph picture during the exposure of the photograph. Since this process does not lead to any blurring of the picture, we may claim good reproducibility and the absence of random errors.

In addition to these errors, the oscillograph method may involve errors arising from electromagnetic induction and various electrical interferences. However, by judicious screening, a proper choice of conditions, and compensation of sinusoidal induced currents, all these interferences may also be reduced to a value smaller than the oscillogram line width. Thus, the accuracy of the measurements is determined by the ratio of the width of the trace on the oscillograph screen to the pulse height.

The dynamic double-modulation method described, as developed for the study of secondary processes, has fundamental and technological advantages over static galvanometric methods; it enables us to accelerate the measuring process by hundreds of times and obtain the whole volt—ampere characteristic with hardly any change on the target surface, and also study the dynamics of the phenomena taking place. This considerably reduces the probability of chance experimental errors and introduces an element of objectivity into the measurement.

It should be noted that, in relation to the adsorption of residual gases on the target surface, the conditions of measurement by the oscillograph method with a vacuum of 10^{-6}-10^{-7} mm Hg are equivalent to conditions only attainable in the static method at a vacuum of 10^{-9}-10^{-10} mm Hg.

As we shall show in the following chapters, the double-modulation method has opened the fundamental possibility of experimentally separating the inertial and inertia-free processes taking place on bombarding a metal surface with atomic particles. In particular, scattered ions may be separated from ions evaporating from the surface of the metal by surface ionization, which of course depends greatly on temperature.

The oscillographic double-modulation method also enables us to obtain the time characteristic of rapidly varying processes, for example secondary processes, from films of variable density on a solid surface [319]. The method allows visual observation on the oscillograph and ordinary or motion photography of the whole evolution of these complicated processes.

§ 7. Double-Modulation Study of Secondary

Processes at Low Target Temperatures

In the preceding sections we have described the application of the double-modulation method to the study of secondary electron and ion emission. Now we shall consider the data obtained by Arifov and Ayukhanov [342] by experiments based on this method at low target temperatures.

Figure 2.33 shows an oscillogram of the volt—ampere characteristic obtained from an adsorbed film of rubidium on a tantalum target bombarded with Rb^+ ions at an energy of 400 eV. We see from the oscillogram that the electron and ion components of secondary emission are clearly represented.

If we clean the target by heating and then, after cooling, take the volt—ampere characteristic, the picture changes sharply. Figure 2.30 shows such an oscillogram, obtained on bombarding a clean, cold tungsten target with 170-eV Rb^+ ions. We see from the oscillogram that there is no secondary electron emission from the clean surface ($\gamma = 0$); the secondary ions have a continuous spectrum of energies, from thermal to limitingly large values.

In Fig. 2.34 we have two oscillograms representing the integral picture of the energy distribution of the secondary ions on bombarding a cleaned, cold tantalum target with Li^+ ions. The oscillogram of Fig. 2.34a is obtained for 170-eV primary ions and the oscillogram of Fig.

2.34b for 570 eV. The amplitude of the alternating voltage between the target and collector was the same in each case. The form of the ion-retarding curves (right-hand branch of the envelope) indicates an increase in the limiting energies of the scattered ions with increasing primary ion energy.

Comparison of the oscillogram (Fig. 2.34a) for Li$^+$ ions with that of Fig. 2.29 for Rb$^+$ ions obtained at the same primary ion energy (170 eV) and amplitude of the sawtooth voltage between the target and collector shows that the value of the limiting energy depends on the mass of the bombarding ions.

Measurements of the alternating sawtooth voltage between the target and collector based on the earlier-calibrated horizontal deflection of the oscillograph beam showed that the limiting energies of the secondary ions approximately agreed with the value deduced from relation (1.4) (within the limits of experimental error for the particular geometry of the apparatus) for the elastic central collision of the bombarding ions with individual target atoms. On considering this fact, and also the way in which these limiting energies depend on the energy of the primary ions and the colliding particle mass ratio, and noting the approximate agreement with the results derived from relation (1.4), we see that there is no doubt of the existence of elastic collisions between the bombarding ions and individual target atoms.

Figure 2.35 presents oscillograms taken at unequal time intervals during the continuous action of a beam of primary Rb$^+$ ions on a previously well-cleaned and cooled tungsten surface, i.e., taken during the formation of a film, comprising rubidium atoms and residual gas atoms present in the vacuum, on the clean target surface. The oscillograms show that there is hardly any secondary electron emission from the clean tungsten surface (Fig. 2.35a); as time passes, however, such emission appears and increases with increasing density of the film formed (Fig. 2.35b, c, d).

We may also convince ourselves of the inverse effect of a rising temperature on the pictures obtained. If, after obtaining an oscillogram similar to Fig. 2.35d, we slowly raise the target temperature, then analogous pictures of the integral secondary-particle energy-distribution will appear in the reverse order, finally giving oscillograms corresponding to a clean target surface.

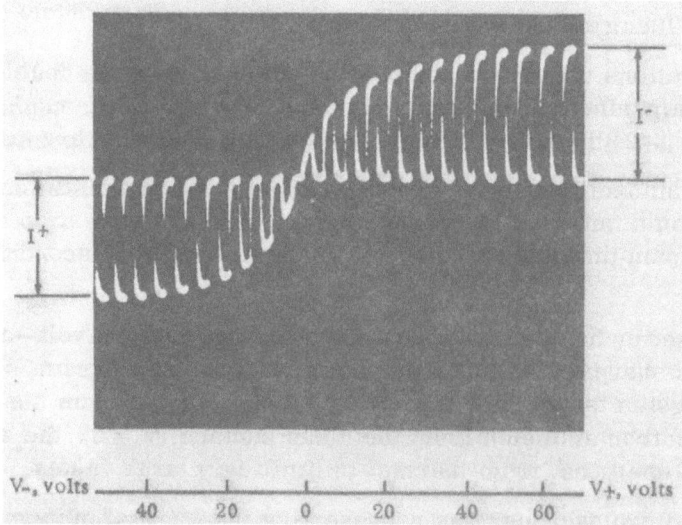

Fig. 2.33. Volt—ampere characteristic of secondary emission from an equilibrium adsorbed rubidium film deposited on a clean, cold (300°K) tantalum target.

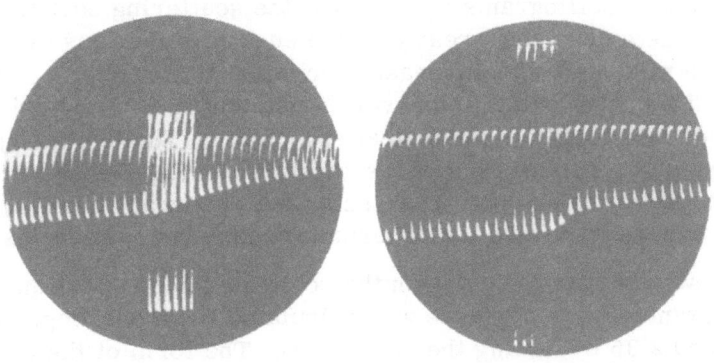

Fig. 2.34. Volt—ampere characteristics of secondary ion emission from a clean cold (300°K) tantalum target bombarded with Li^+ ions at energies of: a) 170; b) 570 eV.

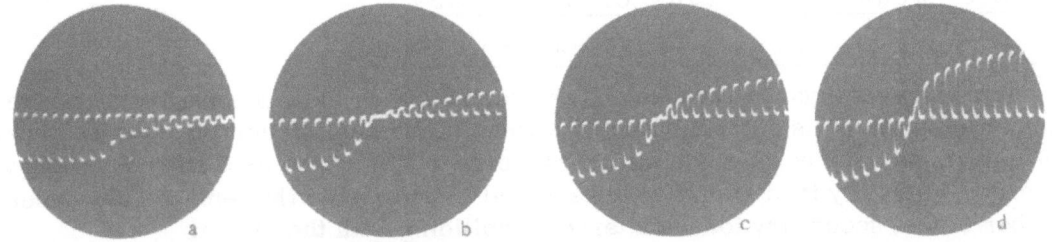

Fig. 2.35. Variation in the shape of the secondary emission V/A characteristic as the primary beam ions are adsorbed on a clean cold (300°K) tungsten target bombarded with 300-eV Rb^+ ions.

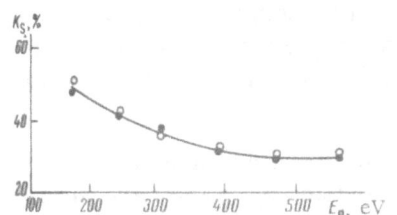

Fig. 2.36. Relationship $K_s (E_0)$ for the Na^+ ion on tantalum at 300°K.

Thus by using the high-speed method of studying secondary processes, we find that the same laws hold at low target temperatures as were obtained for high target temperatures by the galvanometric method, subject to the target surface being reasonably free of foreign adsorbed atoms or molecules.

If various contaminants build up on the cold surface with time, the form of the resultant oscillograms will undergo sharp changes.

The oscillograms just considered show that the appearance of intense secondary electron emission from the surface of metals of the tungsten and tantalum type is exclusively due to the formation of contaminant films deposited by the adsorption of some of the primary ions and also residual gases from the vacuum on the surface under examination. In view of this, the observed temperature dependence [61, 147, 221] may be given a simple explanation. The rise in secondary electron emission on reducing the temperature is mainly due to the rise in the density of the adsorbed films.

Careful study of the oscillograms showed that the scattering coefficient of the ions on striking clean, cold targets depended greatly on the energy of the ions over the range $E_0 \leq 600$ eV, in that the coefficient K_s fell with increasing primary beam energy. This may be seen from the oscillograms of Fig. 2.34a,b, in which several pulses of modulated primary ion current are shown on the large scale to facilitate numerical estimation. The scattering coefficient K_s equals the ratio of the ordinate corresponding to the saturation of the left-hand branch of the volt—ampere characteristic I_s to the ordinate representing the primary current pulse I_0. We see from the figure that this value differs for different primary beam energies.

Figure 2.36 shows the relation between the scattering coefficient and the energy of the primary Na^+ ions on tantalum. The curve was calculated from oscillograms similar to those shown in Figs. 2.34 and 2.35 by finding the ratio I_s / I_0. The form of the curve shows a considerable drop in K_s on raising the energy of the primary beam. It is characteristic that with increasing E_0 the coefficient K_s falls gradually and becomes less dependent on E_0. Analogous results are obtained for Li^+, K^+, and Rb^+ ions on the surface of a clean tungsten target.

The oscillographic method of double modulation offers the possibility of studying secondary processes on cold, clean metal surfaces quantitatively, whereas this is very hard to do by the galvanometric method under similar vacuum conditions.

§8. Double-Modulation Study of Secondary Processes at High Target Temperatures [342]

When studying secondary processes on the surface of an incandescent metal surface, the target temperature, which is determined by the value of the heating current, was kept constant. The conditions of primary beam modulation and the operation of the oscillograph sweep were the usual ones. Naturally it was primarily essential to determine the temperature dependence of the number of fast secondary ions scattered on colliding with the target.

A study of the oscillograms obtained at different target temperatures showed that the number of scattered ions in fact depended very little on temperature. In Fig. 2.37, for convenience of comparison, we show the oscillograms of Figs. 2.30, 2.31, and 2.32 [for Rb^+ ions on clean tungsten, taken at room temperature (a), 1100°K (b), and 1350°K (c), and with a frequency of generator 6 only a few times greater than that of generator 8]. The right-hand sides of these oscillograms, constituting the integral representations of the energy distribution of the secondary (scattered) ions, clearly remain almost constant at the temperature rises by over 1000°. Thus, for any temperature of the clean target, there is a group of back-scattered ions from the primary beam.

At room temperature (on the target) some of the primary beam ions are neutralized. Regarding the neutralized primary beam ions we may make two suppositions: 1) they are scattered without inertia in the same way as (unneutralized) ions; 2) after losing their energy completely, they are adsorbed on the target surface or penetrate further down into the target.

Leaving the question as to the possible scattering of the neutralized portion of the ions on one side for the moment, let us suppose that some of the bombarding ions bounce off the target without inertia, while the others are adsorbed on the surface.

Let us consider the left-hand side of the oscillograms, corresponding to secondary ions. These oscillograms show that on bombarding a clean, cold target, when there is no evaporation of adsorbed ions, the secondary ions consist only of ions of the scattered type, which escape inertialessly with considerable energies, exactly reproducing the shape of the rectangular pulses of the modulated primary ion beam (Fig. 2.37a). On raising the temperature, the total secondary ion current rises and the peculiar lagging described earlier appears at the same time. Ion

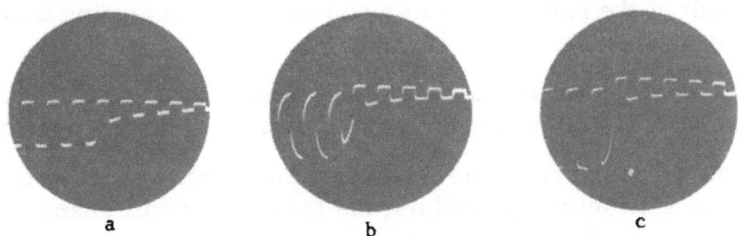

Fig. 2.37. Group of V/A characteristics of secondary emission at various target temperatures.

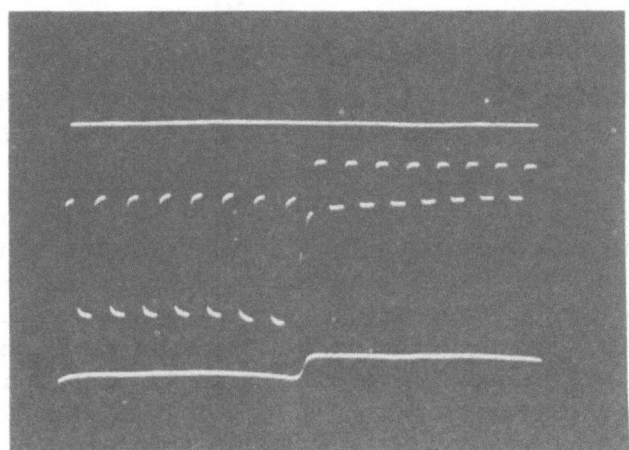

Fig. 2.38. Three successively taken oscillograms of secondary currents from a clean tungsten target bombarded with K^+ atoms at an energy of 600 eV at a temperature of 1350°K.

currents also exist on switching the primary beam off, and these then vary exponentially (Fig. 2.37b).

At high temperatures the life of the adatoms on the surface of an incandescent target becomes quite short, and the whole process of evaporation ends so quickly that the curves become rectangular again (Fig. 2.37c). However, the height of the ordinates of the pulses at this temperature is greater, the rise being due to ions with thermal velocities; this clearly indicates the existence of a considerable group of secondary particles formed as a result of the evaporation of adsorbed atoms in the form of ions, i.e., by surface ionization.

The secondary ions formed as a result of surface ionization only occur at high temperatures, and only appear in large numbers on condition that $V_i < \varphi$. For example, in the case of Li^+ and Na^+ on tantalum, for which $V_i > \varphi$, the picture obtained at high target temperatures is no different, within the limits of the accuracy of the method, from the analogous picture at room temperature. However, back-scattered secondary ions occur at all target temperatures.

The evaporation of adsorbed atoms in the form of ions at high target temperatures may also be observed in another way. When an alternating voltage is applied between the target and the collector, the unhindered evaporation of ions is only possible for a negative (extracting) collector potential. If the collector has a positive (cutoff) potential, evaporation may only take place in the form of neutral atoms, which will lead to a certain excess buildup of adatoms on the surface as a result of the return of scattered ions by the retarding field and the absence of evaporation in the form of ions.

Thus, modulation of the voltage between the target and the collector leads to modulation of the density of the adsorbed film and hence to modulation of the ion current to the collector.

On changing the direction of the field between the target and the collector from positive (retarding) to negative (extracting) potentials, the ion current will change from the value $(I_e)_+ = n_+eW_+$ to $(I_e)_- = n_-eW_+$, where n_+ and n_- are the surface concentrations of the adsorbed atoms for positive and negative potentials on the collector, respectively, W_+ is the probability of the evaporation of an adatom in the form of a positive ion, and e is the charge on the electron. This transformation may easily be shown to take place in accordance with the law

$$I_e = (I_e)_+ \exp\left[-\frac{t}{\tau_a}\right], \tag{2.30}$$

where τ_a is the life of the adatoms on the target surface.

This kind of nonstationary process is used to determine the coefficient of surface ionization and the heats of adsorption of ions and neutral atoms [192, 193, 195], and is considered in §§2, 3, and 4 of this chapter.

When the target temperature is such that the evaporation taking place after switching the primary beam off has not quite ended before the next pulse of primary ions arrives, i.e., when there are still a certain number of adsorbed atoms on the target surface, the integral picture of the energy distribution of secondary particles is superimposed on that of the nonstationary processes just described. However, for fairly high target temperatures, when the life of the adatoms on the surface is vanishingly short, the process of evaporation after the removal of the primary beam ends well before the arrival of the next pulse, and the target surface may be regarded as clean in advance of every pulse. Hence, in this case, each pulse of secondary current starts from the zero line, thus showing the absence of adsorbed atoms from the target surface.

However, when working under conditions corresponding to the transition from attractive to repulsive collector potentials, for high target temperatures we always observed a certain current of different origin, evidently due to an adsorbed film of equilibrium thickness. The ion current always gave a jump (displacement) on passing from negative to positive collector voltages, showing the presence of an unmodulated ion current at negative collector potentials.

Figure 2.38 shows three oscillograms taken successively on one plate. The upper oscillogram was taken before applying the ion beam to the target; it indicates the absence of foreign impurities from the target. The middle oscillogram was taken after applying a beam of K^+ ions to the target. A break was formed in the volt—ampere characteristic, indicating the presence of an unmodulated ion current associated with ions which had penetrated into the target slowly diffusing to the surface. The lower oscillogram was taken 2-3 sec after the removal of the primary beam. The break remains on the curve owing to the continuing diffusion of primary-beam ions which had penetrated deep down into the target.

Thus, for high target temperatures, evaporation of ions with an unusually long lifetime in the absorbed state takes place. This phenomenon cannot be identified with the evaporation of ordinary adsorbed atoms, since at high target temperatures these have lifetimes of the order of fractions of a microsecond. In addition to this, this phenomenon only occurs for fairly high primary-ion energies and a high target temperature ($T > 900°K$).

In Fig. 2.39 we present two oscillograms taken after bombarding a tantalum target heated to about 1300°K with Cs^+ ions at energies of: a) 170; b) 570 eV.

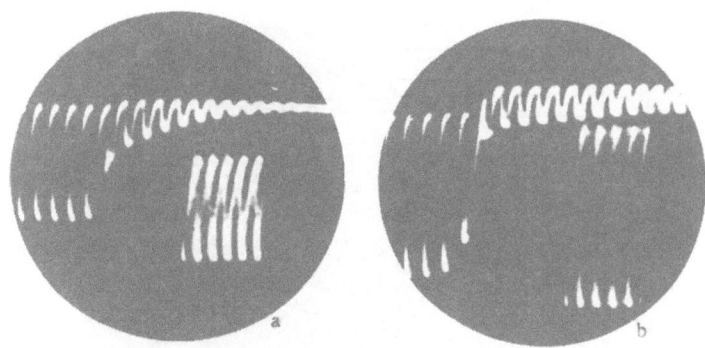

Fig. 2.39. Volt—ampere characteristics of secondary emission from clean tantalum heated to 1300°K on bombarding with Cs^+ ions at energies of: a) 170; b) 570 eV.

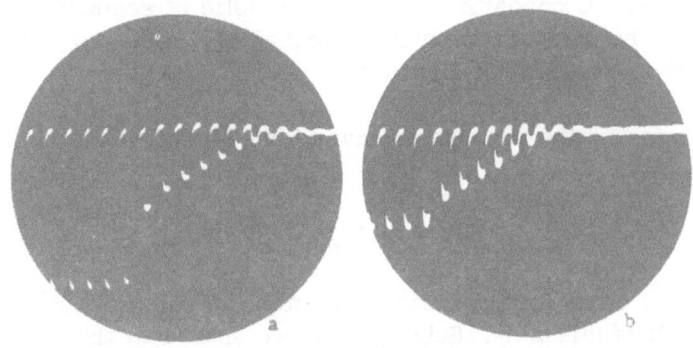

Fig. 2.40. Volt—ampere characteristics of secondary emission from clean tantalum on bombarding with Rb^+ ions at energies of 170 eV and a temperature of: a) 1300; b) 300°K.

Despite the different target temperatures, the oscillograms of Fig. 2.39a and Fig. 2.40a,b only show scattered and evaporated ions. There is no break in the V/A characteristic analogous to that seen in Fig. 2.38. This indicates that, owing to the low energy of the primary ions (170 eV), these cannot penetrate into the target.

In addition to the scattered and evaporated ions, the oscillogram of Fig. 2.39b clearly shows a characteristic break indicating the presence of a special component of secondary ions. In order to explain this phenomenon, it is useful to consider the overall picture of the energy distribution of the secondary ions. The energies of the scattered secondary ions have all possible values, from thermal energies to those determined by elastic collisions between the incident particles and the target atoms; this strongly suggests that the scattering of the ions takes place not only as a result of a single collision with the surface atoms of the target, but also as a result of multiple collisions with atoms lying at a certain depth. Then the primary ions will have lost energy not only at the target surface but also at a depth of several atomic layers. Hence, the evaporation of these ions will be determined not only by the life of the ordinary adatoms on the surface, but also by the rate of diffusion of these ions from inside the target to the top layer. The emergence of these at the surface is determined by ordinary thermal diffusion, which is distinguished by extreme inertia. The value of this ion current is controlled

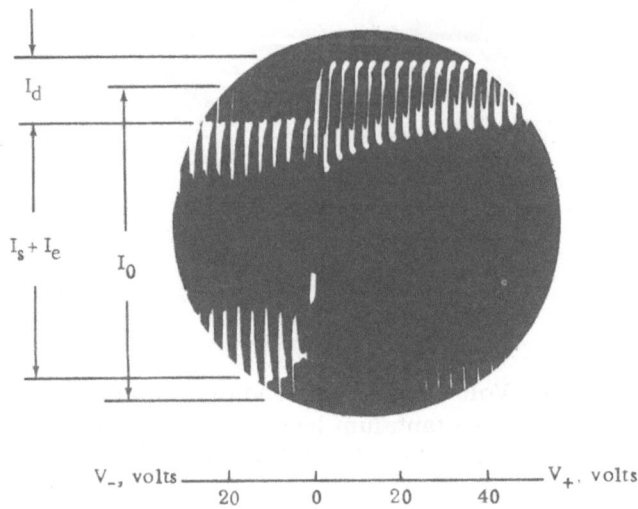

Fig. 2.41. Composite V/A characteristic of secondary emission. The primary ion current and all the secondary currents are shown on the oscillogram.

by the target temperature and the average primary ion current; it may be determined from the relation

$$I_d = \beta \frac{dn}{dt},\qquad (2.31)$$

where β is the coefficient of surface positive ionization, and dn/dt is the rate of access of atoms to the surface as a result of diffusion. Subsequently we shall call this ion current (which is characterized by great inertia and may be measured by determining the height of the break on the zero line of the V/A characteristic) the "diffusion-ion" current (or current of diffusing ions).

Figure 2.41 shows an oscillogram obtained after bombarding a clean tantalum target with 400-eV Rb^+ ions at a target temperature of about 1300°K. On this oscillogram we notice the absence of electron emission and the presence of three very different groups of ions in the secondary ion emission: scattered ions (I_s), evaporated ions (I_e), and ions diffusing from well inside the target (I_d). On the same oscillogram we have the primary ion current I_0 taken in an independent photograph. The total-current picture appears on the oscillograph screen if we short-circuit the target with the collector to the input of the oscillograph. From this oscillogram we may determine the numerical values of the secondary-ion emission coefficients K_s, K_e, and K_d and calculate the energy distribution of the particles forming individual components of secondary emission.

Thus, on bombarding incandescent metals with positive ions at energies of $E_0 > 200$ eV with $V_i < \varphi$, three groups of ions are observed:

a) Ions scattered by metal atoms, inertialessly reproducing the shape of the primary ion pulse ("scattered" ions);

b) ions appearing as a result of the surface ionization of primary beam particles adsorbed on the target surface ("evaporated" ions);

c) ions appearing as a result of the surface ionization of some of the primary beam particles, which first penetrate into the target and then diffuse back to the surface ("diffusion" ions).

Secondary electron emission under the influence of positive ions from pure, hot metals (Ta, W, Mo, and Ni) is negligibly low in the energy range studied; it lies outside the limits of sensitivity of the measuring apparatus.

CHAPTER 3

ION BOMBARDMENT OF INCANDESCENT METALS

The principal laws governing the secondary processes taking place on bombarding a metal with positive ions at high target temperatures have not been studied very extensively. In addition to this, such work as has been done has been carried out without any quantitative checking of the state of the target surface. In this chapter we shall present the results of some investigations into secondary-ionic processes at high target temperatures. The state of the surface of the incandescent metals is checked as necessary by reference to the corresponding work function.

§1. Method of Studying Secondary Processes
on Incandescent Metals

The investigations were based on the double-modulation method which, as indicated in Chapter 2, enables us to make separate measurements of the currents from scattered (I_s), evaporated (I_e), and diffusing (I_d) ions.

Figure 3.1 shows the apparatus in which the experiments were carried out by Arifov, Ayukhanov, Starodubtsev, and Khadzhimukhamedov [418, 457]. Mounted on the ground-glass joint A is the receiving part of the apparatus, consisting of three cylinders: a measuring cylinder 1 and two guard-ring cylinders 2. The measuring cylinder had a cut on the side opposite to the entrance slit to facilitate the free exit of the ions passing around the filament target 3, which was stretched along the axis of the cylinders by means of a small spring 4. In order to prevent foreign ions from falling onto the collector 1 and the target, these latter were surrounded with a screening cylinder 5 having slots to facilitate the entrance of the primary ions and the escape of the ions bypassing the target. Opposite to the cut for the escaping ions was a grounded trap 6 formed by a box blackened with a thick layer of soot, having a grid in front of it. The filament target was fixed to an independent pin 7, which enabled it to be moved in a plane perpendicular to the direction of the primary ion beam. This enables us to direct a focused ion beam rapidly and easily onto the target. The beam cross section at the target surface was about 1×10 mm. If necessary, a positive potential (relative to ground) was applied to the trap 6 in order to hold the secondary electrons from the soot. Mounted on the ground-glass joint B was an ion source with a focusing system, as described in Chapter 1. The use of a filament-type target enabled us to determine its temperature from the heating current and the filament diameter very ac-

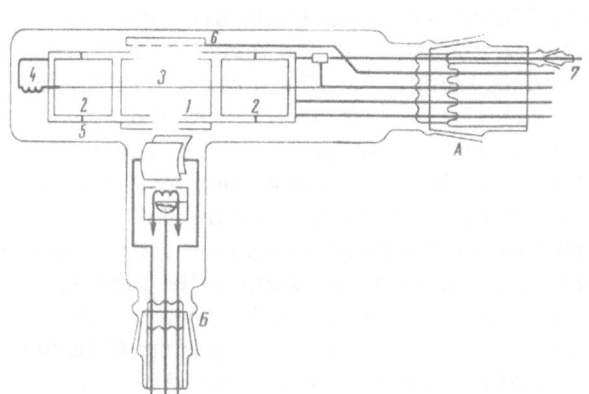

Fig. 3.1. Vacuum apparatus.

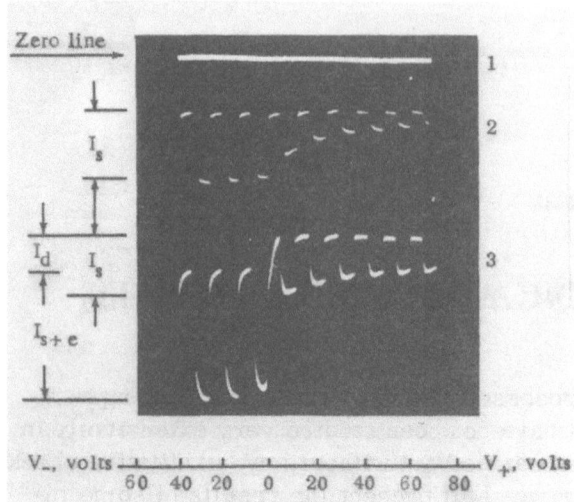

Fig. 3.2. Oscillograms of the volt—ampere characteristics of the secondary ion currents from cold (300°K) and hot (1500°K) metal targets.

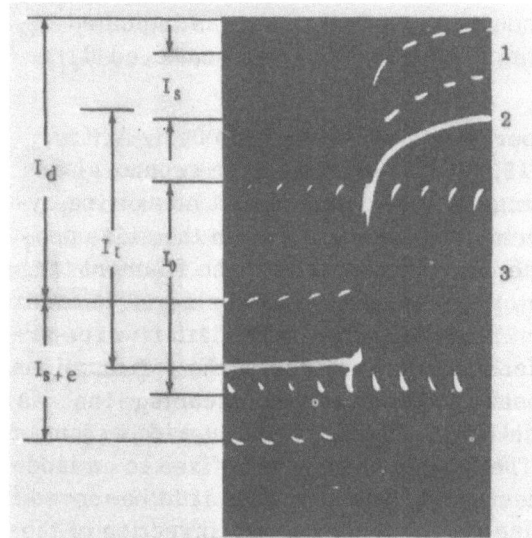

Fig. 3.3. Oscillograms of the volt—ampere characteristics of the primary ion current (I_0), the currents of the scattered (I_s), evaporated (I_e), and diffusing (I_d) ions, and also the thermoelectron (thermionic) current (I_t) at 1800°K from a tungsten target.

curately. The accurate temperature determination, in turn, enabled us to measure the work function of the metal and thus check the cleanness of the target surface. Replacing a strip target by one of the filament type enabled us to carry out measurements at higher target temperatures, thanks to the low value of the thermionic (electron) current.

We studied secondary processes from targets composed of tungsten, molybdenum, tantalum, etc., bombarded with positive alkali ions of various energies between 300 and 1900°K.

Figure 3.2 shows an example of an oscillogram (2) representing the secondary emission from a clean, cold tungsten target bombarded with K^+ ions at an energy of $E_0 = 840$ eV. In the absence of an ion beam on the target, the horizontal zero line 1 appears on the oscillograph screen. We see from the oscillogram 2 that there is no secondary electron emission from the cold target; the secondary scattered ions have a continuous energy spectrum, from thermal values to the limitingly high levels corresponding to the law of elastic collisions.

Oscillogram 3 was obtained for a target temperature of 1500°K. We see that on passing from positive to negative potentials on the collector the zero line of the V/A characteristic gives a displacement indicating the presence of an unmodulated positive ion current; as indicated earlier (Chapter 2, §6) this is a current of diffusing ions.

Let us consider Fig. 3.3, which contains three successively photographed oscillograms (1, 2, 3) obtained from a tungsten filament target at 1800°K bombarded with 840-eV K^+ ions. Let the temperature of the target first be considered as being below the threshold of thermionic (electron) emission (1500°K) and let us apply a sawtooth voltage between the target and collector, with the primary beam switched off. In this case, we shall have a horizontal sweep (zero line) on the oscillograph screen. If at this instant we apply the ion beam to the target, oscillogram 2 of Fig. 3.2 will appear on the oscillograph screen. Then, without applying the ion beam, we raise the heating current in the filament, increasing its temperature to 1800°K. On passing from negative to positive collector potentials the zero line of the V/A characteristic will experience an upward leap, indicating the presence of an unmodulated electron current. It is clear that the height of the ordinate I_t

(Fig. 3.3) determines the thermionic (electron) current from the filament. If now we apply an intensity-modulated primary ion beam to the target, the oscillograph screen will show the V/A characteristic 1; the height of the leap in the corresponding zero line I_d' is determined by the sum of two quantities: the thermionic emission and the diffusing-ion current.

By using the thermionic current measured from these oscillograms at various temperatures we plotted Richardson straight lines and deduced the work function of the target from these.

To the left of the oscillogram in Fig. 3.3 is a schematic representation of the three components of secondary ion currents (I_s, I_d, and I_{s+e}), the thermionic (electron) current (I_t), and the primary ion current (I_0). The contributions of the individual components of secondary currents are related by the following equations:

$$I_e = I_{s+e} - I_s, \quad I_d = I_d' - I_t.$$

By using such oscillograms we may study the laws governing the contributions of scattered, evaporated, and diffusing ions to the secondary-ion emission current at high target temperatures.

The relations

$$K_s = \frac{I_s}{I_0}, \quad K_e = \frac{I_e}{I_0}, \quad K_d = \frac{I_d}{I_0}$$

(3.1)

determine the proportion of each secondary-emission ionic component relative to the primary ion current I_0. The sum

$$K_\Sigma = K_s + K_e + K_d$$

(3.2)

gives the coefficient of secondary ion emission for the particular conditions of the experiment in question. On the oscillogram of Fig. 3.3, $K_s \simeq 30\%$, $K_e \simeq 40\%$, and $K_d \simeq 10\%$. The coefficient K_Σ depends on the physical parameters and the conditions covering the interaction of the colliding particles, for example,

$$\text{for} \quad V_i < \varphi, \quad T \gtrsim 1200^\circ K \quad K_\Sigma = K_s + K_e + K_d;$$
$$\text{for} \quad V_i < \varphi, \quad T \lesssim 1200^\circ K \quad K_\Sigma = K_s;$$
$$\text{for} \quad V_i > \varphi, \quad T \gtrsim 1200^\circ K \quad K_\Sigma = K_s.$$

§2. Temperature and Energy Dependence of the
Components of Secondary Ion Emission

Arifov and Khadzhimukhamedov [458] have investigated the temperature and energy dependence of secondary ion emission at target temperatures of 300–1900°K in the ion-energy range 100–1000 eV. Filament-type tungsten targets were bombarded with Na^+, K^+, Rb^+, and Cs^+ ions after prolonged degassing so as to achieve work functions of $\varphi \approx 4.51\text{-}4.57$ eV.

Figure 3.4 shows the temperature dependence of the coefficients K_s, K_e, K_d, and K_Σ for a tungsten target bombarded with 560-eV K^+ ions. We see from the curves that over the temperature range studied and for the ion energy specified, K_s is almost independent of target temperature, being approximately equal to 30%. The absence of any temperature dependence of the scattering coefficient was also noted in the case of the other alkali ions.

Since there is no marked evaporation for T < 1200°K, none of the alkali ions yield any I_e or I_d currents. Above this temperature, K_e increases, and saturation sets in at 1800°K. For Rb^+ and Cs^+ the corresponding temperature at saturation is roughly 1600°K. The value of K_s

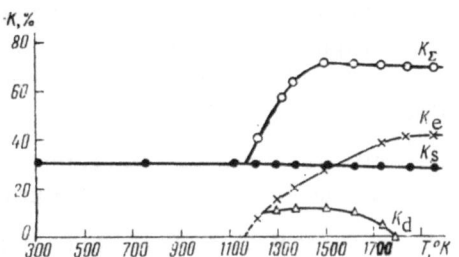

Fig. 3.4. Dependence of the coefficients K_s, K_e, K_d, and K_Σ on the temperature of a tungsten target bombarded with K^+ ions with $E_0 = 560$ eV and $\varphi = 4.57$ eV.

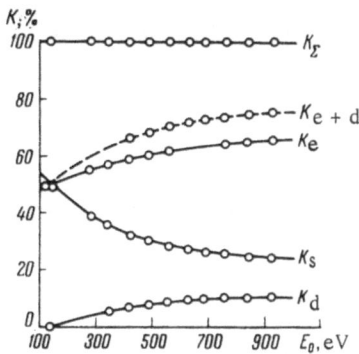

Fig. 3.5. Dependence of the coefficients K_s, K_e, K_d, and K_Σ on the energy of Cs^+ ions bombarding a tungsten surface at 1500°K.

for the surface studied is the greater, the smaller the ionization potential of the ion V_i. For example, whereas for Cs^+ on tungsten the value of K_s reaches 70%, for Na^+ on tungsten under the same conditions K_s only reaches 3 or 4%.

The diffusion ions appear for T > 1200°K in the same way as the evaporated ions. In the case $V_i > \varphi$ (Na^+ on tungsten) no diffusion ions occur (within the limits of sensitivity of the measuring apparatus). For $V_i < \varphi$, the coefficient K_d reaches a maximum at T ~ 1300°K (K^+ on tungsten). On further raising the temperature K_d falls to zero at about 1800°K for K and 1600°K for Cs and Rb. The temperature at which K_d falls to zero coincides with the saturation temperature of K_e. We note that in this respect the currents of the evaporated and diffusing ions supplement one another. The total current of these ions is described by the law of positive surface ionization.

For comparatively low temperatures (T < 1600°K) the behavior of the diffusion ions differs from that of the evaporated ions in that the life of the former (τ_d) is greater than that of the latter (τ_e). For T > 1600°K, the value of τ_d falls and becomes comparable with τ_e. Then the diffusion ions are indistinguishable from the evaporated ions.

The curves giving the dependence of K_s, K_e, K_d, and K_Σ on E_0 (Cs^+ on tungsten) at T = 1500°K are shown in Fig. 3.5. The character of the K_s (E_0) curves for Cs^+ ions and other alkali ions is identical. We see from Fig. 3.5 that the value of K_s first falls sharply with increasing E_0 up to about 600 eV, after which there is a slow fall to about 1000 eV. The coefficients K_e, K_d, and their sum K_{e+d} increase with rising E_0 and tend to saturation at about 1000 eV. The value of K_Σ (within the limits of experimental error) is independent of E_0. An analogous result is obtained for Rb^+ ions on tungsten, except that the value of K_e, and hence K_Σ, is different. The results of these studies of the relationships K_Σ (T) and K_Σ (E_0) obtained on bombarding tungsten targets with Cs^+ ions served as standards for subsequent considerations. It is well known that in the case of Cs^+ on tungsten at T > 1300°K, all the secondary particles should leave the surface of the target in the form of positive ions, i.e., we should have $K_\Sigma = 100\%$. In actual fact, experiment showed that, for $E_0 = 700$ eV and T = 1500°K, $K_s = 26.5\%$, $K_e = 63.5\%$, and $K_d = 10\%$.

On the basis of the foregoing data, we may assert that for all types of ions and targets, the region above the curves of K_Σ (E_0) and K_Σ (T) represents the neutralized proportion of the ions from the bombarded targets. We also note that for T ≤ 1200°K and $V_i > \varphi$, some of the primary beam ions are scattered elastically at individual target atoms, while the rest are neutralized.

Under the conditions T ≥ 1200°K, $V_i < \varphi$, some of the primary beam ions are elastically scattered, some leave the target surface in the form of evaporated and diffusing ions, while the rest are neutralized.

Thus the oscillographic method of studying high-temperature targets offers the possibility of quantitatively determining the integral value of the neutralized part of the primary ion current at a function of the E_0 and V_i of the ions and the T and φ of the target.

Regarding the neutralized part of the primary beam ions, we may say that these also comprise three components; scattered, evaporated, and diffusion-evaporated neutral atoms.

§ 3. Temperature and Energy Dependence
of the Coefficient of Positive-Ion Neutralization
on a Metal Surface

Among the various phenomena taking place during the ion bombardment of a metal surface, the least studied is that of the neutralization of the ions. This is because of the extreme difficulty experienced in recording the number of secondary neutral particles directly. Up to the present time there has been no reliable method of measurement such as would enable us to study the neutralization of primary ions as a function of the parameters and conditions of interaction of the colliding particles. The small number of investigations on this subject [59, 139, 236, 463] also bears witness to the difficulty of studying such processes.

In this section we set out the results of an investigation by Arifov and Khadzhimukhamedov into the relation of the integral coefficient of neutralization of Na^+, K^+, Rb^+, and Cs^+ ions on tungsten, molybdenum, and the tantalum targets to the target temperature and to the energy of the incident ions [495, 544].

In order to study the dependence of the positive-ion neutralization coefficient (K_0) on the target temperature and primary-ion energy, we took oscillograms analogous to those shown in Fig. 3.3. On the basis of these oscillograms we plotted graphs of K_s, K_e, K_d, and K_Σ on T and E_0. The curves of K_Σ(T) and K_Σ(E_0) show the behavior of the sum of the coefficients K_s, K_e, and K_d.

It is clear that the sum of the secondary ion-emission coefficient K_Σ and the integral neutralization coefficient K_0 should be, in %:

$$K_\Sigma + K_0 = 100\%. \tag{3.3}$$

Then the region above the curve of K_Σ(T) (Fig. 3.4) may be ascribed to the neutralized fraction of the primary ions. From the curves in Fig. 3.4 and expression (3.3) we see that, depending on the physical parameters and the conditions of interaction of the colliding particles, the relation between the coefficients and K_Σ and K_0 will vary.

The integral value of the positive-ion neutralization coefficient K_0 was determined for various experimental conditions from the curves representing the temperature and energy dependence of K_Σ and from expression (3.3).

Figure 3.6 gives the temperature dependence of the coefficients K_s, K_Σ, and K_0, obtained on bombarding a tungsten target with 560-eV Cs^+ ions. We see from the figure that the coefficient K_s, as in the case of K^+ on tungsten (see Fig. 3.4), is independent of target temperature for a constant energy of the ions. The value of K_Σ for T < 1200°K equals K_s, owing to the absence of the currents I_e and I_d, while above 1300°K, within the limits of experimental error, it reaches 100%. This shows that all the ions falling on the target surface at T > 1300°K leave it in the form of scattered, evaporated, and diffused ions, while the emission current contains no neutral atoms. In other words, the numbers of primary and secondary ions balance completely.

We see from the K_0(T) curve that, in the temperature range 300–1100°K, the neutralization coefficient is almost constant at about 77%; then between 1100 and 1400°K it falls sharply to zero

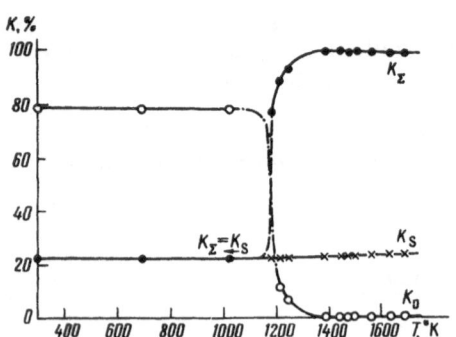

Fig. 3.6. Dependence of the coefficients K_0, K_s, and K_Σ on the target temperature for Cs^+ on tungsten (E_0 = 560 eV, φ = 4.51 eV).

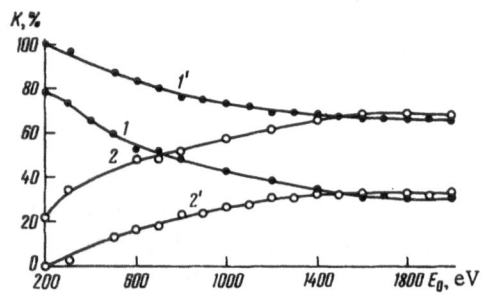

Fig. 3.7. Curves of $K_0(E_0)$ (light circles) and $K_\Sigma(E_0)$ (dark circles) for K^+ ions on molybdenum (1, 2) and Rb^+ ions on tantalum (1', 2') at T ~ 1500°K.

owing to the surface ionization which starts at these temperatures. The difference between the coefficients K_Σ and K_s in Fig. 3.6 characterizes the current of surface ionization (evaporated and diffusing ions).

An analogous result is obtained with Rb^+ ions on tungsten. In this case, in contrast to the previous, the coefficient K_0 differs from zero (~23%) for target temperatures of T > 1200°K; the reason for this is the difference in the ionization potentials of rubidium and cesium.

It is well known that on bombarding a cold, clean target with positive ions under the condition $V_i < \varphi$ the secondary emission will comprise only scattered atoms and ions, while at high temperatures there will be three components of neutral particles in the secondary emission: atoms leaving the bombarded surface by scattering, evaporation, and thermal diffusion, in parallel with the emission of scattered, evaporated, and diffused ions. Hence, the integral ion–neutralization coefficient may differ considerably for cold and hot targets (see Fig. 3.6).

Evaporated and diffused ions appear as a result of surface ionization [342]; on the basis of the Saha—Langmuir formula it should therefore be possible to find the numbers of evaporated (n_e^0) and diffused (n_d^0) atoms. Putting the coefficients of positive surface ionization for sorption-evaporated (β_e) and diffusion-evaporated (β_d) particles equal to the theoretical coefficient of surface ionization (β_t) calculated from the Saha—Langmuir formula, we may find the number of scattered atoms n_s^0, determine the coefficient of surface ionization for the scattered particles β_s [769], and compare this with the value of β_t.

In order to find n_s^0 we suppose that for target temperatures of T ≥ 1500°K there is no buildup of particles on the target, and we shall consider that the number of primary particles N equals the number of secondary particles:

$$N = n_s^+ + n_e^+ + n_d^+ + n_s^0 + n_e^0 + n_d^0, \qquad (3.4)$$

where n_s^+, n_e^+, and n_d^+ are, respectively, the numbers of scattered, evaporated, and diffused ions. Equation (3.4) enables us to calculate n_s^0, since all the other quantities may be determined.

Knowing n_s^0 and n_s^+, we find the coefficient of surface ionization of the scattered particles:

$$\beta_s = \frac{n_s^+}{n_s^+ + n_s^0}. \qquad (3.5)$$

The values of β_s for a number of alkali ions and a tungsten target at E_0 = 560 eV are given in Table 3.1, which also gives the values of β_t for comparison.

We see from Table 3.1 that for Cs^+ ions the values of β_s and β_t coincide, while for K^+ and Rb^+ ions they differ considerably. However, for the case of Na^+, presumably because

Table 3.1

	Cs+	Rb+	K+	Na+
β_s, %	100	69	64	57
β_t, %	100	92	74	0.5

Table 3.2

Ion	Metal	$V_i - \varphi$, eV	K_0, %
Na+	W	0.57	76
K+	Mo	0.17	62
Rb+	Mo	0.00	35
Rb+	Ta	—0.05	30
Cs+	W	—0.63	0

$V_i > \varphi$, the value of β_s is incomparably greater than β_t. In a general way the law relating β_s to the ionization potential of the alkali atom V_i agrees qualitatively with the law of surface ionization, i.e., β_s agrees with β_t in falling as the difference $V_i - \varphi$ increases.

It may be concluded [315, 678] that scattering in the form of positive ions depends not on the charged state but on the ionization potential of the bombarding particle. However, this phenomenon only qualitatively agrees with the laws of surface ionization. This is to be expected, since the secondary scattered ions have a wide energy spectrum, from thermal values to the level determined by single and multiple elastic collisions between the bombarding particle and the metal atoms. The best agreement of the results with the laws of surface ionization may in all probability only be obtained for a group of slow ions with energies of under 10 eV.

The energy dependence of K_Σ and K_0 for K^+ and Rb^+ ions on molybdenum, Rb^+ on tantalum, and Na^+ on tungsten was studied at a target temperature of 1500°K in a primary-ion energy range of 0.2-2 keV.

Figure 3.7 presents curves of K_Σ and K_0 plotted against the energy of the primary ions for K^+ on molybdenum and Rb^+ on tantalum. We see from the picture that K_Σ falls while K_0 increases with rising energy; both reach saturation at fairly high energies of the ions.

Analogous results were also obtained for Rb^+ ions on molybdenum and Na^+ ions on tungsten. The curves corresponding to these cases are similar in shape to the K^+ curves for molybdenum in Fig. 3.7, differing from these only in respect to the values of the coefficients.

Let us consider the experimental data obtained for K_0 (Table 3.2) as a function of the values of $V_i - \varphi$ corresponding to various target materials and types of ion at $E_0 = 1200$ eV and $T = 1500°K$.

We see from Table 3.2 that with increasing values of $V_i - \varphi$, K_0 increases. This evidently explains the almost 100% neutralization of inert-gas ions on the same targets [345, 420], where the value of $V_i - \varphi$ exceeds 10 eV.

For the total current of secondary ions we may write

$$I_\Sigma = I_s + I_e + I_d. \tag{3.6}$$

If we use β_t to denote the coefficient of positive surface ionization of the atoms, then for the sum of currents $I_e + I_d$ we obtain

$$I_e + I_d = \left(I_0 - \frac{I_s}{\beta_s}\right)\beta_t. \tag{3.7}$$

Substituting this expression in (3.6) we find

$$I_\Sigma = I_s \left(1 - \frac{\beta_t}{\beta_s}\right) + I_0\beta_t. \tag{3.8}$$

Dividing both sides of Eq. (3.8) by I_0, we have

$$K_\Sigma = K_s \left(1 - \frac{\beta_t}{\beta_s} \right) + \beta_t. \tag{3.9}$$

From formulas (3.3) and (3.9) we find an expression for K_0:

$$K_0 = 1 - K_\Sigma = K_s \left(\frac{\beta_t}{\beta_s} - 1 \right) - (\beta_t - 1). \tag{3.10}$$

We see from (3.9) and (3.10) that the total coefficient of secondary ion emission and the integral neutralization coefficient of the ions depend on the coefficients of ion scattering and positive surface ionization. If we suppose that the values of β_t and β_s are independent of the energy of the primary ions, then the variation in K_Σ and K_0 with ion energy will be determined by the dependence of K_s on E_0.

Thus, the results of the experiments show that the integral neutralization coefficient depends in a complex manner on the temperature and work function of the target and the energy and ionization potential of the bombarding ions.

§4.   Dependence of the Ion-Scattering Coefficient

on the Nature of the Ion and the Target Material

In order to discover the dependence of the ion-scattering coefficient on the nature of the ion and the target material, the author and some of his colleagues carried out some special experiments [542, 543, 545]. We used positive ions Na^+, K^+, Rb^+, and Cs^+ with energies E_0 between 75 and 1600 eV. The targets were W, Ta, Mo, Ni, and Ti strips. The experiments were carried out at target temperatures of 300 and 1500°K.

Figure 3.8 shows a series of $K_s(E_0)$ curves obtained by bombarding a clean molybdenum surface with positive Na^+, K^+, Rb^+, and Cs^+ ions at T = 300°K. We see from the figure that K_s falls with increasing primary ion energy for all the ions studied and reaches a minimum for large E_0. In the energy range E_0 = 450-900 eV, the curves corresponding to different ions intersect. For $E_0 \geq 1000$ eV, K_s falls with increasing mass of the ion; below 400 eV the reverse is the case, K_s increasing as the ion becomes heavier.

The fall in K_s with increasing ion energy is clearly explained by the character of the scattering at the repulsive screened Coulomb potential acting between the ion and the metal atom [541]. The corresponding calculation shows that for fairly high ion energies it is only by allowing for the energy dependence of the ion-scattering cross section at the atom that we can obtain the experimentally observed energy and mass dependence of K_s, namely, its initial rapid and subsequent slow fall with increasing E_0 and its variation with the mass of the colliding particles [541, 662]. In order to understand these laws more clearly, the $K_s(E_0)$ curves shown in Fig. 3.8 are reexpressed as a relation between K_s and the velocity of the ion v_0 in Fig. 3.9. In this figure, if we discount the anomalous case of Cs^+ ions on molybdenum ($m_2 > m_1$), the curves for the Na^+, K^+, and Rb^+ ions no longer intersect, the value of K_s being larger for smaller m_2.

The presence of scattered ions at very low primary ion energies in the case of $m_2 > m_1$ may be explained by the fact that here the primary ion interacts not with a single free target atom but with groups of atoms having a certain effective mass $m_{eff} > m_1$ (due to interatomic coupling). In interpreting the experimental data for K we thus take the equation

$$K_s = 1 - \frac{m_2}{m_{eff}} \tag{3.11}$$

In the case of Cs^+ ions on a molybdenum target, calculations based on formula (3.11) for the values of K_s taken from Fig. 3.9 and Brunnee's paper [383] show that for energies of 0.14

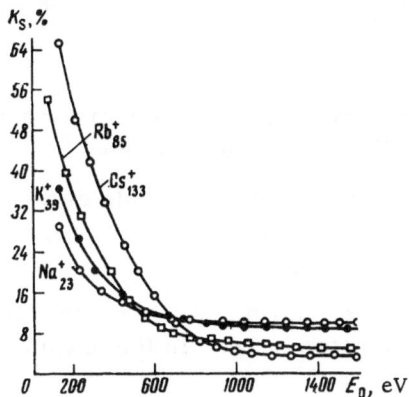

Fig. 3.8. Dependence of K_s on the energy of Na^+, K^+, Rb^+, and Cs^+ ions striking a molybdenum target.

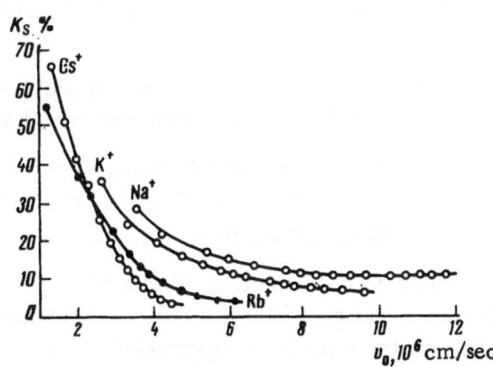

Fig. 3.9. Dependence of K_s on the velocity and mass of primary Na^+, K^+, Rb^+, and Cs^+ ions striking a molybdenum target.

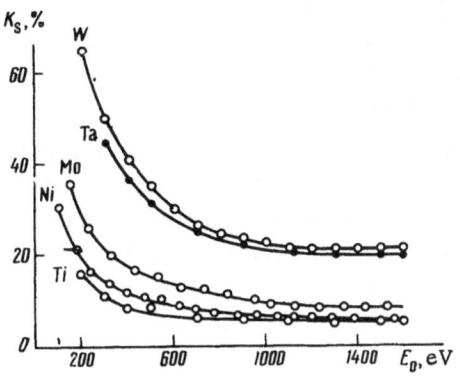

Fig. 3.10. Dependence of K_s on the energy of K^+ ions striking W, Ta, Mo, Ni, and Ti targets.

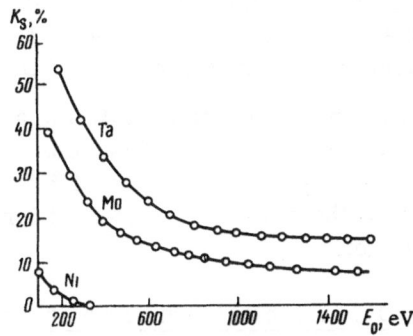

Fig. 3.11. Dependence of K_s on the energy of Rb^+ primary ions striking Ta, Mo, and Ni targets.

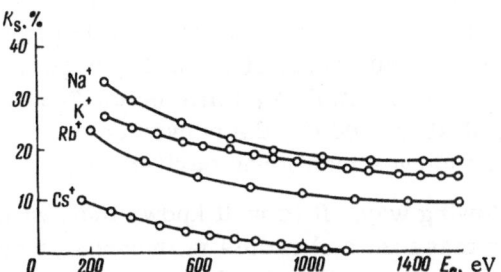

Fig. 3.12. Dependence of K_s on the energy of Na^+, K^+, Rb^+, and Cs^+ ions on a tantalum target at T = 1500°K.

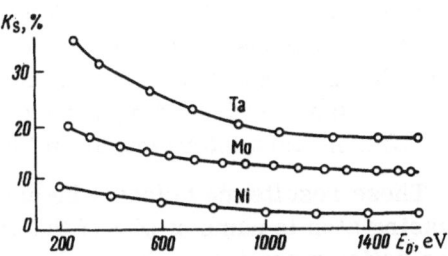

Fig. 3.13. Dependence of K_s on the energy of Na^+ ions for Ta, Mo, and Ni at T = 1500°K.

and 5 keV the value of m_{eff} respectively equals 380 and 133 m_H, where m_H is the mass of a hydrogen atom. The value of m_{eff} approximately equals four times the mass of the target atoms for $E_0 = 0.14$ keV.

The values of m_{eff} given by Petrov [504] for Cs^+ on molybdenum with primary ion energies of 0.1 and 1 keV are not greater than twice the mass of the molybdenum atoms.

An analogous calculation may be carried out for Cs^+ on nickel [494]. In this case, the value of $K_s = 90\%$ ($E_0 = 43$ eV) at the maximum of the $K_s(E_0)$ curve corresponds to $m_{eff} = 1330$ (for $m_1 = 58$). Thus, $m_{eff}/m_1 = 23$.

The values of m_{eff} obtained in the latter case are extremely large. It is true that this relates to a range of ion energies in which the energy of the interatomic bond in the crystal lattice of the metal exerts an appreciable influence.

The results of our study of K_s as a function of the mass of the target atoms at $T = 300°K$ are presented in Figs. 3.10 and 3.11. We see from these figures that for a particular ion K_s falls as E_0 rises and the density ρ and mass m_1 of the target atoms fall. This dependence of K_s on ρ and m_1 is better expressed in the range of low ion energies ($E_0 < 600$ eV), while in the range $E_0 > 600$ eV the curves for the same ion and for targets with similar m_1 values (for example, K^+ on tungsten and tantalum, K^+ on nickel and titanium) almost merge with one another.

Figures 3.12 and 3.13 show $K_s(E_0)$ curves obtained at $T = 1500°K$ for various ions and targets. Here the values of K_s are determined with a collector potential very slightly retarding the ions. In all cases the value of K_s falls with increasing E_0 and ion mass, as in Fig. 3.12, and rises with increasing mass of the target atoms (Fig. 3.13).

In contrast to the results for a cold target, for a hot target (1500°K), under the condition $m_2 > m_1$, no scattering of fast ions was observed (within the accuracy of the experiments). In the case of $m_2 < m_1$ the value of K_s at 1500°K was smaller than at 300°K. There was no intersection of the curves for different ions (Fig. 3.12), in contrast to the situation in the case of the cold target.

§5. Diffusion Component of Secondary

Ion Emission from Incandescent Metals

As shown in §2 of the present chapter, on bombarding the surface of metals with positive alkali ions the scattered and evaporated ions are accompanied by another group of secondary ions known as "diffusion" (or diffusing) ions. Evidently the emission coefficient of the diffusion ions (K_d) is in general determined by a whole series of parameters characterizing both the incident ion and the material and surface state of the target [546, 547].

Figure 3.14 shows the relation between K_s and E_0 for K^+, Rb^+, and Cs^+ ions bombarding a molybdenum target at 1500°K. We see that K_d increases with energy for these ions, but for $E_0 > 1000$ eV it tends to saturation. For energies $E_0 > 700$ eV, the value of K_d is the larger, the greater the mass m_2 of the ion and the smaller its ionization potential V_i while below 700 eV the relation between K_d and m_2 reverses. For energies of about 700 eV, the curves corresponding to different ions intersect, as in the case of the $K_s(E_0)$ curves mentioned earlier.

These results may clearly be explained in the following way. It is well known that, with increasing atomic number (Z_2) of the alkali atoms, their mass m_2 and radius r_2 increase, while the ionization potential V_i diminishes. Of these three parameters, the rise in m_2 clearly leads to an increase in the number of primary ions penetrating into the target and thus to an increase in the contribution of diffusing ions to the emission. At the same time, an increase in the radius of the ion clearly reduces the number of ions penetrating deep down into the target. Thus, the

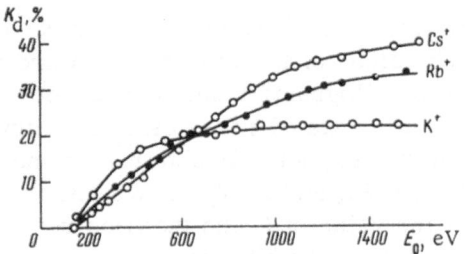

Fig. 3.14. Dependence of K_d on the type and energy of primary K^+, Rb^+, and Cs^+ ions on Mo at T = 1500°K.

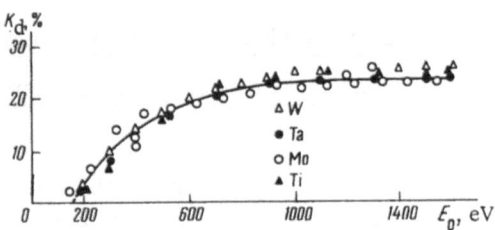

Fig. 3.15. Dependence of the coefficient of emission of the diffusing ions on the energy of the bombarding K^+ ions for W, Ta, Mo, and Ti targets at 1500°K.

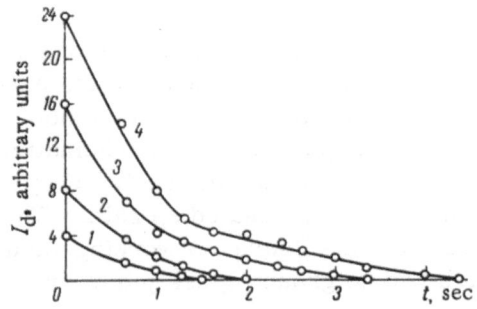

Fig. 3.16. Time dependence of the current I_d for K^+ ions striking tungsten targets at 1500°K.

number of primary ions penetrating into any particular target (the number of absorbed atoms) depends on the mass and radius of the ion.

The number of absorbed atoms (adatoms) diffusing to the surface of the metal at high temperatures and leaving it in the form of positive ions in accordance with the law of surface ionization will be determined for any particular target by the value of V_i. It is probable that the smaller the value of V_i, the greater will be the number of diffusing ions.

The emission coefficient of diffusing ions depends in a complicated manner on the three primary ion parameters mentioned and on the target material. The greatest value of K_d for $E_0 > 700$ eV occurs in the case of Cs^+ ions and the smallest in that of K^+ (Fig. 3.14). The curve corresponding to the Rb^+ ion lies between those of Cs^+ and K^+ in the energy range studied. This is evidently associated with the fact that Rb lies between Cs and K both in mass and in ionization potential.

Let us consider the results of an investigation into the dependence of K_d on the target material; these are shown as $K_d(E_0)$ curves in Fig. 3.15. This figure relates to the bombardment of tungsten, tantalum, molybdenum, and titanium targets with K^+ ions. We see from the figure that the coefficient of the diffusing ions increases with energy after about $E_0 = 150$ eV for all targets, and tends to saturation at $E_0 > 1000$ eV.

A reduction in the density of the target and a rise in the mass of the ion tend to increase the number of ions penetrating into the target. Since the diffusion ions appear as a result of surface ionization, a reduction in the ionization potential of the ion or a rise in the work function of the target should lead to an increase in the emission coefficient of the diffusing ions.

The resultant effect of all these parameters is that the emission coefficient of the diffusing ions is very much the same for different targets and the same ion (Fig. 3.15).

Thus the processes governing the penetration of alkali ions into metals and their back diffusion to the surface with subsequent ionization depends in a complicated manner on a number of properties of the bombarding ions and target surface.

In order to gain a better understanding of the mechanism giving rise to the current of diffusing ions I_d, it is important to know the time characteristics of this current, after the ion beam has ceased reaching the target, as functions of the energy of the bombarding ions and the temperature of the metal. In Fig. 3.16 we present a series of curves giving the dependence of

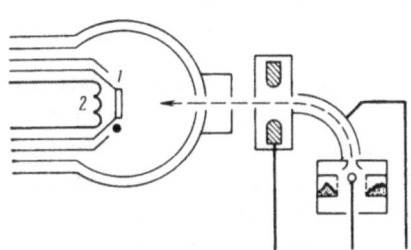

Fig. 3.17. Arrangement of the vacuum apparatus.

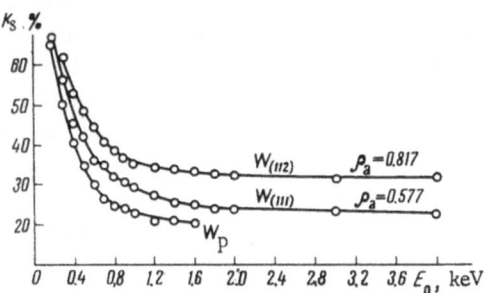

Fig. 3.18. Relation between K_s and E_0 for the bombardment of the (111) and (112) faces of tungsten with K^+ ions.

I_d on t for tungsten targets bombarded with K^+ ions at 1500°K. The values of I_d on the curves were determined from oscillograms taken with a motion picture camera at a rate of 8 frames/sec. Curves 1, 2, 3, and 4 were respectively obtained at energies of 640, 960, 1280, and 1600 eV. We see from the figure that the value of I_d and the time required for the latter to fall off both increase with increasing E_0. We shall arbitrarily call this time the life of the diffusion ions (τ_d). Analogous $I_d(t)$ curves obtained for various target temperatures (E_0 = const) show an inverse temperature dependence of τ_d.

§6. Scattering of Alkali Ions by the Surface of Tungsten and Molybdenum Single Crystals

A few papers have been written on the scattering of ions by single-crystal targets. These have included investigations into the scattering anisotropy [692, 693, 697] and the energy spectra [698, 699, 779, 736, 737] of scattered ions on bombarding the surface of a copper single crystal with Ar^+ ions at energies of up to 80 keV. Some of these papers [692, 693, 697] have been of a qualitative nature.

In this section we shall present the results of an investigation into the scattering of alkali ions by the surface of tungsten and molybdenum single crystals. In order to measure the currents and determine the values of K_s, we used the oscillographic double-modulation method (Chapter 2, §6). The energy dependence of K_s was studied in the apparatus shown schematically in Fig. 3.17. The single-crystal target 1, 11 × 5 × 0.7 mm in size, was heated to 2500°K by electron bombardment on the reverse side, the current coming from the heated spiral 2. The receiving part of the apparatus and the ion source are clear from the figure. A working vacuum of about 10^{-7} mm Hg was maintained in the apparatus (which had no ground-glass joints or taps) by means of two mercury-vapor pumps working in parallel.

Energy Dependence of the
Ion-Scattering Coefficient

In order to discover the influence of the crystal structure of the target and the geometrical factors on the value of K_s for alkali ions, Arifov, Khadzhimukhamedov, and Yunusov [738] studied the relations $K_s(E_0)$ and $K_s(v_0)$ for various faces of tungsten and molybdenum single crystals.

The $K_s(E_0)$ curves for K^+ ions and the (111) and (112) faces of a tungsten single crystal are shown in Fig. 3.18. For comparison, the same figure shows results obtained for polycrystalline tungsten W_p [542]. We see from the curves that the scattering process depends on the crystal structure of the target bombarded; the value of K_s in the E_0 range studied is greater for the single crystal than for polycrystalline material. The form of the curves is the same for both faces. However, the values of K_s differ in the two cases. Naturally, the number of scat-

Table 3.3

	Face			
	100	110	111	112
n	1	2	0,5	1
S_{hkl}	a^2	$\sqrt{2}a^2$	$\dfrac{\sqrt{3}}{2}a^2$	$\dfrac{\sqrt{3}}{\sqrt{2}}a^2$
$\rho_a = n/a^2$	$1/a^2$	$1,414/a^2$	$0.577/a^2$	$0.817/a^2$
$\rho_0 = \rho_a/\rho_a(100)$	1	1.414	0.577	0.817

Table 3.4

Face	Ion				
	Li+	Na+	K+	Rb+	Cs+
110	—	—	0.38	—	—
111	0.43	0.42	0.40	0.31	0.29
112	(0,27)	0.40	0.39	0.29	0.31

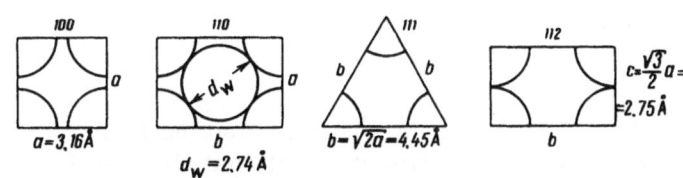

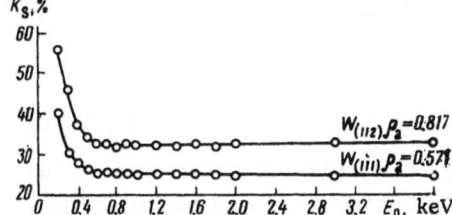

Fig. 3.19. Cross section of the (100), (110), (111), and (112) faces in the unit cell of the bcc tungsten lattice.

Fig. 3.20. Dependence of K_s on E_0 for the bombardment of the (111) and (112) faces of tungsten by Na+ ions.

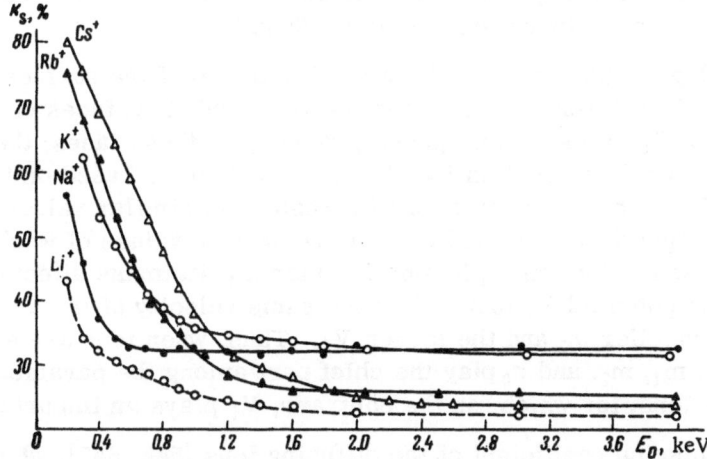

Fig. 3.21. Dependence of K on the energy and type of bombarding ions for the (112) face of tungsten.

tered ions is proportional to the probability of a collision between the primary ion and surface atoms on the target. This probability is the greater, the greater the density of atomic packing ρ_a [738] on the face of the crystal:

$$\rho_a = n/a^2,$$ (3.12)

where a is the lattice constant of tungsten, equal to 3.16 A, and n is the number of atoms associated with unit area of the face (see Fig. 3.19 and Table 3.3).

We see from Table 3.3 that

$$(\rho_a)_{(112)} > (\rho_a)_{(111)},$$ (3.13)

hence we should expect

$$(K_s)_{(112)} > (K_s)_{(111)},$$ (3.14)

as experiment indeed shows; for $E_0 = 4$ KeV,

$$(K_s)_{(112)} \approx 0.32, \quad (K_s)_{(111)} \approx 0.23$$ (3.15)

$$\left(\frac{K_s}{\rho_0}\right)_{(112)} \approx 0.4, \quad \left(\frac{K_s}{\rho_0}\right)_{(111)} \approx 0.4.$$ (3.16)

It follows from (3.16) that the value of K_s is proportional to ρ_0, and to a first approximation in the range $E_0 \sim 1.5$–4 keV, we may put

$$K_s = A\rho_0,$$ (3.17)

where A is a proportionality factor.

An analogous result is obtained for Na^+ ions (Fig. 3.20). In this case also the form of the curves is the same for both faces; the value of K_s almost coincides with that obtained for K^+ ions striking the same face in the high-energy (E_0) range. Hence, the value of A is practically the same for Na^+ and K^+ ions. The experimental values of A for various ions and tungsten faces are shown in Table 3.4 for the energy range $E_0 = 1.5$–5 keV.

For light ions (Li^+, Na^+, and K^+) the value of A, except for the bracketed case in the table, equals ~0.40, while for heavy ions (Rb^+ and Cs^+) it equals ~0.30. The figure in brackets relates to an anomalous case to be considered later (Fig. 3.21).

As regards the dependence of K_s on the work function φ of the surface of the face, we may say the following. The difference in K_s for the (111) and (112) faces of tungsten is mainly due to the dependence of K_s on the atomic packing density on these faces; the work functions of the faces clearly have a weaker effect on K_s. This proposition is based on the fact that, as shown in [543], the $K_s(E_0)$ curves are grouped with respect to similar values of the atomic mass m_1 (see Fig. 3.10) or target density ρ, and not with respect to values of work function. On the other hand, when the mass of the ion m_2 is smaller than m_1, its mass affects K_s more (see Fig. 3.12) than the ionization potential V_i [542]. For the same velocity of the alkali ion the value of K_s is the greater, the smaller m_2 and the larger V_i. Thus, when ions are scattered by the surface of a single crystal m_1, m_2, and ρ_0 play the chief part among the parameters. However, this does not relate to slow ions, for which, on the contrary, V_i plays an important part [702].

A study of the emission coefficient of the diffusing ions [546, 547], which characterizes the degree of penetration of the primary ions into the depths of the target, showed that, in contrast to K_s, the value of K_d for a single crystal was smaller than for polycrystalline tungsten,

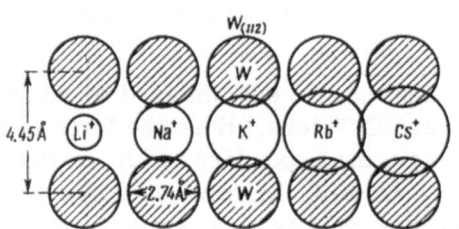

Fig. 3.22. Arrangement of the tungsten
atoms on the (112) face.

i.e., larger values of K_s corresponded to smaller values of K_d. Thus, the coefficients K_s and K_d are mutually related and depend on the crystal structure of the target material.

The dependence of K_s on the energy and type of bombarding ions is shown in Fig. 3.21 for the (112) face of a tungsten single crystal.

For high ion energies ($E_0 > 2$ keV) the value of K_s increases as the mass of the ions m_2 diminishes: $(K_s)_{Cs} < (K_s)_{Rb} < (K_s)_K < (K_s)_{Na}$. We note, however, that the lightest ion Li^+ has the lowest value of K_s. This is probably because the diameter of the Li^+ ion (1.36 Å) is smaller than the width of the channel (1.70 Å) in the (112) face (see Fig. 3.22 for the arrangement of the tungsten atoms on this face); hence, as a result of the possible penetration of a large number of Li^+ ions into these channels, the value of K_s in this case is low $(K_s)_{Li} < (K_s)_{Cs}$. Hence, for the (112) face of a tungsten single crystal we cannot speak of a particular dependence of K_s on m_2 for high ion energies ($E_0 > 2$ keV). For this reason, the value of A corresponding to Li^+ on the (112) face of tungsten appears too low (Table 3.4). In the range $E_0 < 1$ keV, the value of K_s rises sharply with falling E_0 and with increasing mass of the bombarding ion.

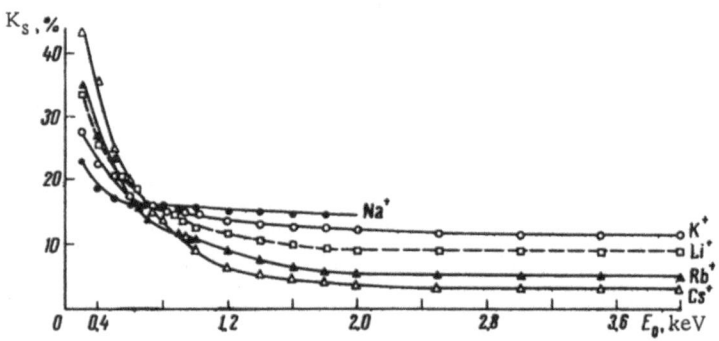

Fig. 3.23. Dependence of K_s on the energy and type of ion
for the (001) face of molybdenum.

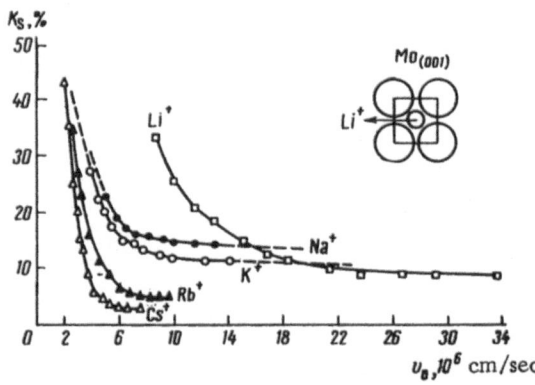

Fig. 3.24. Dependence of K_s on the velocity and type of ion
for the (001) face of molybdenum.

We extended these investigations to a molybdenum single crystal, which like the tungsten had a bcc lattice and a similar lattice constant (3.14 Å). The results of the experiments are shown in Figs. 3.23 and 3.24.

Figure 3.23 gives the $K_s(E_0)$ curves for the energy range 0.3–4.0 keV obtained by bombarding the (001) face of molybdenum with Li^+, Na^+, K^+, Rb^+, and Cs^+ ions. Here the form of the $K_s(E_0)$ curves is the same as in the case of tungsten. However, the dependence of K_s on the mass of the ion is of a specific character. In the range $E_0 > 0.8$ keV, the value of K_s for Li^+ is smaller than for Na^+ and K^+ and larger than for Rb^+ and Cs^+. This is apparently associated with the relative dimensions of the ion and the lattice of the crystal (Fig. 3.24).

Figure 3.24 shows the dependence of K_s on the velocity of the ion v_0. We note that the dependence of K_s on m_2 for all the ions remains the same as in the case of the polycrystalline material [542, 383] up to velocities of $v_0 \approx 1.5 \cdot 10^7$ cm/sec. Above this velocity [possibly because the diameter of the Li^+ ion is smaller than the interatomic distance on the (001) face of molybdenum), Li^+ ions start penetrating more strongly into the depths of the target in a direction of motion coinciding with the crystallographic [001] direction. As a result of this, the values of K_s for Li^+ become smaller than for Na^+ and K^+.

For the looser, (112) and (111) faces of the crystal, as in the case of tungsten, we may expect a sharply expressed anisotropy of alkali-ion scattering.

CHAPTER 4

INTERACTION OF SLOW IONS
WITH A SOLID SURFACE

In order to develop our investigations into the interaction between atomic particles and solids, it is particularly important to understand the mechanism of this phenomenon. Many and various hypotheses have been advanced regarding the mechanism underlying the scattering of atomic particles by a solid surface. Thus, a number of research workers attempted to relate the direction of maximum intensity of the scattering of ions by solids with the angle of specular reflection. Later it was shown that the ions were scattered predominantly forward, and that any reflection at the specular angle or variation in the position of this angle with the energy of the incident ions (which had been considered as a proof of the diffraction of ion beams) were only apparent. Even in Gurney's treatment [49] an attempt was made to explain the scattering of ions by a solid as being similar to elastic scattering in a gaseous medium. However, the limiting energy E of the scattered ions measured in this investigation was much lower than would follow from Eq. (1.9). This was due to the insufficiently clean conditions on the target surface. Agreement would also clearly not be achieved in the case of an ideally clean target, since, for the low energy range of the incident ions employed (20-60 eV) the atoms in the target could not be regarded as completely free owing to the bond forces between them. This would make the energy of the scattered ions greater than it should be according to Eq. (1.9). Much later, Arifov and Ayukhanov [219] and Eremeev [228] reliably established that only for incident energies E_0 of more than a few hundred eV would the ions be scattered from individual target atoms which could justifiably be regarded as free.

In certain theoretical treatments [228, 501] the colliding particles are considered as solid, impenetrable spheres. This view implies spherical symmetry of the ion scattering and energy independent of the elastic-scattering coefficient K_s and thus fails to agree with experiment. Theories using a variety of screened potentials for the interaction between the colliding particles [662, 700, 747] (the model of mutually permeable particles) show that the scattering is directed predominantly forward. This is supported by the majority of experimental investigations into the angular distribution of secondary ions [622, 624, 698].

In a case in which the mass of the target atom is greater than that of the approaching ion $(m_1 > m_2)$, the energy spectrum of the secondary ions contains a clear peak representing singly elastically scattered ions; the energy E of this peak agrees quite well with that calculated from formula (1.9) [228, 323, 598, 708]. The spectrum for the case of $m_1 < m_2$ with scattering angles of $\beta < \beta_{lim} = \arcsin m_1/m_2$ is qualitatively similar to the spectrum for $m_1 > m_2$. However, on increasing β above β_{lim} the maximum of the singly scattered ions gradually vanishes. Under these conditions, i.e., after the maximum has vanished, the ions with greater energies observed are evidently multiply scattered [662, 708, 717].

The presence of ions with energies smaller than the maximum energy E of the secondary ions is explained by many authors as being due to multiple scattering. Regarding ions with energies greater than E, it has been suggested that these are singly scattered in a single impact with a complex of crystal-lattice atoms having a certain effective mass $m_{eff} > m_1$ [323, 500, 658]. This assertion is valid if the atoms in the crystal lattice are considered as rigidly linked together. It is well known, however, that for an energy E_0 greater than a few hundreds of electron volts the atoms in the crystal lattice may be regarded as practically free. It is thus not clear whether one may regard them as rigidly connected for an energy of a few tens of electron volts.

A theoretical study of the influence of the coupling between atoms in the crystal lattice on pair collisions, using potentials of the Born-Meyer type in the calculations, showed that the maximum energy of the scattered atoms corresponding to the group interaction of several particles was smaller than that corresponding to paired collisions [700].

Thus, as a result of the coupling between the atoms of a solid we may expect a departure from the gas model of ion scattering in the low-energy range.

The author and his colleagues considered the interaction of slow ions with solids in a number of papers [494, 702, 710, 711]. The same question was treated by Veksler [614, 615, 659, 709] and Gruich et al. [696, 740] and earlier in [48, 49, 59, 94]. However, in our present view, all this research constituted merely the first steps in studying the interaction of slow ions with solids. The effect of the coupling of the atoms in the target on secondary processes has still not been explained. There are no reliable measurements of the secondary ion- and electron-emission coefficients in the low-energy range. The theoretical treatments cannot explain the experimentally established high values of the coefficients of secondary ion emission in the cases $m_1 < m_2$ and $\beta > \beta_{lim}$. For various targets and ions, the energy spectra of secondary ions have not been studied in sufficient detail. Questions of the interaction of slow ions with solids require further theoretical study.

In this chapter we present a method of studying phenomena taking place during the interaction of slow ions of the alkali and alkaline earth atoms with a solid body, and some of the results obtained. Measurements based on this method enabled us to establish the extent to which the coupling of the atoms in the solid and the sorption forces at the surface affected the energy spectra and coefficients of secondary ion emission. We were also able to separate out a group of slow secondary ions formed by a mechanism differing from that of the elastically scattered type.

§1. Method of Investigation when Working with

Slow Ions

Formation of a Beam of Slow Ions

When studying the interaction of slow ions with a solid surface a particular difficulty is that of obtaining well-focused ion beams of sufficient density. The reason for this is the mutual electrostatic repulsion of the charged particles.

The spreading of charged-particle beams under the influence of space charge has been studied in [114, 153, 189, 278, 336, 520]. The degree of beam spread for a beam of rectangular section at various distances l in drift space is described by the expression

$$\frac{d}{d_0} = 1 + 2.04 \frac{l^2 I}{d_0} \frac{A}{U_0^3}, \tag{4.1}$$

where d_0 and d are the initial characteristic linear thicknesses of the beam and its dimensions at a distance l, A is the atomic weight of the ion, U_0 is the voltage accelerating the ions, and I is the current per unit beam width in amperes [189].

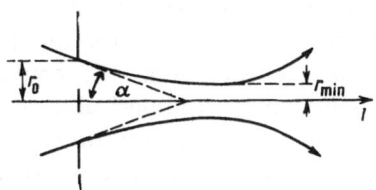

Fig. 4.1. Spread of a beam of charged particles due to space charge.

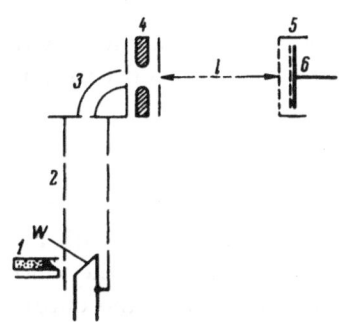

Fig. 4.2. Source and ion-optical system for obtaining focused beams of slow ions.

The radius of an ion beam with a circular cross section at various distances in drift space is determined from the equation

$$l = 1.75 \cdot 10^{-3} r_0 \frac{U_0}{Al^2} \int_0^{\ln r/r_0} e^{v^2} dy, \qquad (4.2)$$

where r_0 and r are the initial beam radius and the radius at a distance l in drift space, respectively [114, 189]. The values of the integral in formula (4.2) are tabulated in [203].

We are interested in the behavior of converging ion beams in drift space. The radius of the beam r at any l may be determined from the following equation [114, 336, 529]:

$$\frac{l}{r} = \sqrt{\frac{2eU_0}{m_2}} \int_1^{r/r_0} \frac{dx}{\sqrt{a \ln x + 2\frac{e}{m_2} U_0 \tan^2 \alpha}}, \qquad (4.3)$$

where

$$a = \sqrt{\frac{8 l_0^2 e}{U_0 m_2}}.$$

At the point of minimum cross section $dr/dl = 0$. Then the radius r_{min} is determined by the following equation (see Fig. 4.1):

$$r_{min} = r_0 e^{-\frac{v_e^2 \tan^2 \alpha}{a}}, \qquad (4.4)$$

or

$$\ln \frac{r_{min}}{r_0} = -\frac{v_e^2 \tan^2 \alpha}{a} = -\frac{U_0^{1/2}}{I_0} \sqrt{\frac{e}{2m}} \tan^2 \alpha. \qquad (4.5)$$

It was shown theoretically and experimentally in [336, 529] that the effect of space charge ceased when, as a result of the reduction in the density of the ion current, the radius of the minimum cross section of the beam became smaller than a certain critical value (of the order of the distance between the charged particles in the beam). Then we were dealing with a converging beam free from any aberration (positive or negative).

In order to obtain well-formed beams of slow ions, ion guns analogous to electron guns in construction (254, 362] were tested. The main disadvantage of such constructions is the sharp fall in beam intensity on reducing the energy of the incident ions.

Our own version of the ion-optical system (Fig. 4.2) was free from this defect [696, 740]. In the ion source employed the ions of alkali and alkaline earth metals were obtained by the surface ionization (on a hot tungsten strip) of the halides of these metals coming from a tantalum-foil evaporator tube 1 filled with salt and heated by the direct passage of an electric current. The ions so formed were shaped into a beam by means of the electrostatic immersion lens 2. Then the ions fell into the cylindrical condenser 3, where they were separated from neutral atoms and directed into the immersion-type electrostatic lens 4. By means of this lens the ion beam was finally focused and retarded to the required energy E_0.

The cross section of the beam of slow ions obtained in the ion-optical system described was checked experimentally as a function of U_0, I_0, and l. For this purpose we set a fine mesh 5 with 75% transparency in the drift space at a distance l from the electrostatic lens 4 (l could be varied from 5 to 70 mm), and behind the mesh at a distance of about 1 mm, a nickel plate 6 covered with a phosphor. By applying a potential difference between the mesh 5 and plate 6 the beam of slow ions was accelerated further to an energy of 1 keV. Under the action of these ions a sharp spot appeared on the phosphor, coinciding in shape and size with the cross section of the slow-ion beam. For a current of $I_0 \sim 10^{-6}$ A, $E_0 < 30$ eV, and $l = 70$ mm, the beam tended to spread, and this could not be compensated by focusing (changing the potential on the middle diaphragm of the lens 4). However, the dimensions of the spot diminished on reducing I_0 or l. For a current of $I_0 \sim 5 \cdot 10^{-9}$ A and $l = 70$ mm, and an energy of up to $E_0 = 10$ eV, the diameter of the luminescent spot on the target was under 2 mm.

The phenomena taking place during the interaction of the slow ions with the surface of the metal were studied in the fields of spherical and cylindrical condensers and recorded with an oscillograph.

Experimental Apparatus of the Spherical-Condenser Type

It is well known that the accuracy with which charged particles may be analyzed with respect to energy in a spherical condenser is determined by the excess of the energy of the analyzed particles over the retarding potential required in order to make them fall into the outer collector sphere. Lukirskii showed [88] that for a 1 : 10 ratio of the diameters of the inner and outer spheres the collector received all charged particles with initial kinetic energies of

$$E = \frac{m_2 v^2}{2} \gg 1.01 \, V_\kappa e, \qquad (4.6)$$

where V_K is the retarding potential on the collector sphere. It was shown in [488] that for a 1 : 10 ratio of the target and collector diameters, condition (4.6) was also satisfied when the target had the form of a disc or a disc-like box.

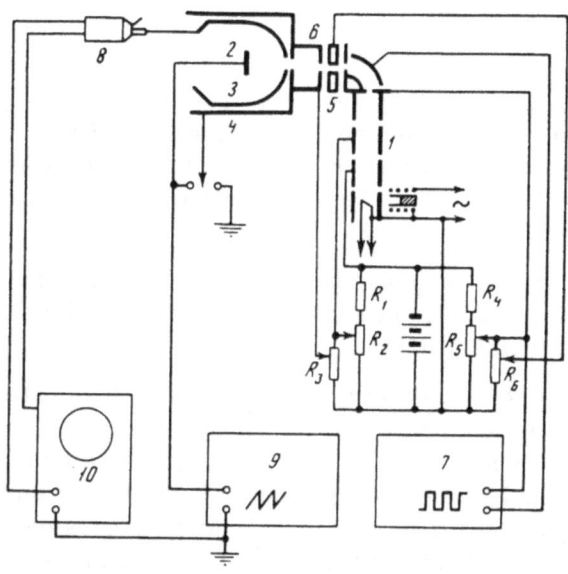

Fig. 4.3. Principal electric circuit of the apparatus for working with slow ions.

Figure 4.3 shows a schematic representation of the experimental apparatus used by Gruich, a component part of which is a spherical condenser satisfying condition (4.6), and the main electrical measuring circuit [696, 740]. The apparatus consisted of a source of slow ions 1 (described at the beginning of the section), a target 2, a hemispherical collector 3, and a guard cylinder 4. Between the lens 5 and the guard cylinder 4 was a drift space 6 about 15 mm long preventing the field of the lens 5 from bulging out into the collector sphere. The focused ion beam did not graze the guard cylinder 4 or the collector 3; this eliminated secondary emission from the edges of the apertures, which might otherwise distort the results of the measurements.

The energy distribution of the back-scattered ions in the field of the spherical analyzer was studied by the double-modulation method described in detail in Chapter 2, §5. In this

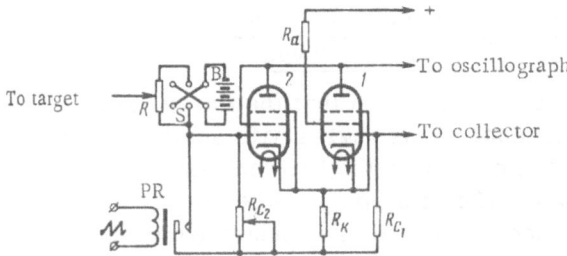

Fig. 4.4. Electric circuit of the amplifying attachment to the oscillograph for the simultaneous recording of the primary and secondary ion currents.

case the electric circuit consisted of a rectangular-pulse generator 7, an amplifying attachment 8, a sawtooth-pulse generator 9, an oscillograph 10, a battery B, and a voltage divider R_1-R_6.

Work with small ion currents (about 10^{-9} A) required an increase of two or three orders in the measuring sensitivity. This was achieved by connecting the amplifying attachment 8 to the oscillograph 10.

In order to improve the accuracy of measuring the coefficients of secondary ion emission, the electrical circuit shown in Fig. 4.3 was considerably modified by adding an attachment to the oscillograph (see Fig. 4.4) so as to be able to record the primary and secondary ion currents simultaneously [696, 740]. The attachment constituted an amplifying stage based on two pentodes with a common plate load R_a and a bias resistance R_K. The control grid of tube 1 was joined to the collector and the control grid of tube 2 to the target. By adjusting R_{C_2} we were able to secure identical amplification factors for the tubes. In parallel with R_{C_2} was a polarized relay PR which enabled us to switch the circuit into two measuring modes.

With the relay PR open, amplified collector and target currents were taken from the plates of the tubes (together these made up the primary ion current I_0); with the relay closed, the target current was short-circuited to ground, and only the amplified secondary ion current from the collector was taken from the tube plates.

Figure 4.5 shows an oscillogram taken with this device and representing the simultaneous recording of the currents of the primary ion beam I_0 and the secondary emission I on bombarding clean, cold tungsten with 360-eV Na^+ ions. For this purpose the tube plates were connected to the vertical input of the oscillograph; the sawtooth voltage feeding the relay and the oscillograph sweep were held at a frequency of about 20 cps, and the primary ion beam was modulated by rectangular pulses at about 160 cps.

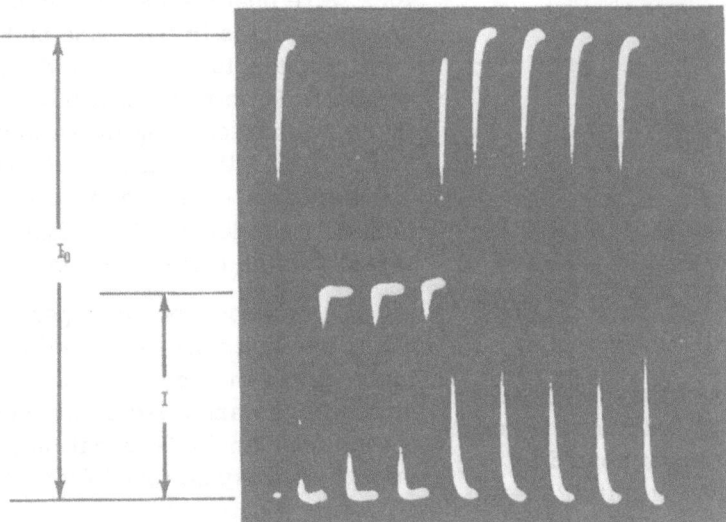

Fig. 4.5. Oscillogram representing the simultaneous recording of the intensity of the primary and secondary ion currents.

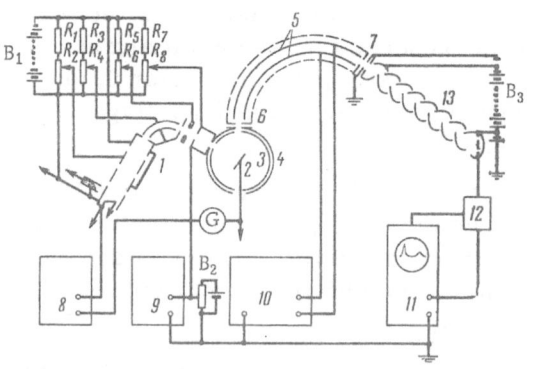

Fig. 4.6. Main electric circuit and apparatus of the cylindrical-condenser type.

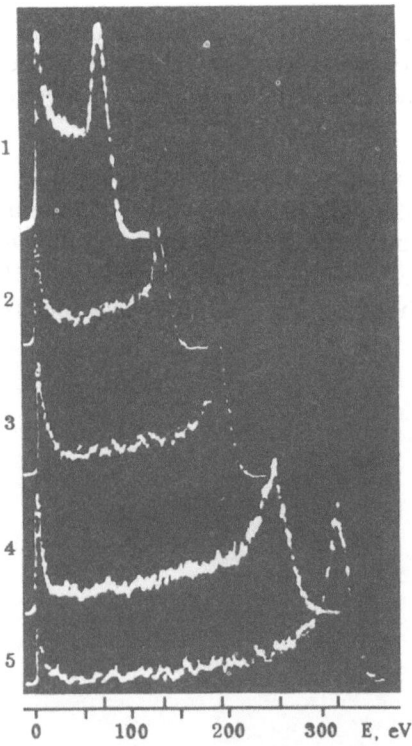

Fig. 4.7. Oscillograms of the energy spectra of the secondary Na$^+$ ions obtained on bombarding a tungsten target heated to 1700°K with Na$^+$ ions having initial energies E_0 of: 1) 90; 2) 190; 3) 290; 4) 395; 5) 500 eV.

Experimental Apparatus of the Cylindrical-Condenser Type

It is well known that the energy spectrum of the secondary ions varies with scattering angle β while the energy of the incident particles remain constant. Hence a study of the energy distribution of the back-scattered secondary particles (integral distribution) will not enable us to carry out any deep study of the nature of the interaction between the bombarding ions and the target atoms. In view of this, it seemed interesting to study the energy spectra of ions scattered at a particular angle β.

In order to analyze the secondary ions with respect to energy we used a 127° cylindrical condenser of the Hughes–Rojansky type. The parameters of the analyzer used in the experiments (Fig. 4.6) were as follows: mean radius of curvature of the copper plates 5, 100 mm, distance between the plates 7 mm, width of the plates 45 mm, dimensions of the entrance and exit slits 6 and 7 both equal to 0.8 × 18 mm. The resolving power of the apparatus determined experimentally was about 1.5% [702, 740]. The ion source 1 was distinguished from that described at the beginning of the section simply in that the ion beam had a rectangular cross section. The target 2 lay in the center of the cylindrical condenser 3 and guard cylinder 4. The distance from target 2 to the entrance slit of the analyzer 6 was 24 mm. The beam fell on the target normally, and the scattered ions were observed at an angle of $\beta = 135°$. The secondary ions in the plane of Fig. 4.6 were cut out at an aperture angle of 7°, determined by the width of the entrance slit 6 of the analyzer. Since there was no focusing in the plane perpendicular to Fig. 4.6 in the cylindrical condenser, the angle of divergence of the beam, about 4°, was determined by the length of the exit slit 7 of the analyzer, and the distance from the latter.

As a result of the dependence of the energy E of the elastically scattered ions on the scattering angle β, the accuracy of measuring E is determined by the angle $\Delta\beta$ at which the secondary ion beam under analysis is cut out. The spread of energy ΔE may be found from the expression

$$\frac{\Delta E}{E} = -\frac{2\sin\beta}{\sqrt{\mu^2 - \sin^2\beta}}\,\Delta\beta, \qquad (4.7)$$

obtained by differentiating (1.9). For $\Delta\beta \approx 7°$, $\Delta E/E \approx 1.5\%$, i.e., of the same order as the resolving power of the apparatus.

The receiver for the secondary ions passing through the analyzer 5 (accelerated to 5 keV following the exit slit 7) was the first dynode of an ion—electron multiplier 13 made of a copper—beryllium alloy. The amplification factor of the multiplier 13 was about $5 \cdot 10^6$. The multiplier, together with the preamplifier 12 and oscillograph 11, enabled us to measure currents of $\sim 10^{-14}$–10^{-15} A. The energy of the primary ions was measured with a tube voltmeter 8, and the constant bias of the sawtooth-pulse generator 10 with the tube voltmeter 9.

The principle of operation of the analyzer 5 was described in Chapter 1, § 2. In order to secure automatic recording of the energy spectrum of the secondary ions on the oscillograph screen 11, a sawtooth voltage from the generator 10, varying linearly with time, was applied to the analyzer plates 5. If the frequency of the sawtooth voltage fed to the analyzer plates and the oscillograph sweep were kept the same (about 25 cps), a stationary oscillogram of the energy spectrum appeared on the oscillograph screen.

Figure 4.7 shows a series of such oscillograms of energy spectra obtained with a tungsten target heated to 1700°K bombarded with Na^+ ions at various values of E_0.

The secondary ion emission in the low-energy range was extremely sensitive to the state of the target surface. Hence, in choosing materials for study, we decided on refractory metals which could be brought to an atomically clean state fairly easily. Targets of tungsten and molybdenum were degassed at about 2200°K, tantalum and zirconium at about 1700°K, and nickel at about 1400°K. Prolonged heating of a tantalum target at T > 1700°K led to its recrystallization, as a result of which the polished surface became rough. Secondary ion emission from such a surface was almost a factor of two lower than that from a polished surface.

The alkali halides and the $BaCl_2$ used in the work were chemically pure. The composition of the ion beam was analyzed with a mass spectrometer. Foreign impurities in the ion beam never exceeded 1%. Evacuation was effected with glass mercury-vapor pumps. The measurements were made at a vacuum of about $5 \cdot 10^{-8}$ mm Hg.

§ 2. Energy Spectra of Secondary Ions on

Bombarding Metals with Slow Ions

Using the apparatus shown schematically in Fig. 4.6 and an angle of $\beta = 135°$, we studied the energy spectra of secondary ions from cold and high-temperature targets of tungsten, tantalum, molybdenum, and nickel bombarded with 20- to 500-eV alkali ions in a direction normal to the surface [702, 710, 711, 740]. We found that the secondary ions had a wide energy spectrum. There were two groups of ions: elastically scattered, with the maximum of the singly scattered ions in the high-energy part of the spectrum for the case $m_1 > m_2$, and slow ions grouped around a maximum in the low-energy part of the spectrum for the case $m_1 \gtrsim m_2$.

The energy of the ions determined from the position of the high-energy maximum of the spectra indicated in Figs. 4.7–4.9 for an incident energy of $E_0 \geq 400$ eV agrees closely with the E calculated from Eq. (1.9) for singly elastically scattered ions. However, we see from the oscillograms of Figs. 4.7–4.9 and 4.11 that the energies of the elastically scattered ions may be either smaller or greater than E. It has usually been considered that the secondary ions with an energy exceeding E appear either as a result of single elastic scattering from several target atoms at the same time, or else as a result of the appearance of coupling (chemical bonds) between the atoms in the crystal lattice. It is more natural (at least for an energy of $E_0 \gtrsim 400$ eV) to explain the development of a wide spectrum of secondary ions by multiple scattering [662, 691, 702, 740, 747].

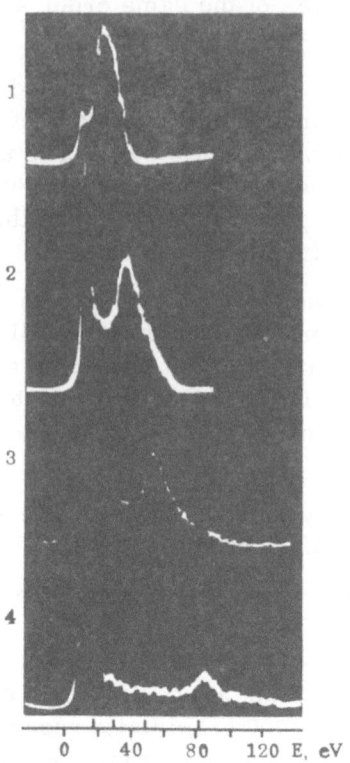

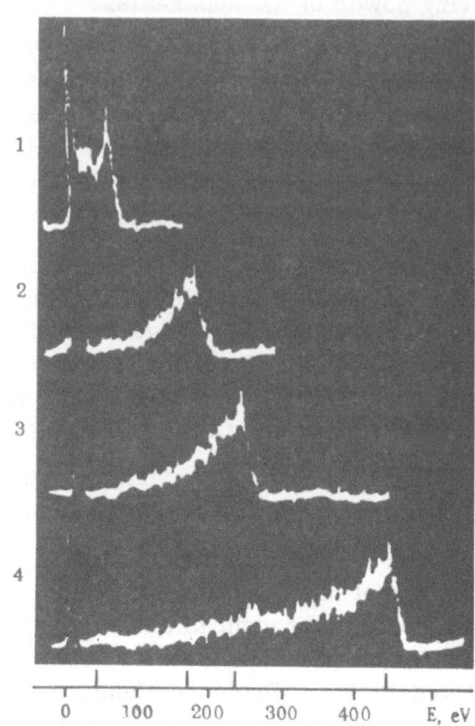

Fig. 4.8. Oscillograms of the energy spectra of secondary K^+ ions obtained on bombarding a molybdenum target heated to 1700°K with K^+ ions at an energy E_0 of: 1) 42; 2) 89; 3) 186; 4) 370 eV.

Fig. 4.9. Oscillograms of the energy spectra of secondary Li^+ ions obtained on bombarding a tungsten target heated to 1700°K with Li^+ ions at an energy E_0 of: 1) 40; 2) 185; 3) 250; 4) 500 eV.

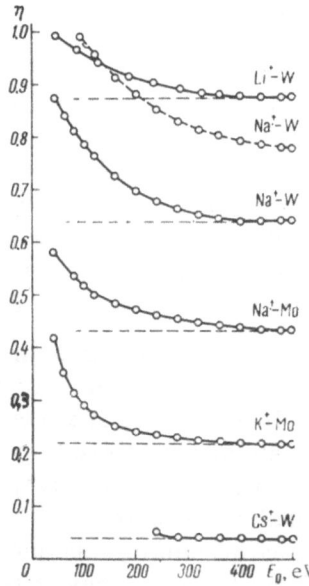

Fig. 4.10. Experimental and calculated values of $\eta(E_0)$ for W and Mo targets bombarded with alkali ions.

The energy of a multiply scattered ion may be calculated from Eq. (1.9) by using the latter several times in succession. The minimum energy E_{min} will correspond to an ion suffering $n-1$ deflections through 180° and one through the angle β under consideration. The maximum energy E_{max} will be retained by a secondary ion if in a twofold collision it is scattered twice in succession through an angle $\beta/2$, in a threefold collision three times through $\beta/3$, and in an n-fold collision n times through β/n. The minimum and maximum energies of the ions scattered as a result of an n-fold collision through an angle β are given by the relation

$$E_{min} = \left(\frac{\mu-1}{\mu+1}\right)^{2n-2} \frac{E_0(\mu-1)^2}{(\cos\beta \pm \sqrt{\mu^2-\sin^2\beta})^2} \leqslant E \leqslant$$

$$\leqslant \frac{E_0(\mu-1)^{2n}}{\left(\cos\frac{\beta}{n} \pm \sqrt{\mu^2-\sin^2\frac{\beta}{n}}\right)^{2n}} = E_{max}. \quad (4.8)$$

If we suppose that n may take very large values, then the spectrum of secondary ions will lie within the range $0 \leq E \leq E_0$. However, the realization of this assumption is not very likely.

For the cases studied, with $\beta = 135°$ and a recording system sensitivity of about 10^{-15} A, the spectrum of secondary ions contained ions with energies E_{max} corresponding to not more than two- or threefold scattering. This agrees with the theoretical calculations of the scattering cross section [662, 747]. For $E_0 \geq 400$ eV, the number of ions to the left of the maximum of singly elastically scattered ions falls to zero much more slowly than the number of ions to the right. The lower limit of the spectrum extends to $E = 0$. On reducing E_0 below 400 eV, the number of secondary ions on the part of the spectrum between the two peaks increases (see Figs. 4.7-4.9). This may possibly be associated with the smaller depth of penetration of the incident ions into the target and with the increased scattering cross section for the less energetic ions. For low energies of the incident ions ($E_0 < 400$ eV) the scattering process is complicated by the influence of the coupling between the target atoms.

It follows from inequality (4.8) that the value of E_{max} is variable, depending on n and the intermediate scattering angles. However, since the number of scattered ions falls gradually to zero with increasing n, the experimental determination of E_{max} is a difficult matter. In individual investigations [500, 614, 615], E_{max} was equated to the retarding potential V_K for which the smallest current of scattered ions measurable by the particular apparatus fell on the collector. The values of E_{max} so defined, of course, depend on the sensitivity of the measuring apparatus and constitute a function of E_0, since, as E_0 changes, so does the number of secondary ions. The essence of the matter remains the same if we even specify that E_{max} coincides with the value of the retarding potential for which the current of secondary ions falls by a specific number of times, since the exact law governing the relation between the ratio of the number of secondary ions with an energy of E_{max} to the total number of secondary ions and the energy E_0 is unknown. In addition to this, since, in the case under consideration we are concerned with measuring very small currents (about 10^{-14} A), the values of E_{max} determined for a particular type of ion may be severely distorted if the primary beam contains a very small number of lighter ions.

The energy of the singly elastically scattered ions E, in contrast to E_{max}, is independent of the sensitivity of the apparatus, and also of the number of scattered ions, which varies with E_0. The energy of the singly elastically scattered ions E in the primary-ion energy range of present interest (20-500 eV) should clearly be defined, as in the case of $E_0 \gg 500$ eV [323, 598, 708] by reference to the position of the maximum in the high-energy part of the spectrum [702, 710, 740].

For Li^+, Na^+, K^+, and Cs^+ ions scattered from tungsten and molybdenum targets heated to 1700°K at an angle of $\beta = 135°$ we plotted the curves of $\eta(E_0)$, where $\eta = E/E_0$, shown in Fig. 4.10. The broken horizontal lines correspond to the values of η calculated from Eq. (1.9) for $\beta = 135°$.

We see from Fig. 4.10 that the experimental curves deviate from the computed ones. The nonlinear rise in η as E_0 falls below 400 eV [instead of the constant value indicated by Eq. (1.9)] may be explained as being due to the influence of the coupling between the target atoms. The deviation of the measured value of η from the calculated values starts at larger values of E_0 for lighter atoms than for heavier atoms. This is clearest for Li^+ and Cs^+ ions striking tungsten targets. For the same displacement of the atoms in the crystal lattice the light ions should have a greater energy than the heavy ones. In fact, for the Li^+ ion to give an energy of about 50 eV to a tungsten atom, it must itself possess an energy of about 400 eV, whereas an energy of about 50 or 60 eV is sufficient in the case of the Cs^+ ion. Since the recoil energy is smaller for light than for heavy ions in the case $m_1 > m_2$, it is natural to suppose that the coupling between

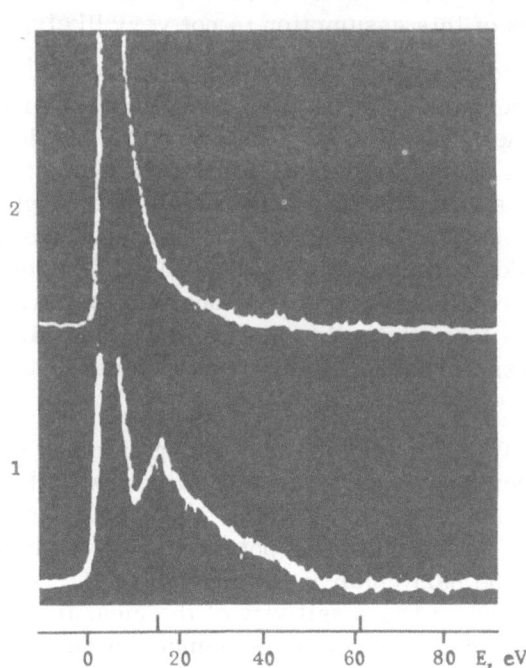

Fig. 4.11. Oscillograms of the energy spectra of secondary Cs$^+$ ions on tungsten for E_0 equal to: 1) 360; 2) 210 eV.

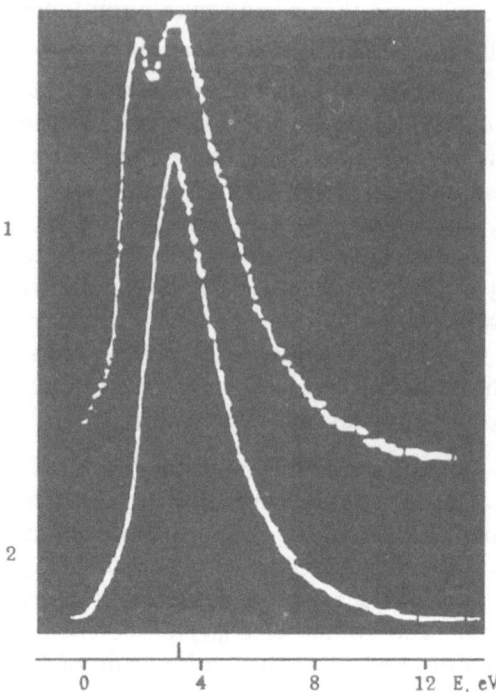

Fig. 4.12. Oscillograms of the low-energy part of the spectra of secondary Cs$^+$ ions on tungsten for $T = 1700°K$ (1) and room temperature (2).

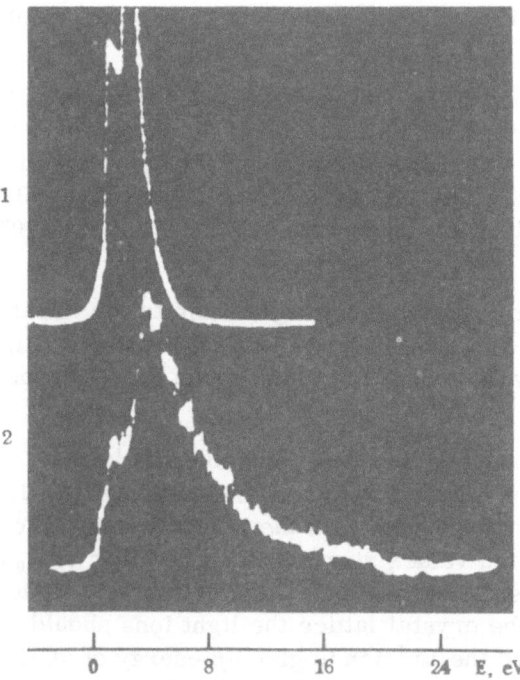

Fig. 4.13. Oscillograms of the spectra of slow Cs$^+$ ions on a Mo target heated to $T = 1700°K$ (1) and at room temperature (although heated to $T = 1000°K$ before measurements) (2).

the target atoms starts affecting the energy of the light ions at larger E_0 values than it does for the heavy ions.

The $\eta(E_0)$ relation for Cs$^+$ ions striking tungsten targets is only shown as far as $E_0 \approx 240$ eV. For lower values of E_0, as may be seen from Fig. 4.11, the peak of the singly scattered ions merges with that of the slow ions. In fact, on the oscillogram 1 of Fig. 4.11, taken at $E_0 = 360$ eV, we clearly see the peaks of the slow and singly elastically scattered ions. On oscillogram 2, obtained at $E_0 = 210$ eV, the peak of the singly scattered ions merges with that of the slow ions, so that the latter broadens, while the multiply scattered ions with $E_{max} > E$ still extend beyond the limits of the slow-ion peak. There is no doubt that for still lower values of $E_0 \approx 80$ eV the elastically scattered ions with $E_{max} > E$ also merge with it; and, since the slow-ion peak is independent of E_0, we may gain the impression that the incident ion interacts with the target as if with an absolutely solid body ($m \rightarrow \infty$). An analogous phenomenon is observed for multiply scattered Cs$^+$

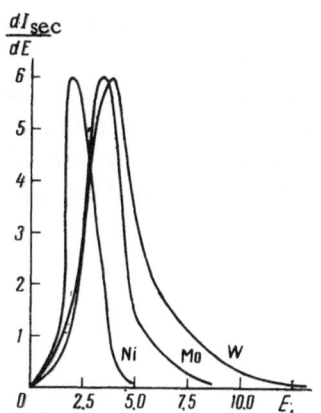

Fig. 4.14. Energy spectra of slow Cs^+ ions on nickel, molybdenum, and tungsten.

ions on bombarding a molybdenum target with ~ 80-eV Cs^+ ions, since the recoil energy for Cs^+ on molybdenum is the same as for Cs^+ on tungsten. At the same time, for Li^+ and Na^+ on tungsten and K^+ on molybdenum, this kind of merging does not take place, owing to the low recoil energy even for $E_0 \approx 40$ eV (see Figs. 4.7-4.9).

The curve of $\eta_{max}(E_0)$ shown in Fig. 4.10 (dotted and dashed line) was deduced from the maximum energy E_{max} of Na^+ ions multiply scattered at $\beta = 135°$ from tungsten; it lies much higher than the $\eta(E_0)$ curve for singly scattered Na^+ ions. The similar shape of these curves indicates that in the range $E_0 < 400$ eV, the coupling between the target atoms also affects the energy of the multiply scattered ions.

The curves of $\eta(E_0)$ shown in Fig. 4.10 disagree with the calculated values in the range $E_0 < 400$ eV, but nevertheless obey the same sequence. Only for Li^+ ions on tungsten at $E_0 \approx 40$ eV is η close to unity, whereas, for heavier ions at the same energy η is much smaller than unity. It follows from this that, despite the growing influence of the coupling of the target atoms on the energy of the scattered ions as E_0 falls below 400 eV, this energy is mainly determined by a pair collision, and even for an energy $E_0 \approx 40$ eV the incident ion never interacts with the crystal lattice as with a solid body.

Figures 4.12 and 4.13 present oscillograms of the spectra of slow ions obtained on bombarding clean tungsten and molybdenum targets, respectively, with 460-eV Cs^+ ions. The oscillograms 1 were taken at a target temperature of about 1700°K and the oscillograms 2 at room temperature. The peak on the left of oscillogram 1 of Fig. 4.12 corresponds to thermionic emission, and the peak with the maximum around $E_0 \approx 3.5$ eV to a group of slow ions. In oscillogram 2 of this figure, only the peak of slow ions occurs; the thermionic emission is absent. The shape of the spectrum of slow ions to the right of the maximum and the position of the maximum are independent of the energy E_0 and the target temperature. In oscillogram 1 (Fig. 4.13) the thermionic and slow-ion peaks are also separated. If, during the continuous ion bombardment, the target remains for a long time at room temperature or in a slightly heated condition, the slow-ion spectrum expands (see oscillogram 2, Fig. 4.13). This takes place as a result of the sputtering (in the form of positive ions) of the film of residual gases and primary-beam ions which have been adsorbed on the surface of the target in the course of time. A similar phenomenon occurs for any target—ion pair if the target surface acquires a substantial covering such that interaction between the incident ions and adsorbed atoms becomes reasonably probable. In oscillogram 2 (Fig. 4.13) we also see a small thermionic peak, since the target was heated to 1000°K before measurement.

Figure 4.14 shows the energy spectra of slow Cs^+ ions obtained from clean nickel, molybdenum, and tungsten targets (of the type of oscillogram 2 in Fig. 4.12) referred to the same height. We see that the width of the spectrum is different for the different targets and increases with increasing energy of the bond between the nickel, molybdenum, and tungsten atoms.

The series of oscillograms of energy spectra presented in Figs. 4.7-4.9 were obtained with different intensities of the primary ion beam I_0, which made comparison difficult. Despite this, it was noted that the relative number of slow and elastically scattered ions in these spectra varied with varying E_0. For example, in the case of K^+ ions from a clean molybdenum target (see Fig. 4.8), the areas of the two peaks increase as E_0 falls below 500 eV, the number of slow ions reaching a maximum for $E_0 \approx 160$ eV (we note that at this energy the threshold corresponding to the emission of diffusing ions also occurs), while the number of elastically scattered ions continues growing; at an energy of $E_0 \approx 40$ eV there are very few slow ions, while the number of elastically scattered ions reaches a maximum.

The rise in the number of slow ions with falling energy E_0 is associated with the reduction in the depth of penetration of the primary ions. The sorption of ions for small values of E_0 (≤ 160 eV) takes place mainly on the surface of the target, from which they are more easily removed than from the lower regions by our assumed mechanism, involving the intensive vibration of that part of the lattice excited by the ion bombardment. The passage of the number of slow ions through a maximum and the rapid fall for low values of E_0 may be explained by the competing rise in the number of elastically scattered ions.

§3. Energy Dependence of Secondary

Ion Emission Coefficients

We studied the energy dependence of the coefficients of secondary-ion emission $K(E_0)$ and $K'(E_0)$ from clean, cold, and high-temperature tungsten, tantalum, molybdenum, nickel, and zirconium targets, respectively, bombarded with alkali and Ba^+ ions with primary ion energies of 20–500 eV [696, 702, 740], using the apparatus illustrated in Fig. 4.3.

The coefficient K takes very high values indeed (60–95%) for alkali ions in the low E_0 range. For Ba^+ ions the values of K are lower than the corresponding values for alkali ions by a factor of 2 or over for the whole energy range studied, as indicated in Fig. 4.15. The reason for this is evidently the large mass $m_2 = 138$ and the high ionization potential (5.21 eV) of barium.

According to theory, the number of elastically scattered ions increases with falling E_0 and m_2 and with increasing radius of the incident ion a. Our $K(E_0)$ curves clearly also depend on the ionization potential V_i of the ions, in that K falls as V_i rises.

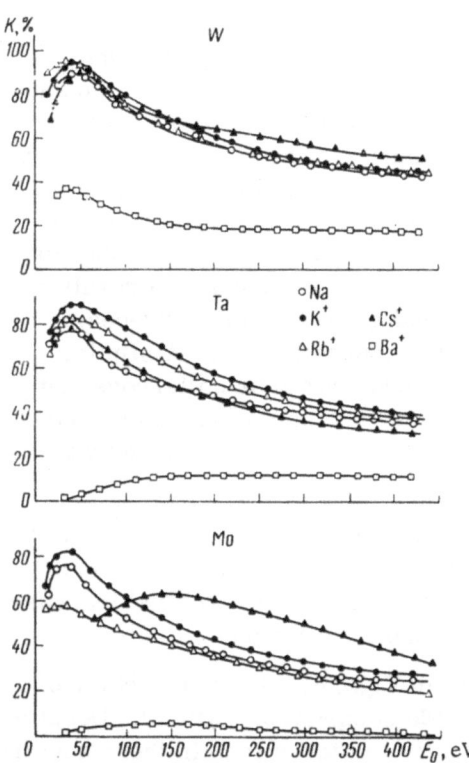

For the alkali ions from Li^+ to Cs^+, the rise in m_2 is accompanied by an increase in a and a fall in V_i. However, since in this case changes in a and V_i influence the number of secondary ions in the opposite sense to m_2, the $K(E_0)$ assumes fairly close values for all the alkali ions.

Results obtained for Ba^+ ions introduced a certain clarity. As compared with alkali ions, Ba^+ has a very large mass and ionization potential. Here, both factors act in the sense of a relative reduction in the number of secondary ions, and the $K(E_0)$ curve for Ba^+ on tungsten, tantalum, and molybdenum passes much lower than that of the alkali ions.

When $m_1 > m_2$, (Na^+, K^+, Rb^+, Cs^+, and Ba^+ on tungsten and tantalum and Na^+, K^+, and Rb^+ on molybdenum), the $K(E_0)$ curves pass through a maximum at $E_0 \approx 40$ eV (see Fig. 4.15), while for $m_1 < m_2$ (Cs^+ on molybdenum, nickel, and zirconium, and Ba^+ on molybdenum) the maximum becomes flat and moves in the high-energy direction (Fig. 4.16).

Information regarding these maxima may be obtained from Fig. 4.17a,b, which represents $K(E_0)$ and $K'(E_0)$ values obtained on bombarding a clean tantalum target (either cold or heated to 1500°K)

Fig. 4.15. The function $K(E_0)$ for alkali and Ba^+ ions on tungsten, tantalum, and molybdenum.

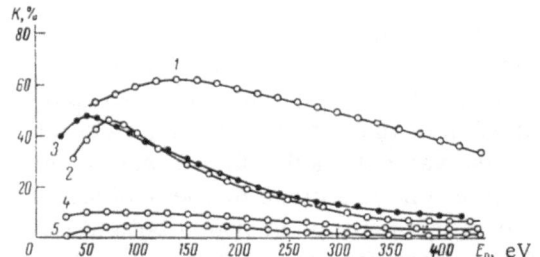

Fig. 4.16. Functions $K(E_0)$ obtained for the case $m_1 < m_2$ with Cs^+ on Mo (1), on Ni (2), or Zr (3); Rb^+ on Ni (4); and Ba^+ on Mo (5).

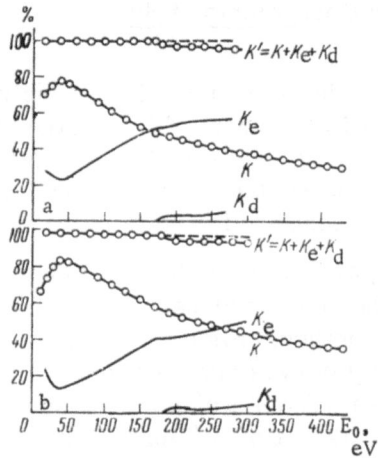

Fig. 4.17. Functions a) $K(E_0)$ and b) $K'(E_0)$ for Cs^+ and Rb^+ ion on tantalum.

with Rb^+ and Cs^+ ions. The functions $K(E_0)$ are analogous to those shown in Figs. 4.15 and 4.16. The $K'(E_0)$ curves consist of two components: secondary ion emission $K(E_0)$ and the emission of evaporated ions $K_e(E_0)$. From $K'(E_0)$ we may also estimate the coefficient of the emission of diffusion-type ions $K_d(E_0)$ which appear for $E_0 \geq$ 180 eV. In the energy range 20–180 eV, for Cs^+ ions on tantalum (Fig. 4.17a), $K' \approx 100\%$, i.e., there is a charge balance of the interacting particles. The current of secondary ion emission and the current representing the emission of evaporated ions are equal to the current of primary ions. For Rb^+ on tantalum (Fig. 4.17b), $K' \approx 98\%$, i.e., only a few percent of the primary-beam ions leave the target as neutral atoms. The break on the $K'(E_0)$ curves starting at $E_0 \approx$ 180 eV is due to the appearance of diffusing ions, which have a considerable inertia and are not recorded by the oscillograph. If we continue the linear part of $K'(E_0)$ into the region $E_0 > 180$ eV (broken lines in Fig. 4.17a,b), then the difference in ordinates between the experimental curve and the broken one will give the energy dependence of the coefficient of emission of the diffusing ions $K_d(E_0)$.

If we subtract $K(E_0)$ from $K'(E_0)$, we obtain the relation between the emission coefficient of the evaporated ions and the energy of the primary ions $K_e(E_0)$. The $K_e(E_0)$ curves pass through a minimum at $E_0 \approx 40$ eV, the position of this coinciding with the maximum on the $K(E_0)$ curves. Hence, the maximum on the $K(E_0)$ curves in the case of $m_1 > m_2$ is due to the increase in the adsorption of secondary ions by the target on reducing the energy E_0 below 40 eV.

The reason for the rise in the adsorption of the ions is evidently the increase in the influence of sorption forces at the target surface as the energy spectrum of the secondary ions contracts. It is clear that on reducing E_0 below 40 eV the number of secondary ions with an energy sufficient to overcome these forces diminishes, and the curve $K(E_0)$ passes through a maximum.

We note that for any target—ion pair in which $m_1 > m_2$, the appearance of maxima on the $K(E_0)$ curves at approximately the same value of $E_0 \approx 40$ eV (see Fig. 4.15) may possibly be explained by the influence of the binding energy between the target atoms, i.e., on reducing the energy E_0 below 400 eV there is a slower fall in the energies of the secondary ions that would follow from expressions (1.9) and (4.8). As a result of this, the energy spectra of the secondary ions come close together for all masses and the maxima on the $K(E_0)$ curves appear at approximately the same value of $E_0 \approx 40$ eV.

In the case of $m_1 < m_2$ (see Fig. 4.16), the secondary-ion spectrum in the low E_0 region is affected all the more strongly by the existence of the limiting scattering angle $[\beta_{lim} = \arcsin(m_1/m_2)]$; this is possibly the reason for the displacement of the maxima on the $K(E_0)$ curves in the direction of larger E_0.

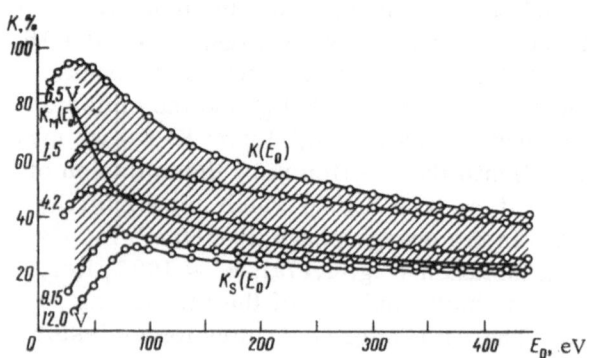

Fig. 4.18. The function $K(E_0)$ for Rb^+ ions on a tungsten target for various collector potentials.

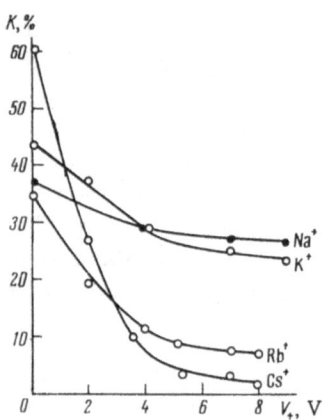

Fig. 4.19. The function $K(V_{K^+})$ for the ions Na^+, K^+, Rb^+, and Cs^+ striking a molybdenum target.

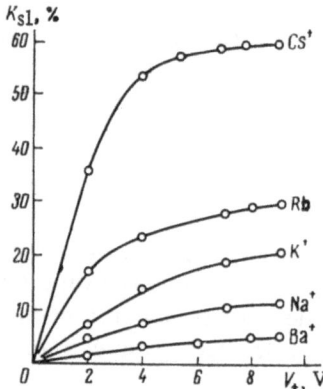

Fig. 4.20. The function $K_{sl}(V_{K^+})$ for the ions Na^+, K^+, Rb^+, Cs^+, and Ba^+ striking a molybdenum target.

The resultant $K(E_0)$ curves (Figs. 4.15 and 4.16) cannot show us which of the factors m_1, m_2, V_i, φ, ρ, a_1, and a_2 play the main part in the formation of the secondary ions, since all the $K(E_0)$ curves obtained with different targets and different ions are close to one another in the range of incident-ion energies between 20 and 500 eV. However, as we have seen, in the case of Ba^+ the dominating role is taken by the ionization potential of the incident ion.

Separate Study of the Coefficients of Elastically Scattered Ions $K_s(E_0)$ and the Emission Coefficients of Slow Ions $K_{sl}(E_0)$

The fact that the width of the spectrum of slow ions was independent of E_0 and the nature of the incident ions (see Chapter 4, §2) was used in order to separate the group of slow ions from the elastically scattered variety [702, 740].

The separation was effected in an apparatus of the spherical-condenser type by the retarding field method. Figure 4.18 shows a family of $K(E_0)$ curves obtained for different collector potentials from a clean, cold tungsten target bombarded with Rb^+ ions. When a slight negative potential with respect to the target was applied to the collector ($V_K \approx -6.5$ V), all the positive secondary ions were collected on the latter and a function $K(E_0)$ analogous to those shown in Fig. 4.15 was obtained. For a collector potential of $V_K \approx 9$ V, the group of slow ions was almost completely retarded, and the curve then obtained gave the relationship $K_s(E_0)$. We see from Fig. 4.18 that an increase in the retarding potential on the collector from 9 to 12 V produces no substantial change in the number of slow ions, particularly at high values of E_0. In order to obtain $K_{sl}(E_0)$, we must subtract $K_s(E_0)$ from $K(E_0)$. In this way, the shaded portion of Fig. 4.18 corresponds to the relation $K_{sl}(E_0)$, represented by the continuous line.

Families of curves similar to those shown in Fig. 4.18 were also obtained from tungsten, tantalum, molybdenum, nickel, and zirconium targets bombarded with Na^+, K^+, Rb^+, Cs^+, and Ba^+ ions. The functions $K(V_K)$ and $K_{sl}(V_K)$ were also plotted for $E_0 = 200$ eV (Figs. 4.19 and 4.20 for the case of a molybdenum target).

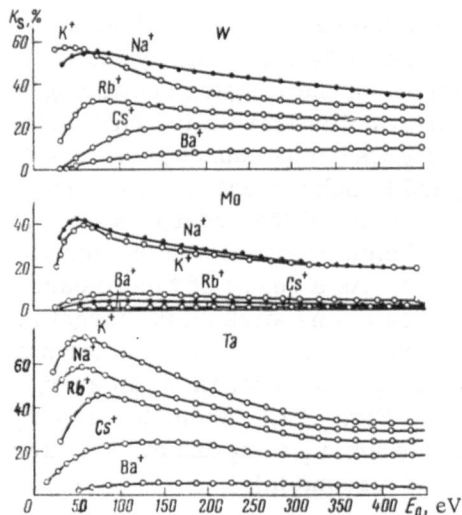

Fig. 4.21. Values of $K_s(E_0)$ for alkali ions and Ba$^+$ ions striking W, Mo, and Ta targets.

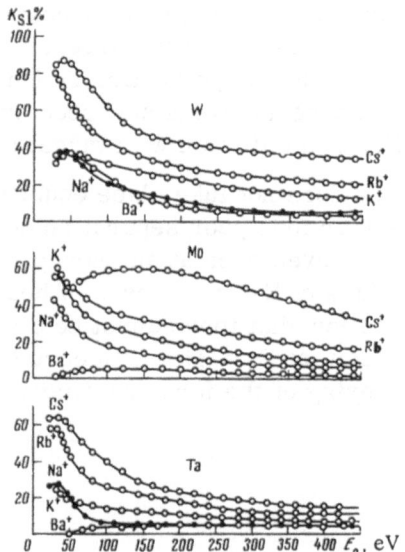

Fig. 4.22. Values of $K_{s1}(E_0)$ for alkali ions and Ba$^+$ ions striking W, Mo, and Ta targets.

We see from Fig. 4.19 that, for retarding potentials V_K close to zero, when a large number of slow ions are still falling on the collector, the coefficient K shows no special dependence on any of the parameters of the colliding particles (m_1, m_2, ρ, a_1, a_2, Z_1, Z_2). Starting from certain positive values of V_K (Fig. 4.19), the coefficient K increases with falling mass of the incident ion m_2. This is expressed most clearly for the value of V_K for which the group of slow ions is stopped completely and only elastically scattered ions fall on the collector ($K = K_s$). These potentials may be established for tungsten, tantalum, molybdenum, nickel, and zirconium targets by reference to the width of the energy spectrum of the slow ions; they are equal to 9, 6.5, 7, 4, and 5 V, respectively. The values of the coefficient K_{s1} in Fig. 4.20 are also set in a specific sequence: they rise with falling ionization potential V_i of the incident ions and show no marked dependence of the mass of the ions m_2.

Figure 4.21 shows the relationship $K_s(E_0)$ for tungsten, molybdenum, and tantalum targets bombarded by Na$^+$, K$^+$, Rb$^+$, Cs$^+$, and Ba$^+$ ions in the energy range 20-500 eV. We see that K_s rises with falling E_0, passing through a maximum for small values of the latter. In the case of tungsten and molybdenum, K_s clearly depends on the mass of the incident ions m_2 [throughout almost the whole of the E_0 range studied, the $K_s(E_0)$ curves lie in a higher position, the lighter the incident ion]. In the case of tantalum, the $K_s(E_0)$ curve for the K$^+$ ions is higher than that for the Na$^+$. A possible reason for this infringement of the m_2 law lies in the complex dependence of the coefficient K on many parameters (m_1, m_2, ρ, a_1, a_2, Z_1, Z_2).

For the same ions and targets, the $K_{s1}(E_0)$ curves are shown in Fig. 4.22. We see that over the whole of the energy range studied the $K_{s1}(E_0)$ curves are situated in a higher position, the lower the ionization potential of the incident ions V_i.

The dependence of K_{s1} on eV_i and φ are shown more clearly by the $K_{s1}(eV_i)$ curves of Fig. 4.23. Individual points on these curves correspond to the values of K_{s1} determined from the results of Fig. 4.22 for $E_0 = 200$ eV. We see that for each target K_{s1} is the greater, the lower the ionization energy of the incident ion eV_i. While $eV_i \gtrsim \varphi$ (Ba$^+$, Na$^+$, and K$^+$ on tungsten, tantalum, and molybdenum), K_{s1} is relatively small. When $eV_i < \varphi$ (Cs$^+$ and Rb$^+$ on tungsten, tantalum, and molybdenum), K_{s1} is much larger.

From the curves of Fig. 4.23 we may also estimate the dependence of K_{s1} on φ. The value of $K_{s1}(eV_i)$ is the greater, the higher the work function of the target ($\varphi_W \approx 4.54$, $\varphi_{Mo} \approx 4.25$,

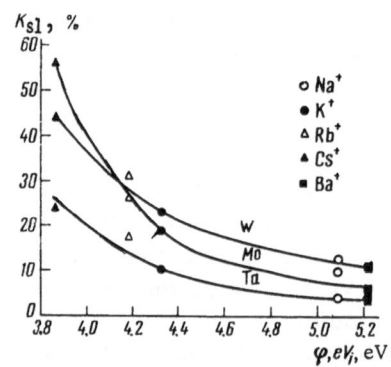

Fig. 4.23. Values of $K_{sl}(\varphi)$ for alkali ions and Ba^+ ions striking W, Ta, and Mo targets.

$\varphi_{Ta} \approx 4.11$ eV). There is a discrepancy for Cs^+ ions on molybdenum; this may evidently be explained by the limitation imposed by the existence of the limiting scattering angle β_{lim}. After their first collision with individual molybdenum atoms ($m_1 < m_2$), the Cs^+ ions cannot pass outside the limits of the target, and in subsequent collisions only a negligible proportion of the ions, of the order of a few percent, possess energies sufficient to overcome the sorption forces on the target surface. As a result of this, clearly, some 95% of the primary beam ions stick on the target surface, and this creates favorable conditions for the formation of slow ions. Thus, for Cs^+ on molybdenum at an energy of $E_0 = 140$ eV, $K_{sl} \approx 60\%$ (see Fig. 4.22). Anomalously large values of K_{sl} always occur when $m_1 < m_2$ and $eV_i < \varphi$ (for example, Cs^+ on nickel and zirconium).

Thus the quantitative data here presented enable us to establish still further properties of the slow ions distinguishing them from elastically scattered ions, namely, the dependence of the number of slow ions on eV_i and φ, and the absence of any dependence on m_2.

The large values of K_{sl} on the range of E_0 values studied cannot be explained by cathode sputtering, i.e., the direct collision of the incident ion with primary ions adsorbed on the target surface. For Cs^+ on tantalum, for example, at $E_0 = 200$ eV, some 40% of the primary beam ions stick to the target surface (see Fig. 4.17). Simple calculations show that for our current density of $j \approx 3.5 \cdot 10^{-8}$ A/cm^2, at an instant 5 sec after disconnecting the target heating circuit, the probability of a direct collision between a Cs^+ ion and primary beam atoms adsorbed on the target surface is roughly three orders lower than the probability of a collision with target atoms.

The separation of slow ions from elastically scattered ions does not take place equally successfully for all energies E_0 and ratios m_1/m_2. For the heavier ions poor separation occurs even for $E_0 \leq 200$ eV, while for lighter ions separation is achieved even at an energy of $E_0 < 100$ eV. This is associated with the fact that, for small values of E_0, in the case of heavy particles the energy of the elastically scattered ions becomes so low that they cannot be distinguished from the group of slow secondary ions. For this reason, at $E_0 \approx 40$ eV, in certain cases the $K(E_0)$ curves of Figs. 4.16 and 4.17 describe the scattering of the ions quantitatively better than the $K_s(E_0)$ curves shown in Fig. 4.22.

§4. Limitation of the Escape of Secondary

Ions by Sorption Forces on the Surface

of a Metal

As has been shown experimentally, the relation $K(E_0)$ reaches a maximum in the case of $m_1 > m_2$ at $E_0 \approx 40$ eV and starts falling for lower E_0 values as a result of the influence of sorption forces forming a potential barrier of height λ on the surface of the metal [696, 740]. There is no doubt that this barrier has a particularly strong influence on the shape of the low-energy part of the secondary-ion spectrum [711, 740]. For the spectrum of secondary ions in a direction making an angle β with the normal to the target, we may write the following expression:

$$I(E, \beta) = \frac{dI(E', \beta')}{d\Omega_{\beta'}} \frac{d\Omega_{\beta'}}{d\Omega_{\beta}} \Omega_0, \qquad (4.9)$$

where $E' = E + \lambda$ is the energy of the secondary ion inside the metal and Ω_β and $\Omega_{\beta'}$ are the solid angle in the vacuum and the corresponding solid angle inside the metal, respectively. The

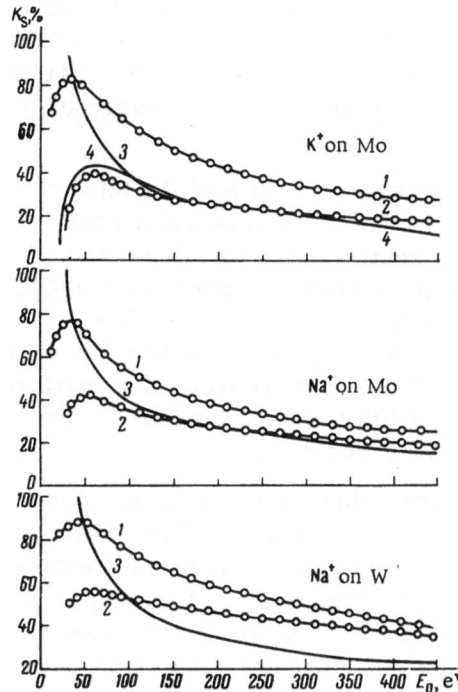

Fig. 4.24. Comparison of the relationships $K(E_0)$, $K_s(E_0)$, and $K_s^T(E_0)$ either with or without allowing for the influence of the potential barrier at the target surface.

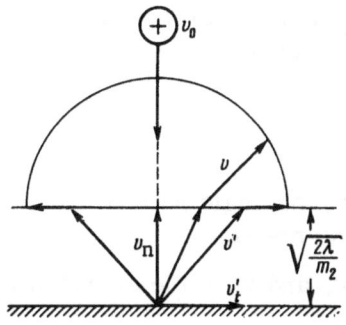

Fig. 4.25. Limitation of the escape of secondary ions by the potential barrier at the target surface.

angle β in the metal corresponds to an angle β' in the vacuum, and the following relation exists between these:

$$\frac{\sin \beta'}{\sin \beta} = \sqrt{\frac{E' - \lambda}{E'}}, \qquad (4.10)$$

whence

$$\frac{d\Omega_{\beta'}}{d\Omega_{\beta}} = \frac{\dfrac{E' - \lambda}{E'} \cos \beta}{\sqrt{1 - \dfrac{E' - \lambda}{E'} \sin^2 \beta}} = \frac{\dfrac{E}{E + \lambda} \cos \beta}{\sqrt{1 - \dfrac{E}{E + \lambda} \sin^2 \beta}}. \qquad (4.11)$$

The first factor in expression (4.9) determines the energy spectrum of the secondary ions propagating in unit solid angle inside the target. The second factor in expression (4.9), as shown by (4.11), is close to unity for very large E', while for $E' \approx \lambda$, this factor is close to zero. Physically, this means that the experimentally measured secondary ion spectrum in the region in which E' is large will differ very little from the energy spectrum beyond the barrier, inside the target, while for $E' \to \lambda$ this difference will be large.

Our experimental energy spectra of the group of slow Cs^+ ions from nickel, molybdenum, and tungsten (see Fig. 4.14) and those corrected for the potential barrier by means of formula (4.11) were compared with the Maxwellian energy distribution. On plotting the results in the form of a graph on a semilogarithmic scale, only the right-hand branches of the corrected spectra lay satisfactorily on straight lines in the downward direction. A characteristic feature is the fact that the slopes of these straight lines have the same sequence as the bond energies of the atoms in nickel, molybdenum, and tungsten.

The fact that the width of the spectrum of the group of slow ions is independent of E_0 and of the type of incident ions, and the existence of a correlation between the slopes of the straight lines (or on the ordinary scale the half-width of the energy spectra of the group of slow ions) and the heat of sublimation of the target atoms strongly suggest that the slow ions acquire their energy at the expense of intensive vibrations of the atoms in that part of the target excited by ion bombardment.

The group of slow ions thus separated out must not be attributed to elastically scattered ions. The subtraction of the group of slow ions $K_{s1}(E_0)$ from the total number of secondary ions $K(E_0)$ offered the possibility of separating out the number of elastically scattered ions $K_s(E_0)$.

Our experimental $K_s(E_0)$ relationships were compared with theory [228, 392, 501, 541]. There was poor agreement with the theory of [228, 501], in which the solid-sphere method was used to calculate K_s. Better agreement was obtained with the theoretical calculations [541] in which we used the Firsov [435] interaction potential, extended to the low-energy range ($E_0 <$ 400 eV) [740].

In each part of Fig. 4.24 three curves appear: 1) our measured values of the total secondary ion emission $K(E_0)$; 2) the experimentally distinguished values of the elastically scattered ions $K_s(E_0)$; 3) the theoretical relationship for the elastically scattered ions $K_s^T(E_0)$ obtained for the Firsov interaction potential [541]. Good agreement is obtained between curves 2 and 3 for K^+ ions (Fig. 4.24) in the incident-ion energy range 120-500 eV. On reducing E_0 below 120 eV, the theoretical curve 3 moves away from the experimental curve 2 and tends toward the maximum of curve 1, continuing to rise as E_0 diminishes beyond 40 eV. The theoretical curve 3 for Na^+ ions on molybdenum (Fig. 4.24) with $E_0 > 250$ eV passes a little below, and that for Na^+ on tungsten with $E_0 > 100$ eV considerably below the experimental curve 2.

Evidently the difference between curves 2 and 3 for large values of E_0 in the case of Na^+ on molybdenum and tungsten, and in all cases for $E_0 < 100$ eV, may be explained by the incompleteness of the theory. The theory fails to allow for multiple scattering (which may be considerable for the comparatively small sodium atoms), the energy dependence of the depth of penetration of the ions into the crystal lattice (on which the number of scattering centers depends), the influence of coupling between the target atoms on the back-scattering cross section of the ions, and the effect of the potential barrier at the target surface (which limits the escape of low-energy scattered ions).

It may be shown by fairly simple calculations that the presence of a potential barrier of height λ due to the sorption forces at the surface of the metal leads to a limitation of the secondary-ion emission. Emission is absent when the component of secondary-ion energy normal to the barrier in the metal $E_n' \leq \lambda$.

The relation between the normal component of velocity v_n' and the total velocity v' of the secondary particle under the barrier in the metal is given (Fig. 4.25) by

$$v_n' = v' \cos \beta'. \tag{4.12}$$

For normal incidence of the primary beam on the metal surface, the angle $\beta' = \pi - \theta'$, where θ' is the scattering angle of the particle behind the barrier in the metal.

If in expression (4.12) we transform from the velocities v_n' and v' to the energies E_n' and E', we obtain

$$E_n' = E' \cos^2 \beta'. \tag{4.13}$$

If, however, neglecting the effect of the interatomic coupling in the metal on the energy of the secondary ions which, as we saw in §3 of the present chapter, increases as E_0 falls below 400 eV, we replace E' by the value derived from Eq. (1.9), we obtain

$$E_n' = \frac{E_0'(\mu - 1)^2 \cos^2 3'}{(\cos 3' \pm \sqrt{\mu^2 - \sin^2 3'})^2}, \tag{4.14}$$

where $E_0' = E_0 + \lambda$.

The condition for the cessation of secondary ion emission $E_n' = \lambda$ determines the boundary angle

$$\beta_b' = \pi - \theta_b' = \arccos \sqrt{\frac{\mu + 1}{\frac{E_0 + \lambda}{\lambda}(\mu - 1) - 2\sqrt{\frac{E_0 + \lambda}{\lambda}}}}. \tag{4.15}$$

We see from (4.15) that β_b' falls with diminishing E_0.

The coefficient of essentially scattered ions, allowing for the effect of the potential barrier, is given by the expression

$$K_s = \sigma_p(\theta_b')n, \tag{4.16}$$

where σ_p is the back-scattering cross section of the ions, allowing for the effect of the potential barrier, θ_b is the boundary scattering angle of the ions under the barrier in the metal, and n is the number of atoms per unit area of the target surface. If σ_p is expressed in terms of the collision parameter

$$\sigma_p = \int_\pi^{\theta_b'} d\sigma(\theta') = 2\pi \int_0^{\rho(\theta_b')} \rho(\theta')d\rho(\theta') = \pi\rho^2(\theta_b'), \tag{4.17}$$

expression (4.16) takes the form

$$K_s = \pi\rho^2(\theta_b')n. \tag{4.18}$$

The value of K_s reaches a maximum for the condition $dK_s/dE_0 = 0$, and hence for

$$\frac{d\rho(\theta_b')}{dE_0} = 0.$$

Since ρ is a function of E_0 and θ_b', while θ_b' is a function of E_0, we have

$$\frac{d\rho}{dE_0} = \frac{\partial\rho}{\partial E_0} + \frac{\partial\rho}{\partial\theta_b'}\frac{\partial\theta_b'}{\partial E_0} = 0, \tag{4.19}$$

whence, finally, we obtain the conditions for the curve $K_s(E_0)$ to pass through a maximum:

$$\frac{\partial\rho}{\partial E_0} = -\frac{\partial\rho}{\partial\theta_b'}\frac{\partial\theta_b'}{\partial E_0}. \tag{4.20}$$

Curve 4 in Fig. 4.24 constitutes the values of $K_s(E_0)_T$ calculated from expression (4.18) for the case of K^+ on molybdenum. In calculating ρ from Eq. (4.18), we used the Firsov potential [435], and in determining θ_b' from Eq. (4.15) we took the height of the potential barrier λ as ≈ 5 eV, which approximately equals twice the heat of evaporation of the K^+ ions from the molybdenum target.

The theoretical curve 4 agrees closely with the experimental curve 2. For an energy of $E_0 \approx 55$ eV, curve 4 passes through a maximum in the same way as the experimental curve, while for an energy of $E_0 \approx 20$ eV, a threshold of elastically scattered ions appears.

The slight difference between curves 4 and 2 is probably due to the failure to allow for multiple scattering, the energy variation of the number of scattering centers, and the effect of coupling between the atoms in the crystal lattice on the scattering cross section.

CHAPTER 5

ANGULAR RELATIONSHIPS
OF SECONDARY ION EMISSION

§1. Brief Review of Investigations into the
Angular Distribution of Secondary Ion Emission

In order to understand the mechanism responsible for the interaction of atomic particles with solid surfaces, it is extremely important to study the angular distribution of the scattered particles and the angular dependence of the energy spectra. There have been a number of investigations into the angular distribution of scattered ions [83, 84, 122, 293, 341] and atoms [148, 159, 167] in gas targets. At the same time, the distribution of ions scattered by the surfaces of solid bodies cannot be regarded as thoroughly understood. Some authors have studied the angular distribution of the number and energy of scattered alkali ions as a function of the nature, initial energy, and incident angle of the bombarding ions, and also of the material, surface state, and temperature of the metal target bombarded. However, the results have varied widely, either because of differences in the conditions of the surface of the materials studied or inadequacies in the method employed.

In describing the angular relationships, the authors have used different terminology and different notations. For convenience of subsequent analysis we shall specify the terminology and notation more precisely. The angle between the direction of incidence of the primary ion beam and the normal to the target surface will be called the angle of incidence of the primary ions, and we shall denote this by φ (Fig. 5.1). The angle between the normal to the target surface and the direction of escape of the particle, we shall call the escape angle (angle of reflection) θ. The angle between the direction of incidence of the beam and the escape direction we shall call the scattering angle β. The sum of the angle of incidence of the primary ions and the escape angle of the secondary ions we shall denote by α ($\alpha = \theta + \varphi = \pi - \beta$).

In accordance with the generally accepted terminology, by "scattering" we shall understand any deviation of the particle from its original direction as a result of interaction with other particles. This concept includes both the altered motion of the particle inside the solid and its motion outside the latter. In order to distinguish these two parts of the flow of scattered particles, we sometimes use the terms "reflection" (for $\varphi = \theta$, "mirror" or "specular" reflection) and "penetration" (or "insertion"). One often speaks of the "scattering of particles by the surface of a solid" meaning just that part of the flow directed outside. We shall use both terms, giving preference, however, to "scattering" as the most widely used.

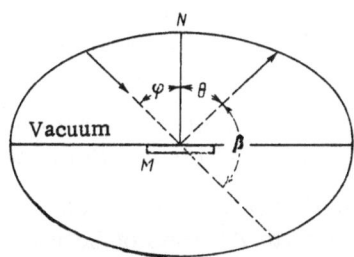

Fig. 5.1. Schematic representation of the angles. φ) Angle of incidence; β) scattering angle of the primary particles; θ) escape angle of the secondary particles.

Read [48] was the first to describe (in 1928) the angular distribution of ions scattered from a platinum target on bombarding with positive Li^+ and K^+ ions in the energy range 20-260 eV. The scattered ions were received on a collector which could be moved relative to the target. The target was cleaned from contaminants by pulsed heating. The measurements, however, were carried out on a cold target. The author found that the maximum number of ions escaped at approximately the specular angle, the angle of maximum reflection varying slightly with the energy of the incident ions E_0. The energy of the scattered ions increased with increasing angle of incidence φ.

Gurney [49] used a magnetic analyzer to study the energy spectra of Li^+, K^+, and Cs^+ ions scattered from a platinum target as a function of incident angle φ and energy E_0 in the range 21-60 eV with a constant escape angle θ. The author found that the scattered ions were approximately homogeneous in energy; the number of scattered ions varied with varying φ. The distribution of scattered ions was also studied for various φ and θ as a function of the energy E_0. The angle of maximum ion escape θ was independent of the energy E_0 and was always greater than the specular value, i.e., the ions were reflected predominantly forward. The $K(E_0)$ curves, independently of φ and θ, passed through a sharp maximum at energy $E_0 \approx 40$ eV.

Longacre [94] studied the angular and energy distribution of the scattered ions as functions φ, β, and E_0 between 125 and 450 eV for a nickel target bombarded with Li^+ ions. It was found that the ratio of the most probable energy of the scattered ions to the energy of the incident ions fell with increasing scattering angle β; however, this ratio, and also the direction of maximum ion scattering, were independent of E_0 and φ.

In Longacre's opinion these results failed to agree with Eq. (1.9); he considered that, to a first approximation, the energy distribution of the scattered ions was similar to the distribution of macroscopic particles elastically reflected from a rough surface.

Sawyer [59] studied the angular distribution of scattered Li^+ ions from an incandescent Pt—Ir target and from nickel deposited on a tungsten substrate over the energy range $E_0 = 20$-700 eV; he found that the scattering at small angles of incidence φ was more diffuse than that indicated by Read and Gurney. On increasing the retarding potential, the maxima on the distribution curves of the scattered ions obtained from a platinum target moved in the direction of larger escape angles θ. According to Gurney, this shows that the ions only slightly deviated from their original direction of motion retained a high energy. Sawyer found no dominant direction of scattering for Li^+ ions striking a nickel target deposited on a tungsten substrate for low retarding potentials on the collector. However, for $E_0 \gtrsim 200$ eV, on increasing the retarding potential, two directions of maximum ion scattering began to stand out: one at an escape angle greater than the specular value, and another near the normal to the target surface; the energy of the ions escaping near the normal was $E > 0.8E_0$. Since the direction of this maximum was independent of E_0, it could not be explained by ion diffraction. An attempt to relate the effect to specular reflection from the (110) plane of nickel single crystals also gave no positive results.

Seeliger and Sommermeyer [101] found that the angular distribution of the scattered ions on bombarding a platinum target with K^+ ions at energies of up to 1000 eV obeyed the cosine law.

Eremeev and Zubchaninov [165] studied the angular distribution of secondary particles as a function of φ and θ on bombarding hot and cold tantalum targets with K^+ ions of between 2 and 4 keV. These authors emphasized that there was a direction of maximum secondary-ion intensity close to the angle of mirror reflection. With falling φ, the intensity maximum of the scattered ions became diffuse and deviated in the direction of θ values smaller than specular, while the number of slow secondary ions increased. At $\varphi < 50°$, the secondary ions consisted mainly of sputtered ions from the target. The authors considered that a convincing proof in favor of specular reflection was the retention of a considerable proportion of their original energy by ions scattered in the direction of large θ, or the maximum of the scattered-ion intensity. In the opinion of these authors, for a glancing angle of incidence the interaction of the ions falling on the target surface involved not single atoms of the crystal lattice, but a specific surface element; they concluded that the angular distribution of fast scattered ions, in contrast to that of slow secondary ions, did not obey the cosine law.

Eremeev [323] used a mass spectrometer to study the angular and energy distributions of secondary ions on bombarding tantalum, molybdenum, nickel, and graphite targets with Li^+, K^+, and Cs^+ ions for E_0 values between 1 and 4 keV; he also studied their dependence on the state of the surface, the target temperature (300-1450°K), and the energy and incident angle of the primary ions.

For a clean, cold target there was only a slow-ion peak in the secondary-ion spectrum. A change in the energy and incident angle of the primary ions did not lead to any marked change in the character of the distribution. For a high target temperature the distribution had a sharply expressed maximum in the high-energy range; the position of the peak closely corresponded to the energy of ions scattered as a result of single collision with individual target atoms.

On bombarding a graphite target with K^+ and Li^+ ions there was also only a slow-ion peak in the secondary-ion spectrum. The absence of ion back-scattering on bombarding the graphite with Li^+ ions was explained by Eremeev as being due to the considerable penetration of the ions into the graphite. Analogous results were obtained on bombarding molybdenum and nickel targets with Cs^+ ions. It was shown that on reducing the scattering angle (45 to 20°) neither the position of the maximum nor the maximum energy of the secondary ions altered; there was only a reduction in the height of the single-collision peak.

Analogous investigations were carried out by Petrov [583] with a magnetic analyzer. Targets of tantalum and tungsten, heated to 2000°K, were bombarded with Li^+ and Ca^+ ions having energies between 10 and 1100 eV; the working vacuum in the apparatus was of the order of $(2-3) \cdot 10^{-7}$ mm Hg. Principal attention was paid to determining the energy distribution of the ions scattered in the direction of mirror reflection. The angles of incidence φ and escape θ were accordingly made equal, and the scattering angle was varied in accordance with these.

Petrov observed a certain angular dependence of the energy distribution of the secondary ions. With increasing scattering angle, the height of the maximum on the distribution curve diminished, while the half width increased as a result of an increase in the number of slow ions. The energy of the secondary ions corresponding to the maximum on the distribution curve coincided with the energy E obtained from formula (1.9) for an elastic single collision between the particles. The appearance of scattered ions with a maximum energy $E_{max} > E$ was explained by the author as being due to the possibility of a collision between the ion and two target atoms at the same time, and also to the influence of the bond energy of the metal atoms. The angular distribution of the secondary ions was also studied. The maximum number of ions occurred close to the direction of specular reflection though not always exactly at this point.

Mashkova and Molchanov [622] measured the angular distribution of fast scattered particles on bombarding copper, tungsten, and graphite targets with Ar^+ ions over the energy range $E_0 = 0-30$ keV. A special attachment enabled the authors to record not only the ionic, but also the neutral component of the emission of fast scattered particles. The particles scattered by the target passed through an opening in the aperture diaphragm of the ion receiver and fell onto an inclined first collector situated inside a Faraday cylinder. The secondary electrons ejected from the first collector by these particles were collected by a second collector. It was considered that the intensity of the secondary electron current was proportional to the intensity of the scattered particles. In the course of the experiment, the secondary-electron emission current I^- was measured at each scattering angle for the same primary ion current I_0. On comparing the secondary-ion and secondary-electron currents, the authors came to the conclusion that a large proportion of the Ar^+ ions were neutralized on the target surface. The experimental curves agree closely with those of the differential effective scattering cross sections associated with a screened Coulomb potential. The slight disagreement between the theoretical and experimental curves with increasing scattering angle and the absence of any limiting scattering angle on bombarding beryllium and carbon with Ar^+ ions was ascribed by Mashkova and Molchanov to the existence of multiple as well as single collisions, even for glancing angles of incidence. It should be noted that the recording of scattered atoms and ions was based on the emission of electrons from another target initiated by these particles; the authors neglected the fact that this emission depended greatly on the energy of the bombarding particles.

Molchanov and Soshka [708] used an electrostatic analyzer with double curvature to study the dependence of the energy spectra on the angle of incidence. Copper and graphite targets were bombarded with 25-keV Ar^+ ions at an angle of $\varphi > 70°$. The energy spectrum of the ions scattered by the copper target had a maximum in the high-energy region; as the scattering angle β increased, this moved in the low-energy direction, which corresponded to single scattering at the target atoms.

Fluit and Kistemaker [697] studied the distribution of secondary ions and atoms with respect to both angle and energy on bombarding polycrystalline and single-crystal copper targets with Ar^+ ions in the energy range 5-20 keV. The secondary ions and neutral emissions were measured by reference to the secondary electron emission which these produced. It was shown that the energy distribution of the secondary particles had a sharp peak in all cases. The number of scattered particles in general increased with increasing angle of incidence φ and had a minimum for $\varphi = 45°$. The latter corresponded to the [110] crystallographic direction of the copper crystal. Fluit and Kistenmaker considered that the secondary emission contained not only scattered, but also sputtered particles; with increasing E_0 the number of the former increased much faster than that of the latter. The orientation of the crystal affected reflection much less than sputtering; this was because the sputtering of neutral copper atoms took place in deeper layers of the crystal than the emission of fast scattered particles. The maximum scattering angle was determined by the roughness of the surface.

Yurasova, Brzhezinskii, and Ivanov [692] bombarded the (100) face of copper with 1.5-keV Ar^+ ions. The direction of the maximum yield of sputtered particles, corresponding to the crystallographic [110] direction, coincided with the direction of minimum ion scattering. The experimental results agreed closely with calculations carried out by Yurasova [701]. In these calculations the collision of the incident ion with an atom in the lattice was considered as the impact of elastic spheres, while the radius of the impact was determined from a partly screened Coulomb potential of the Born-Meyer type. It was also considered that the greater part of the ions were reflected after two collisions with lattice atoms.

Veksler [709] studied the relation between the maximum energy E_{max} of K^+, Rb^+, and Cs^+ ions scattered by a molybdenum surface and the primary-ion energy over the range 30-260 eV for various scattering angles. The value of $\eta = E_{max}/E_0$ increased with decreasing β. The presence of a broad energy spectrum among the scattered ions was explained by the geometrically asymmetrical collision of the ion with four lattice atoms, and also by the collision of the ion with one lattice atom or two or three simultaneously.

The angular distribution of the secondary ions approximately coincides with the cosine law. The shape of the $\eta(E_0)$ curve is independent of the scattering angle β. However, it should be noted that the variation in the $(E_{max}/E_0)(E_0)$ and $(m_{eff}/m)(v_0^2)$ curves with scattering angle β

was only studied between 120 and 90°, the maximum energy E_{max} being found by the method of the retarding field, and the accuracy of this method was not very high. At the same time, Gay and Harrison [700] calculated on an electronic computer the interaction between atomic beams of Cu and the (100), (110), and (111) faces of a copper single crystal between 0.025 and 10 KeV, and came to the conclusion that, for normal incidence of the ion on the surface of the target, the probability of a collision with several lattice atoms simultaneously was low, such collisions not leading to the back-scattering of the ion but rather to penetration into the target. The calculations show that the conception of the "effective mass" is not a suitable model for describing the scattering of atomic particles by the lattice of a metal.

Abroyan, Lavrov, and Titov [751] studied the angular dependence of the secondary-emission coefficient $K(\varphi)$ on bombarding a silicon single crystal with K^+ ions in the energy range 1-5 keV; they found that the minima on the $K(\varphi)$ curve coincided with angles corresponding to the crystallographic directions [110], [111], and [112]. These authors indicate that the depths of the minima on the curve depend sharply on the state of the target surface. Experiments by Datz and Snoek [698] showed that the bombardment of even an atomically smooth surface leads to the formation of rough places, which hold back a certain proportion of the scattered particles.

An optical study of the rough places created on the (100) face of a copper crystal [705] showed that the furrows formed on bombarding with Ar^+ ions along the [110] axis had a much smaller slope (15°) than the furrows formed along a high-index direction (for example, for $\theta = 37°$ the slope was 45°).

Mashkova and Molchanov [736, 737] used an electrostatic analyzer to study the effect of the crystal structure of a copper target on the energy spectra of the scattered ions. These authors bombarded the (100) and (114) faces of a copper single crystal with Ne^+ and Ar^+ atoms at an energy of 30 keV. The working vacuum was about 10^{-7} mm Hg. Leaving the angles φ and θ constant, the authors rotated the copper single crystal around the [110] axis. It was found that in the directions of the principal axes of the crystal, the peaks of the scattered ions in the energy spectra were sharper than those in the transparent directions. On bombarding the (114) face of the copper single crystal, the experimental conditions were chosen so that, for $\varphi = 10°$ and $\theta = 28°$, the [110] axis coincided with the axis of the electrostatic analyzer. In this case, the energy spectrum clearly showed a peak of recoil atoms, while the position of the Ar^+ peak remained unaltered, as when bombarding the (100) face under the same experimental conditions ($\varphi = 10°$, $\theta = 28°$). This may be explained by the fact that the direction of the primary ion beam coincided with the more transparent directions of the copper single crystal. It was found that on bombarding a copper target with Ne^+ ions the Cu^+ and Ne^+ peaks were resolved better than the Cu^+ and Ar^+ peaks in the case of bombardment with Ar^+ ions.

Veksler [752, 786] studied the energy spectra of secondary ions on bombarding molybdenum and tungsten single crystals with slow (20-250 eV) K^+, Rb^+, and Cs^+ ions at various angles of incidence φ and with a fixed β. The targets were in an incandescent state during the measurements (1500-1600°K). Veksler observed anisotropy of the maximum energy of the secondary ions E_{max} and a displacement of the maxima on the $E_{max}(\varphi)$ curves in the direction of higher values of φ as the scattering angle β diminished. The appearance of peaks in the high-energy region was associated with unpaired collisions between the bombarding ions and lattice atoms.

Dahl and Magyar [753] used an electrostatic analyzer to study the angular dependence of the energy spectra of secondary ions on bombarding the (100) face of an aluminum single crystal with Ar^+ ions at an energy of $E_0 = 50$ keV. In spectra obtained for angles of incidence $\varphi = 5, 25, 45°$, and a scattering angle of $\beta = 50°$, these authors found some peaks of recoil atoms corresponding to the secondary ions Al^{3+}, Al^{2+}, Al^+, and O^+. The peaks for $\varphi = 25°$ were higher than for $\varphi = 45$ and $5°$. This may be explained by the coincidence between the direction of the primary ion beam and the closely packed faces of the single crystal, which leads to intensive sputtering of the target material in the form of ions. The absence of any peak attributable to Ar^+ ions singly scattered from aluminum atoms is due to the fact that the analyzer lies at an angle of $\beta = 50°$ greater than the limiting value. The secondary-ion spectra obtained for $\varphi = 22°$ and $\beta = 25°$, showed not only Al^{3+}, Al^{2+}, H^+, and O^+ peaks, but also those of Ar^+ and Ar^{2+} The presence of an Ar^+ peak is explained by the single scattering of Ar^+ ions at individual aluminum atoms, and the Ar^{2+} peak by multiple ionization due to stripping. The probability of the scattering of Ar^+ and the sputtering of Al atoms was also determined as a function of β. The probability of sputtering Al particles increased with increasing scattering angle β, while the probability of scattering Ar^+ particles had a threshold at $\beta \sim 40°$, and rose with diminishing scattering angle.

Mashkova, Molchanov, Parilis, and Turaev [749, 779] sought to observe the structure of the energy spectrum of ions reflected from a single-crystal face predicted in [699] by bombarding the (100) and (114) faces of a copper crystal with 30–keV Ar^+ ions. The incident and scattered beams lay in the (110) plane. The scattering angle was taken as $\beta = 50°$, φ varying between 10 and 30°. It was found that on reflection from the single crystal the energy spectrum of the ions possessed a fine structure. Calculations carried out by the method of Parilis and Turaev [699] with due allowance for inelastic losses showed that this structure was due to twofold collisions of the incident ion with the surface-layer atoms. The main part of the spectrum studied in these experiments corresponded to twofold scattering in the [110] direction.

Analysis of the various investigations considered in this review shows that the angular and energy distributions of secondary ions in many earlier investigations were studied under insufficiently clean surface conditions, inertial apparatus was used for recording the currents, and the measurements were principally carried out at low target temperatures. When investigations were carried out at high target temperatures, the surface-ionization phenomena accompanying the secondary processes were not always taken into account. The bombarding ion beams also contained neutral particles. The result of all these factors was that different authors were working with different experimental conditions and their results failed to supplement each other, or in some cases were contradictory. For example, certain authors [48, 165] considered that the existence of specular ion reflection was well established, while others [49, 59, 94] denied this, asserting that the ion was scattered predominantly forward. Whereas Seeliger and Sommermeyer [101] asserted that the cosine law was satisfied for the angular distribution of secondary ions, Eremeev and Zubchaninov [165] rejected this law.

In many investigations the energy distributions of secondary ions were obtained from retardation curves obtained in an ordinary apparatus of the spherical-condenser type. It is well known, however, that the method of the graphical differentiation of experimental curves is insufficiently sensitive for the reliable revelation of the desired characteristic features in the distribution functions.

In recent years the advantages of the electrostatic analyzer of the Hughes–Rojansky type have been convincingly demonstrated [54, 352, 644] in studying the energy distributions of secondary particles. However, the detailed study of the angular and energy distributions of scattered ions based on this exact method is still in its early stages.

Up to the present, the angular dependence of the energy distribution of secondary ions on bombarding light targets with heavy ions has hardly been studied at all.

In this chapter we set out the results of investigations based on the use of the static and dynamic (oscillograph) methods of recording, incorporating an electrostatic analyzer for studying the angular and energy distributions of the ions scattered by the surface of a solid body for the cases of direct $m_1 < m_2$ and inverse $m_1 > m_2$ ratios of the masses of the colliding particles.

§2. Static and Dynamic Methods of Studying
the Angular and Energy Distributions of
Secondary Ions

Galvanometric Method

The scattering of ions as a function of incidence angle was studied by Arifov and Ayukhanov [282] by the galvanometric method using the apparatus with two ground-glass joints shown in Fig. 5.2 (one of the joints is perpendicular to the plane of the sketch). The apparatus consisted of an ion source 4, 5, and 6, a 127° cylindrical condenser, a receiving section (in the form of a nickel cylinder with slits for the entrance of the primary beam and for receiving the secondary ions on the collector 3, placed opposite to one of the slits), and a tantalum target 1 drawn out along the axis of the cylinder. The cylinder had eight slits of exactly the same size (6 × 0.8 mm) arranged around the circumference of the cylinder at 14°18' intervals; one the slits was used for receiving and studying the secondary ions, while the rest served as exits for the primary ion beam at a certain fixed angle. The whole rigidly fixed receiving section, including the target, the receiving cylinder 2, and the secondary-ion receiver (collector 3), could be rotated around an axis coinciding with the axis of the cylinder 2. This enabled us to use a primary beam traveling at various angles to the target surface, the angle being determined by the number of the slit through which the primary beam passed. In studying the escape of secondary ions as a function of the angle of incidence of the primary beam, the slit receiving the secondary particles (with the collector 3 behind it) was arranged so that its middle point lay on the normal to the target surface. Between the slit of the receiving cylinder and the collector was an electric field retarding the positive ions.

In order to study the distribution of secondary ions with respect to escape angle, an additional ground-glass joint was attached to the apparatus. In this case the receiving cylinder 3 and the target 1, rigidly attached to it, remained stationary with respect to the ion source, i.e., the angle of incidence was always constant (in our case, the primary beam was directed perpendicular to the target surface, $\varphi = 0$). The angle θ was varied by rotating the collector 3 around the receiving cylinder, starting with the angle 14°18'.

The measuring circuit is illustrated in Fig. 5.3. The secondary ion current to the collector 3 was measured with the electrometer 4 and the total current to the target 1 and the receiving cylinder 2 with the ordinary mirror galvanometer 5. The remaining components of the circuit are clear without explanation.

In order to study the limiting energies of the secondary ions as a function of the angle of incidence of the primary beam we obtained the volt—ampere characteristics of the secondary ions, i.e., we measured the ion currents passing to the collector every 14°18' as a function of the voltage of the retarding field created between the collector 3 and the receiving cylinder 2.

The secondary-emission measurements were carried out at a target temperature of 1300°K after brief heating (flashing) to 2500°K. The ion beam was only applied to the target surface during the measurements for a period not exceeding a few times the time constant of the electrometer. The width of the ion beam at the target surface, for an angle of incidence of 14°18', was no greater than 1 mm; the beam passed through the axis of rotation of the receiving part of the apparatus.

The volt—ampere characteristics were taken in succession for each slit separately, using the same electrometer sensitivity.

The results of our measurements of the angular dependence of the secondary-ion energy obtained by this method on bombarding a hot (1300°K) tantalum target with Na$^+$ ions at an initial

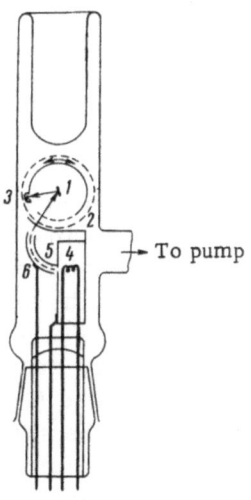

Fig. 5.2. Arrangement of vacuum apparatus for studying the angular and energy distributions of the secondary ions.

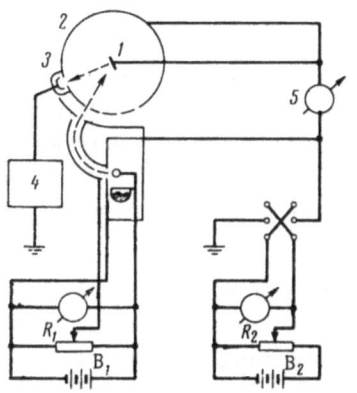

Fig. 5.3. Principal electric circuit of the apparatus for studying the angular and energy distributions of the secondary ions.

energy of $E_0 = 170$ eV and Rb^+ ions at $E_0 = 360$ eV are presented in Figs. 5.4 and 5.5. The abscissas here represent the value of the retarding field and the ordinates the secondary ion currents to the collector 3 (Fig. 5.3), referred to unit total current (in arbitrary units).

Curve 1 in both figures corresponds to the passage of the primary beam through the first slit with $\varphi = 14°18'$; curves 2–5 correspond to passage through the second, third, etc., slits in steps of φ equal to $14°18'$.

The complete stoppage of the ions occurs at different retarding potentials for different angles φ, i.e., the limiting energy of the secondary ions depends on the angle of incidence of the primary ions.

The values of limiting energy agree closely with relation (1.9) for the elastic collision of an ion with one surface atom.

The calculated values

$$\eta_t = \frac{E}{E_0} = \frac{(m_1 - m_2)^2}{m_2^2 \left[\cos\beta \pm \sqrt{(m_1/m_2)^2 - \sin^2\beta}\right]^2} \quad (5.1)$$

and those obtained experimentally, $\eta_e = E/E_0$ for Na^+ and Rb^+ are shown in Table 5.1, where the angle of incidence φ for slit No. 1 equals $14°18'$ for $\theta = 0°$ and $\alpha = \varphi$, while for subsequent slits the angle φ increases by $14°18'$ over the angle φ for the previous slit.

In addition to this, the curves in Figs. 5.4 and 5.5 show that with increasing φ the number of ions possessing high energies becomes relatively greater.

In order to study the limiting energies of the scattered ions propagated at different escape angles, we obtained the V/A characteristics of the ion currents. The V/A characteristics of pure tantalum heated to 1350°K and bombarded with 180-eV Na^+ ions and 360-eV Rb^+ ions are shown in Figs. 5.6 and 5.7, respectively. We see from these curves that the limiting energies of the secondary ions are different for different slits, i.e., they depend on

Table 5.1

Ion	Slit									
	№ 1		№ 2		№ 3		№ 4		№ 5	
	η_t	η_e	η_t	η_e	η_t	η_e	η_t	η_e	η_t	η_e
Na^+	0.61	0.60	0.62	0.62	0.64	0.65	0.67	0.70	0.72	0.76
Rb^+	0,13	0.15	0.15	0.18	0.17	0.20	0.20	0.24	0.26	0.31

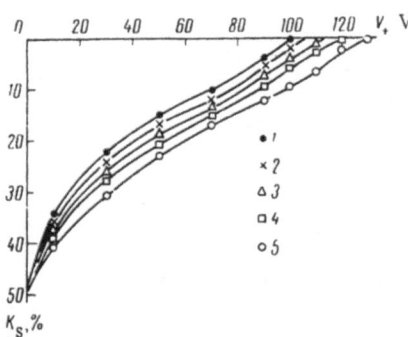

Fig. 5.4. Relation between the V/A characteristics of secondary ion emission and the angle of incidence φ of the primary beam of Na$^+$ ions.

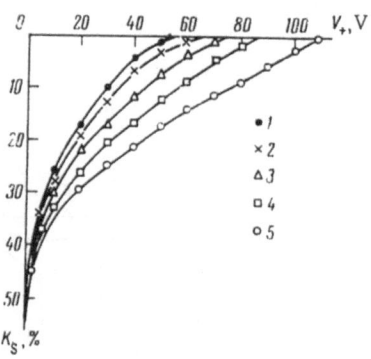

Fig. 5.5. Relation between the V/A characteristics of secondary ion emission and the angle of incidence φ of the primary beam of Rb$^+$ ions.

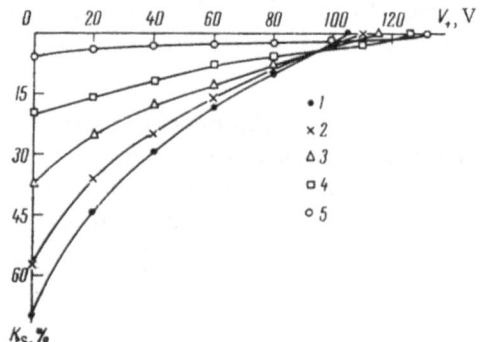

Fig. 5.6. Relation between the V/A characteristics of secondary ion emission for Na$^+$ and the escape angle θ (for $\varphi = 0$). 1) $\theta = 28°36'$; curves 2-5 correspond to successive increases of $14°18'$ in θ.

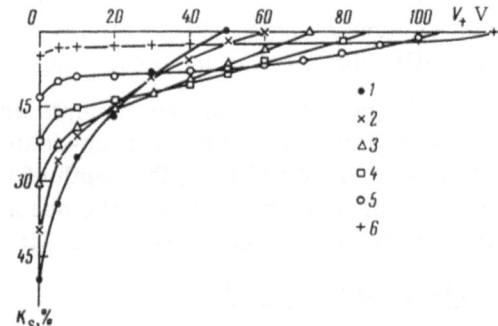

Fig. 5.7. Relation between the V/A characteristics of secondary ion emission for Rb$^+$ and the escape angle θ (for $\varphi = 0$). 1) $\theta = 28°36'$; curves 2-6 correspond to successive increases of $14°18'$ in θ

Table 5.2

Ion	Slit											
	№ 1		№ 2		№ 3		№ 4		№ 5		№ 6	
	η_t	η_e	η_t	η_e	η_t	η_e	η_t	η_e	η_t	η_e	η_t	η_e
Na$^+$			0.62	0.58	0.64	0.61	0.67	0.63	0.72	0.69	0.76	0.74
Rb$^+$	0.13	0.14	0.15	0.17	0.17	0.20	0.20	0.24	0.26	0.30	0.33	0.34

the escape angle of the secondary ions θ. Consideration of the limiting energies of the secondary ions indicates satisfactory agreement with relation (1.9) for elastic collisions. The calculated values of η_t based on (1.9) and the experimental values of η_e for Na$^+$ and Rb$^+$ are given in Table 5.2, where the escape angle θ for slit No. 1 equals $14°18'$ for $\varphi = 0$ and for subsequent slits the angle θ increases by $14°18'$.

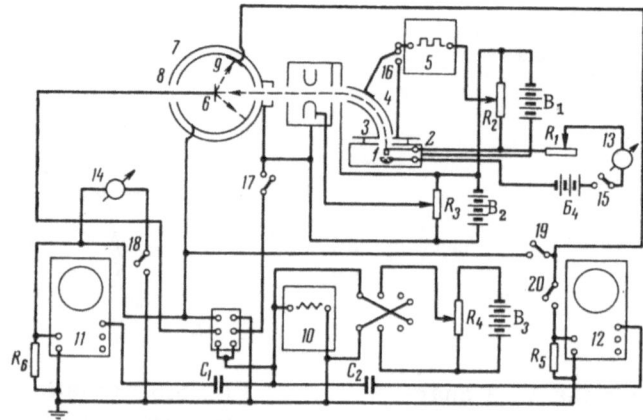

Fig. 5.8. Principal electric circuit of the experimental apparatus for measuring the angular distribution of secondary ions.

It was very important to discover the relationship between the number of scattered ions and the escape angle. We see from the curves in Figs. 5.6 and 5.7 that the number of scattered ions falls with increasing escape angle θ.

The results of the experiments are in agreement with the idea that the processes in which the secondary ions arise are deep-seated and associated with multiple collisions involving target atoms. In actual fact, the depth of penetration of the primary ions diminishes with increasing angle of incidence, so that the secondary ions emerging into the vacuum normally to the target surface will leave the latter with a high energy.

Oscillographic Method

The angular distribution of secondary ions was studied by Arifov, Ayukhanov, and Aliev [553] by the oscillographic method in an experimental apparatus differing from the usual one used in studying secondary processes in that a movable probe was mounted therein in addition to the secondary-particle collector. Provision was also made for changing the orientation of the target with respect to the incident ion beam. The movable probe lay between the target and collector at a distance of about 1 mm from the outer wall of the latter, and could be rotated through 180° to either side of the slit provided for the exit of the primary ion beam. This probe made it possible to measure the intensity of secondary ions proceeding from the target at various escape angles θ, for a fixed angle of incidence φ of the primary ions on the target.

The secondary currents to the collector and to the movable probe were measured by the double-modulation oscillograph method [281]; in this way the screens of a pair of oscillographs automatically reproduced stationary V/A characteristics, on which the secondary currents of scattered ions I_s, evaporated ions, I_e, and diffusion ions I_d, the thermionic (thermoelectron) emission current I_T, and the primary ion current I_0 could all be clearly separated and measured.

The principal electrical circuit of the experimental apparatus is shown in Fig. 5.8. Here, 1 is the ion source, 2 is the body (casing) of the source, 3 is the slit, 4 is the cylindrical condenser, 5 is the rectangular voltage-pulse generator, 6 is the target, 7 is the collector, 8 is a guard cylinder, 9 is the movable probe, 10 is the sawtooth voltage generator, 11 and 12 are oscillographs, 13 is an ammeter, 14 is a microammeter, and 15-20 are switches.

By using the oscillograms of the V/A characteristic, the coefficients K_s, K_e, K_d, and K_Σ could be determined as functions of the primary-ion incident angle φ. The value of K_s measured

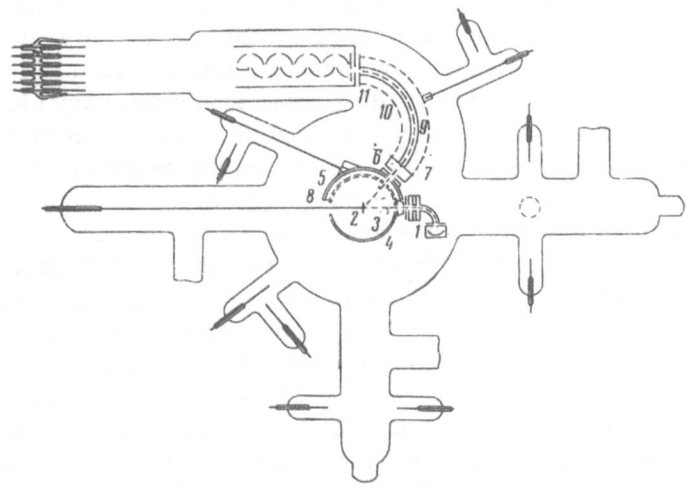

Fig. 5.9. Experimental apparatus used for studying the angular dependence of the energy spectra of secondary ions.

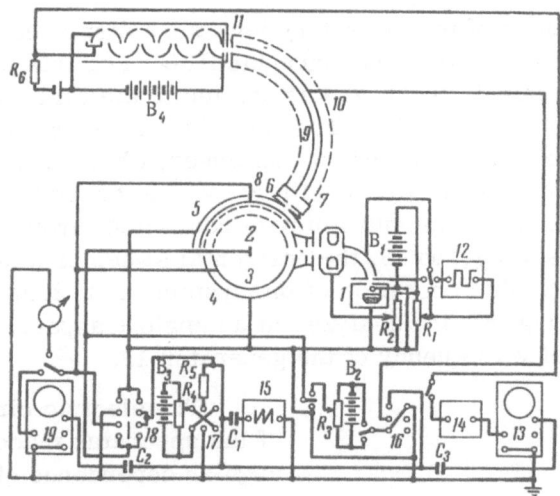

Fig. 5.10. Electric circuit of the experimental apparatus for studying the angular distribution of the energy spectra of secondary ions.

in the circuit of the movable probe for various orientations of the latter relative to the target surface gave the distribution of secondary ions with respect to the escape angle θ.

The angular dependence of the secondary-ion energy spectra was studied in a vacuum system (Fig. 5.9), using the oscillographic method and an electrostatic analyzer of the Hughes-Rojansky type [54, 352, 644]. The main components of the apparatus and the principal electric circuit are shown in Fig. 5.10. The beam of positive ions obtained from the source 1 (the arrangement and operation of which were described earlier) and modulated by the rectangular voltage-pulse generator 12 is directed against the target 2, which constitutes a polycrystalline or single-crystal target of dimensions $0.03 \times 5 \times 30$ or $0.5 \times 5 \times 15$ mm, respectively. The secondary-ion collector is a cylinder 3 of radius R = 30 mm. In the collector 18 identical slits are cut at strictly equal distances from each other; these are 1 × 10 mm in size. The collector is surrounded by a guard cylinder 4 which also has 18 slits, 2 × 12 mm in size, lying opposite to the slits in the collector. Beyond the guard cylinder is a stationary electrode 5 with a slit 6, making an angle $\theta = 50°$ with the direction of the normal to the target surface; the secondary ions leaving the target in the direction of this slit fall on the entrance of the electrostatic analyzer 7, while those passing through the remaining slits are collected by the small collector 8 connected to the main collector 3 (16-19 are switches). The analyzer was a cylindrical condenser 9 of the Hughes-Rojansky type with an aperture angle of 127° machined from copper, with the radius of the equilibrium trajectory equal to 55 mm. In order to protect it from the effects of external electric fields, the analyzer was placed in an enclosure 10 with entrance and exit slits 0.8 × 10 mm in dimension. Calculation showed that the energy resolution of the analyzer was about 2%. The ion—electron multiplier 11, with an amplification factor of about 10^6, served to amplify the currents passing through the analyzer. The ion source, guard cylinder, and collector, were rigidly connected and could be rotated around their own axis by means of an external magnet, which enabled the angle of incidence of the primary ions on the target to be varied.

Thus the apparatus may be used to make an energy analysis of secondary ions having a certain escape angle as a function of the angle of incidence φ of the primary ions. By varying the position of the target with respect to the analyzer, we may also study the dependence of the energy spectra of the secondary ions on the escape angle θ.

The energy distributions of the secondary ions were displayed automatically on the screen of the oscillograph 13 and recorded on photographic film. For this purpose a slight positive potential relative to the collector 3 (20-40 V) was applied to the target and a sawtooth voltage was applied to the outer plate of the analyzer. The secondary ions propagating in the direction of the slit 6 ($\theta = 50°$) fall into a 127° cylindrical condenser, where under the influence of an electric field varying in time they are analyzed with respect to energy and then fall on the first dynode of a multiplier. The anode of the multiplier is connected through the amplifier 14 to the vertical amplifier of the oscillograph 13, the horizontal sweep of which is synchronized with the generator 15. As a result of this operation of the system, a stationary energy-distribution curve of the secondary ions passing into the analyzer at an angle θ appears on the oscillograph screen, repeating itself with the pulse frequency of the generator 15.

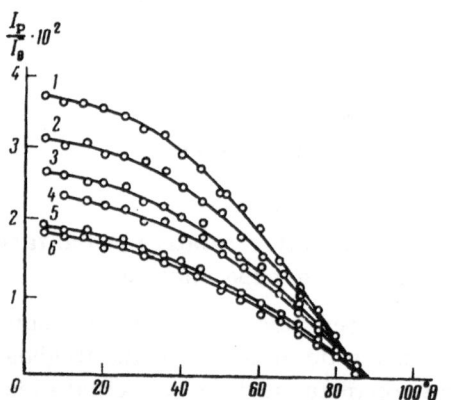

Fig. 5.11. Angular distributions of secondary ions from pure Ta for energies E_0 = 300 (1); 600 (2); 800 (3); 1000 (4); 1500 (5); 1700 eV (6).

The electric circuit and the construction of the apparatus made it possible to study the angular dependence of the coefficient of secondary ion emission at the same time as the angular dependence of the energy spectra of the secondary ions by the double-modulation method [281].

§3. Angular Distribution of Ions

Scattered by a Metal Surface

Bombardment of Heavy Targets with Light Ions

By using the apparatus described in Fig. 5.8, we obtained curves (Fig. 5.11) relating the intensity of the scattered-ion current to the escape angle on bombarding a clean tantalum target heated to 1500°K with Na^+ ions at an angle of $\varphi = 0$.

The number of ions scattered in all directions falls with increasing E_0, while the distribution of the ions with respect to their escape angles corresponds to the cosine law and is independent of E_0 and target temperature.

The angular distribution of the scattered ions was also studied for different orientations of the target with respect to the direction of incidence of the primary ions.

The diagram representing the scattered-ion current obtained on bombarding a tantalum target heated to 1500°K with Na^+ ions as a function of escape angle for angles of incidence of the primary ions equal to 0, 30, 45, and 60° showed that the distribution of the scattered ions with respect to escape angle for 0, 30, 45, and 60° was almost the same as for normal incidence of the primary ions on the target, i.e., there was no preferential direction of scattering. An appreciable deviation of the angular distribution from the cosine law, due to a slight preference for forward scattering of the ions, occurred for angles of incidence $\varphi \geq 60°$ [553].

The angular distribution of secondary scattered ions was also studied by Arifov, Aliev, and Ayukhanov [665] for cases in which $V_i < \varphi$.

On bombarding high-temperature tantalum and molybdenum targets with Rb^+ and Cs^+ ions we find that, in view of the fact that $V_i < \varphi$, the secondary ion emission contains a considerable number of secondary ions resulting from the thermal evaporation of primary ions adsorbed on the surface or penetrating into the target. These secondary ions, of course, have low energies. Hence, in order to obtain the angular distribution of the scattered ions in the case of $V_i < \varphi$, a slight retarding field was applied between the target and the collector; this prevented the collector and movable probe from receiving evaporated ions or ions which had diffused to the surface. The application of this potential in no way distorted the angular distribution of the secondary scattered ions, since these had considerable energies.

Figure 5.12 shows the angular distributions of secondary ions obtained with a retarding potential of 1-2 V on bombarding clean tantalum targets heated to 1500°K with Rb^+ ions having energies of $E_0 = 500, 800, 1100, 1400,$ and 1700 eV with $\varphi = 30°$. The results of the measurements are given in the form of a polar diagram, in which the radius represents the ratio of the current in the movable probe I_p to the primary ion current I_0 on an arbitrary scale, while the angles between the axis of the polar system and the radii correspond to the escape angles of the secondary scattered ions. The broken circle represents the function $2R \cos \theta$, where R is the radius of the circle in question. By analyzing the curves, we may assert that, with increasing energy of the bombarding ions, as in the case of Na^+ ions on tantalum targets, the number of scattered ions diminishes in all directions [665].

The angular distribution of the scattered ions was also studied for various angles of incidence of the ion beam in the case of $V_i < \varphi$. The polar diagrams characterizing the distribution of the scattered ions as a function of θ on bombarding tantalum targets heated to 1500°K with Rb^+ ions at an energy of $E_0 = 1000$ eV are given in Fig. 5.13; curves 1, 2, and 3 are obtained, respectively, for $\varphi = 30, 45,$ and 60°. In order to compare these, curves 2 and 3 are referred to the intensity values for $\varphi = 30°$. We see that all three curves are nearly circular in shape. However, comparison of the polar diagrams obtained experimentally with the function $2R \cos \theta$ (broken circle) shows that, with increasing angle of incidence of the primary ions, the diagrams extend slightly forward relative to the direction of motion of the incident ions.

As indicated earlier, the results obtained in no way contradict the principle of the deep-seated character of the interaction between the bombarding ions and individual atoms of the solid. If the primary ions penetrate very slightly into the surface layers of the solid, the yield of scattered ions will be a maximum in the direction of the normal to the target surface; with increasing escape angle, the path traversed by the scattered ions in the solid becomes longer, and their yield diminishes. Ions which have undergone scattering mainly from the surface atoms

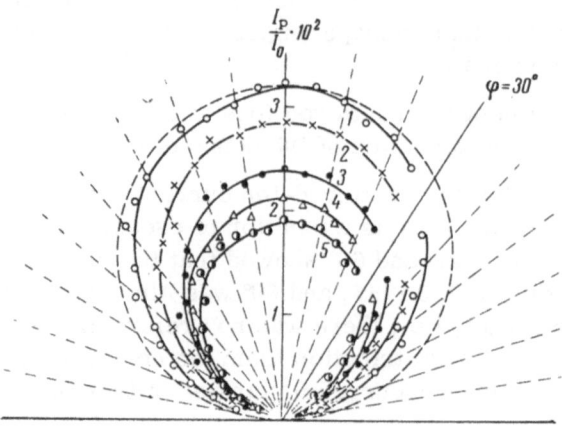

Fig. 5.12. Polar diagrams characterizing the angular distribution of Rb^+ ions scattered by tantalum as a function of energy E_0: 1) 500; 2) 800; 3) 1100; 4) 1400; 5) 1700 eV.

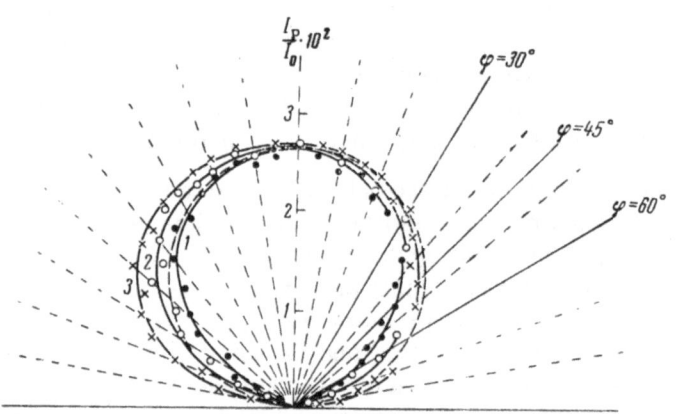

Fig. 5.13. Angular distribution of secondary Rb^+ ions scattered by tantalum.

of the target fall into the collector at large escape angles. This may be seen from the energy distribution of the ions scattered at large angles. It is well known that with increasing θ the relative number of fast ions rises (see §4 of this chapter). This view of the scattering of the bombarding ions from a solid is supported by the $K_s(\varphi)$ relationship. With increasing φ, the coefficient K_s rises. Hence, on bombarding a target with ions at large angles of incidence, the depth of penetration of the primary ions into the solid decreases and the yield of scattered ions becomes greater (see §4 of this chapter) [776].

Bombardment of Light Targets with Heavy Ions

It was shown in the foregoing that, in the case $m_1 > m_2$, the scattered ions were able to leave the surface of the target at any escape angles θ. A slightly different picture should emerge if the mass of the target atoms is smaller than that of the bombarding ions. In this case, relation (1.9) only holds for cases in which $\beta < m_1/m_2$. Hence, singly scattered ions can only leave the surface of the target at $\beta < \beta_{lim}$, where

$$\beta_{\lim} = \arcsin\left(\frac{m_1}{m_2}\right). \tag{5.2}$$

It was shown in [393] that in fact, on bombarding nickel with Cs^+ and molybdenum with Ba^+ ions there was no back-scattering. The same was noted elsewhere [323, 544]. However, certain authors [383, 504, 709] assert that this simple law is not entirely satisfied; they observed the back-scattering of ions in the case $m_1 < m_2$ as well, and explained this as being due to the influence of the bond energy or to the collision of the ions with several target atoms at the same time.

In order to discover the mechanism of the interaction between ions and target atoms in the case of $m_1 < m_2$, Arifov, Aliev, and Ayukhanov [624] studied the angular distribution of the secondary ions produced by bombarding molybdenum and nickel with Cs^+ and Rb^+ ions, respectively.

On bombarding a molybdenum target with Cs^+ ions at angles φ not exceeding a certain limiting value, the currents passing to the probe are small, and their angular distribution agrees with the $\cos\theta$ law. For large angles, however, the angular distribution of the secondary ions has quite a different character. Figure 5.14 shows a polar diagram of the angular distribution of the currents passing to the movable probe on bombarding a molybdenum target heated to 1500°K with 500-eV Cs^+ ions for $\varphi = 50, 60,$ and 70° (curves 1-3, respectively). We see that, for these angles of incidence, in addition to small probe currents distributed in accordance with the cosine law, a considerably larger secondary ion current appears in the direction of large escape angles. On further increasing the angle of incidence, the boundary corresponding to the appearance of these ions moves in the direction of smaller escape angles; however, the limiting scattering angle β remains constant for any angles of incidence of the primary ions, and for Cs^+ on molybdenum it equals about 50°. The ions having an angular distribution in agreement with the cosine law are comparatively slow, their maximum energy not exceeding a few tens of electron volts, although this increases slightly with increasing incident angle φ. The maximum energy of the ions characterized by a limiting scattering angle of 50° for the case of a molybdenum target bombarded with 500-eV Cs^+ ions (determined from the stopping curves) equalled about 75 eV.

It follows from relation (5.2) that, for this case, the limiting scattering angle equals 46°30' and the energy of the ions scattered in the direction of the limiting angle should be about 80 eV. This close agreement of the limiting angle and maximum energy of this group of secondary ions with the values calculated from (5.2) and (1.9) clearly shows that this group of ions appears as a result of the elastic scattering of Cs^+ ions from individual molybdenum atoms.

It was also found that the limiting angle β was independent of E_0 and φ, and coincided with the value derived from relation (5.2) in the case of Cs^+ ions on molybdenum. The number of secondary ions scattered in this direction fell with increasing primary ion energy.

The number of ions possessing an angular distribution coinciding with the $\cos\theta$ law is independent of energy. In order to discover the relation between the slow group of ions and the presence of a surface film on the bombarded surface, the angular distribution was measured for various values of primary ion current. Figure 5.15 shows the angular-distribution curves of the secondary ions measured for the same repetition frequency of the primary ion current pulses, curve 1 being obtained for a pulse length 10 times as long as that corresponding to curve 2. Comparison of these curves shows that the number of ions possessing an angular distribution agreeing with $\cos\theta$ depends very slightly on the primary-ion pulse length. Hence, the reason for the appearance of these ions should clearly not be attributed to the cathode sputtering of the target material (see Chapter 4, §§2 and 3) or contamination created by the primary ion beam. Analogous measurements of the angular distribution were made on bombarding

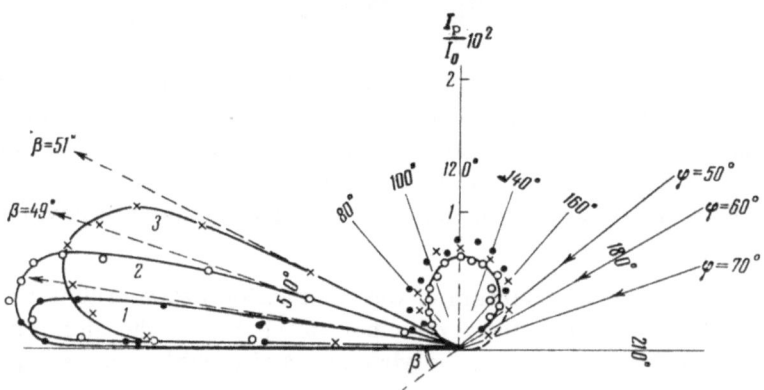

Fig. 5.14. Angular distribution of secondary Cs⁺ ions from molybdenum.

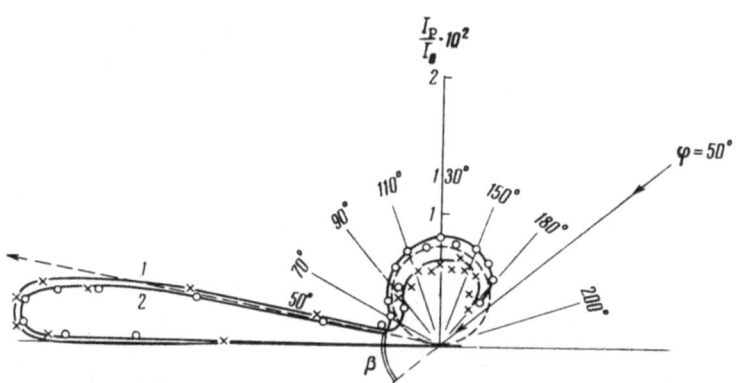

Fig. 5.15. Angular distribution of secondary ions as a function
of the duty factor of the rectangular-pulse generator.

nickel with Rb^+ and Cs^+ ions at an energy of $E_0 = 500-3000$ eV. In this case, the mass ratio of the colliding particles differs from the case of Cs^+ on molybdenum and the value of the limiting scattering angle should accordingly differ.

It should be noted that the limiting scattering angle of the secondary ions slightly exceeds the limiting scattering angle calculated from relation (5.2) in all the cases considered. These results show that in studying the angular distribution for the case $m_1 < m_2$ we may distinguish two groups of secondary ions. One group of ions is fast and only occurs for large incident angles of the primary ions on the target; it is characterized by the fact that it is propagated within a specific limiting scattering angle with an energy $E \geqslant (m_1 - m_2)^2 / m_2^2 \cos^2 \beta$. The other group is slow, many times smaller in intensity, and is observed for all primary-ion incident angles. The angular distribution of these latter is close to the $\cos \theta$ law. The maximum energy of these ions increases with increasing angle of incidence of the primary ions on the target.

The fact that the secondary ion emission contains a group of ions propagating predominantly within a specific limiting angle indicates the existence of elastic paired collisions between the incident ions and individual free target atoms.

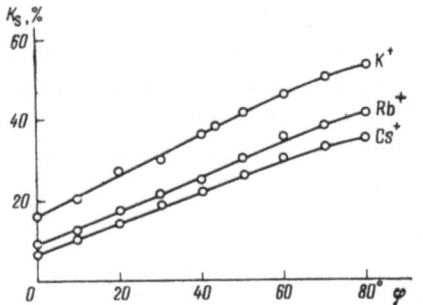

Fig. 5.16. Coefficient K_s as a function of the angle of incidence φ of the primary ions on a Ta target.

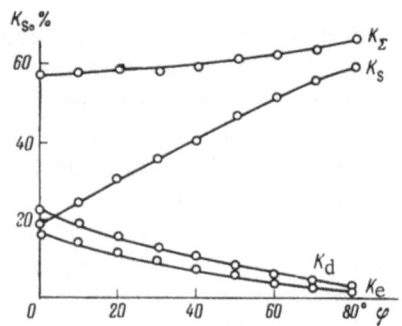

Fig. 5.17. Coefficients K_s, K_e, K_d, and K_Σ as functions of the angle of incidence φ of primary K^+ ions on a tungsten target.

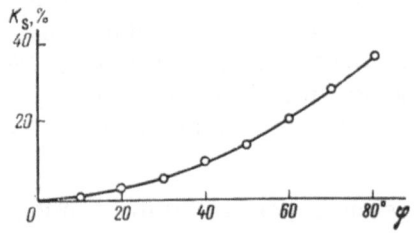

Fig. 5.18. Coefficient K_s as a function of the incident angle φ of primary Cs^+ ions on molybdenum.

§4. Dependence of the Coefficient of Secondary Ion Emission and Its Components on the Incident Angle of the Primary Ions

By using the double-modulation method, Arifov and Ayukhanov [282] found that at high target temperatures the secondary ion emission contained evaporated ions and ions of diffusion origin in addition to scattered ions.

In this section we shall consider the angular dependence of the emission of scattered (K_s), evaporated (K_e), and diffused (K_d) ions, and also the total ion emission K_Σ obtained by Arifov, Aliev, and Ayukhanov [776].

Figure 5.16 shows a series of $K_s(\varphi)$ curves obtained by bombarding a tantalum surface heated to 1500°K with K^+, Rb^+, and Cs^+ ions at an energy of $E_0 = 1000$ eV. We see from the figure that K_s for all three ions increases with increasing incident angle of the primary ions and falls with increasing mass of the ion.

Analogous $K_s(\varphi)$ curves obtained on bombarding heated tungsten and molybdenum targets with Li^+, Na^+, K^+, and Rb^+ ions are not shown here. Analysis of such curves shows that, in the case of tungsten and molybdenum targets, the dependence of the $K_s(\varphi)$ curves on the mass of the ion remains the same as in the case of tantalum. However, the value of the scattering coefficient for the same type of ion differed for different targets.

The dependence of the coefficients K_s, K_e, K_d, and K_Σ on the incident angle of the primary ions on bombarding a target heated to 1500°K with 1000-eV K^+ ions is shown in Fig. 5.17. We see that the coefficients K_e and K_d diminish with increasing φ and tend to zero for $\varphi > 80°$, while the total coefficient K_Σ rises with increasing φ. Analogous results were also obtained for tungsten targets bombarded with Cs^+ ions. Analysis of the resultant $K_s(\varphi)$, $K_e(\varphi)$, $K_d(\varphi)$, and $K_\Sigma(\varphi)$ relationships for K^+, Rb^+, and Cs^+ ions shows that the values of the coefficients K_s, K_e, and K_d not only vary with incident angle and type of bombarding ions, but are also correlated with each other. As shown in the foregoing, the coefficient of the emission of scattered ions K_s increases with increasing angle of incidence of the primary ions. Hence, on bombarding a target with ions at high incident angles, the depth of penetration of the primary ions into the solid diminishes and the yield of scattered ions increases. At the same time, K_e and K_d become smaller, and the variation in $K_\Sigma(\varphi)$ is clearly related to the variation in the relation between the components of secondary emission and the difference in their degree of ionization.

The relation $K_s(\varphi)$ was also studied for $m_1 < m_2$. In this case, by $K_s(\varphi)$ we understand the coefficient of secondary emission of the group of ions having energies above thermal. Figure 5.18 shows a curve of $K_s(\varphi)$ for the case of 1200-eV Cs^+ ions bombarding a molybdenum target. We see from this figure that K_s increases with increasing φ. The energy of this group of ions rises with increasing φ, and for values $\varphi \geq \arcsin(m_1/m_2)$ the secondary ion emission contains ions which have suffered single collisions with free target atoms.

§5. Dependence of the Energy Spectra on the Angle of Incidence of the Primary Ions and the Escape Angle of the Secondary Ions

Bombardment of Heavy Targets with Light Ions at Various Incident Angles

Figure 5.19 presents oscillograms of the energy distribution of secondary ions obtained by Arifov, Aliev, and Ayukhanov [691] on bombarding a tungsten target with 900-eV Na^+ ions at $\varphi = 0°$ and $\theta = 50°$. Oscillogram 1 corresponds to an insufficiently clean target. We see that the distribution curve has a maximum in the low-energy range in the form of a very sharp peak. Subsequently we shall call this the slow-ion peak. A change in the energy and incident angle of the primary ions produces no marked changes in the character of the distribution curve. Oscillogram 2 was obtained with the target at room temperature immediately after cleaning by flashing. We see that in this case, in addition to the slow-ion peak, there is a second peak in the high-energy region. The position of this peak depends linearly on the energy of the primary ions and corresponds to the energy E for the single scattering of Na^+ ions at individual target atoms. Oscillogram 3 was obtained with a target at 1800°K. In this case, owing to the presence of evaporated ions, the slow-ion peak increases sharply, while the character of the high-energy part of the curve remains constant.

The energy spectra of ions scattered by the surface of a metal was studied by Panin [598] for $E_0 = 10-80$ keV. The presence of ions with energies exceeding E is explained by the multiple scattering of the ion at target atoms. Calculation by Parilis and Turaev [747] shows that after the successive multiple collision of an ion with a number of metal atoms the secondary ions may leave the target with an energy greater than that corresponding to a single collision. We see, in fact, from the oscillograms 2 and 3, that on the high-energy side the curve falls off relatively slowly, and the peak broadens as a result of the presence of ions with energies exceeding E.

Figure 5.20 shows spectra obtained on bombarding a tungsten target heated to 1800°K with Na^+ ions having various initial energies E_0. The energies of the primary ions for each oscillogram of the distribution curve are indicated in Table 5.3, which gives the energies corresponding to the peak of the fast scattered ions on the oscillograms. The value of this energy increases with increasing energy of the bombarding ions. The ratio $\eta = E/E_0$, as may be seen from the table, approximately equals 0.63. The value of $\eta_t = E/E_0$ calculated from formula (1.9) equals 0.66. The agreement is quite satisfactory. Thus, the relation between E and E_0 in the range $E_0 > 500$ eV is adequately described by the straight line 1 in Fig. 5.21.

Table 5.3 contains the maximum energies of the scattered ions E_{max} obtained from the oscillograms and also their ratios with respect to E_0. The relation between E_{max} and E_0 also lies on a straight line (line 2 in Fig. 5.21). The height of the peak corresponding to slow ions falls with increasing energy of the primary ions (oscillograms 7 and 8 in Fig. 5.20), and becomes considerable for low values of E_0 [691].'

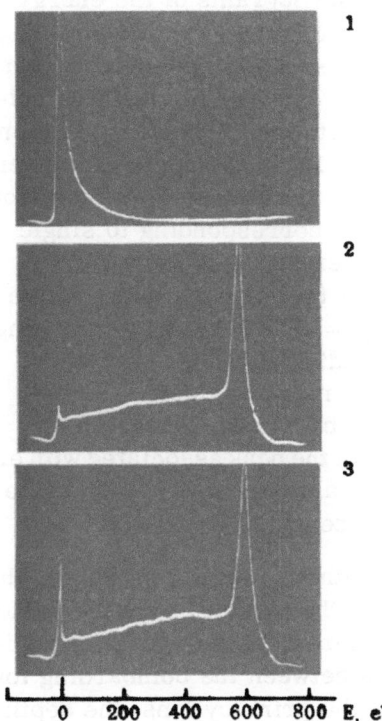

Fig. 5.19. Oscillograms giving the energy distribution of secondary ions.

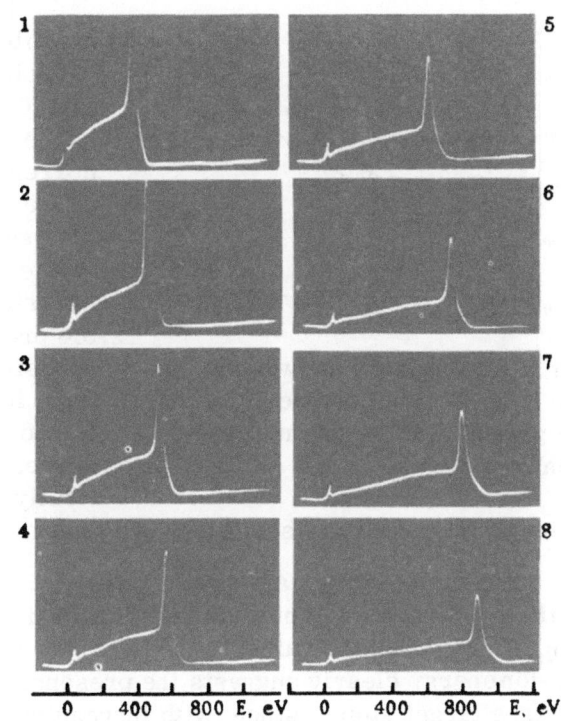

Fig. 5.20. Oscillograms of the spectra of secondary ions for an energy of E_0: 1) 500; 2) 600; 3) 700; 4) 800; 5) 900; 6) 1000; 7) 1100; 8) 1200 eV.

Table 5.3

No. of oscillogram	E_0, eV	E, eV	E_{max}, eV	$\eta_e = E/E_0$	$\eta_{max} = E_{max}/E_0$
1	500	315	380	0.63	0.76
2	600	390	456	0.64	0.75
3	700	459	540	0.63	0.75
4	800	520	610	0.64	0.76
5	900	580	680	0.63	0.76
6	1000	650	770	0.63	0.75
7	1100	700	830	0.63	0.75
8	1200	760	910	0.63	0.75

Table 5.4

	Incident angle φ, deg								
	0	10	20	30	40	50	60	70	80
η_t	0.66	0.68	0.71	0.74	0.79	0.82	0.85	0.90	0.94
η_e	0.63	0.66	0.70	0.72	0.76	0.80	0.82	0.87	0.90
η_{max}	0.75	0.78	0.80	0.82	0.87	0.89	0.91	0.95	0.98

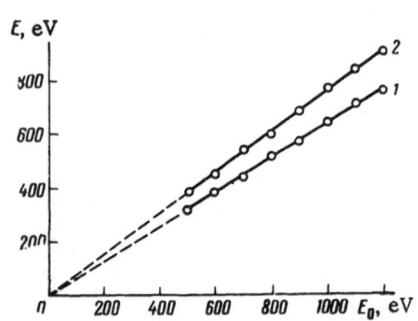

Fig. 5.21. Relation between E (1) and E_{max} (2) and the energy E_0 of the primary Na^+ ions on a tungsten target.

Figure 5.22 presents oscillograms of the energy distribution of secondary ions obtained for different angles of incidence of 700-eV Na^+ ions in a tungsten target heated to 1800°K. Oscillogram 1 was obtained for normal incidence of the beam on the target, and each succeeding one corresponds to an increase of 10° in the angle of incidence. The escape angle θ equals 50° in all cases. We see from the oscillograms that the peak corresponding to single scattering moves in the high-energy direction with increasing angle of incidence of the primary ions. Table 5.4 presents the values of $\eta_t = E/E_0$ calculated for single collisions of Na^+ ions with individual tungsten atoms from formula (1.9), the experimental values of η_e obtained from the oscillograms by measuring the energy corresponding to the position of the peak associated with elastically scattered Na^+ ions, and also the values η_{max} of the ratios of the maximum secondary-ion energies to the primary ion energy.

We see from the table that η_t and η_e are quite close to each other. We also see from the oscillograms and the table that with increasing angle of incidence of the primary ions the values of η_{max} approach the value of η_e for single collisions. This behavior of the maximum secondary-ion energy clearly suggests the presence of multiple collisions between the bombarding ions and free target atoms, since, with increasing angle of incidence of the primary ions, the depth of penetration of the primary ions into the target diminishes, while the probability of multiple collisions with the most favorable angles becomes smaller, which also leads to an approach between η_e and η_{max}.

Table 5.5

E_0, eV	E, eV	$\eta_e = E/E_0$	$\eta_t = E/E_0$	E_{max}, eV	$\eta_{max} = E_{max}/E_0$
\multicolumn{6}{c}{Rb^+ on Ta ($\varphi = 0°$, $\theta = 50°$)}
500	88	0.17	0.19	180	0.35
700	120	0.18	0.19	240	0.35
900	155	0.17	0.19	310	0.35
1000	175	0.17	0.19	340	0.34
1200	210	0.18	0.19	400	0.34
1400	250	0.18	0.19	490	0.35
1500	260	0.17	0.19	520	0.34
1700	290	0.17	0.19	580	0.38
2000	355	0.17	0.19	670	0.33
\multicolumn{6}{c}{Cs^+ on Ta ($\varphi = 0°$, $\theta = 50°$)}
500	20	0.039	0.042	90	0.18
700	29	0.039	0.042	120	0.17
900	36	0.040	0.042	150	0.18
1000	40	0.040	0.042	180	0.18
1200	50	0.039	0.042	205	0.17
1400	57	0.040	0.042	240	0.17
1500	60	0.039	0.042	255	0.17
1700	68	0.038	0.042	300	0.17
1900	80	0.039	0.042	320	0.16
2000	81	0.038	0.042	340	0.17

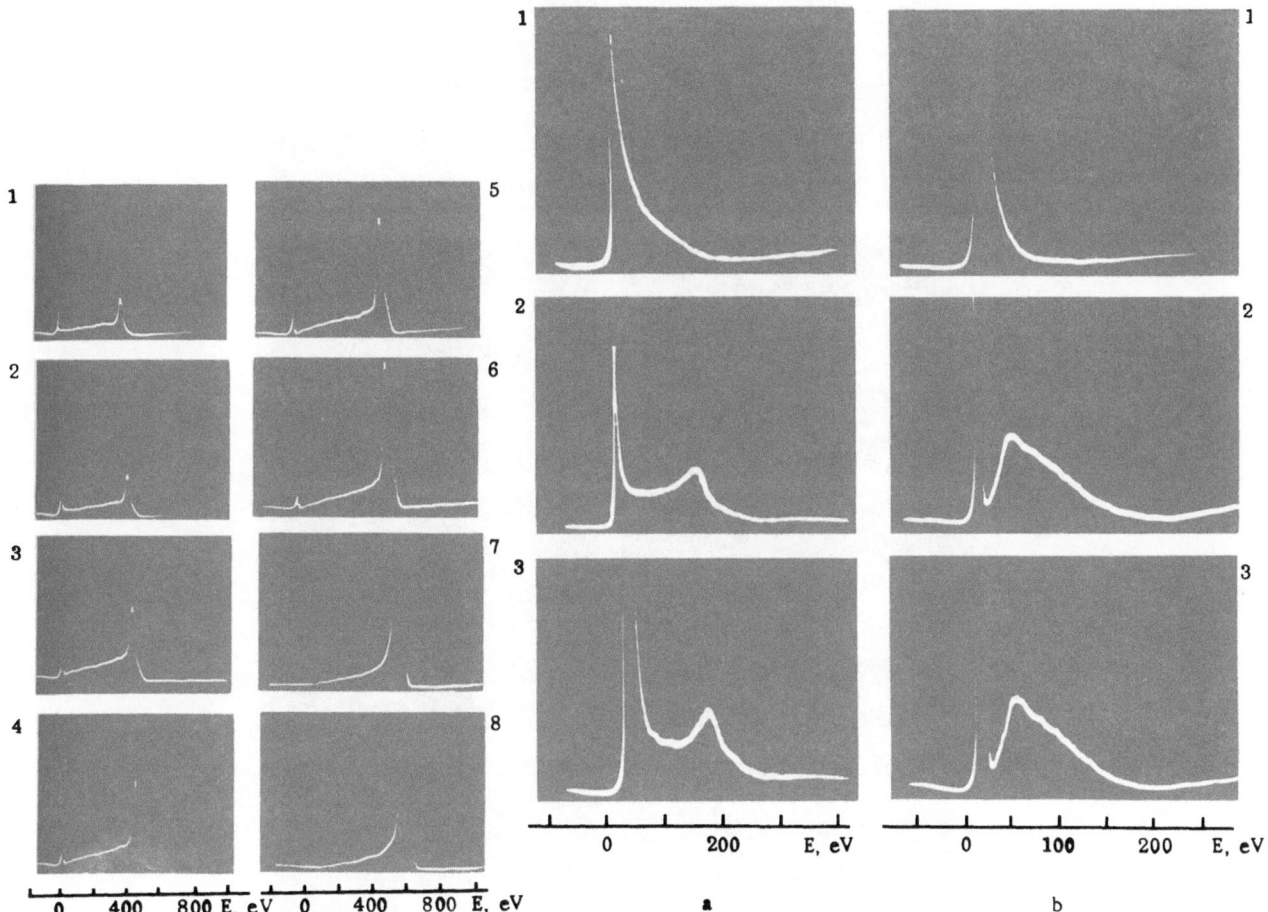

Fig. 5.22. Oscillograms of the secondary-ion spectra for incident angles φ equal to: 1) 0°; 2) 10; 3) 20; 4) 30; 5) 40; 6) 50; 7) 60; 8) 70°; escape angle $\theta = 50°$.

Fig. 5.23. Oscillograms of the secondary-ion spectra of Rb^+ on Ta (a) and Cs^+ on Ta (b) for $\varphi = 0°$, $\theta = 50°$. Targets: 1) Cold and dirty; 2) cold and clean; 3) hot (1800°).

We see from Fig. 5.22 that there is also a considerable variation in the slow-ion peak, according to the angle of incidence. With increasing angle of incidence of the primary ions, the slow-ion peak diminishes, and for $\varphi \geq 70°$ it vanishes completely, indicating the absence of any marked penetration of the primary ions into the depths of the target.

Analogous results are also obtained on bombarding an incandescent tantalum target with Na^+ ions.

It is well known that the energy of the scattered ions depends on the mass ratio of the interacting particles. Naturally, in order to secure a good understanding of the mechanism underlying the interaction between the particles it is of interest to study the angular dependence of the energy spectra of the secondary scattered ions on bombarding the same target with different ions or different targets with the same ions.

Figure 5.23 presents some oscillograms of the energy distribution of secondary ions obtained on bombarding tantalum targets in various states with Rb^+ and Cs^+ ions at an energy of $E_0 = 900$ eV.

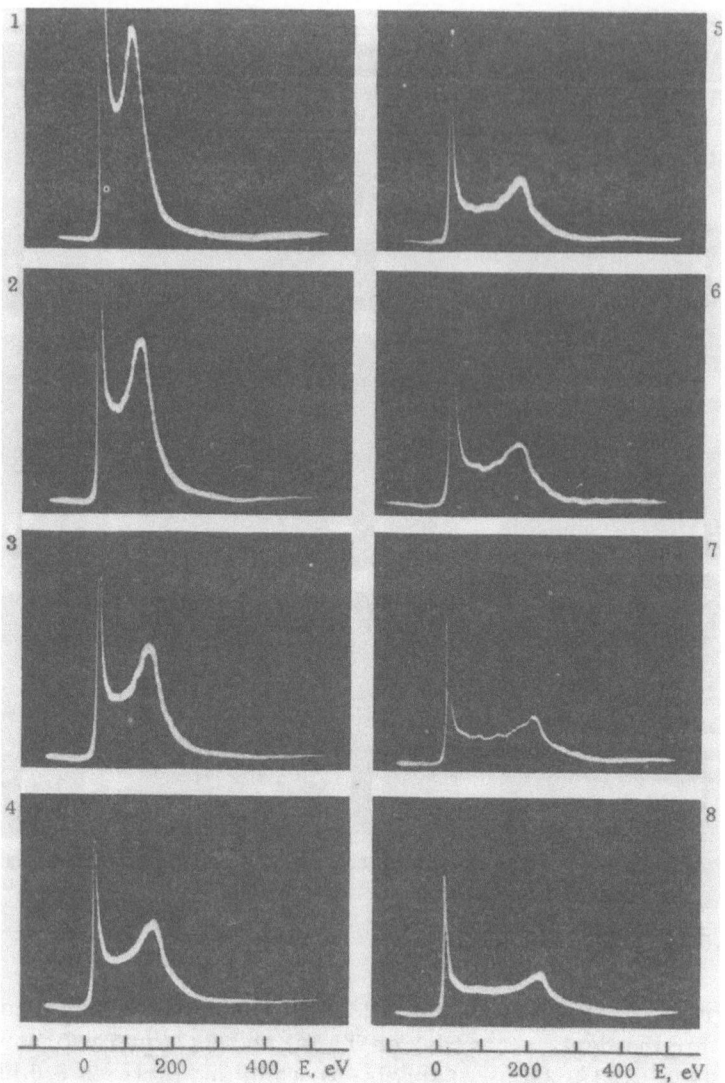

Fig. 5.24. Dependence of the secondary-ion spectra on the primary-ion energy E_0 equal to: 1) 500; 2) 650; 3) 750; 4) 850; 5) 950; 6) 1000; 7) 1200; 8) 1400 eV.

We see from the figures that the character of the secondary-particle distribution from a clean target is in general aspect similar to that obtained for Na^+ on tungsten and tantalum. In this case, first, since $V_i < \varphi$, the slow-ion peak is much greater than for Na^+ on tantalum and tungsten. Secondly, the scattered Rb^+ and Cs^+ ions form a wider peak, extending in the direction of higher spectral energies, thus showing that the difference between the energy characteristic of single collisions and the maximum energy of the secondary ions depends on the mass ratio of the colliding particles.

On bombarding a tantalum target heated to a high temperature with Rb^+ and Cs^+ ions, in view of the fact that $V_i < \varphi$, the secondary emission includes a considerable number of secondary ions obtained as a result of the thermal evaporation of atoms from the primary ion beam adsorbed on the surface and absorbed within the target. Hence, in the case of the hot target (oscillograms 5 and 6 in Fig. 5.23) there is a sharp rise in the slow-ion peak owing to the presence of evaporated ions, while the form of the curve remains unaltered in the high-energy region.

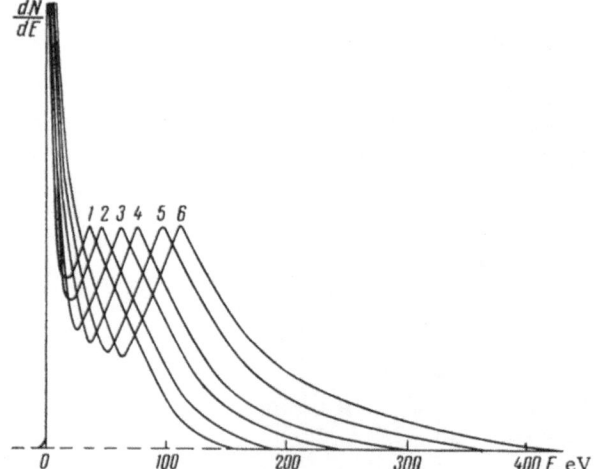
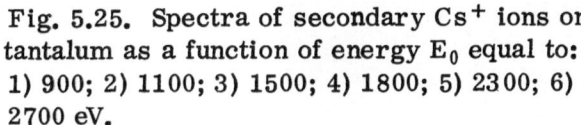

Fig. 5.25. Spectra of secondary Cs$^+$ ions on tantalum as a function of energy E_0 equal to: 1) 900; 2) 1100; 3) 1500; 4) 1800; 5) 2300; 6) 2700 eV.

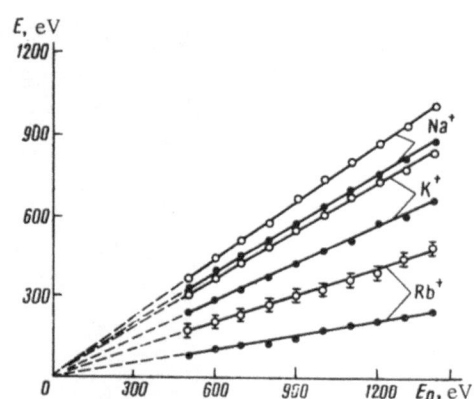

Fig. 5.26. Dependence of E (black circles) and E_{max} (light circles) on the primary-ion energy E_0.

In Fig. 5.24 we have some oscillograms of the energy distribution of secondary ions obtained on bombarding a tantalum target with Rb$^+$ and Cs$^+$ ions at various energies. All the oscillograms were taken with $\varphi = 0°$ and a target temperature of 1800°K. Figure 5.25 shows the corresponding curves for Cs$^+$ on tantalum at 1800°K. We also see from the figures that with increasing energy of the primary ions the maximum on the distribution curves moves in the direction of higher energies, in accordance with the relation for single elastic collisions (1.9). As E_0 rises the peak of the scattered ions falls.

The asymptotic fall in the scattered-ion peak on the high-energy side when the masses of the colliding particles are close together (Rb$^+$ and Cs$^+$ on tantalum) leads to a certain indeterminacy in measuring the maximum energy of the secondary ions E_{max} as a function of E_0 and φ owing to the dependence of E_{max} on the height of the peak. Hence, measurements of the maximum secondary ion energy E_{max} in terms of E_0 and φ were usually carried out for the same height of the scattered-ion peak. This was achieved, not by increasing the primary ion current density at the target, but by changing the sensitivity of the vertical amplifier of the oscillograph, the screen of which automatically recorded the energy distribution of the secondary ions.

A group of E and E_{max} values obtained by the foregoing method appears in Table 5.5, from which we see that the average value of E/E_0 is 0.17 for the case of Rb$^+$ on tantalum and 0.039 for the case of Cs$^+$ thereon. This fully agrees with the values of 0.19 and 0.042 obtained from formula (1.9) for an elastic single collision of Rb$^+$ and Cs$^+$ ions with tantalum atoms on scattering through a specific angle β.

The relation of E and E_{max} to E_0 on bombarding a tantalum target heated to 1800°K with Na$^+$, K$^+$, and Rb$^+$ ions appears in Fig. 5.26. We see from the figure that the dependence of E and E_{max} on the initial energy of the primary ions is analogous to the case of Na$^+$ on tungsten, and in the range of energies studied each relationship falls neatly on a straight line.

Figures 5.27 and 5.28 show oscillograms of the angular dependence of the energy distribution of secondary ions on bombarding a tantalum target heated to T = 1800°K with 600- and 800-eV Rb$^+$ and Cs$^+$ ions, respectively. We see from the oscillograms that the character of the change in secondary-particle distribution with changing angle of incidence is, in general features,

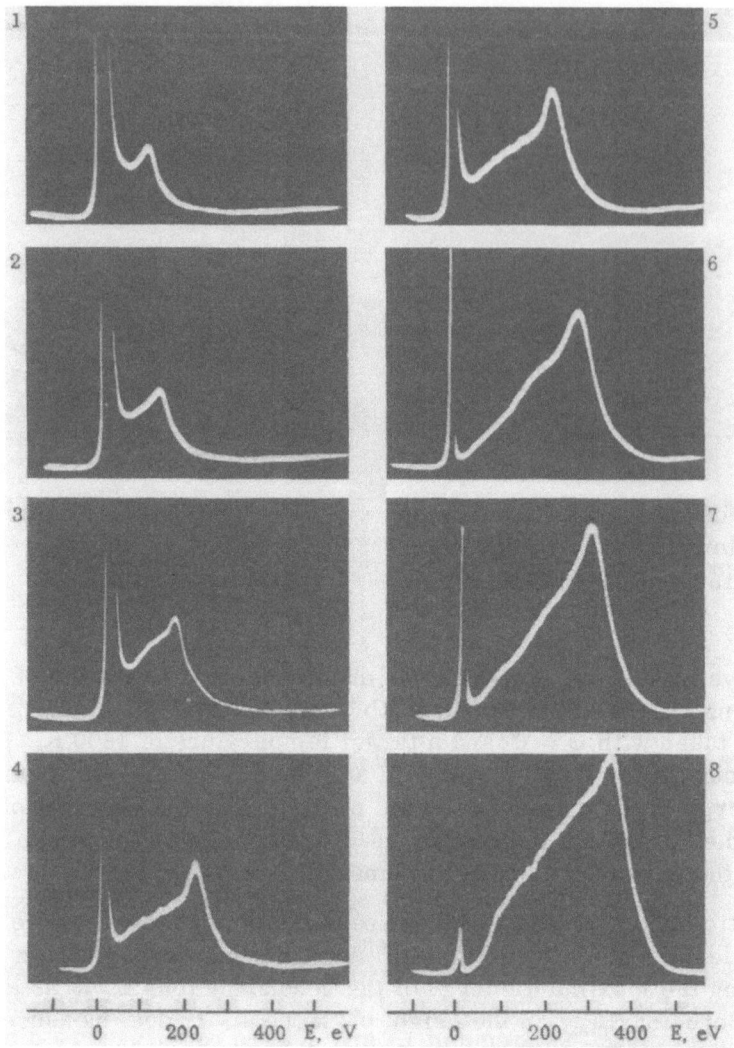

Fig. 5.27. Spectra of secondary Rb$^+$ ions on Ta as a function of the angles of incidence φ equal to: 1) 0°; 2) 10; 3) 20; 4) 30; 5) 40; 6) 50; 7) 60; 8) 70°; escape angle θ = 50°.

Table 5.6

	Angle of incidence φ, deg								
	0	10	20	30	40	50	60	70	80
η_t	0.19	0.21	0.25	0.29	0.36	0.44	0.51	0.60	0.69
η_e	0.17	0.20	0.22	0.27	0.33	0.42	0.50	0.57	0.66
η_{max}	0.35	0.37	0.40	0.44	0.49	0.56	0.63	0.70	0.79

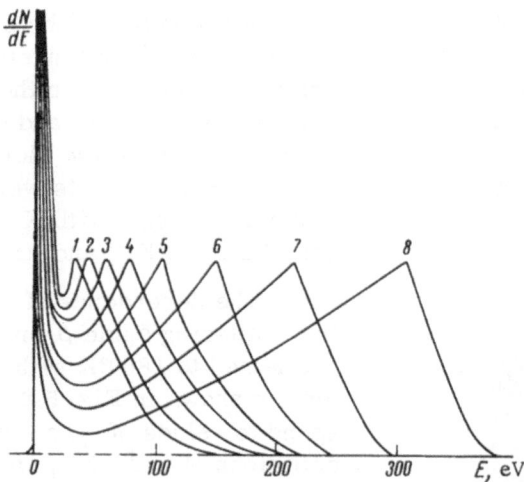

Fig. 5.28. Spectra of secondary Cs$^+$ ions on Ta as a function of the angles of incidence φ equal to: 1) 0°; 2) 10; 3) 20; 4) 30; 5) 40; 6) 50; 7) 60; 8) 70°; escape angle $\theta = 50°$.

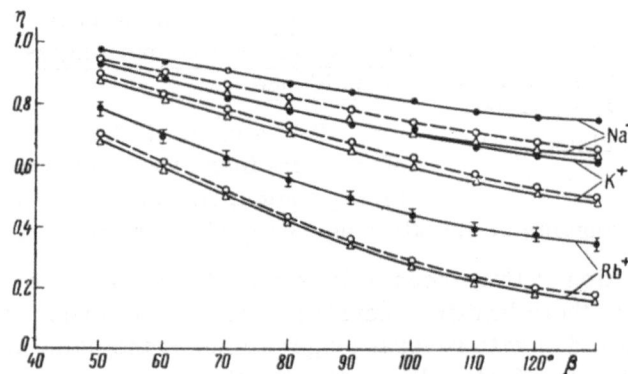

Fig. 5.29. Dependence of the values of η_t, η_e, and η_{max} on scattering angle β.

analogous to the case of Na$^+$ on tungsten (see Fig. 5.22). In this case, in view of the similarity between the masses of the colliding particles, the angular dependence of the position of the single-collision peak is more sharply expressed than in the case of Na$^+$ on tungsten. For the same reason there is a greater difference between the energy of single collisions E and the maximum secondary-ion energy E$_{max}$.

Table 5.6 shows the values of η_t, η_e, and η_{max} averaged over many measurements on bombarding a tantalum target heated to 1800°K with 1000-eV Rb$^+$ ions. From the table we see the good agreement of the η_t and η_e values and the considerable discrepancy between these and the values of η_{max}. With increasing angle of incidence the difference between η_e and η_{max} becomes a little smaller. This occurs, as we showed earlier, as a result of the reduction in the probability of multiple collisions with the most favorable angles. It follows from the oscillograms of Figs. 5.27 and 5.28 that with increasing angle of incidence the relative number of ions leaving the surface with high energies increases, indicating a reduction in the depth of penetration of primary ions with increasing incident angle.

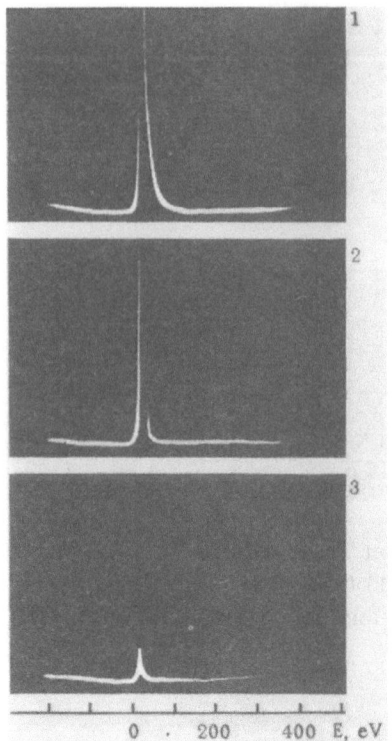

Fig. 5.30. Oscillograms of secondary ion spectra. Targets: 1,2) cold-and-dirty and hot; 3) cold and clean.

A reduction in the slow-ion peak (analogous to the case of Na^+ on tungsten) with increasing incident angle φ also occurs in the case of a tantalum target bombarded with Rb^+ and Cs^+ ions. However, no complete vanishing of the slow-ion peak occurred in this case. Similar results were obtained on bombarding a tantalum target with K^+ ions and a molybdenum target with Na^+, K^+, and Rb^+.

The dependence of the values of η_e and η_{max} on the scattering angle β on bombarding a tantalum target heated to 1800°K with 1200-eV Na^+, K^+, and Rb^+ ions is shown in Fig. 5.29. The broken curves correspond to values of η_t calculated for the case of single collisions of Na^+, K^+, and Rb^+ ions with individual tantalum atoms, using formula (1.9). We see that the $\eta_e(\beta)$ curves obtained for all the ions coincide with the $\eta_t(\beta)$ curves, while $\eta_{max}(\beta)$ always lies higher than the curves of $\eta_e(\beta)$ and $\eta_t(\beta)$ and slightly approaches the $\eta_e(\beta)$ curves with diminishing scattering angle β.

Bombardment of Light Targets with Heavy Ions at Various Angles of Incidence

It was shown in the foregoing that, in the case of $m_1 < m_2$ there was a certain emission at back-scattering angles; in contrast to the case of single scattering, these ions had an angular distribution agreeing with the cosine law.

The energy distribution of this group of ions was studied by Petrov [504] and Arifov et al. [704], and it was found that their limiting energies were high; in order to explain this, one had to assume that the ions collided with several target atoms at the same time or that there was a considerable binding-energy effect. These investigations, however, were based on the retarding-field method, the accuracy of which was of course inadequate.

In order to obtain more accurate results, we used the method described in §2 of this chapter under the same experimental conditions; the energy distribution of the secondary ions on bombarding a nickel target with Rb^+ ions and a molybdenum target with Cs^+ ions at various angles of incidence was studied by Arifov, Aliev, and Ayukhanov [717].

Figure 5.30 shows three oscillograms of the energy distribution of secondary ions obtained on bombarding a molybdenum target with 700-eV Cs^+ ions. The primary ions fall on the target normally. The secondary ions are energy analyzed in the $\theta = 50°$ direction. Oscillogram 1 corresponds to the ion bombardment of a contaminated surface. It is well known that in this case the secondary emission consists mainly of particles arising as a result of the cathode sputtering of the surface film in the form of positive ions. Oscillogram 2 corresponds to the bombardment of a target heated to 1800°K. In this case the target is quite clean and the emission consists mainly of ions evaporated from the molybdenum surface. Oscillogram 3 is obtained on bombarding a cold, clean target. We see that in this case there is a slight emission of positive ions of low energy; this is neither cathode sputtering nor evaporation.

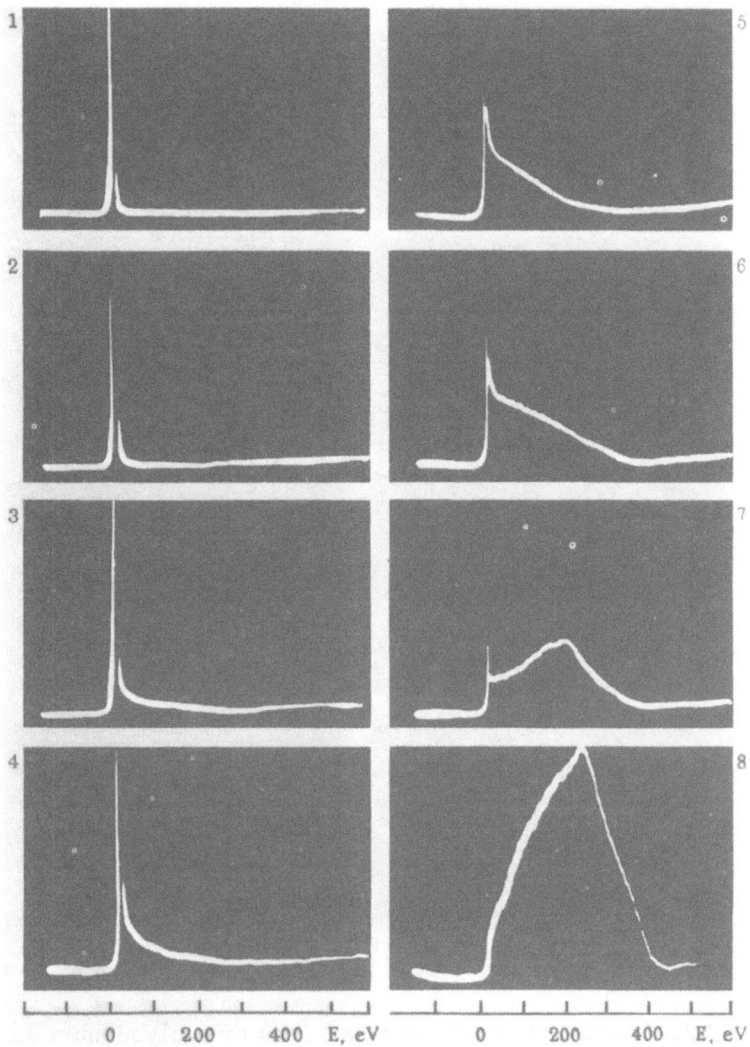

Fig. 5.31. Spectra of secondary Rb$^+$ ions on nickel as a function
of incident angle φ equal to: 1) 10°; 2) 20; 3) 30; 4) 40; 5) 50; 6)
60; 7) 70; 8) 80°; escape angle $\theta = 50°$.

Figures 5.31 and 5.32 show oscillograms of the energy distribution of secondary ions obtained at various incident angles of the primary ions (900 eV) on the heated surface of the target; these are respectively, for Rb$^+$ on nickel (T = 1300°K) and Cs$^+$ on molybdenum (T = 1800°K). Oscillogram 1 (Figs. 5.31 and 5.32) is obtained for $\varphi = 10°$ and the rest with successive 10° increments of incident angle. The escape angle is 50° in all cases.

We see from the oscillograms that for $\varphi < 30°$ the energy distribution of secondary ions remains almost constant, indicating that, as in the case of $\varphi = 0$, only evaporated ions occur. Starting from incident angles of 30° (i.e., for $\beta = 100°$) secondary ions faster than evaporated and sputtered ions appear. On further increasing φ the energy spectrum of this group of ions expands, their number increases, and, finally, for $\varphi = 60°$, $\beta = 70°$ we obtain secondary ions with high energies. On further increasing φ (oscillograms 7 and 8, Figs. 5.31 and 5.32), this group of fast ions becomes dominant and the distribution curve develops a maximum for this group of ions, its peak corresponding to the energy of ions which have suffered single elastic collisions at the limiting angle.

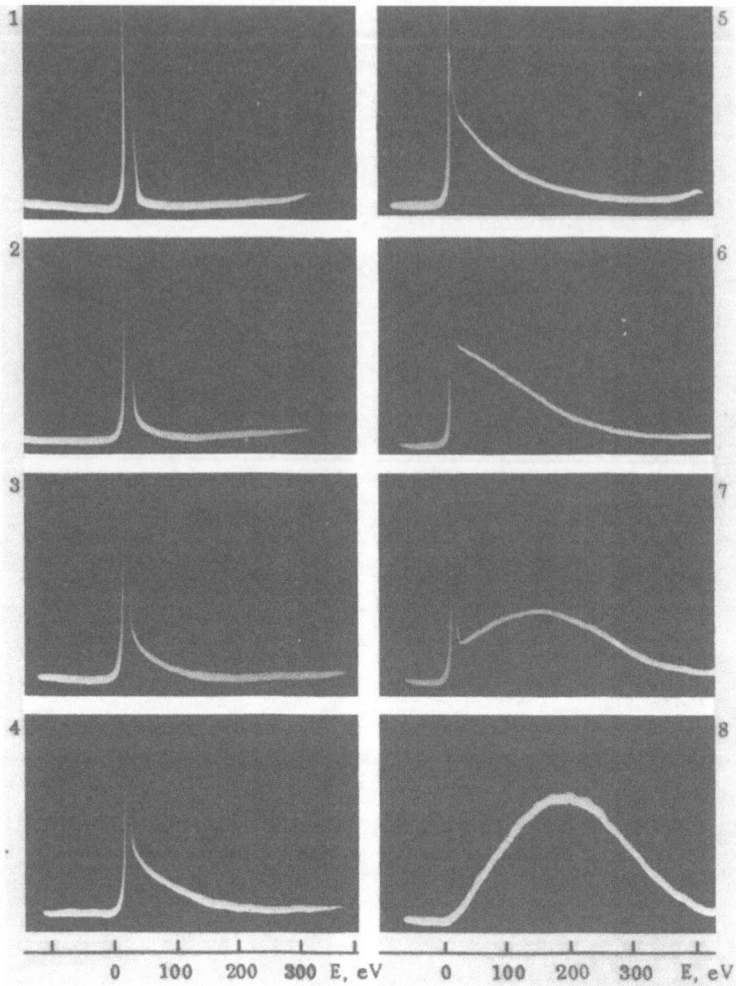

Fig. 5.32. Spectra of secondary Cs$^+$ ions on molybdenum as a function of the angle of incidence φ equal to: 1) 10; 2) 20; 3) 30; 4) 40; 5) 50; 6) 60; 7) 70; 8) 80°; escape angle $\theta = 50°$.

The Cs$^+ \rightarrow$ Mo oscillograms not shown here indicate that, with increasing primary-ion energy, the maximum on the distribution curves moves in the direction of higher energies, in accordance with the relation for elastic single collisions. Analogous pictures are obtained on bombarding nickel with Rb$^+$ ions.

Thus, consideration of the angular dependence of the energy distribution of the secondary ions shows that, both in the case of $m_1 > m_2$ [691] and in the case of $m_1 < m_2$ [717], the secondary emission includes ions which have undergone single collisions with free target atoms. For $m_1 < m_2$, these are only observed within a certain limiting scattering angle. In all cases, the secondary ions which have undergone single collisions have the form of a peak in a more or less narrow spectral range on the secondary-ion energy distribution curves. However, the angular dependence of the energy spectra for $m_1 < m_2$ indicates the scattering of ions through angles considerably exceeding the limiting angles of single scattering. The form of the energy distribution curve of this group of ions and the manner in which it varies with the angle of incidence of the primary ions in no way contradicts the principle that these originate as a result of multiple collisions.

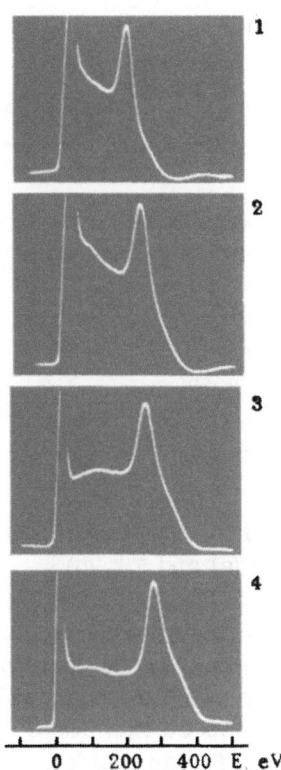

Fig. 5.33. Spectra of secondary K^+ ions on Mo as a function of the escape angle θ equal to: 1) 30; 2) 50; 3) 60; 4) 70°; incident angle $\varphi = 0°$.

Hence, on bombarding metals with ions in the energy range 1-3 keV, the scattering phenomenon is satisfactorily explained on the basis of pair collisions, single and multiple. The appearance of secondary ions possessing energies differing from that of the ions which have suffered single collisions and also scattering at angles greater than the limiting scattering angles, are thus explained in a consistent way.

Angular Dependence of the Energy Spectra of Secondary Ions at Various Escape Angles

In the foregoing sections we have described results obtained by analyzing the energy spectra of secondary ions propagating at a specific escape angle θ as a function of the angle of incidence φ of the primary ions. By varying the position of the target with respect to the analyzer in the apparatus (Fig. 5.9) we could also study the relation between the energy spectra of the secondary ions and the angle of incidence of the primary ions for various values of escape angle [754, 777].

Figure 5.33 presents some oscillograms of the energy distribution of secondary ions propagating at various escape angles on bombarding a molybdenum target with 1000-eV K^+ ions. All the oscillograms are taken at the same angle $\varphi = 0°$ and the same target temperature T = 1800°K. For convenience of comparison, the maxima of all the oscillograms in the figure are normalized to the same height.

Oscillograms 1-4 (Fig. 5.33) are obtained for θ = 30, 50, 60, and 70°, respectively. We see from the oscillograms that the peak corresponding to single collisions moves in the higher-energy direction with increasing escape angle, i.e., the energy of singly scattered ions rises.

The dependence of the secondary-ion energy spectra on the incident angle for 1000-eV K^+ ions bombarding a molybdenum target heated to 1800°K is shown in Fig. 5.34 for θ = 30° and 70°. Oscillogram 1 of Fig. 5.34 is obtained for normal incidence of the beam on the target. Each succeeding oscillogram corresponds to a 10° increment in incident angle. We see that the angular dependence of the secondary-ion energy spectra taken for different escape angles clearly indicates that the scattering angle plays a decisive part. As the relation between the escape and incident angles varies, with constant β, there is only a slight change in the slow-ion peak. With increasing angle of incidence of the primary ions the slow-ion peak becomes smaller. For incident angles of $\varphi \geq 80°$ the slow-ion peak vanishes for all values of escape angle; this indicates that, in this case, there is no appreciable penetration of the primary ions into the target.

A slightly different picture is obtained for $m_1 < m_2$. Figure 5.35 shows four oscillograms of the secondary-ion energy distributions obtained on bombarding a molybdenum target heated to 1800°K with 1000-eV Cs^+ ions. The primary ions fall on the target at an angle of $\varphi = 30°$; the secondary ions with escape angles θ between 30 and 70° are energy analyzed. We see that, for escape angles of under 50°, there is only a group of slow ions. Starting from an escape angle of $\theta \geq 60°$, faster ions appear. The variation in the energy distribution of this group of ions with increasing secondary-ion escape angle is analogous to the variation in the distribution of this group of secondary ions with increasing incident angle.

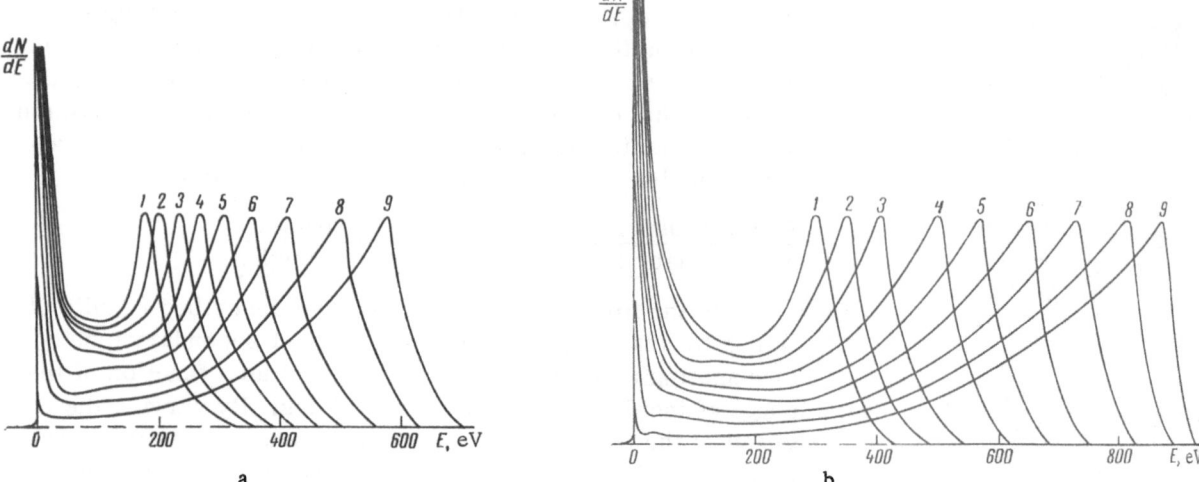

Fig. 5.34. Secondary ion spectra as a function of incident angles φ equal to: 1) 0; 2) 10; 3) 20; 4) 30; 5) 40; 6) 50; 7) 60; 8) 70; 9) 80° for escape angles of $\theta = 30$ (a) and 70° (b).

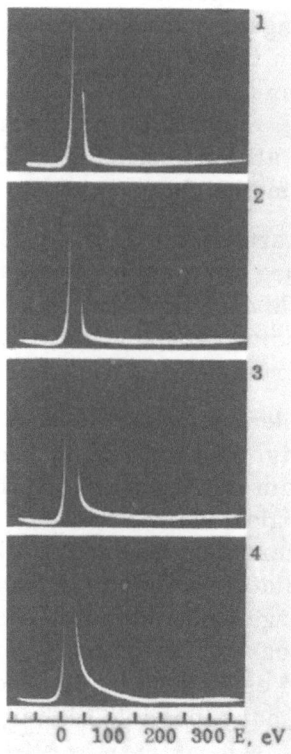

Fig. 5.35. Secondary ion spectra in the case of Cs^+ on Mo, corresponding to escape angles of θ equal to: 1) 30; 2) 50; 3) 60; 4) 70° for an incident angle of $\varphi = 30°$.

Figure 5.36 shows a series of oscillograms of the angular dependence of the secondary-ion energy spectra obtained for various escape angles on bombarding a molybdenum target heated to 1800°K with 1000-eV Cs^+ ions. Oscillogram 1 (Fig. 5.36b) is obtained for an incident angle of $\varphi = 0°$ and the others for successive incident-angle increments of 10°. The escape angle in Fig. 5.36a is $\theta = 30°$ and that in Fig. 5.36b is $\theta = 70°$.

In the case of $\theta = 30°$, for incident angles of under 50°, the secondary-ion energy distribution remains unchanged. As in the case of normal incidence of the primary ions, only evaporated and slow ions with an energy of no greater than 25 eV are observed. Starting from incident angles of 50° (i.e., scattering angles of 100°) we observe secondary ions faster than those of the evaporated and sputtered types.

For $\theta = 70°$, secondary ions faster than evaporated and sputtered ions are observed, starting from normal incidence of the primary ions. With increasing angle of incidence the energy spectra of this group of ions broadens and their number increases, and for angles of incidence of 40° (scattering angle $\beta = 80°$) secondary ions with high energies appear. On further increasing the angles of incidence (oscillograms 6–8, Fig. 5.36), the fast-ion group predominates, and the distribution curves develop a peak, its maximum corresponding to the energy of ions which have suffered single collisions at the limiting angle.

The absence of the peak corresponding to singly scattered Cs^+ ions on molybdenum for an incident angle

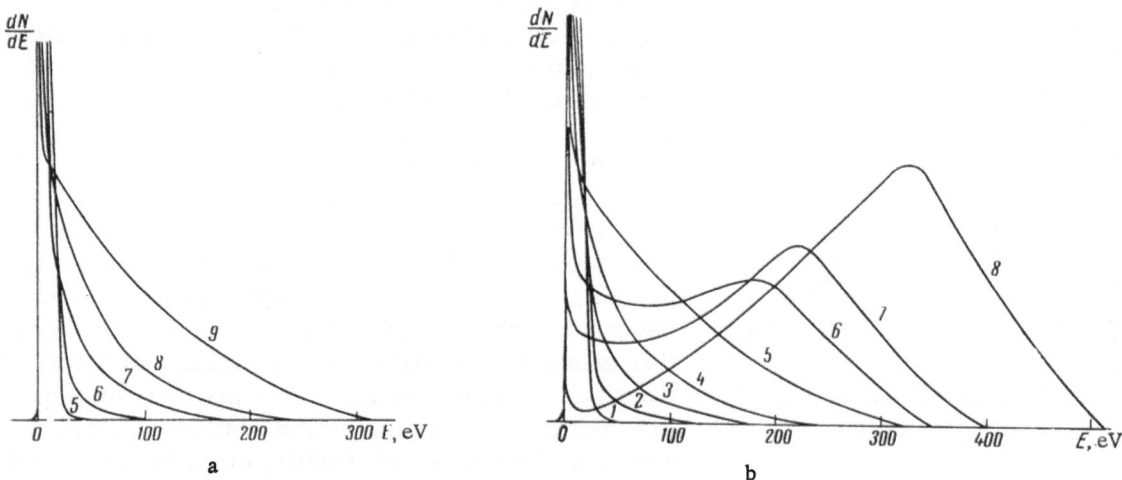

Fig. 5.36. Spectra of secondary Cs$^+$ ions on Mo as a function of incident angle φ equal to: 1) 0; 2) 10; 3) 20; 4) 30; 5) 40; 6) 50; 7) 60; 8) 70; 9) 80° for escape angles of $\theta = 30$ (a) and 70° (b).

of $\varphi = 80°$ in the case of $\theta = 30°$, and the presence of such a peak for an incident angle of $\varphi = 50°$ in the case of $\theta = 70°$ indicate that the limiting angle of the singly scattered ions is determined by the scattering angle for $m_1 < m_2$.

Figure 5.37 shows the dependence of η_{max} and η_e on the scattering angle of the secondary ions on bombarding a molybdenum target heated to 1800°K with Cs$^+$ ions having an initial energy of $E_0 = 1000$ eV. The broken curve corresponds to the value of η_t calculated from formula (5.1) for the case of single collisions of Cs$^+$ ions with individual molybdenum atoms. We see from the figure that the curve of $\eta_e(\beta)$ coincides with the broken $\eta_t(\beta)$ curve, thus showing that, inside the limiting angle, the angular dependence of the energy of singly scattered ions in the case of $m_1 < m_2$ is analogous to that corresponding to the case in which the mass of the bombarding ion is smaller than that of the target atom. However, secondary ions with energies greater than the evaporated and sputtered ions occur in the spectrum of Cs$^+$ on molybdenum starting from scattering angles of ~110°, i.e., from angles greatly exceeding the limiting angle of single scattering.

Analogous results were also obtained on bombarding a nickel target with Rb$^+$ and Cs$^+$ ions.

Consideration of the angular distribution of secondary ions shows that the results obtained may be excellently explained from the point of view of pair collisions between the bombarding ions and free atoms. The rise in the scattering coefficient with increasing angle of incidence clearly indicates that the collisions take place at a certain depth in the target. With increasing angle of incidence, over a large proportion of its range the primary ion runs only slightly below the surface and thus has a high probability of emerging from the solid.

The observation of secondary ions at angles greater than the limiting scattering angle in the case in which the mass of the ion is greater than that of the target atom contradicts the foregoing point of view if we start from the assumption of single pair collisions. The rapid intensification of scattering in directions close to the angles indicates that the view based on paired multiple collisions with a system of free atoms is certainly more accurate; it also indicates the need for further experimental data in order to explain the appearance of secondary ions at angles greater than the limiting values.

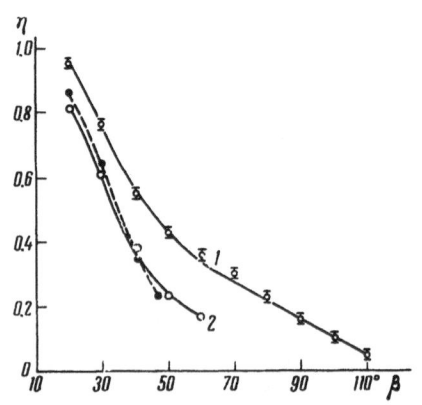

Fig. 5.37. Dependence of η_{max} (1) and η_e (2) on β.

It is clear that the scattering of primary ions takes place as a result of multiple as well as single collisions. The presence of secondary ions with energies smaller than those of ions suffering single collisions has, for a long time, been explained as being due to multiple collisions. Multiple changes in the direction of the primary ion inside the solid (not exceeding the value of the limiting angle on each occasion) may also be used to explain the appearance of secondary ions at angles greater than the limiting scattering angles. The validity of this point of view is supported by the results of investigations into the angular dependence of the energy spectra. The energy spectrum of this group of secondary ions at angles a long way from the limiting scattering angles shows that they only possess low energies. On reducing the scattering angle, the energy spectrum broadens and, finally, close to the limiting scattering angle, we obtain a group of ions which have undergone single collisions with a deviation through the limiting angle.

In view of the observation of a certain number of secondary ions with energies greater than the energy of single collisions, Petrov [583] and Veksler [709] have expressed the view that a certain influence is exerted by the energy of the interatomic bond or that ions collide simultaneously with several target atoms. Consideration of the energy spectra in the case in which the mass of the target atom is greater than that of the bombarding ion shows that this result may also be explained from the point of view of pair collisions with a system of free atoms. In this case, as we saw, the secondary ions have a very wide spectrum, showing that, in the course of scattering, the ions may leave the surface with energies between zero and a value even somewhat exceeding the energy of ions which have suffered a single collision with a target atom. Singly scattered ions form the maximum. The ions which have experienced multiple collisions lie on both sides of this maximum. Parilis and Turaev [747] showed that an ion scattered through a particular angle as a result of multiple successive collisions with scattering through small angles will possess a larger energy than an ion scattered through the same angle as a result of a single collision. In fact, the energy E retained by the ion after scattering through an angle β' as a result of, for example, two successive elastic collisions with atoms of the solid equals

$$E_{\beta'} = \frac{E_0 (1 - \mu)^4}{(\cos\beta_1 - \sqrt{\mu^2 - \sin^2\beta_1})^2 (\cos\beta_2 - \sqrt{\mu^2 - \sin^2\beta_2})^2} , \qquad (5.3)$$

where $\mu = m_1/m_2$, β_1 and β_2 are the scattering angles in the first and second collisions, respectively. Calculations show that $E_{\beta'} > E_\beta$ if $\beta' = \beta_1 + \beta_2$, $\beta' = \beta$. The energetically most favorable case (although also the least probable) when considering n collisions is the case $\beta_1 = \beta_2 = \beta_3 = \ldots = \beta_n = \beta/n$. As shown earlier, the probability of multiple collisions with the most favorable angles, leading to escape into the vacuum, does have an appreciable value.

Thus, by using the screened Coulomb potential for the interaction of atomic particles and the simple model of elastic pair collisions (single and multiple) we may explain all the main features in the scattering of ions (atoms) at the surface of a solid polycrystalline body over the energy range studied (0.5-3 keV) [754, 777].

For primary-ion energies in the region of 500-3000 eV we found no serious effect arising from the coupling between the target atoms although such an effect would be expected for extremely low energies of the primary ions.

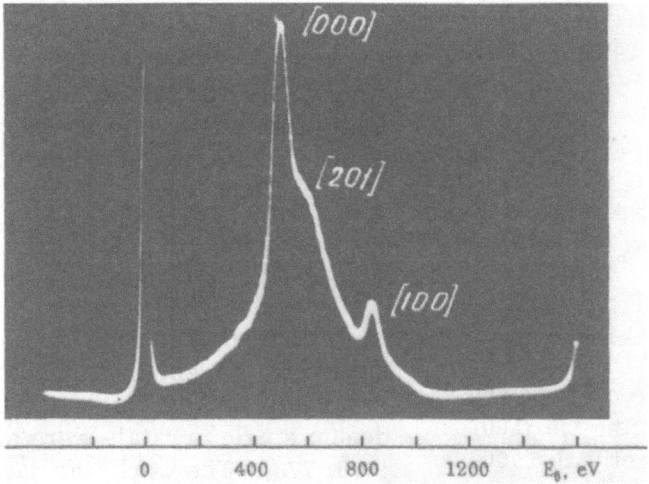

Fig. 5.38. Oscillogram of the secondary-ion spectrum on bombarding the (100) face of tungsten with
1200-eV Rb$^+$ ions.

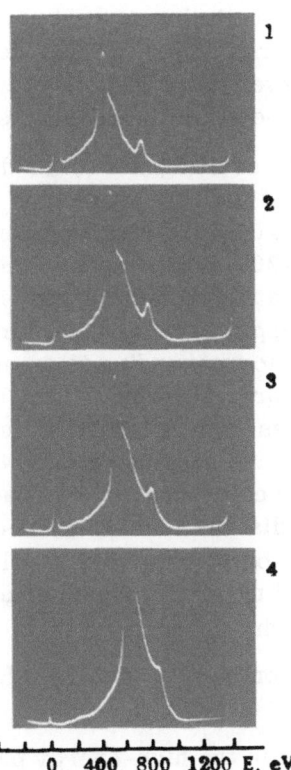

Fig. 5.39. Spectra of secondary
Rb$^+$ ions on the (100) face of tungsten for scattering angles of β
equal to: 1) 90; 2) 80; 3) 70; 4) 60°.

§6. Angular Laws Governing the Scattering of Ions by the Surface of a Single Crystal

From the considerations developed in the last section regarding paired single and multiple collisions between ions and target atoms, it follows that not only may
the angular distribution be anisotropic, but the secondary-
ion energy distribution obtained on bombarding single-
crystal targets with various ions may have a characteristic structure. As indicated in §1 of this chapter, in
certain recent investigations [692, 697, 751] anisotropy
was observed in the scattering of ions by the surface of
a single crystal; the scattering depended on the particular crystallographic direction. Parilis and Turaev [747]
predicted that not only anisotropy of the angular distribution, but also a characteristic structure of the energy
distribution of ions scattered by a single crystal would
be expected. The structure of the energy spectrum of
scattered ions experiencing single or double collisions
with atoms in the surface layers of a particular face of
a single crystal of cubic symmetry was also calculated
by Parilis and Turaev [699]. It was shown, furthermore,
that the probability of a double scattering of the ions was
by no means small, so that for a reasonable resolving
power of the analyzer it should be possible to observe
these experimentally in the form of peaks in the low- and
high-energy parts of the spectrum.

The energy spectrum of secondary Ar$^+$ ions
scattered from the surface of the (100) face of a copper single crystal was studied by Mashkova and Mol-

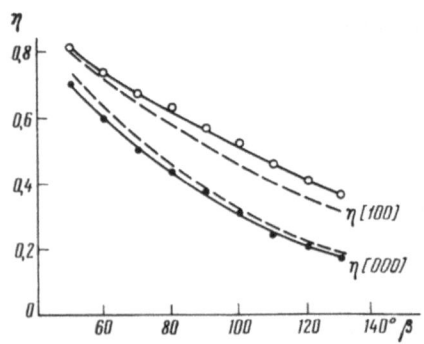

Fig. 5.40. Dependence of $\eta_{[100]}$ and $\eta_{[000]}$ on β on bombarding the (100) face of a tungsten target with 1500-eV Rb$^+$ ions. The broken curves characterize the angular dependence of $\eta_{t[100]}$ and $\eta_{t[000]}$ calculated from the formulas for elastic single and double scattering.

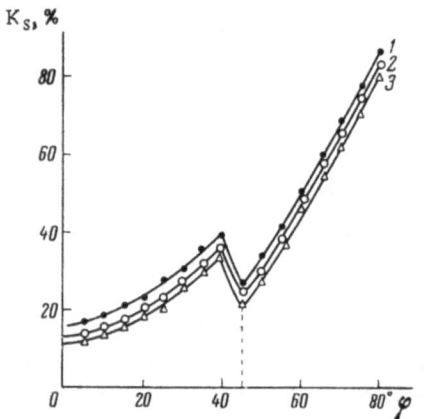

Fig. 5.41. Angular dependence of the scattered-ion coefficient K_s on bombarding the (100) face of a tungsten target with Rb$^+$ ions at energies of $E_0 = 1, 2,$ and 3 keV (figures on curves).

chanov [723]. A peak was found on the falling part of the peak of the singly scattered Ar$^+$ ions in the direction of high spectral energies; this was explained as being due to the multiple scattering of Ar$^+$ ions by copper atoms. It was shown that the structure calculated by Parilis and Turaev [699], of the energy spectrum of the scattered Ar$^+$ ions experiencing single and double collisions with atoms of the (100) copper face, agreed with the experimental curves [749].

The angular dependence of the coefficient of secondary ion emission and the energy distribution of secondary alkali ions scattered by tungsten and molybdenum single crystals were studied by Arifov et al. [770, 778]. The work was carried out with the apparatus described in §2 of this chapter. However, in order to increase the resolving power of the experimental apparatus, the following improvements were made. The width of the entrance and exit slits of the Hughes-Rojansky electrostatic analyzer were changed to 0.4 mm so as to raise the energy resolving power of the analyzer to 1%. The ion-electron multiplier (with an amplification factor of ~10^6) at the analyzer output was replaced by a multiplier with an amplification factor of ~10^8. After preliminary treatment, one particular face of a tungsten single crystal, taken in the form of a molding, served as target. This was cleaned by electron-beam heating from the far side.

Figure 5.38 shows an oscillogram of the differential energy distribution of the secondary ions obtained on bombarding the (100) face of a tungsten crystal heated to 2000°K with 1200-eV Rb$^+$ ions. The primary ions fell on the target at an incident angle of $\varphi = 50°$; energy analysis was applied to the secondary ions propagating in a direction corresponding to an escape angle of $\theta = 50°$ with the normal to the surface. The orientation of the target was such that the incident and scattered beams lay in a plane passing through the [100] axis of the crystal. We see that, in this case, the oscillogram of the energy distribution of the secondary ions has a slightly different form when compared with that obtained on bombarding a polycrystalline target (see Chapter 5, §5). In addition to the peak corresponding to ions singly scattered from individual target atoms, the spectrum shows two peaks in the high-energy part of the spectrum.

Oscillograms of an analogous character are also obtained on bombarding the (100) face of tungsten with K$^+$ and Cs$^+$ ions [244].

Calculations analogous to those of [699] show that the end peak in the high-energy part of the spectrum corresponds to ions repeatedly scattered at an atom in the [100] direction after a first scattering at an [000] atom, while the peak close to the peak of the singly scattered ions corresponds to ions repeatedly scattered at an atom in the [201] direction after first colliding with an [000] atom (Fig. 5.38). The index [000] on the oscillogram corresponds to a single

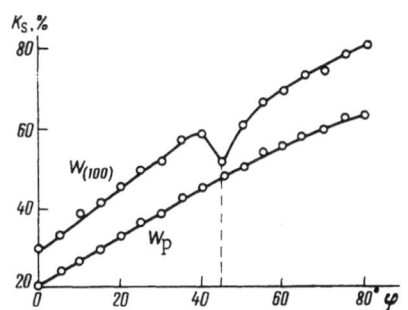

Fig. 5.42. Relationship $K_s(\varphi)$ for a polycrystalline target and for the (100) face of a tungsten single crystal.

collision, since we have arbitrarily taken an atom on the (100) face at which the first collision takes place as origin of coordinates; the remaining indices denote the atom with which the second collision took place after the first collision with an [000] atom. The energy retained by the ion in single and double collisions at an angle β is determined by relations (1.9) and (5.3), respectively.

Figure 5.39 shows oscillograms of the secondary-ion energy distribution obtained on bombarding the (100) face of tungsten with 1200-eV Rb$^+$ ions. Oscillogram 1 is obtained for a scattering angle of $\beta = 90°$ and oscillograms 2-4 for successive 10° reductions in scattering angle. We see that the peaks corresponding to double collisions move in the high-energy direction with decreasing scattering angle, in the same way as the peak of singly scattered ions. As β falls the peaks of the doubly scattered ions predominate and become quite strong, indicating an increase in the number of ions striking atoms in the [100] and [201] directions after first colliding with an [000] atom.

Figure 5.40 shows the relationships $\eta_t(\beta)$ obtained from (1.9) and (5.3) for the case of singly and doubly scattered ions. The same figure shows the experimental values of the [000] and [100] peak energies. We see that $\eta_{t[000]}$ and $\eta_{e[000]}$, and also $\eta_{t[100]}$ and $\eta_{e[100]}$ agree quite closely.

Figure 5.41 shows the curves of $K_s(\varphi)$ obtained on bombarding the surface of the (100) face of a tungsten single crystal, heated to 1500°K, with 1-, 2-, and 3-keV Rb$^+$ ions, respectively. An analogous result (Fig. 5.42) is obtained for 1.5-keV K$^+$ ions. For comparison the figure also shows the curve of W_p for a polycrystalline target. We see from Figs. 5.41 and 5.42 that the general run of the curves is the same for both cases, except that there is a minimum at $\varphi = 45°$ for the single crystal; this evidently corresponds to the crystallographic direction [110]. Thus, the minimum ion scattering occurs when the direction of incidence of the ion beam coincides with the principal crystallographic directions [100] and [110] of the crystal. In these directions the lower-lying atoms are screened by the surface atoms of the crystal; the existence of channels with various cross sections is associated with this. Some of the primary ions pass into these channels and this leads to a reduction in the value of K.

In working with polycrystalline targets one has to consider greater or smaller densities of the material, and in working with a single crystal there are crystallographic directions with different transparencies. The different transparencies for different orientations of the crystal to the incident ion beam lead to anisotropy of the ion-scattering coefficient.

As regards the rise in K_s with increasing φ, this is associated with the reduction in the depth of extraction of the secondary ions and also with the reduction in the number of ions penetrating deeply into the metal.

CHAPTER 6

SECONDARY PROCESSES UNDER THE INFLUENCE OF BOMBARDMENT BY INERT-GAS IONS

§1. Brief Review of Investigations into Secondary Emission under the Influence of Gas Ions

Electron emission under the influence of gas ions (under gas-discharge conditions) was first studied by Penning [50] in 1928. Penning found that in the case of bombarding a copper target with positive Ne^+ ions at $E_0 \sim 100$ eV the value of $\gamma = I^-/I^+$ equalled ~3%. However, the inadequate precision of experiments on ion—electron emission in a gas discharge (inhomogeneity of the ion composition and the large energy spread, complex conditions on the cathode surface, back diffusion of secondary electrons into the target) impeded any explanation of the results obtained. For this reason all subsequent investigations were carried out with independent ion sources by vacuum-separating the source and target spaces with special diaphragms, using differential pumping [257].

Penning also studied ion—electron emission with an independent ion source [60]; he studied secondary phenomena on the surface of a previously degassed molybdenum target bombarded with He^+, Ne^+, and Ar^+ ions at E_0 values between a few tens of eV and 1000 eV, finding that in this range of energies γ varied linearly from 3-8% for Ar^+, 7-17.5% for Ne^+, and 8.5-21% for He^+. For all the ions K was small, indicating the complete neutralization of these ions on the molybdenum surface. Penning explained the rise in γ on passing from an ion with a low ionization potential V_i to one with a large V_i as being due to the extraction of electrons by virtue of the energy of neutralization of the ion.

Oliphant [61], in addition to studying the dependence of γ on E_0, was the first to study the secondary-electron energy distribution on bombarding a molybdenum target with 80- to 1000-eV He^+ ions. The construction of the apparatus made it possible to bombard the target with ions separated from neutral particles in the field of a plane condenser inclined at 30° to the direction of the beam. The experiments were made with a residual gas pressure of $\sim 10^{-5}$ mm Hg. Measurements were made with cold targets and targets heated to 1100°K. It was found that the values of γ were greater for a cold target than for a hot one. The author ascribed this difference to the effect of molecules adsorbed from the residual gas on the value of γ. Measurements of the secondary-electron energy distribution showed that the contamination of the target surface increased the number of slow and reduced the number of fast electrons. It follows from the results obtained on studying the energy distribution of the secondary electrons for a heated target that the main contribution to the secondary electrons comes from slow electrons with almost zero velocities, while very few fast electrons occur. The limiting energies of the secondary electrons were estimated as 21 eV for $E_0 = 80$ eV and 25 eV for $E_0 = 1000$ eV. The author also explained the existence of a component of emission independent of E_0 up to 800 eV as being due to the emission of electrons by virtue of the energy of neutralization of the ions. In addition to this, the dependence of γ on the beam incident angle (φ) was studied; it was found that $\gamma \sim (a - b \cos \varphi)$, where a and b are constants.

Brasefield [80] made comparative studies of the secondary electron emission of nickel bombarded with argon ions and neutral atoms. In these experiments the vacuum was $\sim 10^{-5}$ mm Hg and the target was not specially cleaned; instead of this, it was covered with soot to prevent the reflection of the primary ions. Neutral atoms were obtained by charge exchange between the accelerated ions and gas atoms in a special chamber. The intensity of the beam of neutral atoms was determined, in accordance with Wolf [73], from the relation between the charge—exchange cross section and the energy of the ions. Under these conditions the author found that γ rose from 3.5 to 6.5% on raising the energy of the neutral atoms from 30 to 180 eV; in the case of ions, however, γ fell with increasing E_0. The author concluded from this that the principal part in maintaining the cold discharge was played by the neutral atoms, owing to their greater efficiency in extracting electrons from the metal.

Rostagni [96] studied the secondary electron emission from copper and brass on bombarding with ions and neutral atoms of He, Ne, and Ar; he found that for ions γ was 100 times greater than for neutral atoms in the energy range 6-600 eV, and showed that the values of γ corresponding to the minimal values of E_0 were almost proportional to the ionization energies of the ions. As we see, Rostagni's results contradict those of [80], according to which γ was greater for neutral atoms than for ions.

Healea and Houtermans [147] studied secondary electron and ion emission on bombarding targets with He^+, Ne^+, and Ar^+ ions in the energy range up to 1400 eV. The measurements were made in a vacuum of $\sim 10^{-5}$ mm Hg. In contrast to the authors of [60, 61, 96], Healea and Houtermans explained the increase in γ on passing from Ar^+ to He^+ with a given E_0 as being due to the effect of the mass of the ions on the coefficient γ; in their opinion, a light ion produced greater emission than a heavy one owing to its greater penetration into the metal. The authors therefore concluded that there was no particular dependence of the γ coefficient on the ionization potential of the incident ions. We should note that in this investigation a considerable ionic emission, almost comparable in value with the electron emission, was observed.

Healea and Chaffee [137, 109] studied electron emission from nickel on bombarding with H_2^+ and D_2^+ ions at energies up to 1400 eV. Measurements of the $\gamma(E_0)$ relationship showed that γ increased from 13.1 to 44.7% for H_2^+ and from 12.2 to 35.7% for D_2^+ on bringing E_0 up to 1400 eV. The authors emphasize the difference in the behavior of the resultant $\gamma(E_0)$ curves from those obtained by Hill et al. [135]. Considering that in such cases nickel is always saturated with hydrogen, the authors explained their results as being due to the presence of contaminations on the surface. In addition to this, the absence of any analysis of the H^+ ions with respect to e/m prevents us from comparing the results of these authors for H_2^+ and D_2^+, since atomic ions may also occur in the beams of these molecular ions.

The secondary ion—electron emission of various metallic targets (Be, Mo, Cu, Al, Pb, Au, Ni, Pt, Mg, Fe) on bombarding with light ions (H^+, H_2^+, D^+, He^+, etc.) at high energies (tens and hundreds of keV) has been studied by a number of authors [69, 135, 138, 534, 587, 625, 673]. The results showed that γ depended only slightly on the target material. With increasing energy there was a reduction in the emission of electrons, which was explained by the comparatively extensive penetration of the faster ions into the target, leading to a reduction in the probability of the extraction of excited electrons.

The dependence of γ on the angle of incidence of the beam φ was studied by a number of authors [108, 288, 534, 676]. It was found that $\gamma \sim \sec \varphi$. In order to explain this relationship, we must suppose that the secondary electrons arise in a thin surface layer, so that as φ increases the formation of secondary electrons takes place closer to the surface and thus the probability of their extraction becomes greater. In other words, the depth of the effective generation of secondary electrons (d_{ie}) is much smaller than the depth of penetration of the bombarding ions (d_p). On these lines we may also explain the absence of any substantial dependence of γ on E_0 for a given target material. It is interesting to note that the correctness of this last proposition was confirmed by special experiments made by Gurtovoi and Sosenko [344] to determine d_{ie} and d_p on bombarding a silver target with protons, the target having the form of a wedge so as to facilitate the measurement of γ for various target thicknesses by varying the position of the narrow ion beam.

Bourne, Cloud, and Trump [316] sought to determine the role of the positive gas ions in a high-vacuum breakdown by studying emission for metals and ions typically employed in various electrical vacuum systems. For this purpose, they used aluminum, magnesium, steel, copper, gold, and lead targets, bombarding these with 10- to 120-keV H_2^+, He^+, N_2^+, Xe^+, and Hg^+ ions. The measurements showed that up to energies of 20 keV there was a sharp rise in γ for all cases; after this point the rise became slower.

A number of authors [289, 387, 584, 586, 627, 718] studied secondary ion—electron emission from alloys of the Cu—Be, Ag—Mg, and Al—Mg types and cathodoluminophores (ZnS—Ag, ZnS, ZnSe—Ag, $CaWO_4$, ZnO); this is of interest in view of the use of such materials as sensitive ion detectors.

Okano [263] studied the secondary electron emission of molybdenum bombarded by various ions, including inert-gas ions with energies between 700 and 1000 eV. In order to separate the ion beam a mass spectrometer with double focusing and a resolving power of ~50 was employed. The residual-gas pressure during the experiments remained at $3 \cdot 10^{-5}$ mm Hg. For all the ions Okano found a linear dependence of γ on E_0 in the energy range indicated. However, the resultant values of γ showed no simple dependence on the mass of the ions. There was a certain parallelism between γ and the first ionization potential of the ions. There was also a fall in γ for ions of the heavier isotopes.

All these investigations, which were characterized by a low degree of vacuum (~10^{-5} mm Hg) and the absence of any adequate measures for cleaning the targets, nevertheless led to an important qualitative conclusion, namely, that the secondary phenomena were sensitive to a large number of different factors. It follows from considering the results obtained that the extent of the secondary ion—electron emission depends substantially on the energy of the primary ions, the nature of the bombarding ion, and the state of the target surface. For this reason, later investigations were devoted to studying ion—electron emission as a function of the physical properties of the target and the bombarding ions under better vacuum conditions and with more careful degassing of the target surface.

In a series of investigations, Hagstrum [262, 264, 290, 345] studied the secondary electron emission of molybdenum, tantalum, and tungsten under the impact of singly and multiply charged inert-gas ions for energies of $E_0 = 10$-1000 eV. Before measuring the secondary emission, the beam geometry and the physicochemical conditions on the target surface were specially examined.

The apparatus for studying secondary emission consisted of an ion source based on ionization by electron impact, a mass analyzer, and a receiving section. The ion beam was focused both in front of the mass-spectrometer entrance and after the latter, using electrostatic lenses with previously determined properties [264].

The electrodes of the apparatus were made from tantalum plate, separated by quartz spacers, and fixed on quartz rods. This arrangement ensured a high accuracy in adjusting the electrode system. The tungsten filament of the ion source and the target were fixed each to its own pin in its own individual sleeve, so as to facilitate replacement without harming the overall adjustment. The e/m analyzer was a magnet with a 90° deviation. There were no ground-glass joints or taps with sealing grease anywhere in the vacuum part of the apparatus. The focusing systems made it possible to bring the ion beam onto a section of target 0.2 × 0.8 cm in size without the ions falling on the edge of the entrance aperture in the collector and guard disc in any way. The apparatus was evacuated with three two-stage mercury diffusion pumps. The use of Bayard and Alpert ionization manometers [208] enabled the residual gas pressure to be reduced to $2 \cdot 10^{-10}$ mm Hg. In order to secure atomically clean surfaces, target degassing was therefore carried out by heating to 1750°K in the initial experiments of Hagstrum [262] and to between 2100 and 2200°K later on [264, 290].

The degree of surface cleanness was checked by measuring the number of adsorbed molecules, using the flash method [264]. It was found that the time required for the formation of a monomolecular coating by the adsorption of residual-gas molecules on the target surface equalled 14–15 h at 10^{-10} mm Hg, while under less favorable experimental conditions it equalled a few hours. In Hagstrum's opinion this slow adsorption suggested that, on measuring immediately after heating the target, the surface was atomically clean.

In addition to this, the effect of the magnetic fringing field of the mass analyzer and the effect of the contact potential difference on the secondary electron spectrum were analyzed. In order to reduce the intensity of the magnetic field around the target, screening and magnetic compensation were applied; nevertheless, it was impossible to eliminate the effect of the magnetic field on the results of the measurements entirely.

In order to find the target—collector contact potential difference, a special circuit with half-period heating was amployed.

A study of the processes underlying the ejection of electrons from molybdenum, tantalum, and tungsten on bombarding with He^+, He^{2+}, and He_2^+ in the original experiments [262] and with singly and multiply charged He, Ar, Ne, Kr, and Xe ions in later work [290, 345] led Hagstrum to the following conclusions.

1. The value of γ is almost independent of E_0 for all targets and all singly charged ions in the energy range 10–1000 eV; this indicates that the ejection of the electrons is due to a potential mechanism.

2. The ion-scattering coefficient K is small compared with γ for all singly charged ions; this indicates a high degree of neutralization of the primary ions. The coefficient K increases with E_0.

3. The measured values of γ are smaller than 0.5 for all the singly charged ions. In Hagstrum's opinion, this indicates a partial internal reflection of the excited electrons from the potential barrier at the exit from the metal.

4. There is a relation between the yield of electrons on bombarding with He^+, He^{2+}, and He_2^+ ions and the energies released on their neutralization. The average energies of the electrons ejected by He^+ and He^{2+} differ very little.

5. The maximum energy of the electrons ejected by 40-eV ions is in no case greater than the limit of $V_i - \varphi$, while the minimum limit is zero.

6. In all cases, without exception, γ increases with increasing charge of the ion. In explaining this circumstance, Hagstrum supposes that a multicharged ion is neutralized, on approaching the surface of the metal, not in a single process, but in a whole series of stages, in each of which an electron is excited inside the metal. Another confirmation of the validity of this assumption is the fact that the average energy of the ejected electrons is independent of the charge on the ion.

7. For all the singly charged ions studied except He^{2+}, γ falls with increasing kinetic energy of the ions. The fall is explained by the fact that as E_0 increases the process takes place closer to the surface, which leads to a broadening of the secondary-electron energy spectrum; this causes a reduction in the number of electrons having an energy sufficient to emerge from the metal. The results relating to He^{2+} remain unexplained.

8. The maximum on the electron energy-distribution curve depends on the charge of the ion, but in the majority of cases it is independent of the nature of the latter. Thus, the maxima on the curves for all triply charged ions are close to $16 \cdot 10^{-2}$, for doubly charged ions to $5 \cdot 10^{-2}$, and for singly charged ions to $2.5 \cdot 10^{-2}$. The dependence of γ on the nature of the ions (reduction in the maxima for heavy particles), particularly substantial in the case of singly charged ions, is associated with the fact that a low ionization energy of the particles makes it possible for only a portion of the excited electrons to emerge from the metal.

By using Ne^{20} and Ne^{22} isotopes, the dependence of γ on the mass of the ion was studied. For energies of 200 eV, the values of γ were the same for both ions, within the limits of experimental error.

It should be noted that, whereas Hagstrum was able to degas the molybdenum and tungsten targets to the atomically clean state, this was impossible in the case of tantalum owing to the capacity of this metal to absorb gases when hot; he therefore regarded the data for tantalum as appertaining to a contaminated surface.

The foregoing results were explained by Hagstrum from the point of view of the Auger mechanism of electron ejection by virtue of the ionization energy of the bombarding ions; on the basis of these data he proposed an improved theory for the potential ejection of electrons.

Later, Hagstrum studied the thermionic, sorption, and secondary-emission properties of hafnium [388]. These experiments were set up in order to discover the dependence of γ on the work function of the metal φ. However, the heated hafnium tended to enter into a chemical reaction with the residual gases, and this prevented it from being cleaned to the atomically clean state, whatever the form of heat treatment. Hagstrum therefore considered that the results obtained corresponded to a contaminated surface, as in the case of tantalum, so that it was quite impossible to use these results in order to determine the $\gamma(\varphi)$ relationship.

Hagstrum obtained some interesting information regarding electron emission from molybdenum and tungsten under the impact of metastable noble-gas ions [347]. Metastable singly charged ions were obtained by single electron—atom collisions in an ion source. The dependence of γ and the secondary-electron energy distribution on the energy of the electrons in the ion source was studied. The degree of deviation of the γ values for the same intensity of the ion beam, expressed as a function of electron energy, made it possible to estimate the ratio of the cross sections corresponding to the formation of metastable (σ_m) and normal (σ_n) ions. The latter was determined from the following expression:

$$\gamma_{obs} = (1 - f_m)\gamma_n + f_m\gamma_m,$$

where γ_n and γ_m are, respectively, the coefficients of secondary electron emission under the impact of normal singly charged and metastable ions, while f_m is the proportion of metastable ions in the beam. From this formula we may write

$$f_m = \frac{\sigma_m}{\sigma_m + \sigma_n} = \frac{\gamma_{obs} - \gamma_n}{\gamma_m - \gamma_n}.$$

Putting $\sigma_m \ll \sigma_n$ we obtain

$$f_m \approx \frac{\sigma_m}{\sigma_n}.$$

An estimate of f_m from this expression based on experimental data relating to the dependence of γ on the energy of the electrons in the source gave $f_m = 0.02$, in good agreement with the results of other authors. Hagstrum therefore considered it useful and convenient to use the potential emission of electrons for observing and recording particles in various excited states.

Hagstrum also studied the influence of the adsorption of various gases on the potential emission of electrons under the impact of inert-gas ions [347]. For this purpose N_2, H_2, and CO_2 with purities of 99.9, 99.7, and 96.0%, respectively, were used as adsorbed gases. The degree of purity of both the inert and the adsorbed gases was determined mass-spectrometrically. The experiments were carried out with a residual-gas pressure of 10^{-10}-10^{-11} mm Hg.

The effect of gas adsorption on γ was studied during the formation of coatings at an adsorbed-gas pressure of 10^{-7}-10^{-8} mm Hg. Before admitting these gases into the target chamber, the target was degassed to atomic cleanness by heating at 2100°K. The measurements showed that at these pressures of adsorbed gases a monolayer was formed in 10 min. The effect of the adsorption of these gases was studied by measuring the dependence of γ on the energy of the bombarding ions and obtaining secondary-electron energy-distribution curves after the formation of the monolayer.

It was found that on adsorption of all the gases in question γ fell as the film thickened, saturation setting in after a monolayer had been formed; the effect of the adsorption of these gases on γ was greater for small E_0 than for large. Heating a tungsten target at 2000°K led to the complete removal of N_2 and CO_2, but it was almost impossible to free it from H_2. Hagstrum could not explain the fall in γ on adsorbing H_2, N_2, and CO_2 simply by a change in the work function of the target. For this reason, in addition to a change in work function, Hagstrum admitted the possibility of a change in the energy level of the bombarding ion near the surface covered with gas, and the emission of electrons by some other mechanism, for example, of the kinetic type.

The validity of these conclusions regarding the influence of gas adsorption on the ejection of electrons is substantially impaired by the fact that the work functions of the surface were not measured in the presence of H_2 and CO_2 during the experiments. Comparison was only made for the case of N_2/W, in which $\varphi \sim 5$ eV was taken from Mignolet [317]. From this point of view the assertion of the author that the results could not possibly be explained simply as being due to a change in work function would appear to be premature.

Analogous results were obtained in [285, 668, 669, 680, 719] when studying the influence of gas coverage (O_2, H_2, N_2) on the secondary ion—electron emission of a number of metallic targets (copper, tantalum, platinum).

Tel'kovskii [349] studied the secondary electron emission of various metals on bombarding with neutral atoms and singly and multiply charged ions of various elements, including inert gases, for energies between a few keV and 120 keV. The experiments were carried out in a metal apparatus with a mass spectrometer to effect e/m separation of the primary ions. Principal attention was devoted to processes of target degassing and to the cleanness of the target surface. The target surface was considered as atomically clean on achieving constancy of the γ coefficient as the intensity of the primary beam varied by a factor of 1000, i.e., from 10^{-7} to 10^{-4} A/cm². During the measurements the vacuum was kept constant at the order of 10^{-8} mm Hg. The application of a magnetic field perpendicular to the primary-beam direction in the target area made it possible to separate the secondary electrons from negative ions. Tel'kovskii found that for all the ions and targets studied the coefficient γ increased linearly up to velocities of $2 \cdot 10^8$ cm/sec. On further raising the velocity of the ions, for example, in the case of protons, there was a flat maximum; the author considered that this indicated the deep-seated character of the ejection mechanism. In all cases there was a fairly clear threshold of secondary electron emission. The values of the velocities corresponding to this threshold depended very little on the target material and the type of ions; for the cases under consideration it lay between $0.5 \cdot 10^7$ and $2 \cdot 10^7$ cm/sec, or between 1 and 4 keV. The integral secondary-electron velocity distribution was Maxwellian, its shape being similar to the probe characteristics in the gas discharge. The average energy of the secondary electrons, or, as we may call it, the electron "temperature," was independent of the energy of the incident ions. This led to the conclusion that there was no potential emission of electrons in the case of $V_i > 2\varphi$.

Pradal and Simon [427] used a magnetic spectrograph to study the energy spectra of electrons escaping from various metals after bombardment with Ar^+ ions. The resolution of the spectrograph was of the order of 1 eV for an accelerating voltage of 20 keV. The experiment showed that the energy distribution of the secondary electrons could be excellently described by the formula

$$\ln \frac{i_e}{I_e} = -\frac{\mathscr{E}}{kT},$$

where $\mathscr{E}$ is the energy of the secondary electrons, i_e is the intensity of the secondary electrons corresponding to an electron energy between $\mathscr{E}$ and $\mathscr{E}+d\mathscr{E}$; I_e is the value of i_e at $\mathscr{E}=0$. By T these authors meant the equivalent "emission temperature," which depended on the type of metal.

Riddoch and Leck [423] bombarded a tungsten target with positive inert-gas ions in a vacuum of 10^{-6} mm Hg and then degassed the target by heating to 1600°C; in addition to the desorption of gas from the target, positive ions were emitted. Mass-spectrometric analysis showed that these were Na, K, Rb, and Cs ions. The authors considered that these ions arose from impurities already present in the target before bombardment. In none of the experiments was there any emission of the bombarding inert-gas ions from the target.

Petrov carried out a number of experiments [500, 504-507, 552] aimed at further elucidating the nature of the potential and kinetic emission of electrons from metals under the impact of positive ions. The experiments of [500] were concerned with electron and ion emission from molybdenum, tungsten, and nickel targets bombarded with Cd^+ and Zn^+ and from tantalum bombarded with H_2^+, He^+, and Ar^+. The experiments were carried out in a vacuum of $\sim 10^{-6}$ mm Hg. The work with Cd^+ and Zn^+ ions is interesting because the potential-emission condition $V_i - 2\varphi \geq 0$ was satisfied. For a target which had not been degassed, there was a relatively rapid rise of γ with E_0; this occurred as E_0 varied from 0.3 to 1.0 keV in the case of Cd^+ and 0.3 to 1.6 keV in that of Zn^+. With a cleaned target, γ fell to a certain value and then remained practically independent of E_0, and for molybdenum and tungsten $\gamma \neq 0$ as $E_0 \to 0$. For a nickel target no potential extraction was observed. On bombarding a tantalum target heated to 1400-1500°K with Ar^+ ions at energies up to 1.5 keV, potential extraction of the electrons took place, while for energies above this value and up to 7.5 keV there was an almost linear rise in γ. For ion energies of the order of several keV, the $\gamma(E_0)$ curve was much sharper for He^+ than for Ar^+. The coefficient of ion—ion emission K, measured on bombarding heated molybdenum and tungsten targets with Cd ions, was independent of the energy of the ions, while for a nickel target it increased slightly with rising E_0. The greatest emission of ions occurred for molybdenum and the smallest for tungsten.

By using Ar^+ and K^+ ions (neighbors in the Periodic Table), Petrov [506] tried to determine the part played by the structure of the electron shells in the ions during the kinetic extraction of electrons. It was found that for the energy range $E_0 < E_{min}$ there was no kinetic extraction of electrons for either ion, while for $E_0 > E_{min}$ the $\gamma(E_0)$ relationship was linear over a certain range of energies, which was the wider, the greater the mass of the bombarding particle. In the energy range 0.5-6 keV, the coefficients of the kinetic emission of electrons for Ar^+ and K^+ ions were exactly the same for tantalum and tungsten targets.

The secondary emission (ion—electron and ion—ion) from pure metallic targets was studied by Petrov [507] at temperatures up to 2000°K. In order to separate the secondary currents from the thermionic emission of electrons, the primary ion beam was modulated with a sinusoidal voltage of frequency 600 cps. The secondary electrons were separated by a resonance circuit in the target—collector loop. It was found that, on bombarding tantalum and tungsten targets with K^+ ions, electrons started being extracted at an ion energy of 1.5 keV. The coefficient γ then increased linearly with the energy of the bombarding ions, and the slope of the $\gamma(E_0)$ straight lines was independent of the target temperature and material in the range 1700–2000°K. The coefficient K for energies above 2 keV reached 35% and was independent of ion energy.

The secondary emission from a heated molybdenum target bombarded with 0.2- to 1.4-keV Cs^+ ions was studied by Petrov [504]. There was almost a linear increase in the coefficient γ for ion energies between 0.3 and 1.4 keV; for $E_0 = 0.2$ keV, $\gamma < 0.1\%$. In the reflected-ion spectrum there were fast positive particles with energies up to 110 eV for $E_0 = 1$ keV; the author explained these by collisions between a Cs^+ ion and two molybdenum atoms at the same time. By using a beam of K^+ ions modulated in intensity the secondary emission from tantalum and tungsten targets was studied at temperatures up to 2000°K. It was found that for ion energies of under 1.5 keV there was hardly any electron extraction. Subsequently, γ increased linearly with increasing energy of the ions up to 6 keV, the slope of the straight line not depending on the nature of the target nor its temperature between room temperature and 2000°K. The absence of any temperature dependence in this temperature range was also noted in the case of the coefficient of potential electron extraction.

Petrov explained these experimental results as being due to the excitation of bound electrons in the metal [505]. From this point of view the possibility that an electron may be detached from an incident ion arises, for example, in the case of ions with low V_i values and negative ions. In addition to this, there is a rise in $d\gamma/dE_0$ with increasing second ionization potential of the atoms forming the ion beam directed at the target. However, the meaning of this relationship is not quite clear, and it may be due to a connection between these two quantities and some third parameter.

In order to determine the role of the mass of the ion and the structure of its electron shell in the kinetic emission of electrons, Petrov [552] made a comparative study of the secondary electron emission from the refractory metals (molybdenum, tantalum, and tungsten) under the influence of He^+, Ar^+, and Ca^+ ions, hydrogen ions (H^+, H_2^+), and nitrogen ions (N^+, N_2^+). It was found that for a particular energy the H^+, H_2^+, and He^+ ions caused a greater electron emission than Ar^+ and Ca^+. However, for the N^+ and N_2^+ ions, the $\gamma(E_0)$ relationship had a sharper appearance than in the case of H^+, H_2^+, and He^+ ions. Petrov therefore concluded that the electron shell played a more important part in the kinetic emission of electrons than the mass of the ion. However, it should be noted that this conclusion was only drawn on the basis of data obtained with nitrogen ions, without any adequate checking of the state of the target surface. Checking the state of the target surface by measuring the potential emission of electrons for an energy of $E_0 \sim 1$ keV is not correct, in view of the fact that at this energy there may be a considerable kinetic ejection of electrons [347] from a surface covered with a film of nitrogen molecules.

Chaudhri, Mustafa, and Abdul [353] first studied the emission of electromagnetic radiation on bombarding nickel with Ne^+, He^+, and H^+ ions in the energy range $E_0 = 300$–3000 eV; they found that the number of protons necessary to secure the emission of one photon from nickel was $5 \cdot 10^4$ to $2 \cdot 10^5$, the intensity of the radiation increasing rapidly with the contamination of the surface. The wavelengths of a considerable proportion of the radiation lay between 3300 and 4900 Å. The radiation of light was also studied in [119, 121, 158, 758].

Berry and Abbott [467] studied the angular distribution of secondary electrons struck·
from tungsten targets by He$^+$ ions. A rotating collector made it possible to select secondary
electrons every 9° in an angular range of 0 to 120° for three angles of incidence of the ion beam
on the target (15, 22.5, 30°). The measurements were made for bombarding ion energies of 40,
400, and 825 eV with target temperatures of 300 and 1100°K in a vacuum of $5 \cdot 10^{-5}$ mm Hg. According to the authors themselves the target surface was not atomically clean. For all the experiments a cosine distribution of secondary electrons with respect to escape angles was established.

Fogel', Slabospitskii, and Rastrepin [502] studied the secondary emission of charged particles on bombarding a molybdenum target with singly charged H$^+$, He$^+$, Ne$^+$, Ar$^+$, Kr$^+$, and O$^+$
ions and tantalum, tungsten, copper, and iron targets with H$^+$, Ne$^+$, and Ar$^+$ ions over the energy range 10–40 keV. The use of three plane electrodes (target with screen, grid, and collector) in a magnetic field perpendicular to the electric field as the measuring part of the apparatus, in the opinion of the authors, facilitated the simultaneous measurement of the secondary-emission coefficients of positive and negative ions (K and K$^-$) and electrons (γ) and also
the primary-ion reflection coefficient K$_s$. The (e/m)-separated ion beam was focused on the
target by means of a magnetic lens. Before the measurements, the molybdenum, tungsten, and
tantalum targets were degassed at a temperature of 2300°K for 30–40 min, and the copper and
iron targets at 1200°K for the same period, after which they were subjected to ion bombardment
for 1 h.

It was found that the coefficients K, K$^-$, and K$_s$ were of the same order and varied with
the nature and velocity of the ion between a few tenths of a percent to several percent. The
value of γ was two orders greater than that of the other coefficients.

There was a strong dependence of the coefficients measured on the nature of the ion. The
coefficients K and γ, comparable at similar velocities, increased monotonically with increasing
mass of the ion. For K$^-$ and K$_s$, this monotonic dependence on the mass of the ion only occurred
for ions of the noble gases. Whereas the value of K$^-$ only varied slightly with increasing velocity of the ion, the values of the other coefficients increased rapidly with increasing velocity.
In addition to this, experiments were carried out with H$^-$ and H$^+$ ions in order to determine the
part played by the structure of the electron shell in secondary emission. It was found that K$^-$
was greater for H$^-$ than for H$^+$ ions, whereas K was the same in both cases. The value of γ
for H$^-$ was almost double that of γ for H$^+$.

It should be noted that, in view of the fact that the measurements were made on cold targets, these authors' results relate to contaminated surfaces. This is also indicated by the fact
that on bombarding molybdenum, tantalum, and tungsten targets the authors observed considerable currents of negative ions, which could only arise as a result of contaminant atoms and
molecules. In addition to this, the separation of the total emission of the secondary ions into
two components representing the emission of positive and negative ions is incorrect, since for
such primary-ion energies the secondary ion emission receives a considerable contribution
from secondary electrons leaving the collector as a result of the impact of particles (ions and
neutral atoms) reflected from the target.

In our own opinion, the secondary ion current in such experiments can only properly be
classified after analyzing these with respect to masses and energies.

In 1961–1966 the study of interactions between gas ions and the surface of solid targets
underwent an intensive development, both in extending the number of substances studied (new
targets and new bombarding ions) and in relation to a deeper study of the mechanisms responsible for the observed phenomena (work with single crystals, precision measurements of the
angular and energy characteristics, etc.), with an extremely careful check on the state of the
surfaces bombarded.

Thus, for example, Batanov [531] studied the extraction of electrons with H_2^+, He^+, and Ar^+ ions from a dielectric target (glass). The energy E_0 of the H_2^+ and He^+ ions was varied between 0.2 and 30 keV. There was a nonlinear rise in γ on increasing E_0. For the Ar^+ ions, γ increased linearly from 200–2000 eV. The values of γ and $d\gamma/dE_0$ for the object under examination (glass) were much higher than for high-temperature metals.

Hagstrum [514, 532] studied the Auger mechanism of the extraction of electrons by singly charged inert-gas ions from the (111) and (100) faces of silicon single crystals and from the (111) face of germanium. The targets were cleaned by special heat treatment, and the state of the surface bombarded was checked with optical and electron microscopes. The coefficients of ion—electron emission and the energy spectra of the Auger electrons were studied. Hagstrum associated the shape of the spectral curves and their dependence on E_0 with the function representing the density of electron states in the valence band of the semiconductor. The $\gamma(E_0)$ relationship in the region $E_0 \approx 10$–100 eV was also explained by reference to the electron-state density function in the valence band of the semiconductor.

The emission of electrons from semiconductor targets on bombardment with gas ions was also studied in [577, 626].

Magnuson and Carlston [650, 651] studied the γ from polycrystalline molybdenum and copper targets and from the (100), (110), and (111) faces of a pure copper single crystal on bombarding with inert-gas ions at an energy of 0.5–10 keV. In order to clean the surface, the targets were subjected to preliminary ion bombardment. It was found that γ increased linearly with increasing E_0; the electron yield was a maximum for the (111) face and a minimum for the (110) face.

Mahadevan and Layton [667] bombarded molybdenum with He^+ and Ar^+ ions in the energy range $E_0 \sim 100$–2500 eV and found that γ was independent of the target temperature. It was also found that the threshold energy for the kinetic emission of electrons was greater for the heavier atom.

Klein [628] measured the angular distribution of secondary electrons ejected from clean and gas-contaminated tungsten targets by inert-gas ions at energies of 1300–4000 eV.

Krebs [630] studied the energy spectrum of secondary electrons under the influence of positive ions passing through a mass spectrometer and obtained results in agreement with the Poisson distribution.

Medved [679] considered ion—electron emission and the reflection of primary ions on bombarding metals with K^+ and Ar^+ ions and Ar^0 atoms at energies up to 4 keV. In Medved's opinion, the agreement between the reflection coefficients of K^+ and Ar^0 indicated that the electron structure of the bombarding particles played a dominant part in the emission.

Mashkova and Molchanov [674, 721, 723] studied the dependence of γ on the incident angle. For this purpose, beams of Ar^+ ions with energies of 20 to 36 keV were directed at the faces of nickel, copper, zinc, molybdenum, and tungsten single crystals at various angles. The surface was cleaned by ion bombardment; the pressure in the measuring chamber was $\sim 2 \cdot 10^{-7}$ mm Hg. When the direction of the beam coincided with the principal crystallographic directions, γ passed through a minimum.

An analogous behavior of the $\gamma(\varphi)$ relationship occurred in [693, 762, 764]. On increasing the energy of the bombarding ions the anisotropy became all the more marked. The absolute values of γ increased with energy roughly in accordance with a $\sqrt{E_0}$ law. With increasing temperature the anisotropy of γ diminished [724] and the maxima and minima gradually smoothed out.

In another investigation [757], Mashkova and Molchanov, studying the angular dependence of γ, bombarded the (100) face of a copper single crystal with N_2 and H_2 ions at an energy of 15-25 keV. The anisotropy was sharper for the N^+ ions than for the H^+. It was noted that the values of γ obtained on irradiating the target with 17.5-keV atomic ions (N^+) were just twice the values of γ obtained on irradiating with 35-keV molecular ions (N_2^+). This confirms that the molecules dissociate near the target surface.

Takeischi and Hagstrum [681, 741] studied the total yield of secondary electrons and their energy spectrum on bombarding the (111) faces of nickel and germanium single crystals with low-energy (4-100 eV) He^+, Ne^+, and Ar^+ ions. The pressure in the experiments was $\sim 8 \cdot 10^{-10}$ mm Hg. The sample purity was 99.999%. The experiments were carried out for an atomically clean target surface and for one covered with adsorbed gases. It was found that a rise in the energy of the primary ions caused a fall in the proportion of fast Auger electrons at the expense of an increase in the proportion of the slow electrons. The presence of adsorbed H_2, O_2, and CO_2 gases on the surface of the face led to a fall in the number of fast electrons and a reduction in γ.

Kondrashev and Petrov [759] studied the secondary electron emission of LiF, NaCl, KBr, and CsI on bombarding with He^+ and Ar^+ ions in the energy range 20-600 eV. The coefficient γ, as one might expect, was greater for the He^+ ions. For high energies, kinetic emission occurred. The coefficient γ increased with diminishing forbidden-band width of the sample material.

Jamerson and colleagues [760] bombarded a uranium—nickel alloy (5.7 wt.% proportion of nickel) with fission fragments. The vacuum in the apparatus was 10^{-8} mm Hg. The width of the secondary-electron spectrum was no greater than 20 eV. The number of fission fragments was calculated theoretically; the coefficient of secondary electron emission was 307 ± 10.

Karlston et al. [744] studied γ from the (100), (110), and (111) faces of copper, aluminum, silver, molybdenum, and nickel targets bombarded with He^+, Ar^+, Kr^+, and Xe^+ ions at an energy of 1-10 keV. The working vacuum was $4 \cdot 10^{-8}$ mm Hg. The ratios of γ for different faces were independent of the energy of the incident ions.

Euring [761] measured the coefficient γ for a tungsten single crystal as a function of the energy of bombarding protons. The energy was varied between 50 and 225 keV. The working pressure in the apparatus was $2 \cdot 10^{-10}$ mm Hg, and before the experiments the target surface was carefully roasted; the measurements were made with the target at room temperature 2 min after cleaning. The residual-gas pressure in the apparatus was $1 \cdot 10^{-10}$ mm Hg. The maximum value of γ occurred for proton energies of ~125 keV.

Mahadevan [765] bombarded a cleaned molybdenum target with H^+, H_2^+, H_3^+, O^+, O_2^+, N^+, N_2^+, and H^-, O^-, and O_2^- ions having energies of 40-2000 eV. The value of γ was studied as a function of the kinetic and potential energy of the ions and the degree of surface contamination. The pressure in the apparatus was $\sim 2 \cdot 10^{-9}$ mm Hg; the surface of the molybdenum was cleaned by heating to 2000°K. It was found that the potential emission fell exponentially with time as the contamination of the target surface increased, up to the formation of a monolayer. The kinetic emission due to the positive ions depended very little on the surface contamination. With increasing E_0, the kinetic emission rose the more rapidly, the greater the mass of the incident ion. For negative ions there was no potential emission, and the γ was much greater for a contaminated surface than for a clean one.

Ion—electron emission associated with gas-ion bombardment was also studied in [318, 346, 425, 456, 460, 537, 666, 668, 671, 677, 720, 722, 756, 763]. Secondary processes due to the bombardment of targets with ions were treated in a number of reviews [63, 79, 91, 98, 533, 629, 631].

In recent years a number of special investigations have been published on the penetration of gas ions into targets and there has been a great deal of work on the composition of secondary ion emission due to the bombardment of metals with inert-gas ions.

The penetration of positive He$^+$ ions into a molybdenum target was studied by Varnerin and Carmichael [396]. The number of ions penetrating tended to saturate in the course of time. The authors suggested that saturation set in at a number of He atoms corresponding to several equilibrium monolayers.

The penetration of He$^+$, Ne$^+$, Ar$^+$, and Kr$^+$ ions into a nickel target and their reemission were studied by Carmichael and Trendelenberg [437, 511]. Ions with energies of 100 eV, after penetration into the target, were freed by cathodic sputtering on bombarding the target surface with ions of a different kind. A mass spectrometer was used for analyzing the primary and secondary ions. The penetration of Ne$^+$, Ar$^+$, and Kr$^+$ ions reached saturation for concentrations of less than one monolayer, while for He$^+$ ions there was no saturation effect up to 15 monolayers. The authors explained this latter fact by pointing out that the atomic diameter of He was smaller than the interatomic distance in the nickel lattice.

In the opinion of Trendelenberg and Carmichael [511], if the He atoms are distributed within the metal in accordance with a $1/x$ law (x = distance from the surface of the metal), then the number of He atoms coming to the surface of the metal as a result of diffusion will vary with time in accordance with $1/t$; the authors consider that in the thermal emission and back diffusion of ions the $1/x$ law may be used for the distribution of He atoms in nickel [437].

Bradley [459] used a mass spectrometer to study the secondary positive ions emitted by the surfaces of molybdenum, tantalum, and platinum on bombarding these with inert-gas ions of low energy (<1000 eV) in a vacuum of $< 10^{-8}$ mm Hg.

Bradley observed ions characterizing both the material of the target and the surface contaminations and bulk impurities. The secondary ions corresponding to the target metal only constituted a small part of the total sputtered particles (0.01% for platinum); they were singly charged monatomic particles with an energy of the order of 4–5 eV. Ions could arise either from the sputtering of the target itself, or from the dissociation of surface compounds. Bradley pointed out that the method described could be useful in studying the kinetics of surface phenomena and also cathodic sputtering processes.

Castainy, Jouffrey, and Slodzian [509] studied secondary ion emission from various metal samples. The use of a mass spectrometer made it possible to separate the secondary and reflected ions. A powerful primary ion beam was employed; in the authors' view this ensured the cleanness of the surface owing to the cathode-sputtering cleaning effect. The secondary ion—ion emission was large for silicon, aluminum, and magnesium, and rather smaller for beryllium, copper, and nickel. The authors came to the conclusion that there was a considerable emission of metal ions from the halides of the alkali metals, whereas a graphite target gave no appreciable emission of positive ions.

Using an apparatus furnished with a mass analyzer and an energy analyzer (90° condenser), Stanton [508] studied the composition and energy distribution of positive ions emitted from targets of stainless steel, brass, lead, tin, and beryllium on bombarding with He$^+$, Ne$^+$, Ar$^+$, Kr$^+$, Xe$^+$, N$_2^+$, O$_2^+$, and Co$^+$, ions. The energy of the primary ions varied from 0 to 3 keV. Except for beryllium (heated to 200°C), the targets were not subjected to heat treatment. For the lead and tin targets, positive ions of the principal material were observed. Owing to presence of a high background, no peaks of impurity metal ions (for example, iron, nickel, chromium, copper, and zinc) were observed for the stainless steel and brass targets. The secondary-ion energy distribution was only studied for the beryllium target. It was found that a considerable proportion of the ions ejected had energies of under 5-10 eV, while some ions had energies of over 200 eV for a primary-ion energy of 1000 eV. The energy distribution of the contaminant secondary ions was very similar to a Maxwell distribution; however, the energy distribution curve of the secondary ions of the principal material could not be represented by a Maxwell curve corresponding to a single temperature, but rather by the sum of several such distributions relating to different temperatures. The yield of secondary Pb^+ and Be^+ ions depended very little on the mass of the bombarding ions.

Fogel', Slabospitskii, and Karnaukhov [503] studied the mass-spectral composition of secondary positive and negative emission arising as a result of the bombardment of a molybdenum surface with Ne^+, Ar^+, and Kr^+ ions. The residual-gas pressure was $(3-5) \cdot 10^{-6}$ mm Hg. The current density of the primary ions was no greater than 10^{-8} A/cm^2. The authors studied the dependence of the number of ejected secondary ions on the primary-ion energy (5-40 keV) and the target temperature, and also the change in the mass spectrum of the secondary emission when H_2^+ and $(D_2O)^+$ ions struck a previously degassed surface. It was found that the curve relating the number of ejected secondary ions to the velocity of the primary ions had a maximum, the position of which depended on a number of factors, such as the nature of the primary and secondary ions, the angle of incidence of the primary beam on the target, and the material and surface state of the latter.

Hagstrum [579] studied the scattering of inert-gas ions from tungsten and molybdenum in the range $E_0 = 100-1000$ eV, showing that for these ions the value of K_s was very small and almost independent of E_0, although it fell with increasing mass of the ion. For a tungsten target and He^+, Ne^+, and Ar^+ ions, the value of K_s varied between 0.01 and 0.2%. On bombarding a molybdenum target with He^+ ions a smaller K_s value was obtained (0.08, error 0.1%) than for tungsten (0.14, error 0.18%). Hagstrum found that on bombarding tungsten and molybdenum targets with He^+ ions in the range $E_0 = 0.1-1$ keV, the values of K_s respectively equalled 0.17 and 0.09%.

Medved [679] studied the scattering of Ar^+ and Ne^+ ions from platinum targets for energy values of $E_0 = 0.5-2.5$ keV, recording the flux of scattered atoms by means of a secondary-emission detector. The scattering coefficient K_s was almost the same for both ions, its value increasing from 0.5 to 1.5% on decreasing E_0.

O'Briain, Lindler, and Moore [445] found that on bombarding silver with H^+ and D^+ ions at an energy of 9.25 keV, K_s fell in inverse proportion to the mass of the ions $K_s(H^+) \approx 12\%$ and $K_s(D^+) \approx 7\%$.

Snoek et al. [750a] studied the scattering of Ar^+ ions ($E_0 = 40-90$ keV) from the (110) face of a copper single crystal and a copper gas target at $\varphi = 60°$. The value of K_s was roughly twice as great for the gas target as compared with the solid. The authors explained this as being due to Auger neutralization or resonance neutralization near the surface in the case of the solid target. The value of the inelastic energy losses by the bombarding ions was independent of the state of aggregation of the target.

Borovik, Katrich, and Nikolaev [750a] bombarded aluminum, copper, titanium, and tantalum targets heated to between 300 and 400°C with 35-keV protons in a vacuum of $\sim 10^{-9}$ mm Hg; they studied the penetration of the protons into the metals and the sputtering of the target material. It was found that the penetrating ions tended to saturate at a concentration equivalent to $5 \cdot 10^{19}$ atoms/cm^2.

Analogous work was carried out by Grant and Carter [750a]. These authors studied the penetration of He^+, Ne^+, Ar^+, Kr^+, and Xe^+ ions into glass and their back diffusion on heating. The probability of the penetration of the ions into the target was compared with theoretical data.

Fogel' et al. [660] studied the influence of gases (N_2, O_2, NH_3) and water vapor on the secondary ion—ion emission from platinum bombarded with 22-keV Ar^+ ions.

Mashkova et al. [649] directed a beam of 30-keV Ar^+ ions at the (110) face of a copper single crystal, varying the incident angle from 0 to 75°. After comparing theory with experiment, the authors concluded that for small angles of incidence atomic layers lying deeper than the tenth had a considerable influence on the yield of sputtered particles. Allowance was made for the possibility of energy transfer taking place from the deeper layers by means of plasma waves, propagating most intensely in the directions of highest electron density. This mechanism is important in the case of ion—electron emission.

Recently there have been a number of papers on secondary emission due to ion bombardment [170, 178, 184, 194, 235, 321, 397, 429, 438, 499, 510, 590, 680, 766].

It follows from the foregoing review that the phenomenon of secondary electron and ion emission from metals under the impact of positive inert- and ordinary-gas ions has been studied on all sides by a large number of authors. Bombarding particles have included ions of all the inert gases except radon (He^+, Ne^+, Ar^+, Kr^+, and Xe^+), while atomic and molecular hydrogen in positive-ion form (H^+, D^+, H_2^+) and multicharged ions have been widely used.

In studying secondary phenomena due to the interaction of gas ions with the surface of a solid body, three stages may be distinguished. The first of these includes the early work carried out in a comparatively low vacuum ($\sim 10^{-5}$ mm Hg) without checking the state of the surface. In this period there were considerable differences between individual results. The second stage starts from a long series of investigations by Hagstrum [262, 264, 290, 345, 672] and Petrov et al. [500, 504-507, 533, 552] carried out at the same time as those by Arifov, Rakhimov, et al. [419-421, 493, 495, 496, 646]. In this period considerable advances were made in perfecting the experimental and vacuum technology. As a result of this reliable data were obtained regarding the emission of electrons from a number of metals under the influence of inert-gas ions, and this helped in understanding the nature of the potential and kinetic emission of electrons. However, many aspects of these phenomena require further detailed study. In particular, the relation between these forms of emission and many of the physical properties of the bombarding particles and targets has not been established. In this respect, good prospects are offered by work with semiconducting and dielectric targets, and also by a change from polycrystalline material to single crystals. Such investigations, which have been undertaken increasingly in recent years, constitute the next stage in the study of emission phenomena. Already some information has been secured regarding the role of the structure of the solid surface and the structure of the energy bands in processes of ion—electron and ion—ion emission. We should expect to obtain some new and interesting results in this connection in the near future.

In the following sections we shall set out our own investigations relating to the second and third stages in the study of secondary emission from solids bombarded with gas ions.

§ 2. Influence of the State of a Metallic Surface

on the Potential Emission of Electrons

It has been shown in a number of papers [262, 345, 680, 734] that the potential emission is extremely sensitive to the conditions on the surface of a metal; the adsorption of foreign atoms of molecules greatly changes both the total electron yield and the electron-energy distribution. The way in which the emission parameters change depends on the nature of the adsorbed particles. From this point of view it is interesting to study the behavior of the potential emission of electrons in the various vacuum conditions created by pumping systems using various working substances, since these may penetrate into the vacuum apparatus and affect the phenomenon under consideration. Remembering that, in experiments on emission phe-

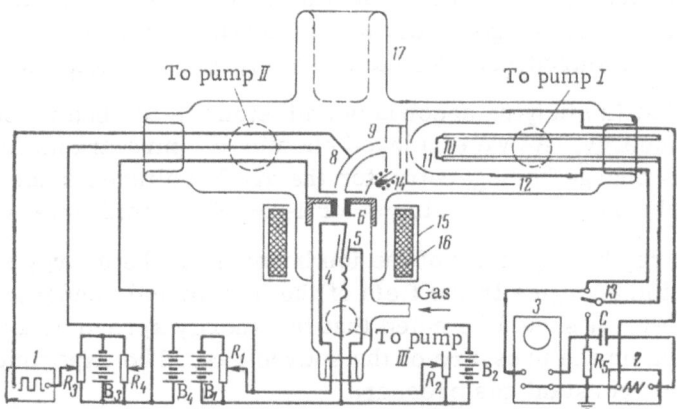

Fig. 6.1. Apparatus and main electrical circuit of the apparatus for studying secondary emission under the impact of gas ions.

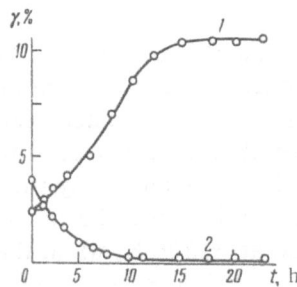

Fig. 6.2. Dependence of γ on the period of heat treatment of the targets.

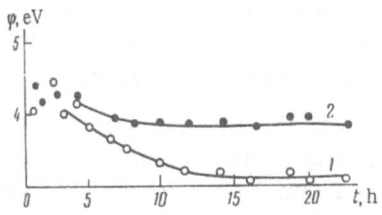

Fig. 6.3. Dependence of the work function on the period of heat treatment of molybdenum.

nomena, mercury vapor and oil diffusion pumps are employed, and that there are differences in the results obtained [262, 349], Arifov and Rakhimov [419-421] studied the behavior of the potential emission of electrons in vacuum systems furnished with these two types of pumps.

Two similarly constructed vacuum systems (A and B) were evacuated with mercury vapor and oil diffusion pumps, respectively (Fig. 6.1). The state of the surface of the strip target was checked by using Richardson straight lines to measure the work function of identical filaments kept under the same conditions. The filaments were sited in special lamps sealed to side tubes of the systems A and B. The experiments were conducted with ∼ 200-eV Ar$^+$ ions for which the kinetic emission of electrons was negligibly small. The material for the target and filament (90 mm long, 0.15 mm in diameter) was molybdenum. Systems A and B were completely identical in construction and consisted of a magnetron ion source [104], a focusing system, and a receiving section. The source was separated from the rest by a vacuum diaphragm 6 (Fig. 6.1). The magnetron consisted of a hollow cylindrical anode 5, a tungsten filament 4 stretched along the axis of the anode, and a solenoid 16 with a screen 15 fitted onto the source outside the bulb. The ions formed in the source were drawn to the diaphragm 6 by a weak field between the filament and the diaphragm. The remaining notation in the figure is as follows: 1 and 2 are "top-hat" and sawtooth pulse generators, 3 is an oscillograph, 14 is a molecular furnace, 13 is a switch, 17 is a liquid-nitrogen trap. The receiving part of the apparatus consisted of a screening cylinder 12 with a double diaphragm for cutting out the required shape of the primary ion beam, a target 10, and a collector 11 for collecting the secondary particles. The target 10 constituted a molybdenum strip 5 × 30 × 0.15 mm in size.

Our apparatus for measuring the work function by the method of Richardson straight lines and the electrical circuit are of the generally accepted type. In order to obtain the ions, we used chemically pure argon introduced into the source through a metal needle valve.

It is well known that in studying secondary emission the ion beam used must be homogeneous in composition and energy. As we employed chemically pure argon, foreign impurities were under 0.1%. As regards charge homogeneity, for the anode voltages taken the proportion of metastable and doubly charged ions was altogether 1-2% of the total intensity [347].

The degree of energy homogeneity of the ion beam, i.e., the energy spread, was determined by analyzing the ion energies in the field of the cylindrical condenser 8. This spread was no greater than ±5 eV. It should be noted that the energy spread of the ions for a given source geometry depends on the pressure of the source gas. The experiments were therefore carried out with the same constant gas pressure.

The glass walls of the apparatus were degassed at a temperature of 350°C. The degassing of the electrodes disposed near the target and filament of the ion source was effected by electron bombardment. Degassing of the target and filament of the source at a temperature of ~1700°K was carried out continuously for several hours until the pressure of the residual gas reached $(1-3) \cdot 10^{-7}$ mm Hg, and then at 2300°K for 20-30 min with breaks so as to avoid overheating the metal—glass seals of the leads.

In order to estimate the degree of cleanness of the surface, the "flash" method [264, 265] was used to determine the time Δt_e required for the formation of an equilibrium coverage of adsorbed molecules on the target surface. It was found that in apparatus B, Δt_e was much shorter than in apparatus A, being about 10-15 sec. This time was much longer than that required to record the secondary-emission V/A characteristics, which appeared almost without any inertia on the oscillograph screen. The secondary currents were measured not only while the target was cooling to room temperature after brief heating to 2300°K, but also at ~1300°K, when the equilibrium coverage was slight owing to the fall in the lifetime of the adsorbed molecules [40, 103].

Curves 1 and 2 in Fig. 6.2 illustrate the variation in γ with the period of heat treatment of molybdenum targets in systems A and B, respectively. The targets were heated for 23 h at 2300°K. During the measurements the temperature was reduced for a short time to 1300°K. We see from the curves that the values of γ are very close to each other for both targets in the first few hours of heat treatment and equal to between 4 and 5%. However, later the value of γ in system A increases, while that in system B falls. After 10-12 h stable values are reached; these are, respectively, 10-12 and under 2%. In subsequent heat treatment, even at very high temperatures (up to the melting point of molybdenum), these values remained unchanged.

At the same time as the γ measurements, we determined the work function of the filaments subjected to the same heat treatment as the targets. The results of measurements made on the work functions of the filaments are shown in Fig. 6.3 (curves 1 and 2 for filaments in systems A and B, respectively). The measurements were carried out in the temperature range 1800-2100°K. It follows from the curves that in the first hours of heat treatment the work functions of both filaments are very much the same; they are unstable and over 5 eV. Over a period of 4-5 h, a fall takes place, and after 10-12 h, the values become stable, equal to 4.2 and between 4.8 and 5.0 eV for filaments in systems A and B, respectively.

Subsequent measurements showed that these values remained stable under prolonged heat treatment at 2500-2600°K. If we compare the behavior of the work-function curves of the two filaments during heat treatment with the γ curves for the two targets, it is not difficult to convince ourselves that these quantities reach stable values in about the same period of time. This

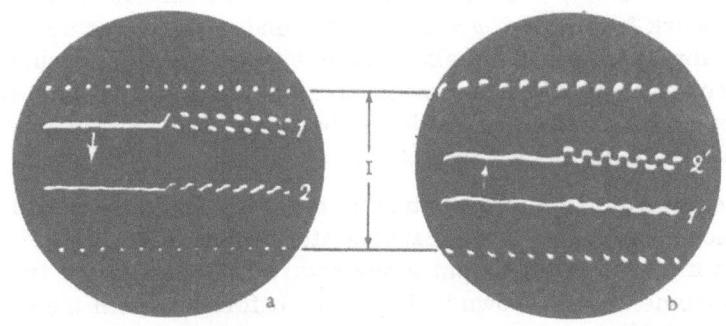

Fig. 6.4. Oscillograms of the volt—ampere characteristics of secondary emission obtained in systems using: a) mercury; b) oil-vapor pumps.

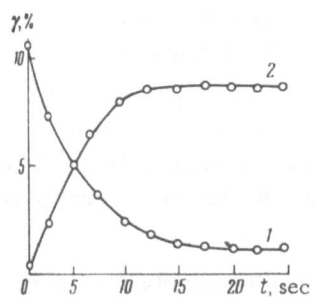

Fig. 6.5. Dependence of γ on the time of target cooling.

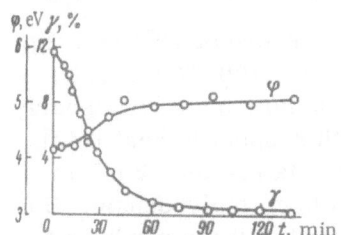

Fig. 6.6. Dependence of γ and φ on the period of heat treatment of the target and filament in system A in the presence of oil vapor at a pressure of $3 \cdot 10^{-6}$ mm Hg.

is evidently a consequence of the identical nature of the phenomena of adsorption and desorption for a filament and a strip of the same material under otherwise equal conditions. Analysis of the time curves of γ and φ during heat treatment shows that there is an inverse relationship between these two quantities. In fact, in the initial stages of heat treatment the work function of the metal is high (curve 1, Fig. 6.3) and the value of γ low (curve 1, Fig. 6.2). Subsequent heating reduces φ to 4.2 eV and increases γ to between 10 and 12%. The form of the γ and φ curves (Figs. 6.2 and 6.3) for the target and filament in system B also shows that for a large value of φ the value of γ remains very low, close to zero.

After reaching a stable state on the surface of the targets corresponding to $\gamma = 10\text{-}12\%$ (A) and $\gamma < 2\%$ (B), we studied the effect of the adsorption of residual-gas molecules on the ejection of electrons. Figure 6.4 presents oscillograms of the secondary-emission V/A characteristics obtained in systems A and B for 1300°K (1, 1') and cold (2, 2') targets after the formation of an equilibrium layer of adsorbed molecules. Figure 6.5 shows the time variation of coefficient γ obtained in the two systems by motion-picture photography. It follows from the curves that as the residual-gas molecules are adsorbed, the value of γ for the target in system A falls while that for the target in system B rises; saturation is reached in 10-15 sec. The time required for the fall and rise of γ to saturation coincides with the time required for the formation of the equilibrium coverage, as earlier determined by the flash method.

Thus, the results indicate that the composition of the residual gas differs for systems A and B. According to the foregoing supposition, this difference was due to the presence of oil vapor in system B, and it therefore seemed interesting to determine the effect of oil vapor on φ and γ under the conditions of system A, evacuated with mercury pumps. For this purpose an ampoule containing previously degassed oil was sealed into system A. After heat-treating the target so as to give a value of γ = 10-12% on bombarding with 200-eV ions (kinetic energy) and reaching a filament work function of φ = 4.2 eV, the end of the ampoule was broken with an armature set in motion by means of an external electromagnet. Communication of the cold oil to the apparatus through an aperture 2 mm^2 in cross section did not lead to any substantial change in φ or γ for an hour; this was evidently because of the low oil vapor tension at room temperatures and the well-degassed state of the oil.

By heating the ampoule to a temperature corresponding to the working condition of the oil in the pump, oil vapor was let into the apparatus at a pressure of $3 \cdot 10^{-6}$ mm Hg and φ and γ were measured with a target and filament temperature of 1300°K after treating at 2300°K. The results of the measurements are shown in Fig. 6.6. It follows from the curves that as heat treatment proceeds, γ falls and φ rises. After 20-30 min, the value of γ remains close to zero, while φ rises to 5.1 eV.

After reaching stable values of γ for the target and φ for the filament, the ampoule was sealed off and these quantities were measured after further heat treatment. However, continued heat treatment above 2500-2600°K would still not reduce φ or increase γ to their original values. In addition to this, prolonged bombardment of the target with 10-keV Ar$^+$ beams having a current density of 10^{-5} A/cm^2 also produced no appreciable change in the values of γ.

Comparison of the curve for γ (Fig. 6.6) with curve 2 (Fig. 6.2) obtained in system B shows that these are identical in character, but the time required to reach the stable value in system A is shorter than in B owing to the slight increase in the oil vapor pressure in A. The stable values of γ and φ reached in A after admitting the oil agree closely with the corresponding values measured in B, which was evacuated with oil-vapor pumps.

Thus the results of measuring γ after prolonged heat treatment of the targets obtained in systems A and B differ sharply from each other. Whereas for the target in system A, γ = 10-12%, for that in B, γ is close to zero. These results are in good qualitative agreement with the corresponding data of Hagstrum and Tel'kovskii, and thus prove the existence of the potential extraction of electrons under conditions in which the apparatus is evacuated with mercury pumps and the absence of this effect when evacuation is based on pumps of the oil-vapor type.

It is well known that, under the conditions of apparatus A, at a pressure of 10^{-7} mm Hg, the surface of molybdenum may be quite well cleaned at a temperature of 2100-2200°K [71], even from adsorbed O_2 molecules. As regards mercury atoms, these amalgamate very poorly with molybdenum [391] and the surface bond so formed is unstable with respect to heat [242]. It is therefore no chance matter that the work function in system A is obtained as 4.2 eV, in good agreement with the results of other authors [209, 428] for pure (clean) molybdenum. However, in the presence of traces of oil vapor in the residual gas of apparatus B the conditions for the cleaning of the target by heat treatment may change considerably. It is well known that under such conditions the carbide of the metal tends to be formed (carbiding), and this is very stable with respect to high-temperature treatment [391, 430]. In addition to this, as shown by Stewart [97], a carbide film on the surface of a refractory metal is a dielectric, and this reduces the thermionic (electron) emission of the metal. Hence, the relatively low value of the coefficient of potential emission of the electrons and the increased value of the work function, as well as the stability of the target with respect to high-temperature heating, are due to the formation of molybdenum carbide in apparatus B.

It follows from a consideration of curves 2 in Figs. 6.2 and 6.4 that the adsorption of oil molecules on the cold metal surface leads to an increase in γ, whereas after heat treatment adsorption produces a fall in this quantity. These different effects of the oil vapor on the value of γ are due to the fact that, in the first case, we are dealing with a surface coating of complex molecules of the $C_n H_{2n}$ type, whereas in the second we are concerned with the corresponding decomposition products in which the adsorbed molecules interact with the metal atoms and form carbides.

Finally we note that, whereas the surface of the metal subjected to heat treatment in system A always had a specular luster, under the conditions of system B it became mat. The reverse transformation of the surface from mat to specular was as impossible to produce by prolonged heat treatment as a change in stable values of φ and γ.

Evidently the absence of potential emission from the carbided surface is associated with its dielectric properties. Parilis, in fact, showed theoretically [548] that for the existence of potential emission $V_i > 2\varphi$ is a necessary but not sufficient condition. A second necessary condition is the presence of free electrons near the surface, the Auger neutralization responsible for the potential emission of electrons being associated with these. Hence, the transformation of the metallic surface into a dielectric increases the kinetic emission of electrons but reduces the potential emission to zero.

Thus, the experimental results show that on working with mercury diffusion pumps it is sufficient to subject the metal to heat treatment at appropriate temperatures in order to clean the target. If oil molecules fall into the apparatus, however, their adsorption on the surface and the subsequent chemical reaction during heat treatment leads to irreversible changes in the conditions on the target surface, accompanied by a rise in work function and a fall in the coefficient of potential emission of electrons.

§ 3. Comparative Study of Electron Emission from Metals Under the Influence of Inert-Gas and Alkali Ions

It follows from our review that almost all authors have studied the secondary electron emission of metals on bombarding with inert-gas ions at fairly low energies ($E_0 \sim 1$ keV). The most reliable results were obtained by Hagstrum in a series of investigations [262, 290, 345] with atomically clean surfaces.

It was found that the coefficient of secondary electron emission of molybdenum and tungsten bombarded by almost any of the inert-gas ions was almost independent of the energy of the bombarding ions E_0 between 10 and 1000 eV. This fact, together with the proportional relationship between γ and the energy evolved on neutralization of the ion at the target surface, indicates the potential character of the electron emission.

However, in the case of He^+, the independence of the coefficient of secondary electron emission γ with respect to the kinetic energy E_0 was only preserved up to ~400 eV; starting from this energy, the coefficient increased, and this could not be explained within the framework of the potential mechanism of electron emission. Owing to their relatively small mass, the He^+ ions have the greatest velocity for a given energy E_0. Hence, in order to explain the rise in γ, in this case one has to consider the possibility of emission by another mechanism (kinetic emission). According to this principle we may suppose that for higher energies of the bombarding ions both potential and kinetic electron emission may occur at the same time. It is important to discover to what extent these are additive.

This problem may be solved by studying the emission from the surface of the same metal under the influence of pairs of ions with the same or similar masses but different ionization potentials, choosing these in such a way that in one case potential emission occurs ($V_i > 2\varphi$) while in the other it is absent ($V_i < 2\varphi$). In this case we should expect that propinquity of the masses and similarity of the electron shells would lead to the same kinetic emission (γ_k) for both particles, and hence that the potential emission would be given by simply subtracting from the total emission

$$\gamma_p = \gamma - \gamma_k.$$

In order to determine the additivity or otherwise of the potential and kinetic emission of electrons, Arifov and Rakhimov [420] chose pairs of ions from contiguous inert-gas and alkali elements ($Ne^+ - Na^+$, $Ar^+ - K^+$, $Kr^+ - Rb^+$). In Table 6.1 the masses and ionization energies of these are compared.

The experiments were carried out with the help of an apparatus analogous to that shown in Fig. 6.1. The construction of the ion source was altered, together with that of the collector, for working with alkali-metal ions.

The target 10 was a strip of the metal under consideration, $30 \times 6 \times 0.015$ mm in size in the case of molybdenum and tantalum and $30 \times 6 \times 0.03$ mm in that of tungsten. The apparatus was evacuated with four two-stage glass mercury diffusion pumps, giving a pressure of $\sim 1 \cdot 10^{-7}$ mm Hg.

It is well known that on bombarding a clean target surface with alkali ions the latter are scattered while still retaining a considerable proportion of their initial energy. Hence, in the case of high primary-ion energies we had to apply a large potential difference between the target and collector in order to retard the scattered ions so as to be able to measure the total number of secondary electrons. This arrangement is liable to disrupt the geometry of the ion beam and lead to the appearance of conductivities in the electric circuit. This factor was eliminated by using a split collector [383], which enabled us to measure the secondary currents of positive and negative particles separately for relatively slight potential differences between the collector and the target 10. Analysis of the volt—ampere characteristics of the secondary currents showed that for a constant potential difference between the front collector 8 and the target (—50 V) and between the rear collector 9 and the target (+40 V) all the secondary positive ions were collected on the front collector and 90-95% of the secondary electrons on the rear collector. The complete collection of the secondary positive-ion current on the front collector was checked by measuring the secondary ion current flowing to this collector on bombarding the clean target with alkali ions at energies up to 1 keV, when there was no appreciable kinetic ejection of electrons. The collection of the secondary electrons on the rear collector was checked by measuring the thermionic-electron current or the current of secondary electrons ejected by inert-gas ions in the collector circuit.

The measurements with both inert-gas and alkali ions were made with the same potentials on the target and collectors.

Study of $\gamma(E_0)$ for Inert-Gas Ions

The secondary emission was measured for a target temperature of 1300°K after brief heating (30-40 sec) at 2400-2500°K. The magnetron ion source gave He^+, Ne^+, Ar^+, and Kr^+

Table 6.1

Ion	Mass	Ionization energy, eV	Ion	Mass	Ionization energy, eV
Ne^+	20,187	21.47	K^+	39.096	4.32
Na^+	22,997	5.09	Kr^+	83.7	13.94
Ar^+	39.94	15.68	Rb^+	85.48	4.19

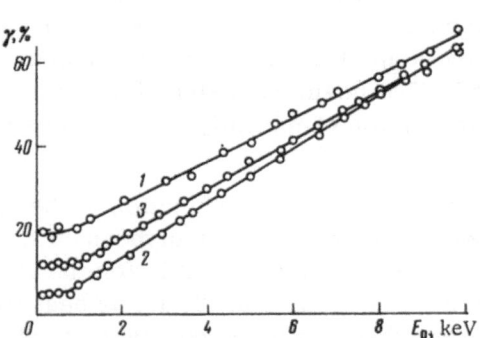

Fig. 6.7. The $\gamma(E_0)$ relationship obtained on bombarding a molybdenum target with Ar^+ ions for various states of the surface.

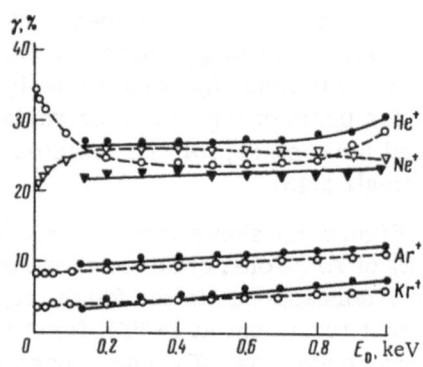

Fig. 6.8. The $\gamma(E_0)$ relationship obtained on bombarding a tungsten surface with inert-gas ions.

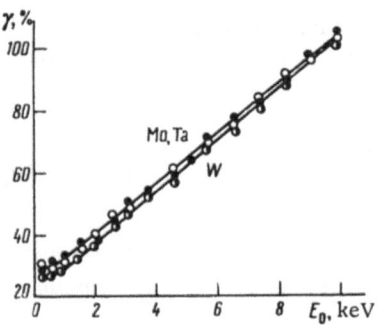

Fig. 6.9. The $\gamma(E_0)$ relationship obtained on bombarding clean molybdenum, tantalum, and tungsten targets with He^+ ions.

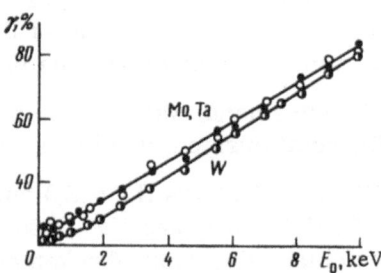

Fig. 6.10. The $\gamma(E_0)$ relationship obtained on bombarding clean molybdenum, tantalum, and tungsten targets with Ne^+ ions.

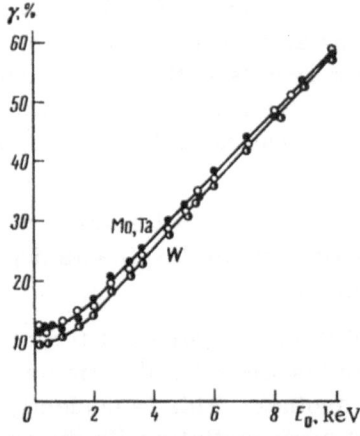

Fig. 6.11. The $\gamma(E_0)$ relationship obtained on bombarding clean molybdenum, tantalum, and tungsten targets with Ar^+ ions.

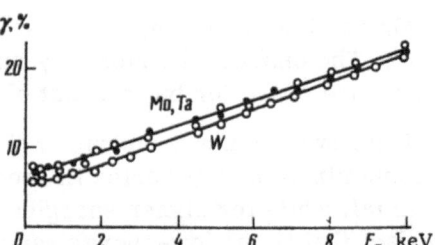

Fig. 6.12. The $\gamma(E_0)$ relationship obtained on bombarding clean molybdenum, tantalum, and tungsten targets with Kr^+ ions.

ion beams with a minimum energy of 110 and 80 eV for anode voltages (V_a) of 120 and 90 V, respectively. The energy spread of the ions was ± 10 eV for He^+ and Ne^+ and ± 5 eV for Ar^+ and Kr^+, as established by velocity analyzing the ions with the help of a cylindrical condenser. Thus, the beam energy could be varied from ~0.1–10 keV. The beam consisted of singly charged ions, since, for the selected values of V_a, the proportion of multiply charged ions was very small [345].

Figure 6.7 shows the coefficient of secondary electron emission as a function of the energy E_0 of Ar^+ ions, obtained at various stages of the heat treatment of the molybdenum target. Curve 1 corresponds to the fresh target before subjecting to heat treatment; 2 is for the target after heat treatment at 1800°K for 4 h; 3 after treatment at 2300°K for 5 h. Thus, as heat treatment proceeded the $\gamma(E_0)$ curve passed from 1 to 3 via 2 and then remained constant. Each point on the curves 2 and 3 was determined from measurements of the secondary currents at a target temperature of 1300°K after briefly heating to 1800 and 2300°K, respectively.

The initial high values of γ (curve 1) were also observed for tantalum and tungsten targets, and they were not reproducible during heat treatment. This fact indicates that curve 1 corresponded to a dirty target. The minimum value given by curve 2 was reproducible when the target remained in the cold state for a long time in vacuum, presumably as a result of the contamination of the target by gas adsorption [285, 347]. Further heat treatment at higher temperatures leads to the removal of the adsorbed gases, so that the emission rises until it reaches curve 3, which characterizes the fully degassed metal.

Measurements were made on all targets after cleaning them to a state corresponding to the $\gamma(E_0)$ relationship given in Fig. 6.7 (curve 3).

Figure 6.8 shows the $\gamma(E_0)$ curves for a tungsten target heated to 1300°K and bombarded with He^+, Ne^+, Ar^+, and Kr^+ ions at energies up to 1 keV, together with the $\gamma(E_0)$ curves of tungsten obtained in [262] (broken curves).

The values of $\gamma(E_0)$ for energies E_0 up to 10 keV obtained by bombarding molybdenum, tantalum, and tungsten targets with He^+, Ne^+, Ar^+, and Kr^+ ions are respectively shown in Figs. 6.9–6.12. We see from the figures that for energies between 0.1 and 1 keV the values of γ are almost independent of the energy of the ions for all targets and types of ion. Together with the rise in γ taking place on passing from an ion with a low V_i to one with a high V_i, this fact indicates the potential character of the emission of electrons at energies up to 1 keV. Quantitatively these data are in good agreement with Hagstrom's results [290, 345], which apparently indicates the cleanness of the metal surface. As regards the relatively high value of γ for tantalum, this may be due to the fact that the tantalum target was here degassed by heating to 2500°K, whereas, in [262], it was only heated to 1750°K.

On raising the energy from 1–10 keV there was a linear rise in γ for all cases considered. The onset of the rise in γ and the slope of the curves relative to the E_0 axis depended on the nature of the ion but not that of the target under bombardment.

It follows from a comparison of the $\gamma(E_0)$ curves in Fig. 6.7 that the effect of target surface contamination on γ varies in accordance with E_0. At fairly low energies this influence is substantial, while for higher energies the curves approach one another. This is evidently due to the fact that in the low-energy range the ion tends to interact preferentially with the surface film of gases, while at high energies the ion penetrates more deeply into the metal and starts interacting with the target itself [285].

The form of curves 1, 2, and 3 (Fig. 6.7) agrees in character with the results obtained in [285, 347] in relation to the effect of adsorbed gases on the potential emission of electrons.

Study of $\gamma(E_0)$ in the Case of Alkali Ions

Owing to the low spread in the energy of ions obtained from sources based on surface ionization (± 3 eV) it became possible to reduce the lower energy limit to 50 eV. The procedure for degassing the target was the same as in the experiments with inert-gas ions.

Figure 6.13 shows the $\gamma(E_0)$ curves for clean molybdenum targets (light circles) and tungsten targets (black circles) bombarded with Na^+, K^+, and Rb^+ ions at energies between 50 eV and 10 keV. It follows from these curves that the value of γ is low for energies up to 1 keV. This agrees with the results of Arifov and Ayukhanov [219, 342] and Eremeev et al. [237], while in the energy range up to 4 keV the value of γ agrees with Brunnee's results [383]. For higher energies (up to 10 keV) there is a linear rise in γ. As in the case of inert-gas ions, for energies of over 1 keV the inclination of the $\gamma(E_0)$ curves to the E_0 axis here mainly depends on the nature of the ions and hardly at all on that of the target.

We shall now try to explain the form of the $\gamma(E_0)$ curves for inert-gas ions in the energy range 1-10 keV. Starting from the experimental fact that E_0 has no effect on the potential emission of electrons from metals for primary-ion energies of up to 1 keV, it is natural to suppose that the potential emission is independent of the energy of the primary ions at higher energies also. Theoretical considerations also confirm this [62, 291].

Let us compare the energy-dependent part of the $\gamma(E_0)$ curves in the case of inert gases with the analogous curves for ions of the alkali elements, for which only kinetic emission of electrons occurs. Figure 6.14 shows the $\gamma(E_0)$ relationships for Ne^+, Ar^+, and Kr^+ ions already recorded in Figs. 6.10-6.12, after subtracting the part of the $\gamma(E_0)$ curve parallel to the E_0 axis, due to the potential extraction of electrons from the molybdenum target. This figure also shows the $\gamma(E_0)$ relations for Na^+, K^+, and Rb^+ ions. It follows from a comparison of these curves that the $\gamma(E_0)$ relationships agree closely for the pairs of ions Ne^+—Na^+, Ar^+—K^+, Kr^+—Rb^+. Since these pairs of inert-gas and alkali ions have almost identical masses and hence the same velocities for a specified kinetic energy, we might expect them to have the same kinetic emission, and this is in fact expressed by the coincidence of the $\gamma(E_0)$ curves for energies of over 2 keV. Thus, the increase in secondary electron emission starting from primary-ion energies of about 1 keV on bombarding these metals with inert-gas ions is due to the kinetic extraction of electrons. The identity between the $\gamma(E_0)$ characteristics of these pairs of ions also indicates that the potential emission depends very little on the energy of the primary ions in the high-energy range.

The $\gamma(E_0)$ relationship found for the alkali and inert-gas ions agrees with Parilis and Kishinevskii's theory of the kinetic emission of electrons [539]. The linear $\gamma(E_0)$ relation corresponds to a quadratic dependence of γ on the velocity v_0, as may be seen from Fig. 6.15. In the case of Ar^+ and Kr^+, and also K^+ and Rb^+, the values of γ coincide for equal velocities. In the case of Ne^+ and He^+ ions, however, there is a considerable deviation of the $\gamma(v_0)$ curves from those corresponding to Ar^+ and Kr^+, the degree of this discrepancy being the greater, the higher the velocity of the ion and the lower its mass.

With increasing primary-ion energy, the effect of the nature of the target on the ejection of electrons diminishes, and for E_0 values over 8 keV the emission is almost identical for molybdenum, tantalum, and tungsten.

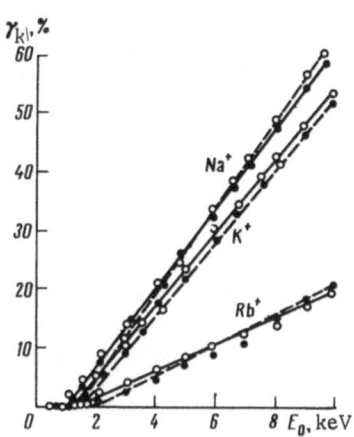

Fig. 6.13. The $\gamma(E_0)$ relationship on bombarding the surface of tungsten and molybdenum targets heated to 1300°K with alkali ions.

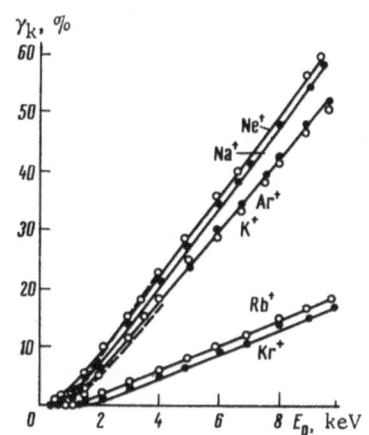

Fig. 6.14. The $\gamma_k(E_0)$ relationship for inert-gas and alkali ions.

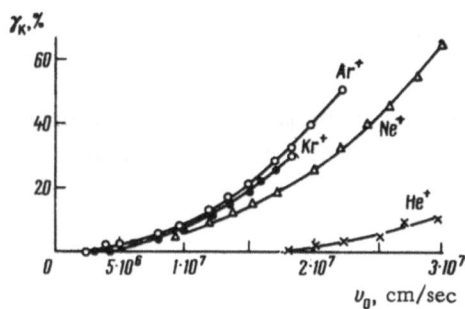

Fig. 6.15. The $\gamma_k(v_0)$ relationship for various ions.

§4. Dependence of the Potential Emission of Electrons on the Temperature and Work Function of the Target [663]

It was found in a number of investigations [61, 138, 147, 221] that the heating of the target led to a change in the coefficient of ion—electron emission from metals. This change consisted chiefly of a reduction in the coefficient of secondary ion—electron emission γ with increasing temperature of the metal. On the other hand, it was shown by Veksler [242] for alkali ions that temperature had no appreciable effect on the value of γ. Other authors considered that the relation between γ and temperature was not established [431]. At the present time there are also no reliable data relating to the dependence of the coefficient of secondary electron emission on the work function of metals. Gurtovoi [157] attempted to elucidate this relationship for thoriated tungsten, and Veksler, Klein, and Shuppe [239] tried it for a number of metals bombarded with mercury ions.

From the results of experiments on tantalum and hafnium [262, 388] Hagstrum was unable to explain the dependence of the coefficient of potential emission of electrons on the work function of the metal. This was because of the impossibility of bringing the tantalum and hafnium to a state of atomic cleanness by heating to 2000°K, these metals being able to absorb gases strongly in the heated state.

In this section we shall set out the results of some experiments by Arifov and Rakhimov [421] on the relation between the coefficient of potential electron emission and the target temperature and work function.

In order to study the $\gamma_p(T)$ relationship the surface of the metal has to be atomically clean. In order to satisfy this requirement, the molybdenum target was degassed at 2300°K and the secondary currents were measured at various temperatures by the double-modulation inertia-free oscillograph method described earlier. In addition to this, as a result of the intensity modulation of the primary ion beam and secondary emission, this method made it possible to study the $\gamma(T)$ relationship over the temperature range 300–1600°K. In order to study the dependence of γ on φ, platinum, nickel, tungsten, tantalum, zirconium, and magnesium targets were taken (work functions between

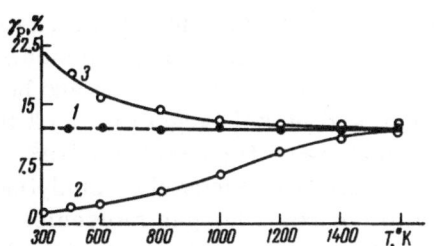

Fig. 6.16. Variation in γ_p with target temperature.

5.29 and 3.46 eV). For the majority of the targets the work functions of similar filaments subjected to the same heat treatment as the targets under the same vacuum conditions were plotted. In order to study the potential emission in pure form and also its dependence on the temperature and work function φ of the target, Ar^+ and He^+ ions with a kinetic energy of 200 eV were used. The experiments were carried out in apparatus A.

After prolonged degassing of the target, the temperature dependence of γ was studied. Measurements of the work function of molybdenum filaments subjected to the same heat treatment as the strip target showed that after 5 h of heat treatment at 2300°K, $\varphi = 4.25$ eV. This fact indicated that the target was reasonably clean. After this, oscillograms of the V/A characteristics and secondary currents were obtained for various target temperatures. The results of an analysis of these oscillograms are presented in Fig. 6.16, which gives the dependence of γ_p on the temperature of the molybdenum target; curve 1 was obtained for a clean surface just after flashing at 2300°K, curve 2 after an equilibrium coating had been formed on the surface, and curve 3 the same in the presence of potassium vapor at a pressure of $6 \cdot 10^{-7}$ mm Hg in the system. Each point on these curves was obtained from individual V/A characteristics on bombarding the target with 200-eV Ar^+ ions.

It follows from a consideration of curve 1 that in the temperature range 300–1600°K the coefficient of potential electron emission is independent of target temperature, while in the presence of surface contamination due to the adsorption of residual-gas molecules there is a γ_p(T) relationship caused by the different densities of adsorbed particles at different temperatures (curve 2). At high temperatures evaporation of the adsorbed gas molecules takes place, and this is expressed by a transformation from curve 2 to curve 1.

Since the residual gas consists mainly of molecules forming part of the composition of air, the fall in γ_p agrees excellently with the results of [285, 347] relating to the effect of gases on potential emission. However, the measurement of γ_p in the presence of adsorbed potassium atoms on the surface showed that in this case there was a rise in γ_p with falling temperature (curve 3, Fig. 6.16).

Thus, if the target is fairly clean, the coefficient of potential electron emission is independent of temperature between 300 and 1600°K.

A variation in γ_p with temperature occurs every time the surface is contaminated with foreign adsorbed particles. The way in which γ_p varies in these cases depends on the nature of the adsorbed particles. Whereas, for example, on reducing the target temperature, the adsorption of residual-gas molecules forming part of the air leads to a fall in γ_p, on adsorbing electropositive contaminants, γ_p rises. We therefore consider that the γ_p(T) relationship observed in [61, 138, 147, 221] was due to the contamination of the target surfaces.

In order to study the dependence of γ on the work function we used targets 6 mm wide, 30 mm long, and 0.015 mm thick in the cases of molybdenum, tantalum, and nickel and 0.03 mm thick in those of tungsten, platinum, and zirconium. Before the measurements the tungsten, tantalum, and molybdenum targets were degassed at 2900, 2500, and 2300°K, respectively, for 8–10 h. In view of the comparatively low melting points of platinum, nickel, and zirconium, these metals could not be heated as strongly as the others, and the highest possible temperatures for degassing these were taken as 100–150° below their melting points.

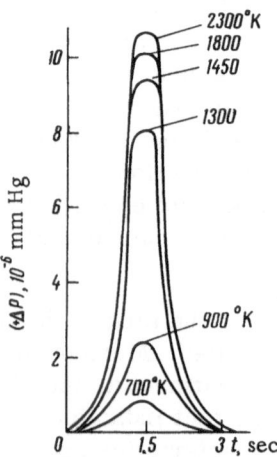

Fig. 6.17. Change in pressure within the apparatus on "flashing" for various temperatures of a molybdenum target covered with an equilibrium layer of adsorbed molecules.

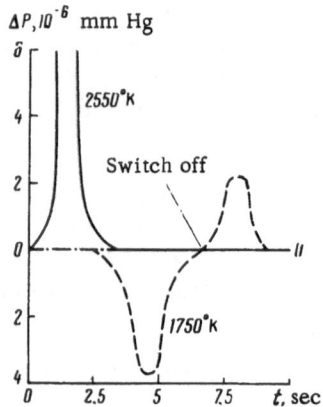

Fig. 6.18. Change in pressure on "flashing" at 1750°K (tantalum and zirconium) and 2550°K (tantalum), and also on disconnecting the heating current.

For the molybdenum, tantalum, and tungsten targets we took the work functions of the corresponding filaments measured under the same conditions of evacuation and heat treatment. For the platinum and nickel targets we took the tabulated values of φ, also measured for filaments [198]. The permissibility of this device was checked by studying the adsorption and desorption of residual-gas molecules on the surface of strip targets and filaments by the "flash" method [345, 388]. It was found that the heating of platinum, tungsten, tantalum, and molybdenum to the degassing temperatures indicated led to evolution of the adsorbed gas, such that the relative amount of gas evolved was proportional to the surface area of the metal and depended on the heating temperature (Fig. 6.17). However, the heating of tantalum and zirconium to 2000°K was always accompanied by the absorption of gas from the volume.

Figure 6.18 shows curves schematically illustrating the variation in residual-gas pressure on heating tantalum and zirconium to 1750°K and tantalum to 2550°K. The horizontal axis indicates the heating period. These curves were obtained after an equilibrium coating of adsorbed molecules had been formed on the surface. It follows from the curves that on heating tantalum and zirconium targets to 1750°K the pressure in the volume of the apparatus fell, while on heating tantalum to 2550°K the gas evolved in almost the same quantity as in the case of molybdenum. The reduction in gas pressure within the volume of the apparatus in the first case can only be explained by the absorption of gas in the tantalum.

On disconnecting the heating current of a target heated to 1750°K the pressure rose, while on disconnecting that of a target heated to 2550°K this did not take place, presumably owing to the reemission of some of the absorbed gas from the metal into the vacuum in the first case.

The evolution of gas from the tantalum on heating to 2550°K and the constancy of the pressure on disconnecting the heating current show that tantalum may be effectively cleaned at this temperature. On heating zirconium up to the melting point the character of the $\Delta p = f(t)$ curve remained the same as in the case of tantalum heated to 1750°K. The zirconium was therefore not clean in our experiments, and for the same reason the tabulated values of φ for zirconium also fail to correspond to the clean surface of this metal. The absorption of residual gases by zirconium and tantalum in the heated state agrees with existing information relating to the adsorption and absorption of certain gases by these metals [210].

A reduction in the residual-gas pressure on heating a metal covered with an equilibrium layer of adsorbed molecules and the reemission of the gas after disconnecting the target-heating current clearly indicate either adsorption as a result of chemical interaction between the gases and the metal or volume absorption. On the other hand, an increase in the residual-gas

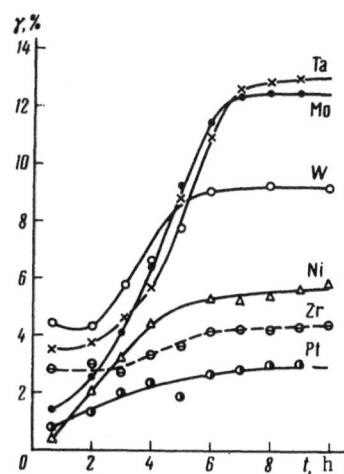

Fig. 6.19. Dependence of γ on the period of heat treatment of various metals.

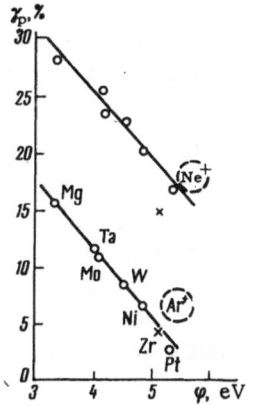

Fig. 6.20. Dependence of the coefficient of potential electron emission γ_p on the work function φ.

pressure on heating a metal covered with an equilibrium layer of adsorbed molecules and the cessation of gas evolution after disconnecting the heater indicate that these processes are not taking place, and that the surface may in fact be cleaned by heating to high enough temperatures.

In the latter case, on further allowing for the possibility of the renewing of the surface by evaporation of the target material on heating to the temperatures in question [391], we may consider that, after prolonged heat treatment, the work function of strip targets will not differ substantially from the values of work function measured in the case of tungsten, molybdenum, and tantalum and tabulated in [198] for platinum and nickel.

We note that the change in pressure before and after disconnecting the heating current from the tantalum depends on the temperature of preliminary treatment, and also on the "flash" temperature. For example, if the preliminary treatment is only carried out up to 2000°K, then Δp is small; if treatment is carried out at 2500-2600°K (tantalum), and the flash at 1700°K, then Δp becomes considerable, and on switching off the current there is a considerable evolution of gas. It would appear that the low temperature of target treatment (1750°K) used by Hagstrum [262] prevented the absorption of gas on heating and the reemission on switching off the current from being observed.

In order to obtain $\gamma_p(\varphi)$ we measured the value of γ_p as a function of the time of heat treatment for various metals (Fig. 6.19) on bombarding with 200-eV Ar$^+$ ions. It follows from the curves that in the initial stages of heat treatment the value of γ_p is small for all metals; as degassing proceeds it becomes greater and tends toward saturation.

The value of γ_p for a magnesium target was measured while depositing metallic magnesium on the surface of pure molybdenum by evaporating from a special furnace. The rate of deposition was such that the curve relating γ_p to the deposition time reached saturation in tenths of a second. This rate of deposition was considered to be sufficient to produce a pure magnesium film since the time required for the formation of an equilibrium layer as a result of the competing adsorption of residual gas when working with magnesium was longer than 20 sec, owing to the formation of a getter on the walls of the apparatus.

Figure 6.20 shows the relation between γ_p and work function for Ne$^+$ and Ar$^+$; each point (apart from that relating to the magnesium target) is taken from the saturation points of the curves in Fig. 6.19. It follows from these curves that γ_p depends greatly on the work function of the surface; as φ changes from 3.46 to 5.29 eV, the γ_p for Ar$^+$ changes from 16 to 2.5% and for Ne$^+$ from 28 to 16.5% in an almost linear manner. The close agreement of γ_p for tantalum and molybdenum targets is evidently due to the similarity of the work functions and the structures of the electron energy bands of these metals.

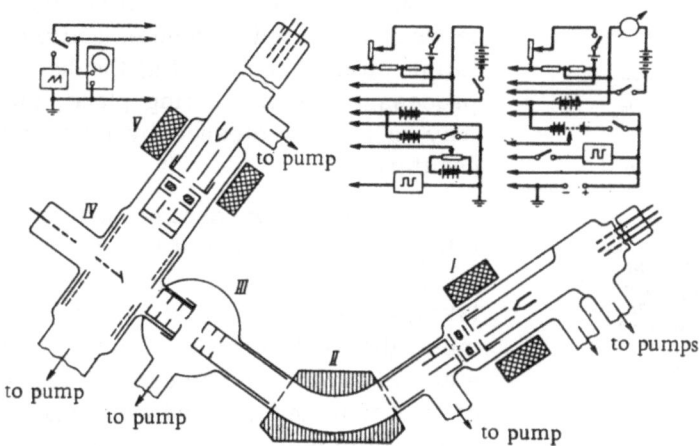

Fig. 6.21. Arrangement of the vacuum apparatus and the electric measuring circuit.

§5. Influence of the Mass of an Ion

on Ion—Electron Emission (Isotopic Effect)

In order to discover the dependence of the coefficient of kinetic emission on the mass of the bombarding particle, it is convenient to carry out experiments with ions constituting different isotopes of the same element.

As mentioned in the review, the isotopic effect was studied for ion—electron emission in [226, 345]. In both cases, neon isotopes (Ne^{20} and Ne^{22}) were employed. For these ions, both the potential and the kinetic emission of electrons were identical for equal velocities. In this section we shall describe another investigation [607] into the isotopic effect in which the ion—electron emission of molybdenum was measured after bombarding with molecular ions of hydrogen isotopes (H_2^+, D_2^+) and He^+ over the energy range $E_0 = 0.1$–45.0 keV. Since the condition $V_i > 2\varphi$ is satisfied for these ions, we should expect both potential and kinetic electron emission to take place for a sufficiently high kinetic energy.

So as to be able to estimate the part played by the potential mechanism of electron emission in relation to the total emission, the measurements were made in two vacuum systems enabling work to be carried out with hydrogen-ion beams in two energy ranges: 0.1–4.0 and 4.0–45.0 keV. The apparatus and operating principles of the first system were described in [419]. Figure 6.21 shows the arrangement of the second system. The apparatus consisted of a magnetron H^+ and He^+ ion source (I), a mass analyzer (II), a principal accelerating section (III), a receiver (IV), and a source of an auxiliary Ar^+ ion beam (V). Ion sources of the Finkelshtein type [150] were used. The source of the additional Ar^+ ion beam was placed in a side tube of the receiving part of the apparatus and was only brought into action in order to verify the cleanness of the target surface. The target was placed in the center of a cylindrical collector in such a position as to enable it to be bombarded by the main and auxiliary ion beams successively. The secondary currents were established (recorded) by the double-modulation method on an oscillograph screen.

The apparatus was evacuated with five mercury diffusion pumps. The residual-gas pressure in the receiving part of the apparatus was of the order of $2 \cdot 10^{-7}$ mm Hg. The target was cleaned by prolonged heating at temperatures above $2000°K$.

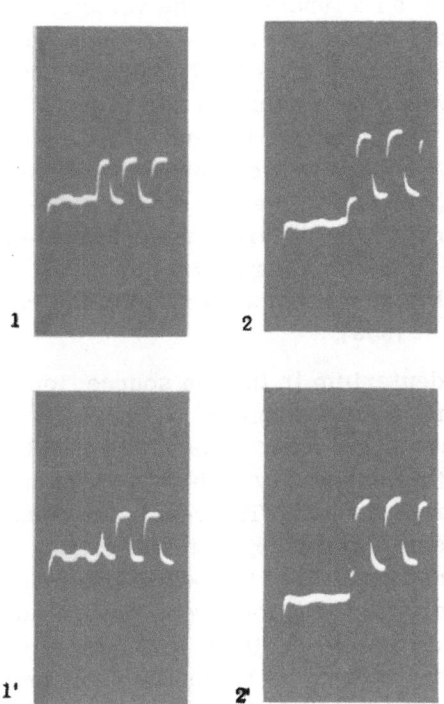

Fig. 6.22. Oscillograms of V/A characteristics on bombarding a molybdenum target with Ar^+ ions at an energy of 400 eV.

The coefficient γ was measured at a temperature of the order of 1300–1400°K after flashing to 2000°K. The vacuum conditions and measuring procedure were identical in both systems.

Whereas in the ion source hydrogen at a pressure of 10^{-4} mm Hg was employed, in the receiving part the hydrogen pressure was ~10^{-6} mm Hg. In view of this, the metal chosen for the target had to be one that would lose all its absorbed hydrogen on heating to fairly high temperatures. The most suitable metal from this point of view was molybdenum, since no interaction took place between the hydrogen and the molybdenum surface up to the melting point [391], while all hydrogen dissolved in the metal could be removed by heating to an appropriate temperature [210].

The state of the molybdenum surface was monitored by reference to the coefficient of potential electron emission on bombarding with 400-eV Ar^+ ions [347].

Figure 6.22 shows oscillograms of the V/A characteristics of potential electron emission obtained before (1, 2) and after (1', 2') admitting hydrogen to the ion source 1 for the same intensity of the Ar^+ ion beam. Oscillograms 1 and 1' are taken at room temperature, and 2 and 2' at 1400°K immediately after the high-temperature "flash" (2000°K).

We see that the potential emission of electrons is much greater for the heated target than for the cold target, both with and without the presence of hydrogen in the system. This fall is probably due to contamination as a result of the adsorption of molecules of the residual gas and hydrogen on the surface at low temperatures of the metal, in quantitative agreement with Hagstrum's observations [347]. However, for the heated target there was the same electron emission in both cases (2, 2'), the value of γ agreeing within the limits of measuring error with the values of γ found by Parilis and Kishinevskii [539]. This indicates that the surface of the heated target was reasonably free from both residual gas and hydrogen. All measurements were accordingly made at target temperatures of 1400°K after a brief "flash."

Investigations within the energy range 0.1–4.0 keV were only carried out with H_2^+ and D_2^+ ions in an apparatus in which the beam was not mass-separated. However, the parameters of the ion source (gas pressure, anode current and voltage) were chosen such that the proportion of atomic ions in the beam should be negligible. This mode of operation was established by special experiments conducted in the same apparatus.

Consideration of the curves in Fig. 6.23 shows that the values of γ for H_2^+ and D_2^+ in the energy range up to 300 eV were almost the same and equal to 6%. On further raising the energy there was a monotonic rise in γ for the H_2^+ ions, a sharp increase starting at 600 eV; the γ value for the D_2^+ ions remained almost constant up to an energy of 1000 eV, after which there was an almost linear rise in γ. We see that the form of the $\gamma = f(E_0)$ curves for the H_2^+ ions is sharper than for the D_2^+.

The faster rise in γ for energies up to 1 keV in the case of the H_2^+ ions may clearly be explained by the fact that, owing to a comparatively low mass of H_2^+, the latter has a higher velocity for a specified kinetic energy. The value of $\gamma = 6\%$ in the energy range < 1 keV may be ascribed to potential emission of electrons from molybdenum by the H_2^+ and D_2^+ ions.

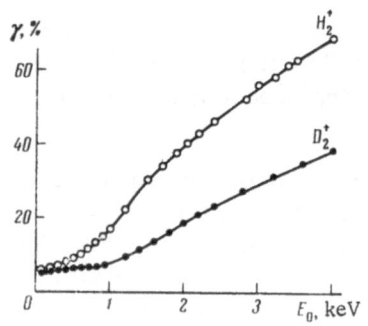

Fig. 6.23. Dependence of γ on the energy of the bombarding particle E_0.

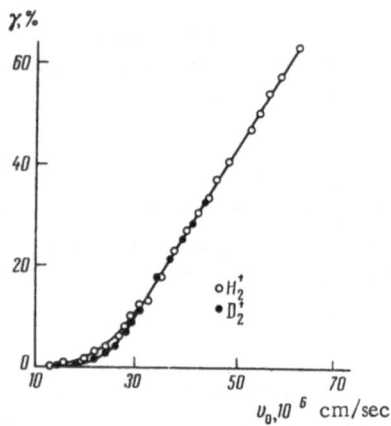

Fig. 6.24. The $\gamma(v_0)$ relationship for molecular H_2^+ and D_2^+ ions.

Figure 6.24 shows γ as a function of the velocity of the ions v_0. Except for the initial part, the experimental points obtained for the H_2^+ and D_2^+ ions lie (within the limits of experimental error) on a single common curve, i.e., the $\gamma = f(v_0)$ curves are the same for H_2^+ and D_2^+. The so-called isotopic effect only occurs in a comparatively low-velocity range. This may be due to the fact that the isotopic effect is associated with the different retardations of ions of different masses on penetrating into the metal; hence, it is most noticeable at low velocities, when the retardation effect is greatest [539].

In the presence of deuterium in the ion source, ions with masses of 1, 2, 3, and 4 were observed. Clearly, a beam with mass 4 consists of D_2^+ ions. Ions with mass 3 are due to DH^+. Ions with mass 2 may be both deuterium ions and ions of molecular hydrogen. The results of measuring γ with beams of mass 2 in the apparatus of Fig. 6.21 gave grounds for considering that the beam with mass 2 was chiefly due to deuterons (if H_2^+ ions were present at all, there were very few of them). On these grounds, we may consider that the beam with mass 1, consisting of protons, was formed as a result of the dissociation of DH molecules rather than H_2. The measurements at 4.0-45.0 keV in the apparatus shown in Fig. 6.21 therefore related to the following ions: H_1^+, H_2^+, D_1^+, D_2^+, DH^+, and He^+.

Figure 6.25 shows curves of $\gamma = f(E_0)$ for all the ions under consideration. We see from these curves that for a given energy the γ coefficient is greater for the light isotope than for the heavy, and that the molecular ion ejects more electrons at a specified energy than the atomic ion. Figure 6.26 shows the coefficient of kinetic ion—electron emission $\gamma_k = \gamma - \gamma_p$ as a function of velocity. Analysis of these curves leads to the following conclusions.

For a given velocity, the molecular ions eject twice as many electrons as the corresponding atomic ions. This is in good agreement with the views of many research workers regarding the decay of molecular ions into their atomic components on striking the surface of a solid.

The experimental points obtained for the molecular ions H_2^+ and D_2^+ fall on a single common curve (Fig. 6.26). The same may be said of the atomic ions (i.e., in the energy range 4.0-45.0 KeV, within the limits of experimental accuracy, there is no mass dependence of γ for the ions of the hydrogen isotopes). The absence of this effect is apparently due to the great energy of the bombarding ions. Clearly, for such energies, the difference in the retardation of the two particles D^+ and H^+ within the layer of material from which secondary electrons are capable of arising is in practice so slight that there is no mass dependence of γ. Thus, the identical values of γ for H_1^+ and D_1^+ and for H_2^+ and D_2^+ at the same velocities indicate that the mass of the ion plays no substantial part in the mechanism underlying the kinetic emission of electrons.

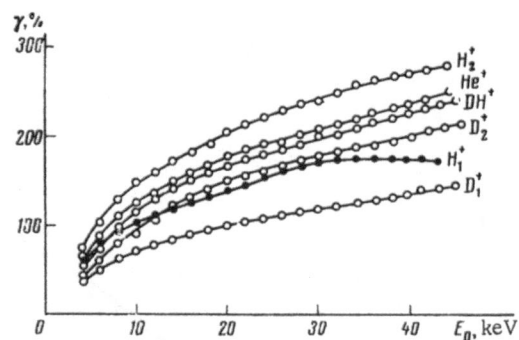

Fig. 6.25. Dependence of γ on the energy
of the bombarding ion.

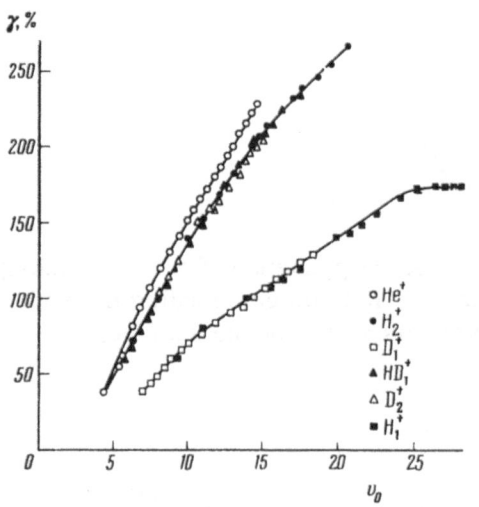

Fig. 6.26. Relationship $\gamma(v_0)$ for ions
formed by the isotopes of hydrogen and
helium. Horizontal axis expressed in 10^7
cm/sec.

§ 6. Energy Dependence of Electron Emission from Metals Under the Influence of Inert-Gas Ions

It follows from our review that in all investigations into the coefficient of kinetic ion—electron emission γ_k as a function of the energy of the bombarding particles there has been a monotonic increase in γ_k with E_0 starting from a certain minimum energy E_{th} (threshold energy), the value of which depends on the nature of the target and the ion. However, the character of the $\gamma_k(E_0)$ curve has not been finally established. Whereas Brunnee [383] and Arifov and Rakhimov [493] reported a linear rise in γ_k with E_0, Petrov [505] observed a deviation of $\gamma_k(E_0)$ from the linear form in the case of light ions. The theoretical calculations of Parilis and Kishinevskii [539] gave a more complicated $\gamma_k(v_0)$ relationship. Whereas, in the energy range up to a few keV, the $\gamma_k(v_0)$ curve had a quadratic form, for higher energies there was a linear rise in γ_k with v_0. It should be noted that in the majority of investigations the $\gamma_k(E_0)$ relationship was studied over different energy ranges. The construction of a curve for a wide energy range based on the results of the various authors is not very helpful, since the vacuum conditions and measuring methods were also different.

We shall now present the results of an investigation into the electron emission of molybdenum, tungsten, and nickel on bombarding with He^+, Ne^+, and Ar^+ ions at energies of $E_0 = 1.0$-50 keV.

The experiments were carried out in the vacuum apparatus shown schematically in Fig. 6.27. The apparatus consisted of an ion source, a region of beam formation, and a measuring section. The ion source used the principle of ionizing the gas atoms by electron impact between the heating filament 1 and the anode 2, placed in the magnetic field of a solenoid. The narrow ion beam was formed by means of a single electrostatic lens incorporating electrodes 4, 5, and 6. The measuring part consisted of a screening cylinder 7 with three successive slits, a spherical collector 8, and a target 9. The apparatus was evacuated with four diffusion pumps connected in parallel, giving a residual-gas pressure of the order of $(2-3) \cdot 10^{-7}$ mm Hg in all parts of the apparatus. The same figure shows (right) the electric circuit used in the measurements, based on the double-modulation method [281]. Intensity modulation of the primary ion beam was effected by applying rectangular voltage pulses from a special generator between the anode 2 and the vacuum diaphragm 3. The second modulation was effected by applying a sawtooth voltage between the target 9 and the collector 8. The principal acceleration of the narrow ion beam was secured by means of an electric field applied between electrodes 6 and 7. The

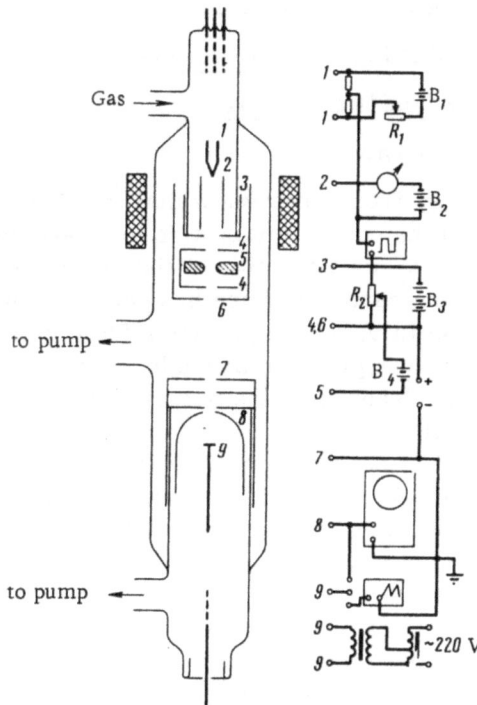

Fig. 6.27. Vacuum apparatus and principal electric measuring circuit.

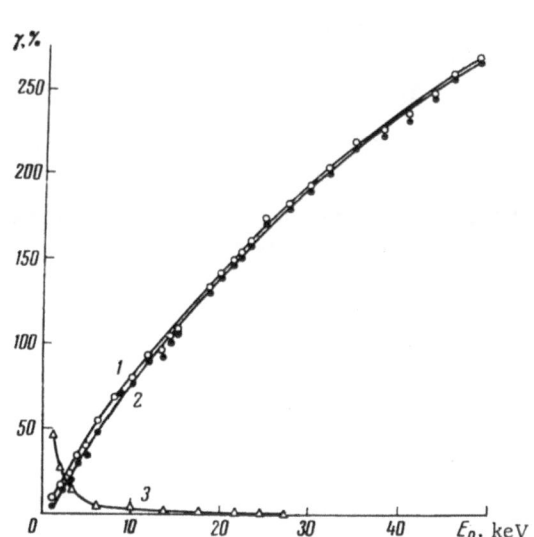

Fig. 6.28. Dependence of γ on the energy of the Ar^+ ions on bombarding molybdenum for various states of the surface.

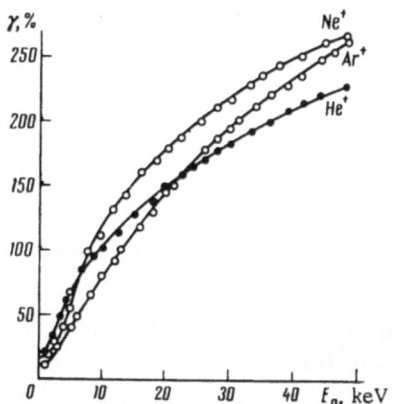

Fig. 6.29. The relation $\gamma(E_0)$ on bombarding clean molybdenum with inert-gas ions.

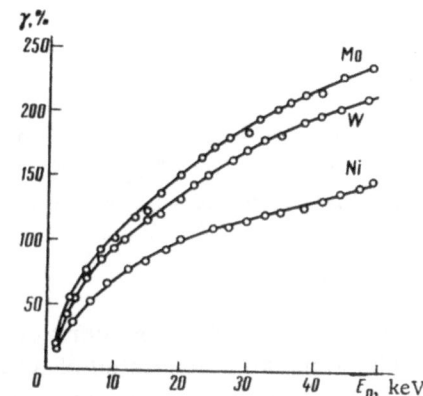

Fig. 6.30. The relation $\gamma(E_0)$ on bombarding molybdenum, tungsten, and nickel with He^+ ions.

construction of the apparatus enabled the energy E_0 to be varied between 1 and 50 keV. Great attention was paid to the cleanness of the surface under consideration. The targets were subjected to prolonged heat treatment at 2500°K in the case of tungsten, 2200°K in that of molybdenum, and 1500–1600°K in that of nickel. The progress of the degassing was checked by reference to the potential electron-emission coefficient γ_p and also by reference to the secondary positive-ion current [262].

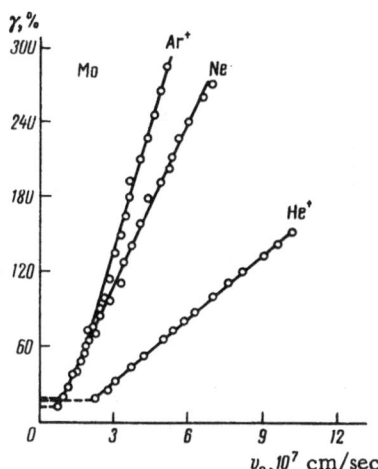

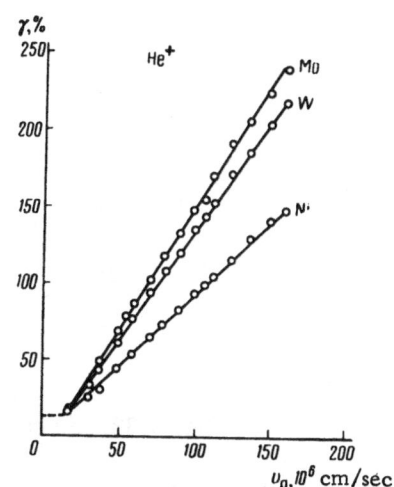

Fig. 6.31. Dependence of γ on the velocity of the ion v_0 obtained from the $\gamma(E_0)$ curves in Fig. 6.29.

Fig. 6.32. Dependence of γ on the velocity of the ion v_0 for various metals.

Measurements were made both at room temperature and at a target temperature of over 1000°K immediately after brief flashing to the temperatures just mentioned.

Figure 6.28 shows the energy dependence of γ with respect to Ar^+ ions bombarding a molybdenum target immediately after high-temperature "flashing" (curve 1) and again after 20 sec, i.e., after the formation of an equilibrium coating by the adsorption of residual-gas atoms (curve 2). We see that up to an energy of 15-20 keV, the value of γ is much greater for a dirty target than for a clean one, while at higher energies the results are almost the same in both cases. In order to show the effect of gas adsorption on γ at various energies the same figure presents the ratio $(\gamma_{cl} - \gamma_{contam})/\gamma_{cl}$ as a function of energy E_0 (curve 3). We see that as the energy increases, this ratio falls sharply and becomes negligible for energies of over 15-20 keV.

Figure 6.29 presents $\gamma(E_0)$ curves obtained by bombarding a clean molybdenum target with He^+, Ne^+, and Ar^+ ions. The value of γ at $E_0 = 1$ keV agrees satisfactorily with the corresponding data from [493, 505, 262] for all the ions; the curves merge with the values of γ in the energy range below 1 keV. On raising E_0 above 1 keV there is a monotonic rise in γ for all cases; the order of the curves and the values of γ in the corresponding energy range agree with the results of [493, 505].

Figure 6.30 shows $\gamma(E_0)$ curves for molybdenum, tungsten, and nickel targets bombarded with He^+ ions. We see clearly that the character of the $\gamma = f(E_0)$ curve established for molybdenum also holds for the other targets. However, the value of γ for a given E_0 falls on passing from molybdenum to nickel, the difference for the different targets increasing with increasing energy of the incident ion.

Figures 6.31 and 6.32 show the dependence of γ on the velocity of the ion v_0 obtained by converting the $\gamma(E_0)$ curves given in Figs. 6.29 and 6.30, respectively. By considering these graphs we may conclude that the form of the $\gamma(v_0)$ relationships is adequately described by a straight line, the inclination of this straight line to the v_0 axis depending on both the nature of the bombarding ions and the target material. The main result of these experiments is the establishment of the following laws in the energy range studied: the coefficient of kinetic electron emission increases almost linearly with increasing velocity of the ion; for a given velocity $\gamma_{Ar} > \gamma_{Ne} > \gamma_{He}$; for a given type of bombarding ion, $\gamma_{Mo} > \gamma_W > \gamma_{Ni}$.

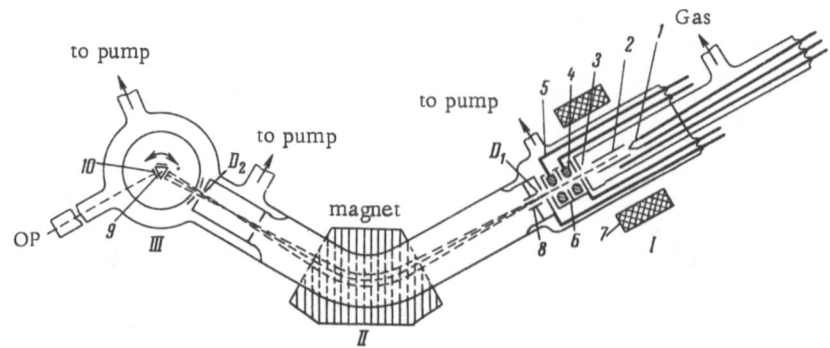

Fig. 6.33. Arrangement of the experimental apparatus.

The rectilinear $\gamma(v_0)$ relationship agrees with that predicted theoretically by Parilis and Kishinevskii [539] for velocities $v_0 > 3 \cdot 10^7$ cm/sec, while the slope of the linear parts of $\gamma(v_0)$ agrees qualitatively with their calculations [539, 618] for Ar^+ and Ne^+ ions; for the He^+ ion there is only qualitative agreement: the order of sequence of the curves is the same, but the calculated slope is double the experimental value.

The difference in the coefficients of kinetic electron emission for the metals in question is evidently due to the effect of the work function on the yield of excited electrons. However, there is no simple proportionality between γ and φ. This circumstance indicates that other factors are also exerting an influence, and further investigations are needed in order to reveal these.

§7. Electron Emission in the Bombardment

of Tungsten Single Crystals by Ar^+ Ions

The study of phenomena taking place on bombarding single crystals with ions was started fairly recently. Magnuson and Carlston [650, 651] investigated the emission of electrons by single crystals. However, in studying ion—electron emission, metallic single crystals with a comparatively low melting point were employed. In order to clean the surface from adsorbed contaminants, sputtering by ion bombardment was therefore applied. This method has the disadvantage that, for the beam densities used, contamination of the target by the beam itself (as a result of the penetration effect) cannot be excluded. In order to study ion—electron emission under clean conditions, the most suitable sample is a single crystal of one of the refractory metals.

In this section we present the results of a study of ion—electron emission from the (111), (100), and (110) faces of a tungsten single crystal on bombarding with Ar^+ and Ar^{2+} ions at energies between 0.6 and 10 keV carried out by Mukhamadiev and Rakhimov [783].

The arrangement of the experimental apparatus is shown schematically in Fig. 6.33. The apparatus consisted of three parts: an ion source (I), a mass spectrometer (II), and a receiving part (III). In the figure, 1 and 2 signify the magnetron ion source while 4 and 5 represent the electrostatic lens. The plates 6 served to modulate the primary ion beam. The individual sections of the apparatus were bounded by plane diaphragms 3 and 8. The 60° mass analyzer enabled us to direct ions of specified mass and a specified charge onto the target 10 and to separate the ions from the fast atoms accompanying them in the beam.

In order to obtain reliable results, the faces being studied should be subjected to the same vacuum conditions and the same thermal action. This requirement was satisfied by mounting

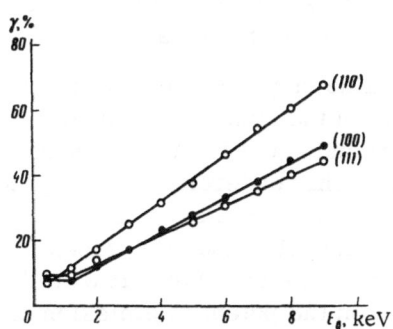

Fig. 6.34. The $\gamma(E_0)$ relationship obtained on bombarding various faces of a tungsten single crystal with Ar$^+$ ions.

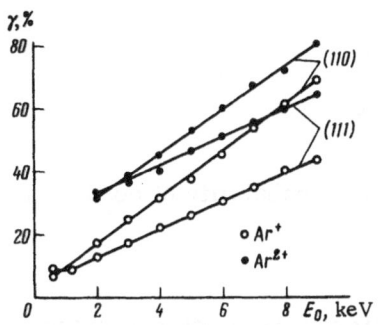

Fig. 6.35. The $\gamma(E_0)$ relationship obtained on bombarding various faces of a tungsten single crystal with Ar$^+$ and Ar^{2+} ions.

the three faces of the single crystal on the sides of a triangular prism 9 fixed on the 00' axis. The prism was made of tantalum plate. Rotation of the objects was effected from outside with the help of a magnet. In this way we were able to place the faces successively in front of the aperture in the collector and direct the ion beam on the desired face. The faces were set perpendicular to the beam. The samples were heated by electron bombardment. Before the measurements the single-crystal target was cleaned by prolonged heating at temperatures of over 2200°K.

Measurement of the secondary currents was effected by the oscillograph double-modulation method. For this purpose, the intensity of the primary ion beam was modulated by applying a rectangular voltage to the plates 6. The low inertia of this method enabled us to make a considerable reduction in the effect of gas adsorption by the target surface on the emission of secondary electrons. In one measuring session several series of $\gamma(E_0)$ curves for each of the three faces of the tungsten single crystal were obtained under the same vacuum conditions.

The apparatus was evacuated by three mercury diffusion pumps; this gave a vacuum in the receiving part of the apparatus of $(2-3) \cdot 10^{-7}$ mm Hg.

The results obtained by measuring the coefficient γ as a function of energy on bombarding with singly charged argon ions are presented in Fig. 6.34. We see that the coefficient γ increases almost linearly with energy. However, the $\gamma(E_0)$ characteristics have different slopes, which indicates anisotropy of the ion—electron emission. In the region of kinetic emission ($E_0 > 1$ keV) there is a sharp anisotropy, i.e., for the same energy the (110) face has the greatest emission and the (111) has the least.

In the case of potential emission the anisotropy of γ is not very sharp; however, γ is greatest for the (111) face and smallest for the (110). A further study of electron emission at lower energies will be required in order to draw final conclusions regarding the effect of crystallographic directions on the emission behavior.

Figure 6.35 shows the $\gamma(E_0)$ relationship for the (111) and (110) faces on bombarding with doubly charged argon ions. For comparison the same figure shows the results for singly charged argon. On considering these curves we see that, as in the case of singly charged ions, the γ factor relating to the doubly charged ions increases almost linearly with rising energy. It is quite clear that the $\gamma(E_0)$ characteristic for Ar^{2+} lies above the $\gamma(E_0)$ characteristics for Ar$^+$; however, for a particular face of the single crystal the inclination of the $\gamma(E_0)$ line to the E_0 axis is almost the same for both singly and doubly charged ions. In agreement with [656], this fact indicates that the velocity of the ion has no effect on the coefficient of potential electron emission γ_p.

In the energy range $E < 1$ keV, in which potential emission predominates, the curves are arranged in accordance with the work functions of the various faces [471]. Since the energy of the ions is not very great in this region, interaction with the lattice atoms occurs in the layer next to the surface, and the predominant influence on the potential-emission coefficient γ_p is exerted not by the density of the target atoms, but by the work function of the surface.

The value of γ for doubly charged ions is greater than the value of γ for those of the singly charged type. It follows from the curves that the difference in the ordinates of every point corresponding to a particular primary ion energy is constant and equal to the difference in the coefficient of potential emission $(\gamma^{2+} - \gamma^+)$ at low energies. This difference is (in agreement with Hagstrum [345]) clearly due to the increased potential emission of electrons in the case of the doubly charged ions. A comparison of this quantity for the (111) and (110) faces shows that for the face with the larger work function the value is smaller than for that with the lower work function. This is due to the effect of the work function on the potential emission of electrons in the case of doubly charged ions also.

In [651], the anisotropy in the kinetic emission was explained from purely geometrical considerations, i.e., by the difference in the packing densities of the atoms in the different faces. The following are the packing density ratios $\delta_{hkl}/\delta_{h'k'l'}$ of the unit cells as compared with the ratios $\gamma_{hkl}/\gamma_{h'k'l'}$, determined from experimental results obtained for ion energies of $E_0 = 5$ keV:

$$\frac{\delta_{110}}{\delta_{100}} = 1.5; \qquad \frac{\delta_{200}}{\delta_{111}} = 1.15; \qquad \frac{\delta_{110}}{\delta_{111}} = 1.64;$$

$$\frac{\gamma_{210}}{\gamma_{100}} = 1.52; \qquad \frac{\gamma_{100}}{\gamma_{111}} = 1.1; \qquad \frac{\gamma_{110}}{\gamma_{111}} = 1.66.$$

Thus the observed anisotropy in γ_k is explained by the change in the probability of collisions taking place between the incident ions and the lattice atoms.

The results lead to the following conclusions.

The coefficient of kinetic electronic emission differs for different faces of a tungsten single crystal. The greatest emission occurs for the face with the greatest packing density; the effect of the work function of the surface on the kinetic emission of electrons is not as great as that of the packing density.

The coefficient of potential electron emission is greatest for the face with the lowest work function.

The coefficient of kinetic emission is independent of the charge and the coefficient of potential emission is independent of the velocity of the ion.

CHAPTER 7

ION AND ELECTRON EMISSION ON BOMBARDING FILMS OF VARIABLE DENSITY WITH POSITIVE IONS

§1. Use of the Double-Modulation Method
for Studying Secondary Processes in Films

It is of great interest to study secondary phenomena taking place in the course of film formation on the surface of a solid body. The double-modulation method offers the best prospects for making a study of this kind. Some changes were introduced into the receiving part of the standard apparatus by Arifov, Ayukhanov, and Starodubtsev [422] and the recording of the currents was specially automated. This modification of the method was necessary because of the rapidly varying nature of the processes taking place in the films. The experiments were carried out in apparatus analogous to that shown in Fig. 1.1 with the addition of an evaporator for depositing the film on the substrate surface. The arrangement of the apparatus is indicated in Fig. 7.1. The evaporator 7 was usually a quartz furnace arranged in such a way that the molecular beam passed through an aperture in the screening cylinder 4 and the collector 3 before falling on the target surface at an angle of 45°. The rate of access of the material to the target surface was regulated by the furnace temperature. Using a slide placed between the furnace and the screening cylinder, access of the beam to the target could be either allowed or prevented. The vacuum in the apparatus was ~$5 \cdot 10^{-7}$ mm Hg.

It is well known that the slow deposition of a film leads to indeterminacy in its composition as a result of the influence of the competing adsorption of residual gases from the vacuum. In order to minimize the effect of the residual gases we must deposit the film at a reasonable velocity. The construction of the evaporator indicated enabled the required rates of film deposition to be achieved.

When studying films of the alkali metals we were able to monitor the rate of access of the material being deposited. For this purpose we stretched a tungsten filament in the space between the target and the collector in the path of the molecular beam arriving at the target. The potential at one end of the filament was equal to the target potential. By applying a negative potential to the collector relative to the filament and heating the latter, the ion current arising from surface ionization from the filament in the path of the beam of molecules could be measured. Knowing the ionization coefficient and dimensions of the section of filament in the path of the molecular beam, we could determine the number of molecules falling on unit surface of the filament, and hence on the surface of the target, in unit time. This method of monitoring is of course only suitable for beams of atoms having reasonably low ionization potentials, i.e., for the case

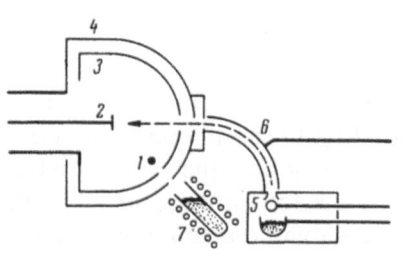

Fig. 7.1. Arrangement of the electrodes of the vacuum apparatus.

$V_i < \varphi$. The rate of access of material to the target (in monatomic layers) per second equals

$$u = \frac{\theta}{t} = \frac{\sqrt{2}}{2} \frac{I_e\, d^2_{alk}\, q_{alk}}{e\beta DL},\tag{7.1}$$

where I_e is the (evaporated) ion current from the filament, d_{alk} is the Goldschmidt diameter of an atom of the alkali element in question, β is the coefficient of surface ionization, D is the diameter of the filament, L is the length of the filament section receiving the beam of atoms, and q_{alk} is the adhesion coefficient of the alkali atoms.

A condition governing the purity of the deposited film is that the ratio of the rate of the competing residual-gas adsorption given by (2.1) to the rate of film deposition should be reasonably small. This leads to the inequality

$$\frac{p q_g e \beta D L d^2_g}{I_e q_{alk} \sqrt{\pi m_r kT}\, d^2_{alk}} \leqslant 1.\tag{7.2}$$

The experiment was carried out in the following way. First the V/A characteristic of the carefully degassed, clean metal substrate (e.g., tantalum) was plotted, then a molecular beam of, for example, sodium was directed against it, and the changes in the V/A characteristics were recorded continuously until a thick sodium layer having all the properties of sodium in bulk form had developed on the tantalum surface.

It was discovered that, in order to obtain a layer possessing all the secondary emission properties of a bulk sample, it was sufficient to deposit a film no more than a few atomic layers thick. Hence, in order to obtain data relating to transitional processes, measurements had to be made very quickly. The measuring and recording method therefore had to be inertia-free and as far as possible automatic.

For the continuous observation of secondary phenomena in transitional processes of this kind, we used the double-modulation method illustrated schematically in Fig. 2.29.

Let us suppose that the conditions on the surface of the target change. In this case there will also clearly be a change in the magnitudes of the secondary (ion and electron) currents and also in the energy distribution of the secondary particles. As a result of this, the V/A characteristics on the oscillograph screen will also change. If the changes in the conditions on the target surface occur less rapidly than the period of variation of the voltage between the target and the collector, then such changes may be observed and recorded.

The transitional processes may be recorded most simply by means of a motion-picture system. The photographing of transitional processes was effected with the help of a motion-picture camera with a speed of 2 frames/sec and an exposure of 0.02-0.2 sec. For a high rate of film deposition on a cold target, these exposures and interframe intervals were inadequate to obtain a reasonable number of intermediate values in the transitional states of the target. For this reason the target heater was switched off at the same moment as the motion-picture system was switched on. Film formation then took place with a cooling target, and this slightly increased the duration of the transitional process.

The series of motion pictures obtained in this way were analyzed from enlarged prints or directly under a magnifier. First the size of the pulse corresponding to the primary ion current was measured; the pulse was measured separately at the beginning and end of the series. Then, on each oscillogram of the series of V/A characteristics of the transitional process the magnitudes of the pulses corresponding to the secondary ion and electron currents were measured, to the right and left of their saturation points, respectively. The ratios of these to the pulses of primary current gave the coefficients K_s and γ for individual stages in the deposition of the film. From the resultant data we plotted curves representing the dependence of γ and K_s

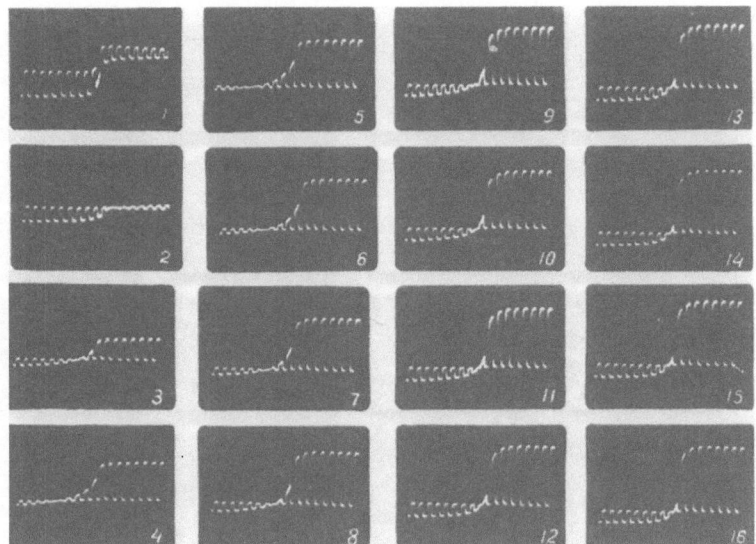

Fig. 7.2. Series of oscillograms of V/A characteristics obtained successively with a motion-picture camera during the increase in film density.

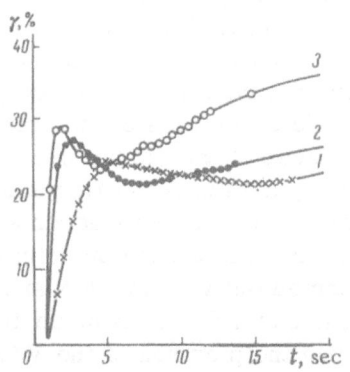

Fig. 7.3. Secondary-emission coefficients of the negatively charged particles as functions of density for various rates of film deposition.

on the time of deposition of the film, and for a constant rate of access of the material to the target, this reflected the dependence of the coefficients on the film density.

The same oscillograms also enabled us to make a qualitative estimate of the energy distribution of the secondary ions and electrons, and to determine the dependence of this distribution on the state of the surface.

§2. Secondary Emission in the Bombardment of Metallic Films with Alkali Ions

In order to elucidate the nature of secondary ion and electron emission, it is of great interest to study secondary processes with metals of the alkali and alkaline earth groups. However, the use of these metals as targets is complicated by the difficulty of cleaning them and ensuring the maintenance of clean surface conditions. The use of films of these metals deposited on metallic substrates offers distinct possibilities.

The most systematic study of secondary emission for the case of films was carried out earlier [422], in which sodium, potassium, rubidium, cesium, magnesium, calcium, antimony, and bismuth films were bombarded with the ions of a number of alkali metals.

In preliminary experiments an attempt was made to create films of alkali metals on refractory substrates by the adsorption of atoms from the beam of bombarding ions.

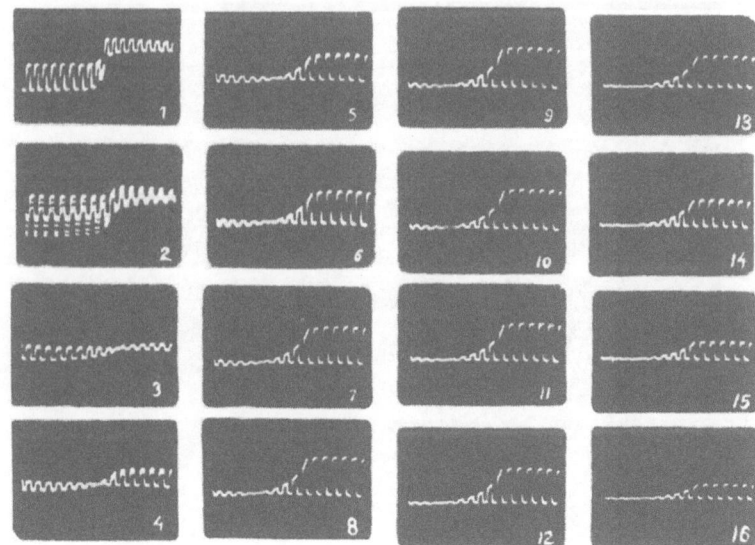

Fig. 7.4. Series of oscillograms of V/A characteristics taken during the increase in the density of a sodium film on a tantalum surface.

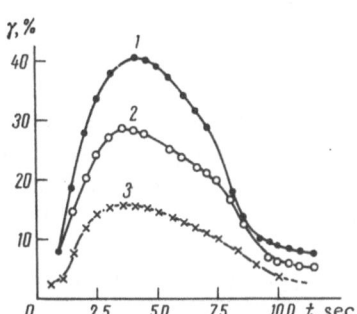

Fig. 7.5. Secondary-emission coefficients of negatively charged particles as functions of the density of a sodium film deposited on tantalum and bombarded with Rb^+ ions at energies of E_0 equal to: 1) 540; 2) 400; 3) 270 eV.

By carefully analyzing the method of obtaining adsorbed films and V/A characteristics it was found that the density of the film varied in a nonuniform manner as a result of cathodic sputtering; for practicable intensities of the primary beam, the rate of access of the atoms to the adsorbed layer was low, and the competing adsorption of residual-gas atoms was considerable for this reason. In view of this, all the principal experiments were carried out with the deposition of the alkali or alkaline earth films from an independent source, while the pictures of the V/A characteristics were recorded with a motion-picture camera. As indicated in the previous section, this method gave films of adequate purity.

It was interesting to discover the way in which the secondary-emission coefficients and the energy distribution of the secondary particles depended on the rate of film deposition. Figure 7.2 shows a series of oscillograms of V/A characteristics successively plotted during the deposition of a sodium film on the surface of a tantalum target at a rate of several monatomic layers per second by bombarding with 500-eV Na^+ ions. Figure 7.3 shows the secondary-emission coefficients of the negative particles as functions of the density of a sodium film deposited on tantalum and bombarded with 540-eV Rb^+ ions for film-deposition rates of 1, 3, and 6 monatomic layers/sec (curves 1, 2, and 3, respectively).

We see from Figs. 7.2 and 7.3 that as the film density increases there is a rise in the coefficient γ, and at the same time the energy spectrum of the secondary negative particles becomes softer.

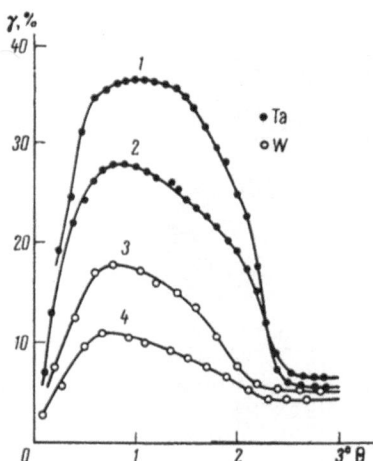

Fig. 7.6. Secondary-emission coefficients of negatively charged particles as functions of the density of a sodium film deposited on tantalum and tungsten surfaces and bombarded with Rb^+ ions at energies of E_0 equal to: 1) 540; 2) 400; 3) 550; 4) 395 eV.

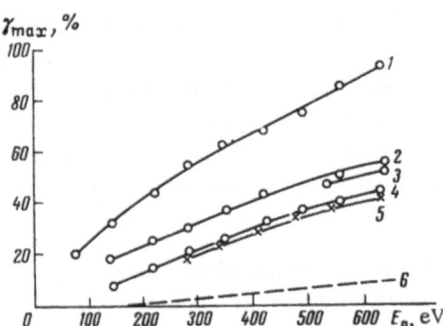

Fig. 7.7. The $\gamma_{max}(E_0)$ relationship obtained by bombarding sodium and potassium films on tantalum substrates with Na^+, K^+, Rb^+, and Cs^+ ions.

However, the change in the secondary-emission coefficient of the negative particles and the energy distribution of the latter expressed as functions of the film density are still determined by the vacuum conditions. If we reduce the influence of the competing adsorption of residual gases, which may be done by depositing the films more rapidly, then these changes take on a completely different character. Figure 7.4 shows a series of oscillograms of V/A characteristics obtained successively during the gradual increase in the density of a sodium film deposited on a tantalum surface and bombarded with 560-eV Na^+ ions for very high deposition rates (20 monolayers/sec), a getter being present in the apparatus. The $\gamma(t)$ curves for sodium films deposited on a tantalum surface and bombarded with 540-, 400-, and 270-eV Rb^+ ions, taken under analogous conditions, are shown in Fig. 7.5. We see from the figure that under these conditions the energy distribution of the secondary negative particles is independent of the film density, and the spectrum is unusually hard. The negative-particle secondary-emission coefficient reaches a maximum (γ_{max}) for a low film density (~1 monatomic layer) and falls sharply as film density increases, reaching a final value of γ_∞. Further improving the vacuum or increasing the rate of film deposition does not lead to any qualitative change in the character of the curve relating the negative-emission coefficient to the film density. This result shows that, for a given rate of film deposition, the competing adsorption of residual gases from the vacuum has no marked effect on the purity of the deposited film. However, study of the numerical values of γ for various ratios between the rate of film deposition and the residual-gas pressure shows that the degree of film purity achieved is not the limiting one, since further increasing the film deposition rate or improving the vacuum conditions still further reduces the value of γ.

In order to discover the reason for the appearance of the maximum value γ_{max} and the meaning of the γ_∞ obtained at high film densities, we studied the dependence of γ_{max} and γ_∞ on the type of film and substrate, and also on the type and energy of the bombarding ions.

Figure 7.6 shows the $\gamma(\theta)$ curves obtained on depositing sodium films on tantalum and tungsten. We see from the curves that the values of γ_{max} are far greater for tantalum than for tungsten, while the values of γ_∞ are almost independent of the nature of the substrate. The values of γ_{max} also depend on the heat treatment of the substrate. The value of γ_{max} may be slightly reduced by treating the target at extremely high temperatures (~2500°K) and depositing the film immediately after this treatment. These results show that the value of γ_{max} charac-

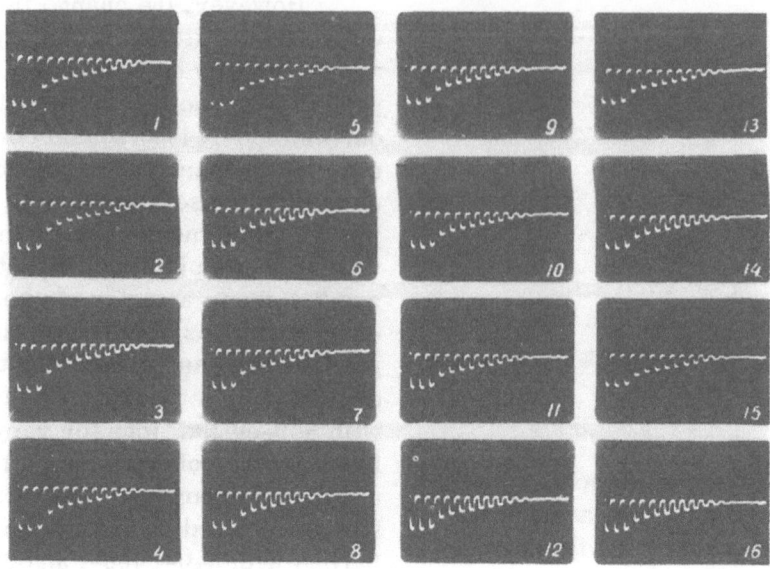

Fig. 7.8. Series of motion-picture oscillograms of the V/A characteristics of the transitional process on passing from clean, cold tantalum (at 300°K) to pure antimony.

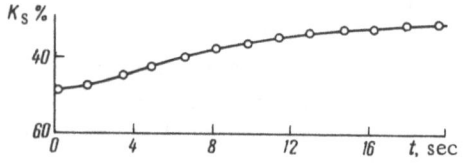

Fig. 7.9. Dependence of K_s on the deposition time (density) of antimony on a clean, cold (300°K) tantalum target on bombarding with 280-eV K^+ ions.

terizes the state of the substrate surface acquired on contamination by the alkali metal. The value of γ_∞ reached at high film densities, on which the substrate has no effect, characterizes the secondary-emission properties of the deposited metal in the bulk state, its purity being governed by the degree of vacuum and the rate of film deposition.

Figure 7.7 shows the dependence of γ_{max} on the energy of Na^+, K^+, Rb^+, and Cs^+ ions bombarding a sodium film deposited on a tantalum surface (curves 2-5, respectively) and the analogous relationship for K^+ ions bombarding a potassium surface (curve 1). The broken line shows the dependence of γ_∞ on the energy of primary Rb^+ ions bombarding a thick sodium film during continuous deposition on a tantalum surface at a rate of 20 monatomic layers/sec (curve 6). It should be noted, however, that these curves only characterize the possible values of γ_{max} and γ_∞ for a certain state of the target, rate of film deposition, and degree of vacuum. As already indicated, different numerical values of γ_{max} may be obtained by changing the method of treating the target; on operating under better vacuum conditions or with higher film-deposition rates, the value of γ_∞ may be somewhat reduced.

A study of magnesium and calcium films on a tantalum surface showed that the bombardment of these with K^+ and Na^+ ions also led to the appearance of the secondary emission of negative particles; the way in which this emission varied was qualitatively the same as that characterizing the results just described for the case of sodium and potassium films. However, results obtained when studying a series of antimony and bismuth films on various targets indicated the possibility of there being another type of transitional process. When films of these metals were deposited on tungsten or tantalum targets, there was no secondary emission of negative particles in any appreciable quantity on bombarding with K^+ ions.

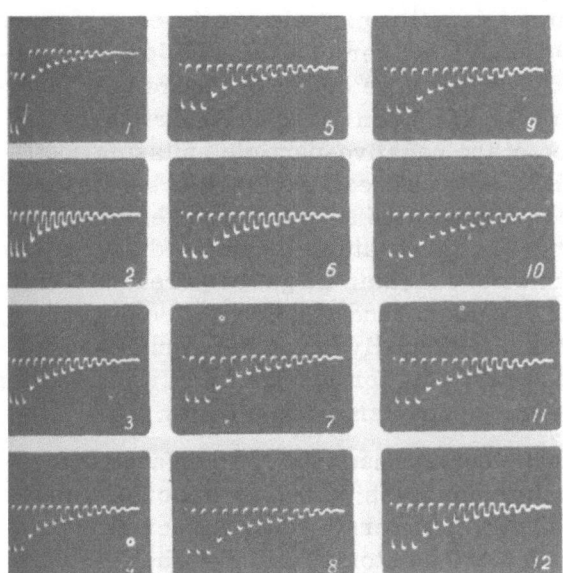

Fig. 7.10. Series of motion-picture oscillograms of the V/A characteristics of the transitional process on passing from pure tantalum to pure bismuth and bombarding with 280-eV K⁺ ions.

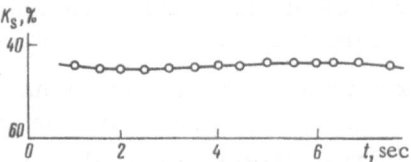

Fig. 7.11. Dependence of K_s on the deposition time (density) of bismuth on a clean, cold tantalum surface on bombarding with 280-eV K⁺ ions.

Figure 7.8 shows a series of motion-picture oscillograms for antimony. We see from the photograph that for clean tantalum (i.e., before deposition) (frame 1) there is no noticeable negative secondary emission; the positive secondary emission corresponds to the ordinary value for clean tantalum. As deposition of the antimony proceeds, the limiting energy of the secondary positive ions falls, the secondary ion emission gradually diminishes, while no emission of negative particles takes place for any degree of coverage in the case of incident K⁺ ions with energies between 200 and 1000 eV. The curve plotted from the oscillograms of Fig. 7.8 is shown in Fig. 7.9.

Figure 7.10 shows an analogous series of motion-picture oscillograms for bismuth on a tantalum surface. We see from Figs. 7.10 and 7.11 that as the deposition of the bismuth proceeds, and its thickness becomes greater, the limiting energies of the secondary scattered ions increase, while the secondary emission of negative particles fails to arise at any degree of coverage, as in the case of antimony.

The difference between the behavior of the secondary-ion emission coefficients in the two examples given (sodium and antimony) is quite easily explained. When a film of sodium has grown on the tantalum surface, the conditions corresponding to the scattering of the bombarding Na⁺ ions will vanish. Since, in this case, we have $m_1 = m_2$, the incident ions will "stick" among the other sodium atoms covering the surface, or will be ejected as neutral sodium atoms as a result of the reduction in the work function of the surface. If we deposit antimony or bismuth on tantalum and bombard with K⁺ ions, however, the sign of the inequality $m_1 > m_2$ remains the same, so that there can only be slight quantitative changes in the coefficient of secondary ion emission, and these accordingly take place, at any rate for antimony (see Figs. 7.8 and 7.9).

The principal result of the foregoing experiments is the conclusion that, on continuously increasing the density of alkali metal films deposited on tungsten, tantalum, molybdenum, and nickel targets and bombarding with alkali ions, the secondary emission of negative particles does not vary monotonically but passes through a certain maximum value corresponding to slight (approximately monatomic) coverages. This variation in γ with increasing film density could not have been obtained before, since on slowly depositing a film by means of the primary ion beam itself, the competing adsorption of residual gases from the main volume of the apparatus was so great that the films were not purely metallic. In this case, as our results indicate, there is a monotonic rise in the negative-particle emission coefficient.

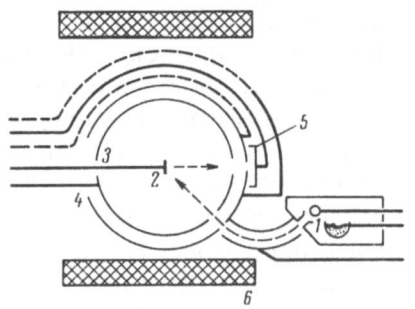

Fig. 7.12. Arrangement of electrodes
in the vacuum apparatus.

For pure alkali metals the secondary-emission coefficients of negative particles in the energy range $E_0 = 0\text{-}600$ eV are not very large. For low film densities, however, we have conditions extremely favorable toward the emission of negative particles. The dependence of the negative-particle emission coefficient on the method of heat-treating the substrate at low degrees of coverage shows that the emissivity of the surface in this state is determined not only by the properties of the main target acquired on "contamination" by the second metal, but also by some other factor. The equally hard spectrum of secondary negative particles for low and high film densities clearly indicates that the negative emission has the same mechanism in both cases.

Thus there is a common factor, associated with the contamination of the surface bombarded, acting on the value of γ for both low and high film densities. The proportion of this contamination is small for high film densities, since then it is determined by the relation between the rate of film deposition and the number of residual-gas atoms falling on the film. The proportion of the contamination may be greater for low film densities owing to the inadequate cleaning of the target and the effect of residual gases on the latter while cooling to the temperature corresponding to the adsorption of the alkali metal film.

The unusually hard energy spectrum of negative particles made it doubtful whether the observed negative emissions were wholly of an electron nature. The negative emission products were therefore subjected to mass-spectrometer analysis and magnetic separation.

Many research workers studying the kinetic emission of electrons by the static method started on the assumption that there were no negative ions in the secondary negative emission.

In order to check the validity of this assumption, experiments must be carried out in such a way as to be able to follow each of these components under controllable conditions for the adsorption of the alkali-metal and residual-gas atoms on the target. With this aim, Arifov and Ayukhanov [545a] studied the composition of the negative emission while increasing the density of the film at various rates of deposition, giving different degrees of purity of the films and creating a great diversity of states on the surface bombarded.

In order to separate the negative-ion component of the secondary negative emission from the electron component, the electrons were taken out to a separate collector by means of a longitudinal magnetic field.

It is well known that an electron leaving a target at an angle α to the normal will move in a longitudinal magnetic field along a spiral line of radius $\rho = (mv \sin \alpha)/eH$. Hence, if the receiving part of the apparatus is placed in a magnetic field coinciding in direction with the electric field, and an aperture is made in the main collector opposite to the point of incidence of the ion beam on the target, with a radius $r \geq (d_i/2) + 2(mv_{max}/eH)$ (d_i is the diameter of the ion beam, v_{max} is the velocity of the electron near the collector), then practically all the electrons will pass through this aperture beyond the main collector. This enables us to measure the integral currents of the ion and electron components of secondary negative emission without any serious changes in the method of measurement. Clearly, the ratios of the currents passing to the collector and the additional collector placed behind the aperture to the primary ion current will give the coefficients of secondary negative ion and secondary ion—electron emission, respectively.

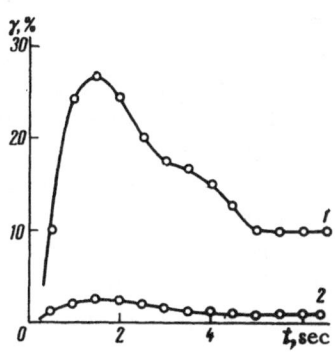

Fig. 7.13. Secondary negative-ion emission (curve 1) and secondary ion—electron emission (curve 2) as functions of the density of a sodium film on tantalum.

Figure 7.12 gives a schematic diagram of the main components of the apparatus. The primary ion beam 1, well focused, modulated in the form of rectangular pulses, and no more than 2.5 mm in cross section, falls on the surface of the target at an angle of 45°. The electric field applied between the target 2 and the collectors 3 and 5, nearly plane in the central region of the receiving section, accelerates the secondary negative particles in the direction of the collector. The secondary negative particles which are not seriously deflected in the magnetic field (400 Oe) created by the solenoid 6 wound onto the apparatus from outside mainly fall on the collector 3. The secondary electrons, under the influence of the magnetic field, pass out through the hole in the main collector and fall on the special collector 5, which is well screened from parasitic currents by the screen 4. The secondary currents to collectors 3 and 5, modulated in accordance with the modulation of the primary ion current, are measured by two cathode-ray oscillographs. An alternating voltage from a sawtooth pulse generator may also be applied between the target 2 and collectors 3 and 5. In this case the measuring method coincides with the double-modulation method, and the oscillograph screens give stationary pictures of the volt—ampere characteristics for the components of the secondary negative emission. The primary ion current passing to the target is measured by connecting the target and the collectors to one of the oscillographs by means of a special switch.

The efficiency of the method used for separating the electrons was checked by placing two or three turns of a tungsten wire with a turn diameter roughly equal to the size of the ion spot on the target at the point of incidence of the ion beam. The thermionic (electron) current arising on heating this filament without applying a magnetic field fell almost entirely onto the main collector. On connecting the magnetic field the thermionic current passed entirely to the special collector 5 (with an accuracy of 2–3%).

We studied the dependence of the coefficients of secondary negative emission on the density of the alkali metal films. For this purpose we first degassed a target by heat treating at ~2500°K and then established a steady temperature of ~1200°K, then simultaneously directed a molecular beam of alkali metal from an evaporator (not shown in Fig. 7.12) and a modulated primary ion beam onto the target. By disconnecting the heater and at the same time connecting the oscillographs, we continuously observed the changes in the components of secondary negative emission as the target cooled and the film density on its surface increased. Rapid changes in secondary currents were recorded with a motion-picture camera every 0.5 sec with an exposure of 0.2 sec. Analysis of the oscillograms thus obtained enabled us to plot the curves relating the secondary negative-ion and secondary ion—electron emissions to the density of the films deposited. The experiments were carried out with a residual-gas pressure of ~10^{-6} mm Hg.

The original experiments with low film-deposition rates showed that the emission of negatively charged particles accompanying the appearance of a film of alkali metal on the surface bombarded contained a considerable number of negative ions. Under these conditions the secondary ion—electron emission becomes considerable and increases as the density of the film becomes greater. As the degree of purity of the film is made higher and higher by increasing the rate of deposition and improving the vacuum conditions, the relation between the negative-ion and electron components of secondary negative emission changes in the sense of reduction of ion—electron emission. For fairly high deposition rates (~20 atomic layers/sec), a getter being present in the apparatus, reproducible relations between the electron and negative-ion components were achieved. The secondary emission of negative ions and electrons obtained under these conditions is shown in Fig. 7.13 as a function of the thickness of sodium films deposited on tantalum and bombarded with 840-eV Rb^+ ions. We see from the figure that the rise in the coefficient of secondary negative emission (curve 1) for low-density pure films to its maximum value (γ_{max}) and the subsequent fall to the stationary value (γ_∞) on further increasing the film density takes place entirely as a result of the emission of negative ions. The slight rise in ion—electron emission for low film densities is evidently associated with the penetration of some of the original ions through the aperture provided for taking electrons to the special collector 5.

It was found that the magnitude of the coefficient of negative ion emission at low film densities depended on the method of treating the target. By heating the target to a higher temperature (~2500°K) and depositing the film immediately after such treatment, the maximum secondary emission of negative ions was slightly reduced. However, under our experimental conditions we were unable to secure a smooth transition of the coefficient of negative-ion emission to its final stationary value. The final values of ion—electron and secondary negative-ion emission achieved for high film densities are not greatly dependent on the nature of the main target. Other conditions being equal, these values only depend on the vacuum conditions and the rate of film deposition.

The results show that the secondary negative emission appearing after bombarding alkali metal films on tantalum and tungsten surfaces with alkali elements is not purely of an electron nature. On the contrary, the main contribution to negative emission comes from the emission of negative ions.

The presence of negative ions in the negative emission is evidently a result of the presence of the film of alkali metal and certain uncontrollable contaminations on the surface bombarded. On the basis of these facts we may consider that the negative ions appear as a result of the cathodic sputtering of contaminant films on the target surface. The film of alkali metal, reducing the work function of the surface, creates favorable conditions for the sputtering of these contaminants in the form of negative ions. The work function may in fact become lower than the electron affinity for certain gases on covering the surface with films of alkali elements. The probability of sputtering these gases in the form of negative ions may be very high indeed. The validity of this assumption was confirmed in experiments involving the mass-spectrometric analysis of the secondary negative ions. These experiments (which will be set out separately later) showed that the negative ions of the secondary negative emission appearing as a result of the bombardment of tantalum and tungsten surfaces covered with thin films of alkali elements were mainly negative oxygen ions. This indicates that atoms with the greatest electron affinity have the highest probability of penetrating the surface in the form of negative ions. Thus, the necessary conditions for the appearance of negative ions within the negative emission are that the work function of the surface should be reduced and that the surface should contain electronegative contaminants with a high electron affinity.

Almost all earlier investigations into the kinetic emission of electrons were carried out with surface conditions such as to favor both factors producing negative ions. As a rule, on bombarding the surfaces under consideration with ions of the alkali elements, the surfaces are contaminated with the primary ion beam. These contaminants create regions of low work function on the surface. In addition to this, for a fairly poor vacuum, adsorption of residual gases from the main volume of the apparatus always occurs, since the instant of measuring is always separated from the instant of cleaning the surface by a short interval of time. On bombarding parts having a low work function with ions, sputtering of the negative adsorbed-gas ions takes place. This occurred particularly in experiments carried out with thoriated tungsten, designed to determine the dependence of the kinetic emission on the work function of the surface [116, 144].

However, even when the dependence of the kinetic emission on the parameters of the bombarding ions (mass, charge, energy) was being studied for constant conditions on the surface bombarded [131, 221, 286], the laws of cathodic sputtering were substantially reflected in the form of negative ions.

Moon [70] and Arifov and Ayukhanov [219] avoided contamination of the surface under study by alkali ions by making the measurements at high temperatures. The substantial reduction in the coefficient of secondary negative emission observed in these cases evidently indicated that there were no negative ions in the negative emission. The coefficients of secondary negative emission here measured really did characterize the kinetic emission of electrons from surfaces with only a little gas coverage.

Thus, in order to obtain the true values of the coefficients of kinetic electron emission from arbitrary surfaces, we must measure the electron emission and the emission of negative ions separately. The kinetic emission of electrons measured by reference to the total negative emission can only give the true value of this quantity if there are no adsorbed gases at all on the surface bombarded.

§3. Secondary Emission in the Bombardment
of Metallic Films with Inert-Gas Ions

A film of alkali metal on the surface of a refractory metal is a convenient specimen for studying the potential emission of electrons in relation to the work function, since the presence of the film reduces the work function of the main target.

Secondary emission from a sodium film on a tantalum target on bombarding with Ar^+ ions was studied by Arifov and Tashkhanova [496, 497]. The measurements were made during the continuous deposition of the sodium film by evaporating metallic sodium from a special evaporator.

Figure 7.14 shows an oscillogram of a V/A characteristic obtained on bombarding a sodium film with 900-eV Ar^+ ions during its deposition on a tantalum surface. This oscillogram differs considerably from the analogous oscillograms obtained when bombarding the same film with alkali ions. It is not hard to see that the composition of the secondary emission of negative particles is divided sharply into two energy groups: one has a comparatively soft energy spectrum and the other is correspondingly hard.

A series of oscillograms of the V/A characteristics plotted successively during the rise in density of the sodium film deposited on a tantalum surface on continuously bombarding with 720-eV Ar^+ ions is presented in Fig. 7.15. The rates of film deposition are relatively high (~20 monatomic layers/sec).

On the basis of a series of oscillograms similar to those shown in Fig. 7.15, curves representing the dependence of the coefficient of secondary emission of the particle groups with the soft and hard energy spectra on the density of the deposited sodium film were plotted; these are shown in Fig. 7.16. We see from the curves that emission of particles with a soft energy spectrum occurs for any arbitrary state of the surface bombarded, while emission of particles with a hard spectrum is absent when bombarding a film-free tantalum surface with Ar^+ ions; however, the latter form of emission appears and intensifies with increasing density of the sodium film on the tantalum surface; after reaching a maximum value (γ_{max}) it falls again to a certain final value (γ_∞) which remains unchanged as the film thickness increases further.

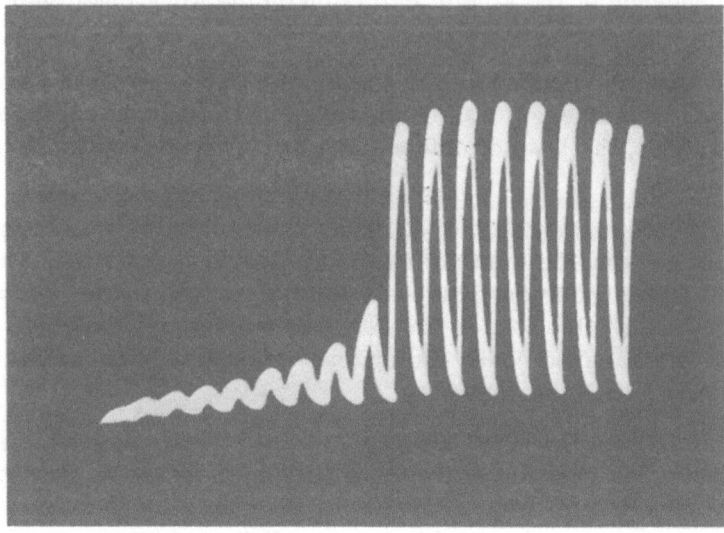

Fig. 7.14. Typical form of the V/A characteristic representing the secondary emission of negatively charged particles on
bombarding a sodium film coating a tantalum target with 900-
eV Ar$^+$ ions.

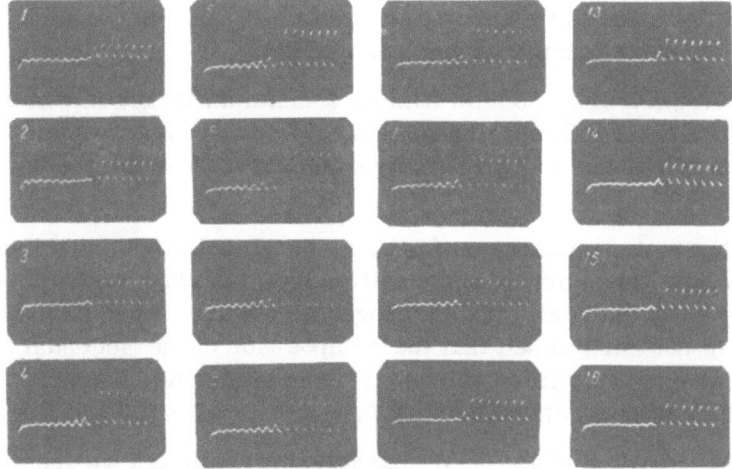

Fig. 7.15. Oscillograms obtained during the rise in the density of a sodium film on a tantalum surface while bombarding
with 720-eV Ar$^+$ ions.

On bombarding the surface of the tantalum substrate with Ar$^+$ ions [289, 345], potential
extraction occurs, since $V_i > 2\varphi$. The presence of potential emission for tantalum is also indicated by the oscillograms obtained before deposition of the film, i.e., from clean metallic targets (Fig. 7.15, 1). On depositing a sodium film on tantalum the relation between V_i and φ can
only change in the sense of increasing the inequality indicated in the foregoing. This means that
the potential extraction of electrons under the influence of the bombarding Ar$^+$ ions should occur for any density of the sodium film on the tantalum (Figs. 7.15 and 7.16).

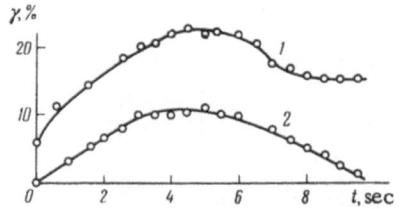

Fig. 7.16. Dependence of the coefficients of secondary emission with soft (1) and hard (2) energy spectra on the density of the deposited sodium film.

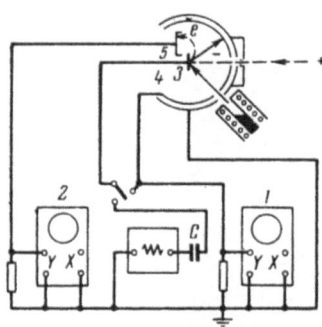

Fig. 7.17. Receiving part of the apparatus with the electric measuring circuit.

Thus the group of particles with the soft energy spectrum consists of electrons emitted by virtue of the potential energy of the ions, while the group of particles having the hard energy spectrum is evidently due to negative ions arising from adsorbed gases. The validity of this assumption was verified later by Arifov, Tashkhanova, and Rakhimov [646-648], who analyzed the secondary emission in a magnetic field.

In order to study the negative ions and electrons separately in the magnetic field, an additional electrode 5 was introduced into the receiving part of the apparatus; this was placed at 2 mm behind the target 3. The electrons were directed onto the collector 5 by means of a transverse magnetic field directed along the axis of the cylindrical collector 4, while the negative ions were taken to the collector 4 itself (the magnet is not shown in Fig. 7.17).

In the absence of the magnetic field, all the negative particles ejected fall on collector 4. On applying the magnetic field, the secondary negative ions, which are not substantially deflected from their original direction, principally fall onto the collector 4. The secondary electrons, however, are rotated by the magnetic field and fall on collector 5. The secondary currents passing to collectors 4 and 5, modulated in accordance with the modulation of the primary ion current, are recorded by two oscillographs 1 and 2.

The efficiency with which the electrons were directed onto the collector 5 was verified by deflecting potential-emission electrons obtained from a clean, heated molybdenum target by bombarding with 540-eV Ne^+ ions. In this case, it is well known [262, 493] that there is no emission of negative ions. Experiments showed that for magnetic fields of 30-40 Oe almost all the secondary electrons fell on the collector 5. The experiments were carried out with a residual-gas pressure of no more than $1 \cdot 10^{-7}$ mm Hg. The ion beam was singly charged; this was ensured by suitably selecting the energy of the ionizing electrons in the ion source.

Figure 7.18 shows oscillograms of V/A characteristics obtained in the circuit of collector 4 on bombarding a sodium film covering a tantalum substrate with 900-eV Ar^+ ions after reaching the density corresponding to oscillograms 4-11 in Fig. 7.15. We see that the group of secondary particles having a soft energy spectrum is in fact taken off by the magnetic field, while the particles with the hard energy spectrum are undeflected in this field and pass to the main collector.

Thus, mass separation confirms the validity of our assumption regarding the possibility of separating the secondary electrons from the negative secondary ions by employing an energy criterion.

Figure 7.19 shows a series of oscillograms of the V/A characteristics of secondary electron emission taken successively during the growth of a sodium film on a tantalum surface while bombarding with 540-eV Ne^+ ions with film-deposition rates of 20 monolayers/sec. As in the case of Ar^+, potential extraction of electrons under the influence of the bombarding Ne^+ ions should occur for all densities of the sodium film on the tantalum substrate (Fig. 7.19, oscillograms 1-25).

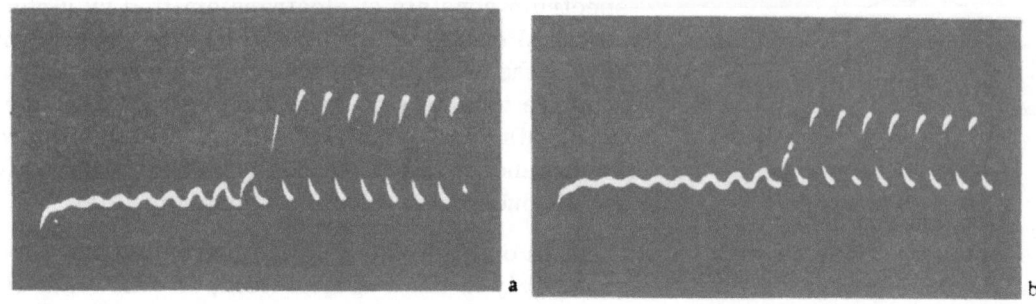

Fig. 7.18. Oscillograms of V/A characteristics taken without any magnetic field (a)
and after applying a magnetic field (b).

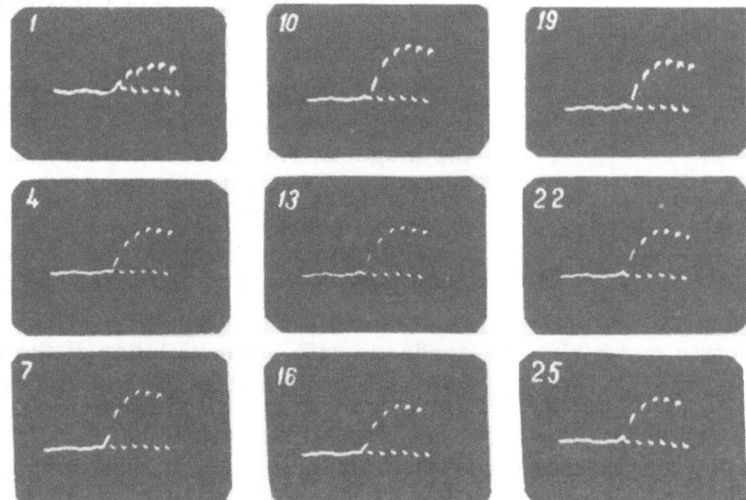

Fig. 7.19. Series of oscillograms taken during the growth of
a sodium film on a tantalum surface.

Figure 7.20 shows a series of oscillograms of the V/A characteristics of secondary
negative-ion emission taken at the same time as the V/A characteristics for the secondary
electrons. It follows from these oscillograms that the emission of negative ions is absent on
bombarding a clean tantalum surface with Ar^+ and Ne^+ atoms; it appears and intensifies with
increasing thickness of the film of alkali metal, reaches a certain maximum value, and then
falls again and almost vanishes. The absence of any emission of negative ions in the final stage
of deposition evidently indicates that the film is reasonably pure.

Figure 7.21 shows the coefficients of secondary electron emission γ and secondary negative-ion emission K^- as functions of the thickness of the sodium film on the tantalum; these
were obtained by analyzing the series of oscillograms presented in Figs. 7.19 and 7.20, respectively. The vertical axis represents the values of γ and K^- in percent, while the horizontal
axis represents the time of film deposition in seconds, corresponding to the thickness of the
deposited layer on a certain nonlinear scale. There is no direct proportionality between the
film thickness and the deposition time, since film deposition occurs during the cooling of the
substrate after preliminary heating to a high temperature.

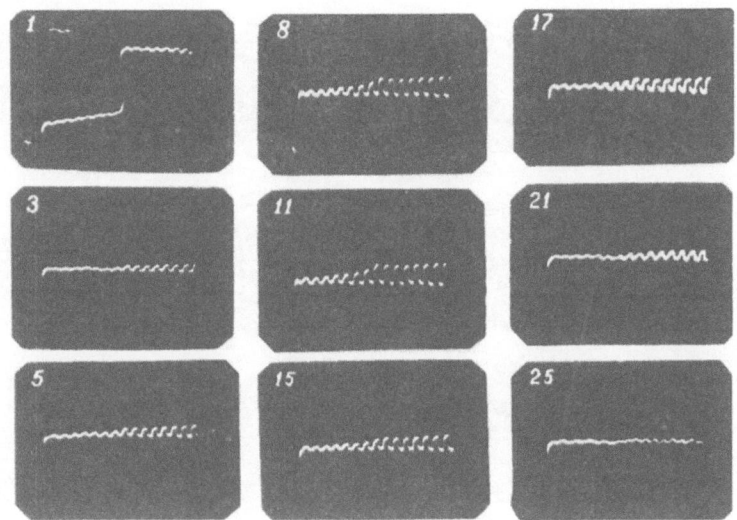

Fig. 7.20.　Series of oscillograms of the V/A characteristics
of secondary negative-ion emission.

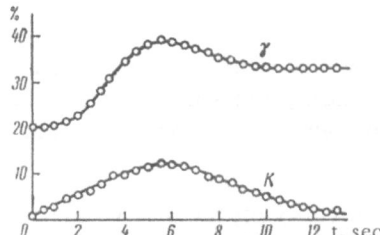

Fig. 7.21.　Coefficients γ and K^- as functions of sodium film thickness.

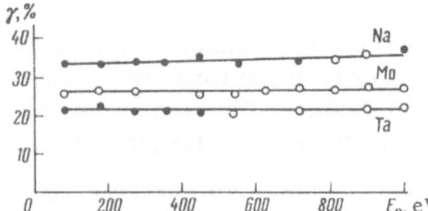

Fig. 7.22.　The $\gamma(E_0)$ relationship on bombarding Ta, Mo, and Na targets with Ne^+ ions.

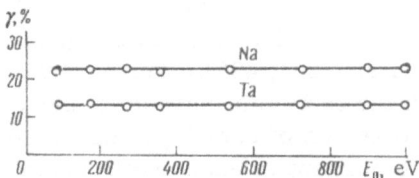

Fig. 7.23.　The $\gamma(E_0)$ relationship on bombarding Ta and Na targets with Ar^+ ions.

We see from Fig. 7.21 that the coefficient γ rises with increasing film thickness to a maximum value γ_{max}, then falls to a certain stationary value γ_∞. The initial value of $\gamma = \gamma_{min}$ relates to the emission of electrons from a film-free tantalum surface, while $\gamma = \gamma_\infty$ relates to a thick film of deposited sodium. The value of K^- also rises with increasing film thickness to a certain maximum value K^-_{max}, and then falls and ultimately becomes negligible (Fig. 7.20, oscillogram 25).

Analogous results were obtained for sodium films on a molybdenum substrate on bombarding with Ne^+ and Ar^+ ions.

The measured values of $\gamma(E_0)$ in the range $E_0 = 100$–1000 eV for Ne^+ and Ar^+ ions bombarding tantalum, molybdenum, and sodium targets are shown in Figs. 7.22 and 7.23, respectively. Before starting the measurements, the tantalum and molybdenum targets were degassed by prolonged heating at 2600 and 2300°K, respectively. Each point on the $\gamma(E_0)$ curves for the tantalum and molybdenum targets was obtained at 1400–1500° K after flashing (brief heating to T > 2000°K), while for the sodium target these points were taken from the γ curves (Fig. 7.21) corresponding to the value of γ_∞.

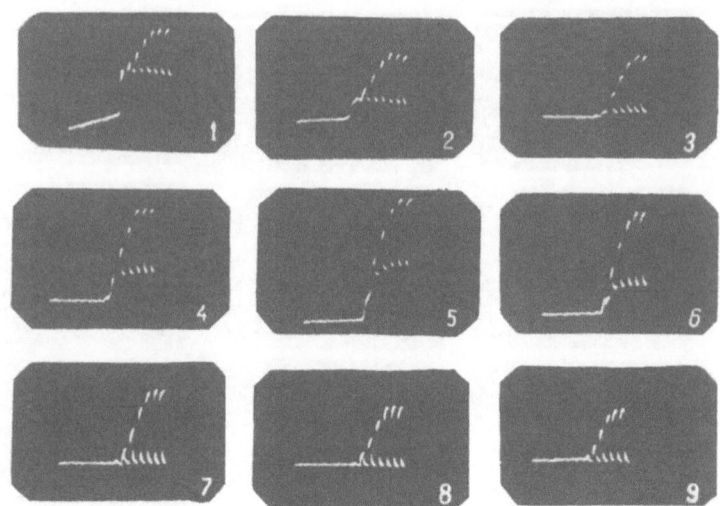

Fig. 7.24. Series of oscillograms of the V/A characteristics
of secondary electron and thermionic currents during the
growth of a potassium film on tantalum.

In the energy range studied (100–1000 eV) the coefficient γ was almost independent of the kinetic energy of the ions for all the targets (tantalum, molybdenum, and sodium) and ions (Ne^+ and Ar^+) studied. This indicates that the emission of electrons was of a potential character in the case of the sodium target as well.

These results all show that the value of γ depends considerably on the work function of the surface; as φ changes from 4.12 (tantalum) to 2.27 eV (sodium), γ changes from 20–35% for Ne^+ ions and from 13–22% for Ar^+. The value of γ obtained for the sodium target confirms the linear dependence of $\gamma(\varphi)$ established in [493].

Figure 7.24 shows a series of oscillograms of the V/A characteristics of ion—electron emission successively plotted during the deposition of potassium (~20 monolayers/sec) on tantalum while bombarding with 400-eV Ne^+ ions. The volt—ampere characteristic 1 (Fig. 7.24) indicates the potential extraction of electrons from tantalum [264, 493]. Subsequently (oscillograms 2–9), as the thickness of the potassium film on the tantalum increased, there was a change in the potential emission of the electrons. In the series of oscillograms presented it is easy to notice the presence of an unmodulated part of the current in each oscillogram (break in the zero line); this characterizes the thermionic (electron) current due to the adsorption of potassium on the tantalum surface during the cooling process. This "adsorption" thermionic emission is well known [22, 42, 43, 295] to have a maximum (Fig. 7.24) for a certain optimum coverage $\theta = \theta_{opt}$. According to Langmuir [22], θ_{opt} corresponds to the monomolecular coverage. However, Gavrilyuk [295] later showed that the greatest thermionic emission or the minimum work function occurred for ~0.7 of a monatomic coverage on the surface of the metal.

Simultaneous observation of the changes in thermionic and ion—electron emission enables us to estimate the thickness of the coating for which a maximum of the coefficient γ_p is attained. Figure 7.25 presents curves relating the coefficient of potential electron emission (γ_p) and the thermionic current (I_T) to the time of depositing potassium on tantalum; these were obtained by analyzing the series of oscillograms in Fig. 7.24. The values of the coefficient γ_p and the thermionic current I_T rise with increasing density of the coating, reach a maximum, then fall until I_T vanishes while γ_p reaches the values characteristic of massive potassium. The maximum of γ_p is displaced slightly with respect to the maximum of I_T in the high θ direction;

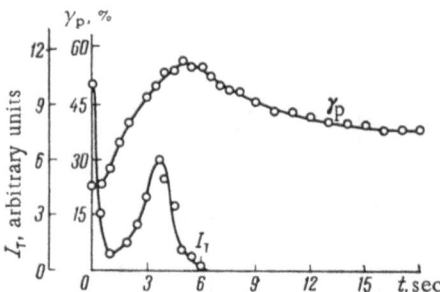

Fig. 7.25. Coefficient of potential electron emission γ_p and the thermionic current I_T as functions of the time of depositing potassium on tantalum.

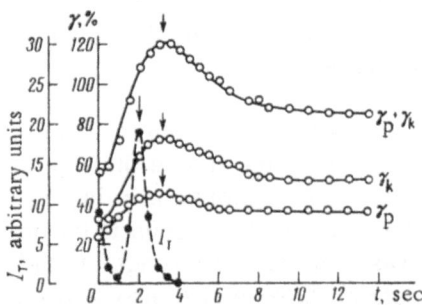

Fig. 7.26. Coefficients γ_p, γ_k, and $\gamma_p + \gamma_k$ as functions of the time of depositing potassium on molybdenum.

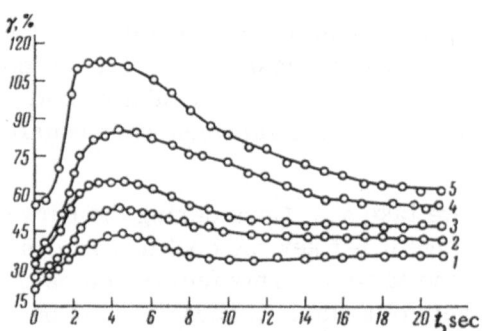

Fig. 7.27. Coefficients γ as functions of the density θ of a sodium film on tantalum, bombarded with Ne^+ ions at energies of E_0 equal to: 1) 360; 2) 1080; 3) 1860; 4) 2620; 5) 4260 eV.

this is evidently because the curves were not obtained under steady-state conditions but during the cooling of the target.

The $I_T(\theta)$ relationship may be explained in the following way. For $\theta = 0$, the value of I_T characterizes the essential thermionic emission of the tantalum, which falls with decreasing temperature. However, as temperature falls below a certain point, the emission starts rising, despite the falling temperature, owing to the reduction in φ resulting from the increased density of the potassium film adsorbed on the tantalum. The rise in the thermionic current continues until at a certain low temperature the emission reaches a maximum, after which it falls sharply to zero as the temperature drops further.

As indicated in [22, 42, 43, 295], the minimum work function determined by the maximum thermionic current corresponds to a coverage estimated as roughly equal to 0.7 of a monatomic layer. This estimate may be extended to our own case. The slight displacement of the maximum of γ_p in the direction of higher coverage relative to the maximum of I_T is evidently due to the fact that the experiments were conducted under nonstationary conditions.

Analogous phenomena were observed on replacing the tantalum by molybdenum and tungsten.

Figure 7.26 shows the coefficient γ_p as a function of deposition time $(E_0 = 0.4$ keV) and also the corresponding relationship for the total coefficient $\gamma = \gamma_p + \gamma_k$ for primary-ion energies of $E_0 = 4.0$ keV. At this energy potential extraction is supplemented by the kinetic emission of electrons, so that the curve $\gamma_k(\theta)$, constituting the difference between curves $\gamma(\theta)$ and $\gamma_p(\theta)$, expresses the relation between γ_k and the deposition time [493, 645] (Fig. 7.26). We see from the figure that $\gamma_k(\theta)$ and $\gamma_p(\theta)$ pass through a maximum at the same film thickness which indicates that the coefficient of kinetic electron emission also depends on the work function.

Thus we may assert that the coefficients of both potential and kinetic electron emission increase with falling φ in the case of film-bearing surfaces. The maximum values of γ_p and γ_k are attained for film thicknesses having the minimum work function.

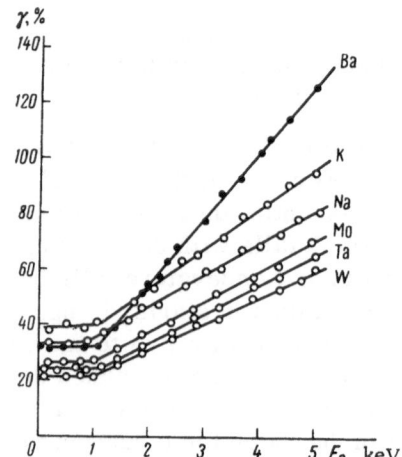

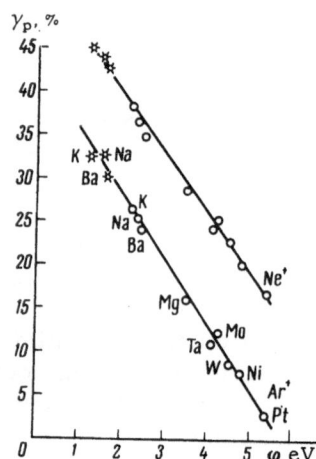

Fig. 7.28. The coefficient $\gamma(E_0)$ on bombarding Mo, Ta, W, Na, K, and Ba targets with Ne$^+$ ions.

Fig. 7.29. The coefficient γ_p expressed in terms of the work function φ for Ne$^+$ and Ar$^+$ ions.

Figure 7.27 shows the coefficient γ as a function of the density θ of a sodium film on tantalum bombarded with Ne$^+$ ions at various energies. We see from the figure that the coefficient γ always has a maximum for a low film density (of the order of a monatomic layer) and falls to a certain final value γ_{max} with further increase in density. An analogous relation between the coefficient γ_∞ and the thickness of a sodium film was obtained with tungsten and molybdenum substrates. The results of experiments by Hagstrum [262] and Arifov and Rakhimov [493] show that for $\theta = 0$, γ_p characterizes the substrate material. The state of the target surface acquired by "contamination" with the alkali metal is determined by γ_{max}. The value of γ_∞ obtained for high film densities is independent of the particular substrate on which the film is deposited, and thus γ_∞ characterizes the secondary-emission properties of the deposited metal, with a purity determined by the degree of vacuum and the rate of film deposition.

The curves relating γ to the energy of the bombarding Ne$^+$ ions in the range 0.1–5.0 keV are given in Fig. 7.28 for tungsten, tantalum, molybdenum, sodium, potassium, and barium targets. We note that our $\gamma(E_0)$ data for refractory-metal targets agree with the results of [262, 493, 505] over the whole energy range studied, indicating that the target surfaces were equally clean in both cases.

As γ for the sodium, potassium, and barium targets we take γ_∞. Considering the curves of Fig. 7.28, we come to the conclusion that in the energy range 0.1–1 keV the coefficient γ is almost independent of the kinetic energy of the ions for all the targets. This fact indicates the potential character of the electron emission from film targets of sodium, potassium, and barium (for ion energies up to 1 keV). On increasing E_0 to 5 keV there is a linear rise in γ in all cases. An analogous variation in γ with the energy of the bombarding ions is obtained for Ar$^+$ ions.

The linear rise in emission with the energy of the ion for $E_0 > 1$ keV may, according to [493], be explained by the kinetic extraction of electrons. We see that on passing along the series W $\rightarrow$ Ta $\rightarrow$ Mo $\rightarrow$ Na $\rightarrow$ K $\rightarrow$ Ba the slope of the $\gamma(E_0)$ curves relative to the E_0 axis increases (Fig. 7.28). This is evidently associated with the difference between the atomic numbers of the elements and the ionization energies of the metal atoms, and is in qualitative agreement with the theory of Parilis and Khishinevskii [539].

Figure 7.29 gives the coefficients of potential electron emission γ_p obtained on bombarding molybdenum, tantalum, tungsten, barium, sodium, and potassium with 200-eV Ne^+ and Ar^+ ions as functions of φ. The asterisks denote the maximum values of γ corresponding to the thickness of the film with the minimum work function. The same figure shows Arifov and Rakhimov's data [493] for platinum, nickel, and magnesium. We see that the value of γ_p varies with the nature of the metal and the relation between γ_p and the work function is almost linear. The work function of barium, sodium, and potassium films estimated from these curves equals 2.3, 2.25, and 2.15 eV for a thick film and 1.58, 1.54, and 1.28 for a coverage giving the maximum potential emission of electrons, respectively. The value of φ giving the maximum γ_{max} obtained from Fig. 7.29 agrees satisfactorily with the results of [22, 42, 43], in which the film work function was determined by a method based on thermionic emission during adsorption.

Thus, a study of the potential extraction of electrons from a metal on bombarding with ions offers the possibility of estimating such an important parameters as the work function, which characterizes the conditions on the surface of the metal.

§4. Study of Secondary Processes Occurring During the Bombardment of Dielectric Films with Positive Ions

A knowledge of the energy losses and range of the ions and atoms in solids is essential when studying many problems of cathode and semiconductor electronics. The passage of low-energy ions through matter in cases in which the velocity of the incident particle is lower than that of the electrons in the atoms of the retarding medium has nevertheless not yet been adequately studied, either theoretically or experimentally.

The traditional method of studying the absorption and retardation of charged particles in a solid is that of firing the particles through foils of the test material. Since the range of the particle in the material decreases with increasing initial energy, samples in the form of very thin films have to be used for low-energy experiments. The difficulties in preparing thin films of uniform thickness and in manipulating these are so great that when using the method of firing ion beams through such films the energy cannot be reduced below 1.5-2 keV, even for light ions.

Nosenko and Strukov [570] irradiated phosphors with ion beams and determined the depth of penetration of various ions from the thickness of the layer destroyed by the bombardment. The thickness of the layer so destroyed was estimated from the change in the relationship between the brightness of the cathodoluminescence and the energy of the exciting electrons. The use of phosphors as absorbents enabled the penetration of heavy ions with energies down to 1 keV to be studied.

For more general application we may use the radioactive method of determining the depth of penetration of ions into solids [394, 571], in which the sample is bombarded with a beam of "tracer" ions, after which the target is dissolved in a special solvent and the change in its content of radioactive impurities is measured in parallel with the dissolution. The sensitivity of this method is low when determining a small quantity of penetrating ions, owing to the fluctuations in natural background; however, it is certainly the best for determining the distribution of penetrating particles on the path of the beam through the material. This method was used to study the penetration of alkali metal ions into semiconducting materials and aluminum at energies between 0.7 and several ions of keV. There are no data regarding the possibility of using the method of "tracer" ions at lower energies.

In order to study the penetration of ions into matter at lower energies we may use the relation between the currents of secondary ion—electron emission and the thickness of the layer bombarded. In passing through a solid, the ion continuously loses energy in interacting with the atoms and electrons of the latter. For an incident-ion energy exceeding the threshold energy E_t for the excitation of kinetic electron emission, secondary electrons are emitted into the vacuum. The maximum depth of generation of the secondary electrons clearly corresponds to the depth at which the ions passing into the emitter, or the target atoms which have received a momentum directed inward from the incident ions, still have an energy equal to or slightly greater than the electron-emission excitation threshold. In substances with a low threshold of kinetic electron emission, the range of an ion after its energy has fallen to a value equal to E_t will be small compared with the total range if the initial energy of the ion is much greater than E_t. In this case we may consider roughly that the limiting depth of origin of the secondary electrons arising under the continuous action of the incident ions corresponds to the maximum depth of penetration of the ions into the emitter. Substances with a low threshold of kinetic emission include, for example, alkali halides; for ions of the alkali metals, E_t is (according to our measurements) under 150 eV.

We may judge the limiting depth of generation of the secondary electrons in alkali halides from the instant of saturation of the curves relating the coefficient of ion—electron emission γ to the thickness of the salt film if the energy of the ion is lower than E_k, the energy of the ions for which the maximum depth of generation of electrons equals the effective depth of emission of electrons into the vacuum. Clearly, the energy spread of the electrons formed near the threshold energies of the ions or atoms should not depend on the nature of the exciting particle. Hence, the conditions for the emission of secondary electrons ejected by different particles into a vacuum should also be identical. Hence, by studying the moment of saturation of the $\gamma(t)$ curves obtained on bombarding a film with different ions we may estimate either the downward penetration of the incident ions or the depth of formation of the "hot atoms" of the target. However, it was pointed out by Lovtsov and Starodubtsev that, on bombarding alkali halide films, negative halide ions are also emitted together with secondary electrons [205].

In order to make a valid study of the effective depth of origin of electrons arising from dielectric films, as well as the depth of penetration of the primary ions into the solid, we must first of all study the secondary electron emission and the emission of negative ions as a function of the film thickness, ion energy, and nature of the bombarding ions separately. An investigation of this kind was carried out by Moroz and Ayukhanov [632, 725] by the double-modulation method, using apparatus similar to that described in §3 (Fig. 7.17). The secondary electrons were brought out to a special collector by applying a transverse magnetic field of 40 Oe. The secondary currents of the electrons and negative ions captured by individual collectors were recorded on the screens of two oscillographs and photographed with a motion-picture camera as the film was deposited on a metal substrate.

Figure 7.30 shows the coefficients of ion—electron emission γ and negative-ion emission K^- as functions of the time of depositing NaCl (the deposition time corresponds to the thickness of the deposited layer on a certain nonlinear scale) on a molybdenum target bombarded with 370-, 100-, and 1500-eV Cs^+ ions. We see from the figure that the negative-ion and electron components of the secondary emission appear simultaneously with the deposition of NaCl on the substrate, but depend on the film thickness in different ways. The coefficient K^- rises very sharply with increasing film thickness to a certain maximum value. On further increasing the film thickness, K^- falls slightly and gradually transforms to a K^- value constant for a specific energy of the bombarding ions. The coefficient γ increases very slowly with increasing thickness of the NaCl film. Saturation of γ is reached for film thicknesses many times greater than those corresponding to the saturation of K^-. Thus, for small thicknesses of the NaCl film, the secondary emission of negative ions is greater than that of electrons.

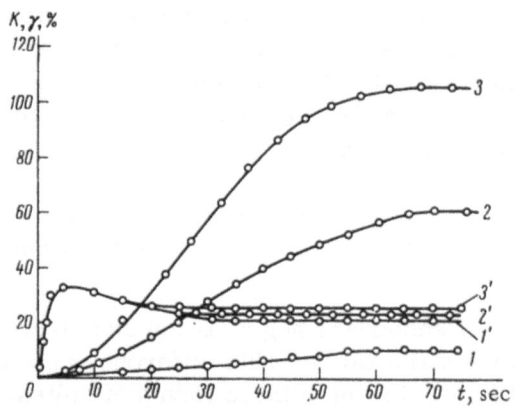

Fig. 7.30. Dependence of K⁻ (curves 1', 2', 3') and γ (curves 1, 2, and 3) on the time of depositing NaCl films on molybdenum while bombarding with Cs⁺ ions: 1, 1') 370; 2, 2') 1000; 3, 3') 1500 eV.

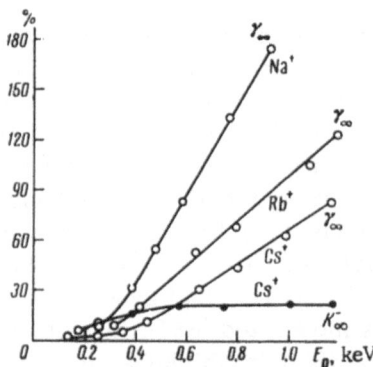

Fig. 7.31. Dependence of γ_∞ and K_∞^- on the energy of the bombarding ions.

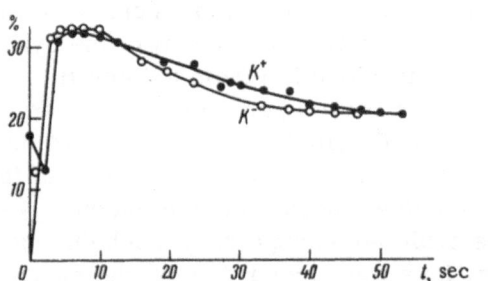

Fig. 7.32. Dependence of K⁺ and K⁻ on the time of depositing NaCl films on molybdenum while bombarding with 580-eV Cs⁺ ions.

In the energy range studied, γ_∞ depends greatly on E_0, while there is little variation in K_∞^- with varying energy for these film thicknesses. The values of γ_∞ are shown as functions of the energy of bombarding Na⁺, Rb⁺, and Cs⁺ ions in Fig. 7.31. Since the coefficient of negative-ion emission is not particularly dependent on the mass of the bombarding ion, as experiment confirms, Fig. 7.31 only shows the K⁻(E_0) relationship for Cs⁺ ions fired through an NaCl film. We see from the figure that for low energies of the primary ions the secondary negative emission may exceed the electron emission even for very thick films. We see that for the same energies of the bombarding ions γ_∞ is greater for light ions than for heavy ones. Hence, the proportion of negative ions in the secondary emission of negative particles on bombarding an NaCl film with ions of different energies increases with increasing mass of the bombarding ions.

In all the measurements we observe similar behavior in the positive- and negative-ion emission. Neither form of emission shows any serious dependence on the energy and mass of the bombarding ions or on the thickness of the NaCl film. There is only a difference between the two forms of emission in the case of very thin films, when the positive emission includes ions reflected from the metal substrate. Figure 7.32 shows the way in which K⁺ and K⁻ vary with time of deposition, i.e., with increasing thickness of the NaCl film on the molybdenum substrate bombarded with 580-eV Cs⁺ ions. We see from the figure that, for a moderate film thickness, after the reflection of ions from the substrate has ceased, the functions K⁺(t) and K⁻(t) are not only similar, but almost coincide with each other for all thicknesses of the NaCl film. This coincidence suggests that both forms of emission from the films in the energy range studied have the same nature and constitute the sputtering of the NaCl film in the form of positive and negative ions, in complete agreement with the conclusions of Lovtsov and Starodubtsov [205], who found no reflected ions among the positive emission of alkali halide films.

The relation between the electron and negative-ion components of the secondary emission of

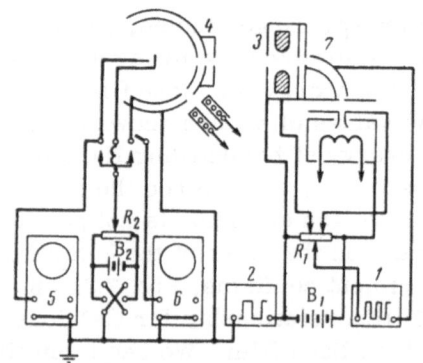

Fig. 7.33. Apparatus and electrical circuit for studying secondary emission of films, involving the separation of the electrons from the negative ions in a magnetic field.

negative particles depends on the energy and mass of the bombarding ions and also on the film thickness. Hence, any measurement of the total emission of secondary negative particles over a wide range of ion energies and film thicknesses will not simply represent the laws of secondary ion—electron emission; on the contrary, in certain extreme cases (small film thicknesses, low ion energies) it will rather reflect the laws of negative-ion emission.

The laws of secondary negative-ion and electron emission differ considerably. This difference appears primarily in the way in which these forms of emission depend on the film thickness. The maximum values of the secondary negative-ion emission are reached for a film thickness of the order of a monolayer. For such thicknesses the electron emission is still very low. Depending on the energy of the bombarding ions, the electron emission rises more or less sharply with further increase in film thickness and reaches saturation for thicknesses many tens of times greater than that at which the maximum of the negative-ion emission occurs. Thus, the secondary negative-ion emission comes almost entirely from the surface, while the secondary ion—electron emission is developed at a comparatively deep level.

It was noted earlier that the curves relating γ to the film thickness obtained for various energies of the bombarding ions were distinguished not only in respect of the values of the coefficients, but also in respect of the point at which they passed to saturation [214, 215, 536]. This evidently indicates that γ depends on the depth of penetration of the primary ions. However, comparison of the $\gamma(t)$ curves obtained by measuring the emission from different films deposited at a high rate was impeded by the impossibility of ensuring identical conditions for the formation of the films. It therefore became desirable to measure γ for several incident-ion energies E_0 simultaneously during the formation of one particular film at a high deposition rate so as not to increase the degree of contamination of the film by residual gases. For this purpose the energy of the primary charged-particle beam was modulated [682]; this enabled the relation between the coefficients of secondary emission and the film thickness to be determined simultaneously for two energies of the bombarding ions while depositing one and the same film.

The arrangement of the electrodes in the apparatus and the electric circuit are shown in Fig. 7.33. On the right-hand side of the sketch is the source part of the apparatus, and on the left-hand side the receiving section, the arrangement of which was described by Moroz and Ayukhanov [657]. The bombarding beam was directed at the target in periodic pulses with a frequency of 50-100 cps derived from the generator 1 connected into the circuit of the cylindrical condenser 7. The primary ions derived their constant acceleration from battery B_1. The energy was modulated by applying rectangular voltage pulses from a generator 2 between the body of the electrostatic lens 3 and the diaphragm of the guard cylinder 4. The depth of modulation could be varied between 0 and 700 V, depending on what difference in the energies of the bombarding particles was desired. The frequency of the pulses modulating the ion energy was 12-15 cps. With this frequency the secondary currents could not change very much during the passage of a single pulse. The electrons and secondary ions were separated for the measurements. The electron current was measured with oscillograph 5 and the negative-ion current with oscillograph 6. The arrangement of Fig. 7.33 also allowed the primary particle current to be measured.

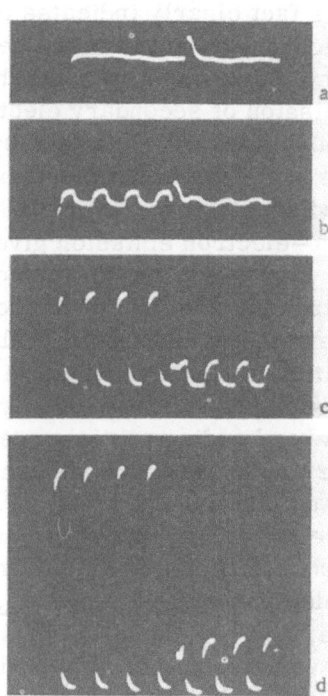

Fig. 7.34. Oscillograms of secondary electron currents obtained while depositing an NaCl film on a molybdenum surface, using "top-hat" (rectangular) modulation of the primary-beam energy.

Figure 7.34 shows oscillograms obtained when measuring the secondary electron current while depositing an NaCl film on a previously cleaned molybdenum target (cleaned by heating). The oscillograph sweep frequency in the present case coincided with the modulation frequency of the primary-beam energy, while the beam-interruption frequency was seven times the frequency of energy modulation. Oscillogram *a* was obtained directly after disconnecting the target heater. Here there is no electron current. The bend in the line halfway along oscillogram *a* , which also occurs in oscillograms b-d, is due to the passage of the modulating voltage pulses through the capacity of the apparatus and the electric circuit. At the time of obtaining the oscillogram a certain salt coating had already been formed on the molybdenum substrate and this led to the appearance of secondary electron emission. Oscillograms c and d illustrate the further growth in electron emission as the thickness of the film deposited on the substrate increased. The height of the step on the left-hand side of the oscillogram was determined by the value of the secondary current arising as a result of bombardment with high-energy ions, while that on the left was attributable to low-energy ions.

The oscillograms were recorded on motion-picture film with a frequency of 2 frames / sec. Analysis of the pictures yielded the dependence of γ on the NaCl film thickness for the two energies of the bombarding ions. In order to be able to compare the moments of passing to saturation more accurately for the $\gamma(t)$ curves when the numerical values of the coefficients γ differed considerably from one another at the same instant of time, it proved convenient to plot the curves in such a way that the values of the ordinates corresponding to the value of γ at saturation γ_∞ should coincide in both cases. This could be done by plotting the time dependence of the ratio γ/γ_∞ for each of the cases being compared.

The ratios $\gamma(t)/\gamma_\infty$ obtained during the formation of one particular film by bombarding this simultaneously with 450- and 920-eV Na^+ atoms are shown in Fig. 7.35. The curve corresponding to the lower energy of the bombarding ions reaches saturation earlier than that corresponding to the higher energy. However, this difference in the moments of saturation only occurs for fairly low energies of the bombarding ions.

Figure 7.36 shows analogous $\gamma(t)/\gamma_\infty$ curves for energies of 1380 and 1840 eV. Here the black circles indicate the values of γ/γ_∞ for an energy of the primary ions equal to 1380 eV and the white circles indicate 1804 eV. We see that both curves reach saturation at the same film thickness, despite the fact that the difference in the energies of the bombarding ions is here roughly the same as in the case illustrated in Fig. 7.35.

A study of curves such as these, obtained for two different values of ion energy in different regions, showed that the film thickness at which saturation set in increased with increasing ion energy E_0 up to a certain critical value E_k and then, as E_0 rose further, i.e., for $E_0 \geq E_k$,

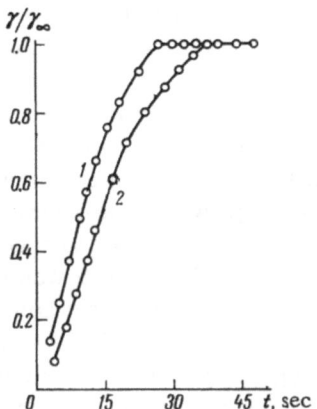

Fig. 7.35. Dependence of γ/γ_∞ on the time of depositing NaCl films on molybdenum while bombarding with Na$^+$ ions: 1) 450; 2) 920 eV.

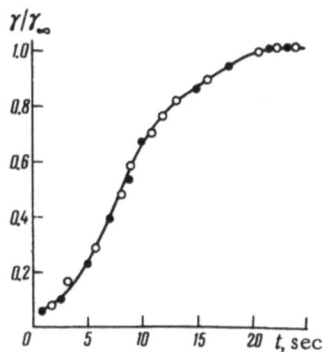

Fig. 7.36. Dependence of γ/γ_∞ on the time of depositing an NaCl film on molybdenum.

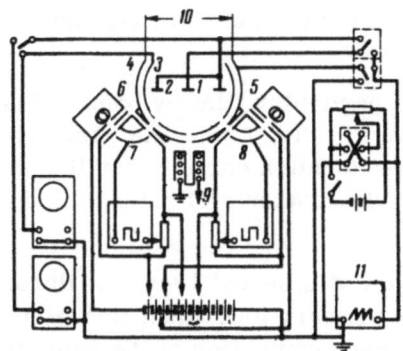

Fig. 7.37. Apparatus and electric circuit with two ion sources.

the moment at which saturation set in ceased depending on E_0. This fact clearly indicates that for $E_0 > E_k$, the secondary electrons are generated in a layer with a thickness exceeding the effective depth of emission of secondary electrons. The value of this thickness was estimated by comparison with values of the "effective depth of emission" of secondary electrons from NaCl in the case of electron—electron emission given by Gomoyunova [470]. For this purpose, the coefficients of ion—electron and electron—electron emission were determined as functions of the time of film deposition while depositing one particular NaCl film, bombarded alternately by electrons and ions with bombarding particle energies $E_0 > E_k$. A comparison of the moments of reaching saturation showed that $d_{ee}/d_{ie} = 1.3 \pm 0.1$. According to [470], the effective depth of emission of secondary electrons from an NaCl film in the case of electron—electron emission was $d_{ee} = 6 \cdot 10^{-6}$ cm. Hence, $d_{ie} = (4-5) \cdot 10^{-6}$ cm, i.e., the secondary electrons formed on bombarding NaCl films with ions are able to emerge from a depth of the order of a hundred monatomic layers.

The results also show that the primary ions starting at energies of $E_0 \sim 1000$ eV penetrate hundreds of monatomic layers into the material.

A comparison of the secondary-emission characteristics obtained on bombarding films of variable thickness with different ions can only be regarded as correct during the formation of one particular film. For this reason, an apparatus with two surface-ionization ion sources simultaneously dispatching different ion beams to the target was built. The arrangement of the electrodes in the apparatus and the electric circuit are shown in Fig. 7.37. The molybdenum substrate-target 1 of width 1.5 and length 40 mm intended to receive the test film was placed on the axis of the cylindrical secondary-ion collector 3. The secondary-electron collector 2 consisted of two strips of blackened tantalum 4 mm wide and 40 mm long arranged in the same plane as the target at a distance of 1.5 mm from the wall of collector 3. The ion beams were obtained in the sources 5 and 6 and directed through a system of diaphragms onto the target 1 with the help of the

cylindrical condensers 7 and 8. The arrangement of the diaphragms and the small width of the target ensured that the ion beams should fall on the same part of the target at the same angles. Particles not falling on the target were caught by the ion trap 10, a hollow metal cylinder with

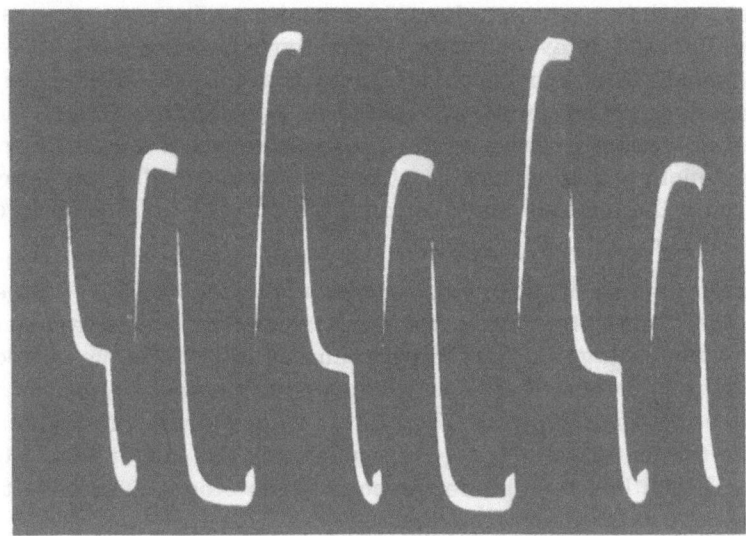

Fig. 7.38. Oscillogram of the secondary electron current in
the circuit of the electron collector of the apparatus shown in
Fig. 7.37, obtained on modulating the intensities of the two-ion
beam with **rectangular-voltage pulses of different frequencies.**

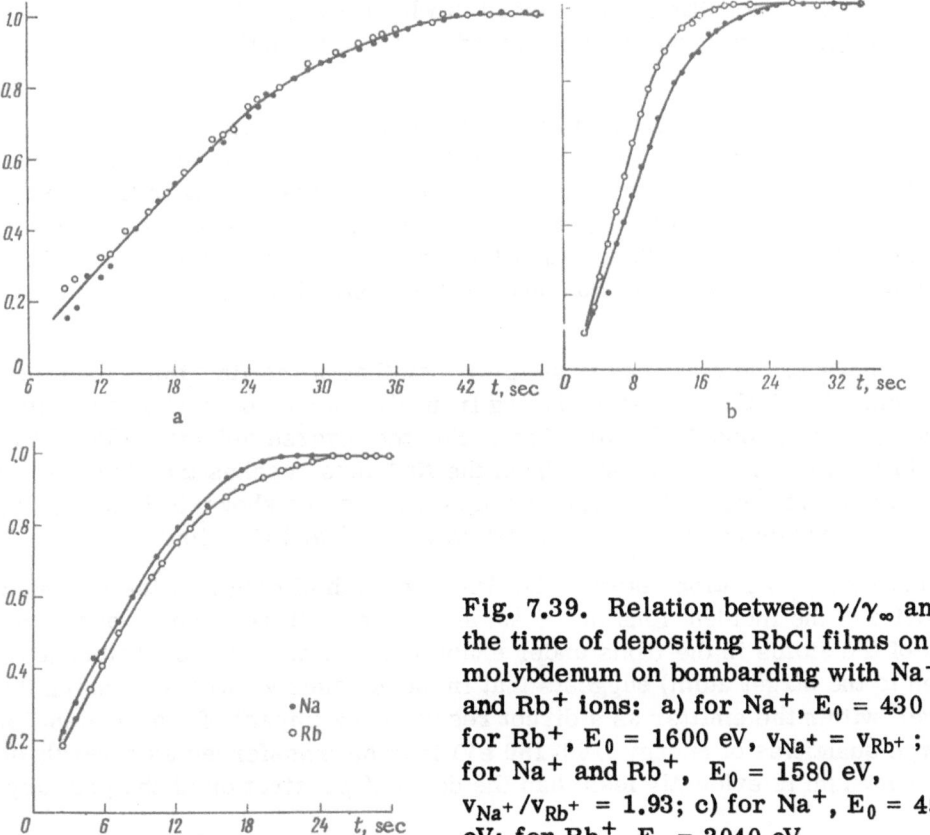

Fig. 7.39. Relation between γ/γ_∞ and
the time of depositing RbCl films on
molybdenum on bombarding with Na^+
and Rb^+ ions: a) for Na^+, $E_0 = 430$ eV,
for Rb^+, $E_0 = 1600$ eV, $v_{Na^+} = v_{Rb^+}$; b)
for Na^+ and Rb^+, $E_0 = 1580$ eV,
$v_{Na^+}/v_{Rb^+} = 1.93$; c) for Na^+, $E_0 = 450$
eV; for Rb^+, $E_0 = 3040$ eV.

blackened walls and a cellular base, on the side opposite to the receiving part of the apparatus. The test films were obtained by evaporation of the corresponding salt (placed in the furnace 9) onto the thoroughly degassed molybdenum substrate immediately after heating. During the measurements a negative voltage of 30–40 V relative to collectors 2 and 3 or a sawtooth voltage from generator 11 was applied to the target. In the presence of a certain optimum magnetic field created by a solenoid placed coaxially with electrodes 3 and 4, the electrons were deflected to the collector 2, while the ejected negative ions, which were only slightly deflected by this value of the magnetic field, fell into collector 3.

The oscillograph method of measuring secondary currents [281] offered the possibility of measuring the secondary emission due to the bombardment of the target by different ion beams separately. If a constant voltage is applied between the target and the collectors and the ion beams are modulated with pulses of different frequency, the oscillogram of the secondary electron current will have the form shown in Fig. 7.38. We see from the figure that the high, narrow pulses correspond to the emission arising on bombarding the target with a beam of ions obtained from one source, while the lower steps correspond to the other. The result of adding the two currents may be seen from the same figure.

As subjects for study in the present investigation, NaCl and RbCl films formed by vacuum deposition on metallic substrates were used. The RbCl films were bombarded with Na^+ and Rb^+ ions and the NaCl films with K^+ and Cs^+ beams.

The results showed that, on bombarding these films with different ions of different energies, the moment of saturation of the γ coefficient occurred for smaller film thicknesses in the case of the heavier ions. On continuously increasing the energy of the heavy ions and reducing that of the light ions, a condition was reached in which γ saturated at the same point for both beams. Calculations showed that this occurred when the velocities of the bombarding ions were equal.

Figure 7.39a shows $\gamma(t)/\gamma_\infty$ as a function of the time of depositing RbCl on molybdenum on simultaneously bombarding with Na^+ and Rb^+ ions at the same velocities. We see from the figure that the values of $\gamma(t)/\gamma_\infty$ for the different ions corresponding to the same film thickness lie neatly on a single curve. The simultaneous achievement of saturation in the $\gamma(t)/\gamma_\infty$ curve for both of these cases indicates that the limiting depths of generation of the secondary electrons on bombarding RbCl with Na^+ and Rb^+ ions striking the target at the same velocities were equal.

If the velocity of the Na^+ ions is smaller than that of the Rb^+ ions, then the $\gamma(t)/\gamma_\infty$ relationship associated with the emission arising from the Na^+ ions reaches saturation earlier than the analogous relationship for Rb^+ ions. For the inverse velocity relationship the $\gamma(t)/\gamma_\infty$ curve associated with emission arising from the Na^+ ions reaches the steady value later, i.e., for greater film thicknesses. The corresponding curves are shown in Figs. 7.39b,c. Similar results were obtained on bombarding NaCl films with K^+ and Cs^+ ions.

The fact that the relation between the limiting depth of origin of the secondary electrons and the velocity of the incident ions is the same for such different ratios of the mass of the ions to the mass of the atoms in the films under examination (i.e., for such different values of the energy given to the target atom) suggests that in these cases we are witnessing the ejection of electrons well within the emitter as a direct result of the impact of the incident ions. The depth to which large quantities of energy (100–150 eV) may be transferred as a result of the estafette ("relay") mechanism is evidently less than the depth of penetration of the primary ions themselves.

Thus for the pairs of ions studied the depths of penetration into the RbCl and NaCl films are the same for equal velocities. For equal energies the lighter ions, which have the greater velocity, penetrate deeper. This latter fact, in general, agrees with the results of [570, 571, 690], in which the penetration of alkali ions was studied.

§ 5. Secondary Emission from Dielectric Films

on Bombarding with Inert-Gas Ions

The kinetic emission of electrons obtained by bombarding alkali halide films on metals with alkali ions was studied in [733]. The comparatively long range of the excited electrons in the dielectric led to certain conclusions regarding the effective depth of emission of the secondary electrons and the depth of penetration of the ions.

Films of alkali halides on metals are convenient subjects for studying the potential emission of electrons also.

The potential emission of electrons from NaCl and CsCl films on molybdenum was first studied by Arifov et al. [755, 782], the targets being bombarded with 100- to 2000-eV Ar^+ and Ne^+ ions. The coefficient of potential electron emission and the energy distribution of the secondary electrons were studied as functions of film thickness. In order to distinguish the potential emission of electrons from the total emission, the electron emission arising from the bombardment of the same films with K^+ and Na^+ ions was studied. The experiments were carried out in a glass vacuum apparatus, the main components of which were an ion source, a focusing system [726], and a receiving section with an analyzer, shown schematically in Fig. 7.40. The analyzer was a 127° cylindrical condenser 1 of the Hughes-Rojansky type. The experimentally determined resolving power of the apparatus was 2%.

It is well known that on bombarding alkali halide films with ions, the emission of electrons is accompanied by the ejection of positive and negative ions [643]. In our experiments the coefficients of electron and ion emission were studied separately with the help of crossed electric and magnetic fields. Special experiments showed that this method ensured the separation of the electrons from the negative ions to an accuracy of 5% in an apparatus of the geometry employed.

The coefficients of secondary electron and ion emission were measured by the double-modulation oscillograph method [281]. The electron and ion components of secondary emission were measured at the same time by oscillographs connected to the corresponding collectors [726]. An alkali halide film was deposited on a molybdenum strip substrate by evaporating from a special evaporator. The rate of adsorption of the salt molecules was maintained at 2-3 monolayers per sec for a residual-gas pressure of $2 \cdot 10^{-7}$ mm Hg. Before the measurements the molybdenum substrate was degassed by prolonged heating at up to 2300°K. The cleanness of the surface was checked by observing the potential emission of electrons. In order to deposit the film on an adequately clean surface under continuous-deposition conditions, the substrate was "flashed" (briefly heated to 2200°K). For this reason, the rate at which salt built up on the surface of the molybdenum in the initial period was not quite constant, since the NaCl molecules were being adsorbed as the substrate cooled.

Measurements continued until reaching deposited film thicknesses such that the coefficient γ or the secondary-electron energy spectrum reached their steady-state values. However, for such thicknesses there were no signs of charging on the film surface.

Dependence of γ on Film Thickness

Figure 7.41 shows curves relating the coefficients of ion—electron emission (curve 1) and positive and negative ion emission (curves 2 and 3) to the time of depositing NaCl on tantalum

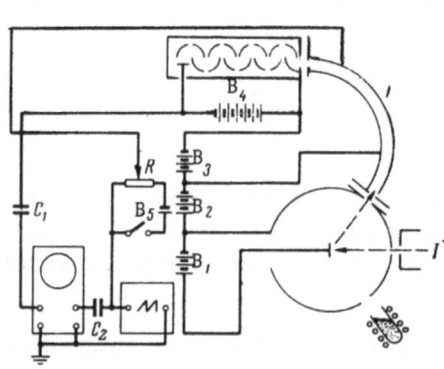

Fig. 7.40. Arrangement of apparatus for studying the energy distribution of secondary electrons.

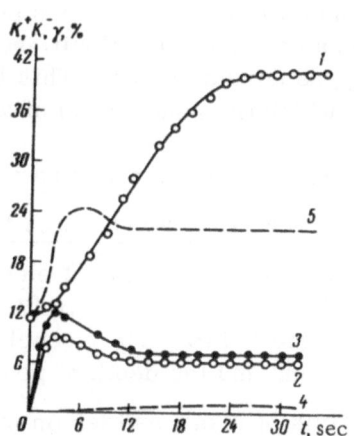

Fig. 7.41. Dependence of $\gamma(t)$, $K^+(t)$, and $K^-(t)$ on the film thickness of NaCl and sodium.

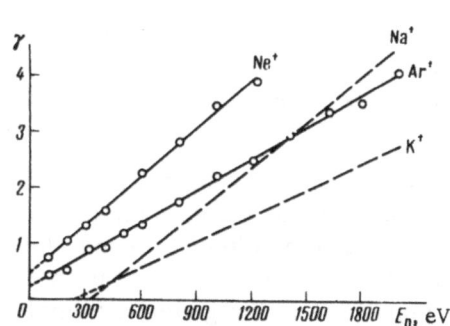

Fig. 7.42. The $\gamma(E_0)$ relation for a "thick" NaCl film.

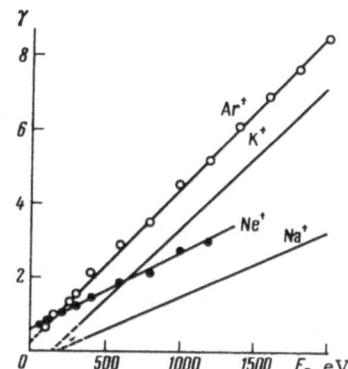

Fig. 7.43. The $\gamma(E_0)$ relation for a "thick" CsCl film.

while bombarding with 100-eV Ar$^+$ ions. Curve 4 characterizes the variation of γ with t for K$^+$ ions of the same energy [733]. Curve 5 represents the variation in γ with the thickness of a sodium film on bombarding with Ar$^+$ ions [734].

We see that before deposition, γ = 11-12% in the case of Ar$^+$ ions, while for K$^+$ ions there is no electron emission. On depositing an NaCl film there is a rise in the electron emission for both ions. However, the rise in emission for the case of Ar$^+$ ions is incomparably greater than for the K$^+$. Whereas, in the case of Ar$^+$, γ increases from 11 to 40%, in the case of K$^+$ the increment is very slight.

Hence the dependence of γ on t in the case of Ar$^+$ ions mainly characterizes the change in the coefficient of potential emission of electrons with varying thickness of the NaCl film.

It is also interesting to compare the $\gamma(t)$ curves for NaCl and sodium films. Whereas saturation of the $\gamma(t)$ curve occurs at thicknesses of about 1-2 monomolecular layers for the sodium film [734], for the NaCl film, saturation sets in at thicknesses of tens of molecular layers.

Before deposition of the film the coefficients K$^+$ and K$^-$ are equal to zero. After deposition of an NaCl film the emission of positive and negative ions begins, increases rapidly, and passes through a characteristic maximum at a comparatively small thickness.

In the case of CsCl films the relationships between γ, K^+ and K^-, and the film thickness are of the same general type.

Dependence of γ on the Energy of the Bombarding Ions

Figures 7.42 and 7.43 show the $\gamma(E_0)$ relationship on bombarding thick films of NaCl and CsCl with Ne^+ and Ar^+, and Na^+ and K^+ ions, respectively [782]. The data for the alkali ions are taken from [733, 268]. We see that in the range of energies studied, the γ associated with all the ions and films increases almost linearly with increasing energy. For the same E_0, the emission of electrons under the impact of inert-gas ions is greater than on bombarding with ions of the corresponding alkali metals of similar mass. The $\gamma(E_0)$ curves for Ne^+ and Na^+, and also for Ar^+ and K^+ ions are almost parallel to each other. The difference in the ordinates of the curves must be ascribed to the potential emission of electrons, since the kinetic emission is the same for the ion pairs $Ne^+ - Na^+$ and $Ar^+ - K^+$ (see Chapter 6, §4).

This conclusion also follows from the fact that the extrapolation of the $\gamma(E_0)$ curves to zero energy leads to their intersection with the γ axis at values of γ = 25 and 55% for Ar^+ and Ne^+ ions on CsCl and 25 and 50% on bombarding an NaCl film. Extrapolation of the $\gamma(E_0)$ to zero values of E_0 in the case of alkali ions, however, as shown in the graph, leads to the intersection of the curve with the E_0 axis, indicating that only kinetic electron emission is taking place. Hence, on bombarding NaCl and CsCl films with Ne^+ and Ar^+ ions in the energy range 100-2000 eV both potential and kinetic electron emission occur.

It is interesting to note that the arrangement of the $\gamma(E_0)$ curves for Ne^+ and Ar^+ and also Na^+ and K^+ ions varies in accordance with the kind of film in question; whereas for an NaCl film the light ions produce kinetic electron emission more effectively, for a CsCl film the reverse is the case. This disagrees with the results obtained for metals, in which case, for a particular E_0, the light ions produce kinetic electron emission more effectively than heavy. This apparently reflects a difference in the mechanism of exciting the electrons in the cases of metals and dielectric films, respectively. It is possible that for films of alkali halides the hot atoms of the target play the predominant part in the excitation of the electrons. This view is favored by the fact that the coefficient γ_k increases as the masses of the colliding particles come closer together, i.e., when the most favorable energy exchange takes place.

The data relating to the dependence of K^+ and K^- on E_0 agree with the results of [268, 733] and may be explained as being due to the sputtering of the film in the form of positive and negative ions. This is also confirmed by the fact that the values of the coefficients K^+ and K^- for pairs of inert-gas and alkali ions with similar masses are identical within the limits of experimental error.

The potential emission of electrons for fairly thick alkali halide films was twice as great as that obtained for films of alkali metals and over three times as great as in the case of molybdenum. An extremely interesting fact is the rise in potential electron emission as the film grows to a thickness of tens of monomolecular layers. This fact cannot be explained as being due to the change in work function only. Indeed, since the work function of a metal surface covered with a thin layer of alkali metal [468] is similar to that of one covered with an alkali halide [354], the potential emission of electrons is almost the same. For large thicknesses, the value of γ_p for a film of alkali metal remains constant as the energy of the primary ions varies from 0.1 to ~1 keV; in the case of a dielectric film, however, γ_p continues rising.

The rise in potential emission up to large film thicknesses may evidently be explained, in analogy with kinetic ion—electron emission from dielectric films, by the comparatively great effective depth from which the excited electrons emerge.

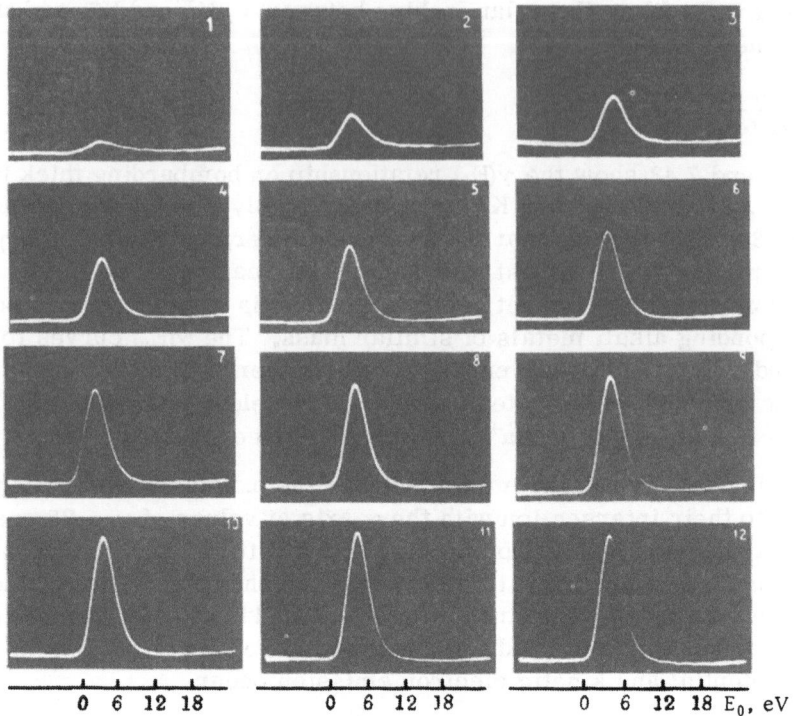

Fig. 7.44. Oscillograms of the energy spectra of secondary electrons obtained while depositing an NaCl film on molybdenum and bombarding with 1000-eV He$^+$ ions.

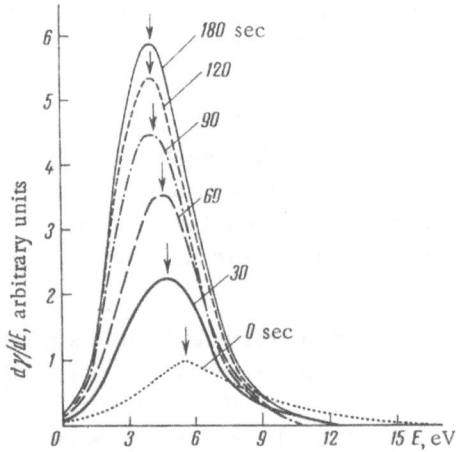

Fig. 7.45. Energy distributions of secondary electrons plotted from the oscillograms of Fig. 7.44.

Energy Distribution of Secondary Electrons

On bombarding NaCl films with Ar$^+$ ions it was noted that negative ions as well as electrons were ejected. In view of the fact that the emission of electrons was an order greater than the emission of negative ions and the electron current was amplified by the multiplier four or five times more efficiently than that of the negative ions, the contribution of the latter to the total current was negligibly low. Hence, when studying the energy spectra of the secondary electrons from alkali halide films no special measures were taken to eliminate the negative ions.

The secondary electron spectra were studied under the same experimental conditions as those used in studying γ.

Figure 7.44 shows a series of oscillograms of the energy spectrum of the secondary electrons obtained during continuous growth in the thickness of an NaCl film on molybdenum bombarded with 1000-eV He$^+$ ions. Individual frames were taken every 15 sec (the first one for pure molybdenum). We see from the oscillograms that as the NaCl molecules were adsorbed on the

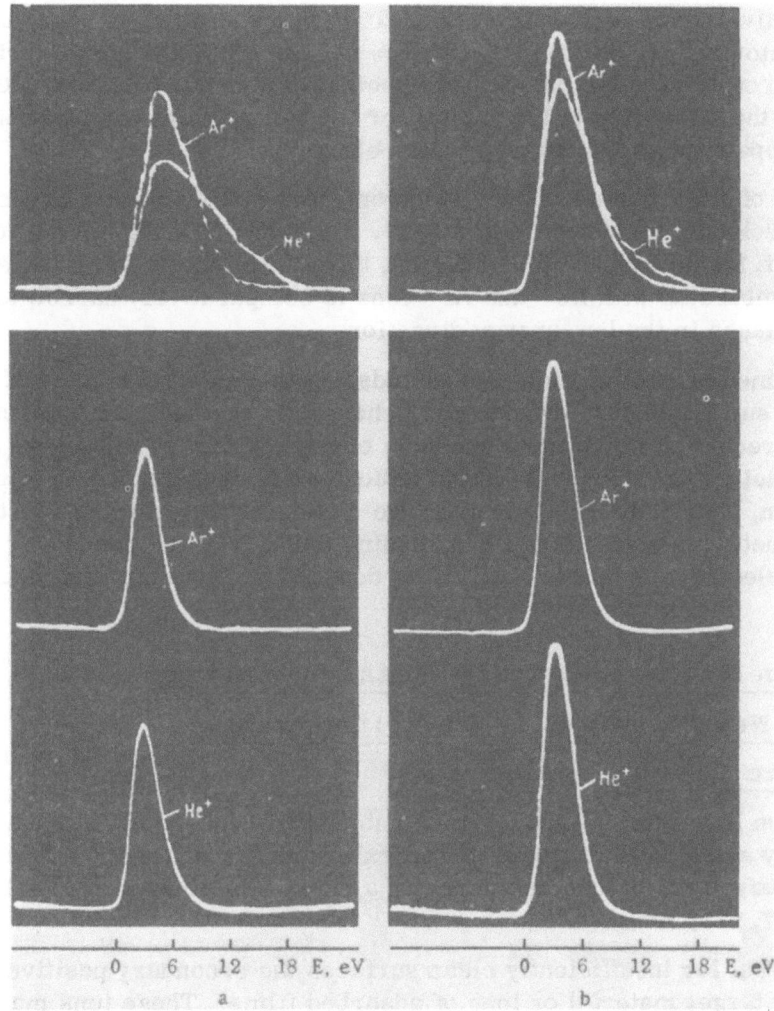

Fig. 7.46. Oscillograms of the energy-distribution curves of secondary electrons, normalized with respect to area. a) $E_0 = 40$ eV;
b) $E_0 = 2000$ eV. 1) Molybdenum film; 2) NaCl film.

molybdenum surface the secondary electron spectrum changed rapidly. For convenience of analysis Fig. 7.45 shows six curves based on the oscillograms of 7.44, which were taken at intervals of 30 sec. The curve characterizing the energy spectrum of the electrons from pure molybdenum is shown with a ten-times magnification in the vertical direction. We see from the
graphs that with increasing film thickness there is a rise in the total area under the curve. The
maximum of the curve moves in the low-energy direction and its half width becomes smaller.
There is a fall in the value of the maximum energy. However, whereas the number of electrons
with the most probable energy increases up to comparatively large thicknesses, the number of
electrons with the maximum energy depends little on the thickness of the film.

Figure 7.46 presents oscillograms of the energy distribution of secondary electrons obtained from molybdenum and an NaCl film on bombarding with 40- and 2000-eV He^+ and Ar^+
ions. The oscillograms are normalized with respect to area. The data for NaCl correspond
to a fairly thick film, such that no change occurs in the spectrum on increasing the thickness
further. We see clearly that on bombarding molybdenum with 40- and 2000-eV He^+ or Ar^+ ions
the energy spectra of the secondary electrons differ considerably. For He^+ ions, the energy

spectrum is relatively rich in fast electrons. The energy spectra for molybdenum at $E_0 = 40$ eV agree satisfactorily in form with Hagstrum's results [345] and characterize the energy distribution of electrons emitted by virtue of the potential energy of the ions. On increasing the energy (2000 eV) the difference in the curves for the two types of ions diminishes. This is probably due to the appearance of kinetic-emission electrons.

In the case of NaCl, the secondary electron energy-distribution curves for Ar^+ and He^+ ions practically coincide for a specified energy. By comparing the curves for molybdenum and NaCl, we see that, for an energy of 40-2000 eV, the energy spectrum of the electrons emitted from NaCl on bombarding with He^+ and Ar^+ ions is comparatively narrow and the maximum of the curve is displaced in the low-energy direction.

The experimental results show that the adsorption and condensation of alkali halide molecules on a metal surface lead to considerable changes in the secondary-electron energy spectrum; there is a reduction in the most probable energy, with a simultaneous growth in the total yield, with increasing film thickness, which indicates the deep-seated character of the ion— electron emission, both in the low-energy range (potential emission) and in the fairly high-energy range (kinetic emission) of the bombarding ions. It is an interesting fact that the energy spectrum of the electrons from the NaCl films depends little on the nature and energy of the bombarding ions.

§6. Mass-Spectrometric Study of the Secondary Emission of Negative Ions from Thin Metallic and Dielectric Films

It was shown in the foregoing that, on bombarding inadequately clean surfaces with positive ions, not only electrons but also a considerable number of negative ions were emitted. The formation of negative ions is also mentioned in a number of other papers [112, 113, 123, 128-130, 181, 276, 277, 528, 703].

It is clear that for insufficiently clean surfaces the secondary positive ions may also be partly ions of the target material or ions of adsorbed films. These ions may be formed on bombarding such surfaces with primary ions as a result of complex processes of dissociation, evaporation, and cathode sputtering.

In order to study secondary-emission phenomena from complex and inadequately clean surfaces, it was essential to use another method so as to facilitate the separate study of the electron and negative-ion components of negative-particle secondary emission, involving a complete mass-spectrometric analysis of the secondary positive and negative particles for different states of the surface bombarded.

A study of these phenomena by means of the ordinary types of magnetic spectrometers was almost impossible in view of their great inertia. In the absence of methods of checking, and in view of the continuous bombardment of the targets by the ions, it was very difficult to secure states of the bombarded surface which were reproducible and constant in time. Under these conditions it is most natural to study the dynamics of secondary processes for specific changes in the state of the surface bombarded; this is only possible, however, by using an inertia-free mass spectrometer. In order to study the composition of the secondary negative-ion emission with a rapid change in the state of the surface, an attempt was therefore made by Ayukhanov and Iskhakov [512] to use a magnetic mass spectrometer, with modulation of the energy of the primary ions. In order to record the ions, an electron multiplier was employed.

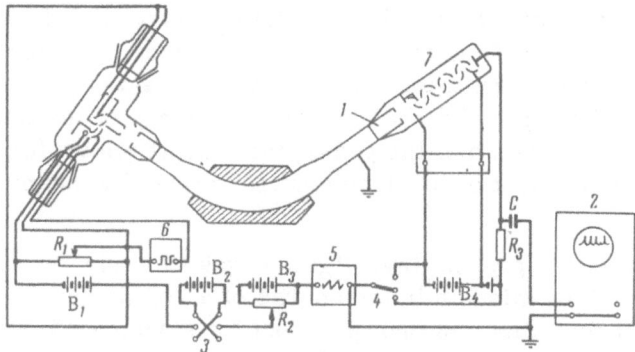

Fig. 7.47. Electric circuit of an experimental apparatus
for analyzing the composition of secondary emission.

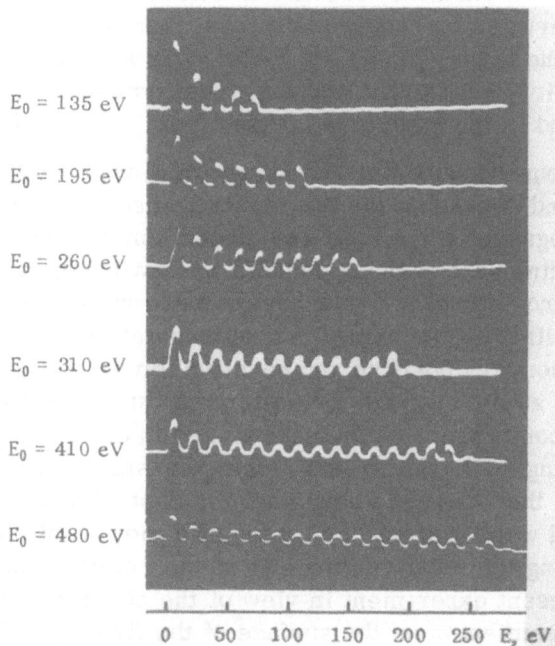

Fig. 7.48. Energy distribution of scattered Na^+ ions on
bombarding a tantalum surface with Na^+ ions.

Mass spectrometers with modulation of the energy of the ions have been described in a
number of papers [178]. The arrangement of the vacuum apparatus and the whole system for
analyzing secondary particles is shown in Fig. 7.47. The secondary-particle analyzer is a
homogeneous sector-type magnetic field with a deflection angle of 78°. The secondary particles
passing through the analyzer traverse the slit 1 and fall on the first dynode of the ion—electron
multiplier 7; they are recorded by oscillograph 2, which has its sweep synchronized with the
generator 5. For a certain value of the magnetic field, the oscillograph screen will reproduce
a part of the mass spectrogram, the extent of this being determined by the depth of modulation
of the ion energy by the generator 5. The system may be used for analyzing both secondary
positive and negative ions. This is facilitated by the switches 3 and 4, and also a switch governing the direction of the magnet supply current, not shown in the figure. In the course of record-

ing, the positive ions receive an additional acceleration in front of the electron multiplier. In order to establish the zero line of the mass spectrogram more precisely, a rectangular-pulse generator 6 is included in the circuit; this modulates the beam of primary ions bombarding the target. In order to vary the depth of modulation of the energy of the ions passing into the analyzer and measure the energy width of the peaks, the circuit allows for the possibility of smoothly varying the constant component of the accelerating voltage by means of the battery B_3 and the potentiometer R_2.

The main problem of the present investigation was to study the composition of the secondary positive and negative ion emission in the course of a rapid change in the state of the surface bombarded. A change of this kind was brought about by depositing various films on a metal surface from a special evaporator, which enabled the rate of film deposition to be regulated. This method of varying the state of the surface bombarded was convenient because if offered the possibility of studying the secondary-emission properties of the substrate target, the dynamics of variations in the secondary phenomena corresponding to transitional states, and the secondary-emission properties of thick films (identical with those of the solid material) under exactly the same conditions of vacuum. The use of refractory metals as substrates made it possible to repeat the experiments a number of times, removing the adsorbed films by heating and redepositing them as the target cooled. Alkali metals and alkali halides were used in order to form the films. In the first case there was a gradual transition from one metal to the other as deposition proceeded, and in the other a transition from the metal to the dielectric.

In carrying out the experiments, the following order was usually observed. After reaching the required vacuum and degassing the target, the ion source and evaporator were switched on, and these were first degassed for a long time under nonlocalized conditions. After restoring the vacuum, the recording part of the apparatus was switched on, and by gradually raising the temperature of the source spiral and evaporator these were slowly brought into the working mode. After reaching the limiting vacuum, the required rate of film deposition was established. The composition of the secondary positive and negative ion emission on ion bombardment of the pure (clean) substrate was studied at high target temperatures both with the evaporator switched on and switched off. The composition of the secondary ion emission from thick films was studied by rapidly depositing a film after establishing constant emission of the secondary ions. For intermediate states of the bombarded surface, the dynamics of the changes in the intensity of the various ions emitted were followed by means of a motion-picture camera. The problem of quantitatively determining the relative intensity of the emission of different types of ions was not included in the present experiment in view of the obvious difference in their coefficients of secondary ion—electron emission on the surface of the first dynode of the multiplier.

Figure 7.48 shows the forms of the peaks representing the secondary positive Na^+ ions, photographed from the oscillograph screen on bombarding a film-free tantalum target with 135- to 480-eV Na^+ ions. These pictures reflect the energy distribution of the secondary ions in differential form for different energies of the primary ions. The photograph clearly shows the wide secondary-ion energy spectrum, with sharply expressed limiting energies, depending linearly on the energy of the primary bombarding ions. The value of these limiting energies agrees with the relation for elastic collisions with individual target atoms.

Secondary positive-ion emission is almost entirely absent from thick films of pure alkali metals, both in the case in which the masses of the atoms in the film are equal to those of the bombarding ions, and also in the case in which the bombarding ions are lighter than the film atoms. Hence, for $m_1 \gtrsim m_2$, there is no appreciable cathode sputtering nor scattering from individual target atoms in the form of positive ions from the surfaces of pure alkali metals.

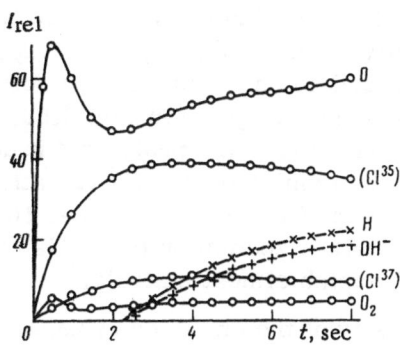

Fig. 7.49. Intensity of a number of negative ions as a function of the density of a sodium film on tantalum after bombarding with 600-eV Na$^+$ ions.

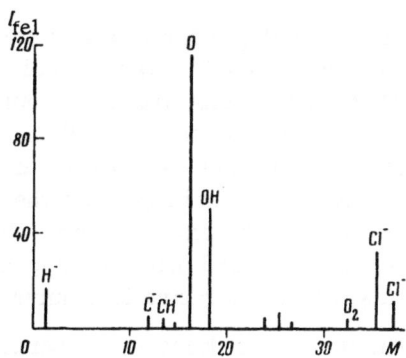

Fig. 7.50. Mass spectrogram of the negative-ion sputtering associated with the bombardment of a sodium film on tantalum by 600-eV Ar$^+$ atoms during the rapid deposition of the film.

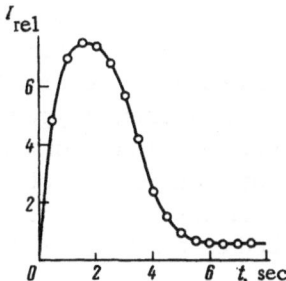

Fig. 7.51. Intensity of negative oxygen ions ejected by 600-eV Na$^+$ ions as a function of the density of a sodium film deposited on a tantalum surface at a high rate of deposition.

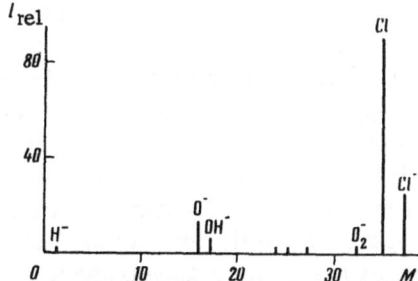

Fig. 7.52. Mass spectrogram of the negative-ion sputtering associated with the bombardment of a KCl film slowly deposited on a tantalum surface by Na$^+$ ions.

In the intermediate states of the bombarded surface, for which the density of the film on the surface of the substrate is comparatively low, both scattering of primary ions from the target and cathode sputtering of the film in the form of positive ions occur. With increasing density of the film the scattering gradually ceases and the cathode sputtering intensifies.

There is no marked emission of any negative ions on bombarding the film-free surface of a tantalum target with alkali ions. However, if the slightest trace of alkali metals occurs on the surface of the target, various negative ions begin to appear, and the intensity of this emission increases up to a specific value as the density of the alkali coating increases.

Figure 7.49 shows the relation between the intensity of negative O$^-$, O$_2^-$, H$^-$, OH$^-$, (Cl^{35})$^-$, and (Cl^{37})$^-$ ions (which appear on bombarding a film of sodium on tantalum with Na$^+$ ions) and the density of the film. We see from the curves that the variation in intensity with film thickness has a different character for each particular ion. In addition to this, the composition of the negative ions and the dynamic variation in the emission intensity, expressed as a function of film thickness for the various ions, depends greatly on the composition of the residual gases in the apparatus and the rate of depositing the film. For slow rates of depositing a sodium film on tantalum (of the order of a few atomic layers per sec) and bombardment with 600-eV Na^+ ions, the negative-ion emission includes, for example, negative ions of H^-, C^-, CH^-, CH_2^-, O^-, OH^-, O_2^-, Cl^-, and ions with mass numbers of 24, 25, 26, 42, and so on. A mass spectrogram taken with the help of the multiplier for a certain fixed film density is given in Fig. 7.50. As indicated earlier, this cannot give any idea of the true relation between the peaks, but it does enable us to say something about the composition of the negative emission and its variation with the conditions of the experiments. For example, at high rates of film deposition (over 10 atomic layers per sec) and after the appearance of a metallic deposit of sodium on the walls of the apparatus, all the negative ions mentioned except O^- and O_2^- vanish almost entirely.

Figure 7.51 indicates the dynamic changes in the intensity of the emission of negative oxygen atoms ejected by 600-eV K^+ ions while depositing a sodium film on tantalum at a high deposition rate in the presence of a metallic sodium deposit on the walls of the apparatus. We see from the figure that the intensity of negative oxygen—ion emission passes through a maximum in the intermediate states of the bombarded surface, but becomes very low for thick films under these conditions.

Thus, in the case of the slow deposition of a film of alkali metal, when the rate of competing residual-gas adsorption is comparable with the rate of film deposition, the secondary negative-ion emission contains a large number of different negative ions. When the film is deposited rapidly, the proportion of gases in the film diminishes and the emission of negative ions almost stops. The presence of negative oxygen ions in low-density films, even for the highest rates of film deposition attainable, is explained by the fact that, despite the measures taken to ensure degassing of the tantalum target, the latter is still not entirely free from oxygen, and that on depositing a thin film of alkali metal this oxygen is ejected very efficiently in the form of negative ions. Hence, on slowly depositing a film of alkali metal the composition of the secondary ion emission provides information regarding the composition of the residual gases within the apparatus, while, on rapidly depositing the film, gases adsorbed on the surface of the main target make their appearance.

Analogous results were obtained on depositing films of alkali halides on tantalum. For the positive-ion emission in this case also, there was a characteristic transition from the scattering of the primary ion beam by the atoms of the main target to the sputtering of positive alkali-metal ions from the alkali halide film forming on the surface, as well as the sputtering of primary ions which had penetrated into the film; the emission intensity of these depended on the rate of film deposition, the intensity of the flow, and the energy of the primary ions.

The composition of the secondary negative-ion emission for the slow deposition of an alkali halide film is in general the same as for the slow deposition of films of alkali metals. Figure 7.52 shows a mass spectrogram of negative ions obtained on bombarding a KCl film deposited on a tantalum substrate with Na^+ ions at a low deposition rate. For rapid film deposition, negative halide ions occur.

Hence, when depositing alkali metal or alkali halide films on tantalum and tungsten surfaces, the ionic component of the secondary emission undergoes very severe changes. The appearance of negative ions is of particular interest. On bombarding a target free from any alkali metal film, the emission of negative ions attributable to the adsorption of residual gases no longer takes place. On the other hand, even a small trace of alkali atoms on the surface leads to the appearance of an intense sputtering of adsorbed gases in the form of negative ions. This may possibly be associated with the electron affinity of the sputtered molecules and the work function of the bombarded surface. Hence, a study of the process of negative-ion sputtering on bombarding targets with various alkali ions should yield a certain amount of information regarding both the sputtering itself and the mechanism responsible for its intensification.

Table 7.1

Mass No.	Ion	Mass No.	Ion	Mass No.	Ion	Mass No.	Ion	Mass No.	Ion
1	H^-	30		45	CO_2H^-	67		128	
12	C^-	31		47		70	Cl_2^-	129	
13	CH^-	32	O_2^-	48	C_4^-	72	Cl_2^-	130	MoO_2^-
14	CH_2^-	33	O_2H^-	49	C_4H^-	73		132	
15	CH_3^- NH^-	34		50		74	Cl_2^-	140	
16	O^-	35	Cl^-	51	ClO^-	108		142	
17	OH^-	36	C_3^-	53	ClO^-	110		143	MoO_3^-
18		37	Cl^-	59		111		144	
19		38		60		112	MoO^-	145	
24	C_2^-	39		61		113		146	
25	C_2H^-	40	C_2O^-	62		114		148	
26	$C_2H_2^-$, CN^-	41	C_2OH^-	63		116			
27	$C_2H_3^-$	42	CNO^-	64		124			
28		43		65		126	MoO_2^-		
29		44	$CNOH_2^-$	66		127			

Table 7.2

Targets bombarded	Negative ions
Aluminum	Al^-, AlO^-, AlO_2^-, Al^+
Beryllium bronze · ·	Cu^-, CuH^-, CuO_2^-, $CuCl^-$, Cu_2^-, Cu_2H^-, Be^-, BeH^-
Copper · · · · · ·	Cu^-, CuH^-, CuO_2^-, $CuCl^-$, Cu_2^-, Cu_2H^-
Molybdenum	MoO^-, MoO_2^-, MoO_3^-
Silicon	Si^-, SiO_2^-, SiO_3^-
Zirconium	Zr^-, ZrO^-, ZrO_2^-, ZrO_3^-
Titanium · · · · · ·	TiO^-, TiO_2^-, Ti_2^-

In order to elucidate this question, Ayukhanov and Abdullaev [784] made a mass-spectrometer study of the composition of sputtering products from a number of solids (aluminum, copper—beryllium alloy, silicon, copper, molybdenum, zirconium, and titanium) in the form of negative ions on bombarding with Cs^+ and Na^+ ions in a mass spectrometer furnished with an ion—electron converter, enabling its sensitivity to be considerably raised. The experiments were made on the principle of the deposition of the alkali film by the primary ion beam itself, i.e., with a slow rate of film deposition.

On bombarding any of the targets mentioned with 700- to 1000-eV Cs^+ ions immediately after cleaning and degassing, a certain emission of various negative ions was established even a few seconds later. A characteristic feature for all the materials studied was the emission of various atomic and molecular negative ions not associated with the material of the target bombarded. The spectra of these ions were very rich in peaks, and the spectra associated with the different targets all had peaks of the same general kind. The density of the peaks was such that, apart from certain exceptions, there was a peak for every mass number. Without exception, all the spectra showed peaks for the following ions: H^-, C^-, CH^-, CH_2^-, CH_3^-, O^-, OH^-, C_2^-, C_2H^-, $C_2H_2^-$, $C_2H_3^-$, O_2^-, O_2H^-, Cl^-, C_2O^-, C_2OH^-, CNO^-, $CNOH_2^-$, CO_2H^-, C_4^-, C_4H^-, ClO^-, Cl_2^-, and also

Table 7.3

Mass No.	Ion	Mass No.	Ion	Mass No.	Ion	Mass No.	Ion	Mass No.	Ion
1	H^-	40	C_2O^-	65	Cu^-	90		131	Cu_2H^-
12	C^-	41	C_2OH^-	66	CuH^-	91	Cl_2OH	133	
13	CH^-	42	CNO^-	67		92		135	
14	CH_2^-	43		68		93		137	
15	CH_3^-, NH^-	44	$CNOH_2^-$	69		95	CuO_2^-	140	
16	O^-	45	CO_2H^-	70	Cl_2^-	97	CuO_2^-	141	
17	OH^-	46		71		98	$CuCl^-$	143	
18		47		72	Cl_2^-	99		145	}
19		48	C_4^-	73		100	$CuCl^-$	147	
24	C_2^-	49	C_4H^-	74	Cl_2^-	101		150	
25	C_2H^-	50		75		102	$CuCl^-$	152	
26	$C_2H_2^-, CN^-$	51	ClO^-	76		103		154	}
27	$C_2H_3^-$	52		77		104		156	
28		53	ClO^-	78		105		160	
29		54		79		106		161	
30		55		80		107		163	}
31		55		81		108		175	
32	O_2^-	57		82		115		178	
33	O_2H^-	58		83		117		180	}
34		59		84		124		182	
35	Cl^-	60		85		126	Cu_2^-	189	
36	Cl_3^-	61		86		127	Cu_2OH^-	191	
37	Cl^-	62		87	Cl_2OH^-	128	Cu_2^-	193	}
38		63	Cu^-	88		129	Cu_2H^-	195	
39		64	CuH^-	89	Cl_2OH^-	130	Cu_2^-		

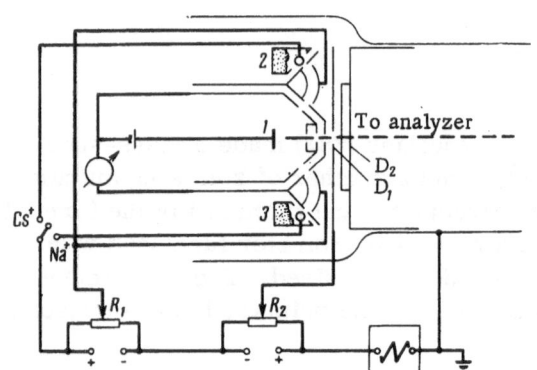

Fig. 7.53. Sputtering chamber with two ion sources.

those of ions with mass numbers 38, 39, 43, 47, 59, 60. These occurred immediately after starting the bombardment for every one of the targets. These peaks evidently constituted various combinations of H, O, N, Cl, and C which were naturally present in the residual gases as water, chlorine, and organic materials arising from the ion sources. There were particularly strong peaks of O^-, OH^-, Cl^-, H^-, C_2^-, C_2H^-, $C_2H_2^-$, C^-, CH^-, CNO^-, and CO_2H^-. By way of example, Table 7.1 shows the composition of negative emission arising on bombarding molybdenum with 800-eV Cs^+ ions.

Negative ions associated with the nature of the bombarded targets themselves were observed for each of the target materials mentioned. The intense peaks attributable to adsorbed atoms and molecules did not prevent the peaks of the materials bombarded from being observed. The list of all observed ions is given in Table 7.2.

Table 7.3 gives the composition of the negative-ion emission on bombarding copper with 800-eV Cs^+ ions. We see from this table that, together with the emission of negative ions not

Table 7.4

Mass No.	Ion	Mass No.	Ion	Mass No.	Ion	Mass No.	Ion	Mass No.	Ion
1	H^-	27	$C_2H_3^-$	45	CO_2H^-	58 }	$NaCl^-$	74	Cl_2^-
12	C^-	32	O_2^-	46		60 }		75	
13	CH^-	33	O_2H^-	47		61		77	
14	CH_2^-	35	Cl^-	48	C_4^-	62		81 }	$NaCl_2^-$
15	NH^-, CH_3^-	36	C_3^-	49	C_4H^-	63		83 }	
16	O^-	37	Cl^-	50		64		86 }	Cl_2O^-
17	OH^-	38		51	ClO^-	65		88 }	
18		39		52		66		90 }	
19		40	C_2O	53	ClO^-	67		92	
23	Na^-	41	C_2OH^-	54		68			
24	C_2^-	42	CNO^-	55		69			
25	C_2H^-	43	$CNOH^-$	56		70 }	Cl_2^-		
26	$C_2H_2^-, CN^-$	44	$CNOH_2^-$	57		72 }			

Table 7.5

Mass No.	Ion	Mass No.	Ion	Mass No.	Ion	Mass No.	Ion	Mass No.	Ion
1	H^-	33	O_2H^-	52		72	Cl_2^-	92 }	Ti_2^-
12	C^-	34		53	ClO^-	73		93	
13	CH^-	35	Cl^-	54		74	Cl_2^-	94	
14	CH_2^-	36	Cl_3^-	55		76		95	
15	NH^-, CH_3^-	37	Cl^-	57		77		96	
16	O^-	38		58		78 }	TiO_2^-	97	
17	OH^-	39		59		79		98	
18		40	C_2O^-	60		80 }		99	
19		41	C_2OH^-	61		81		100	
23	Na^-	42	CNO^-	62 }	TiO^-	82 }		101	
24	C_2^-	43	$CNOH^-$	63		83		102	
25	C_2H^-	44	$CNOH_2^-$	64 }		84		103	
26	$C_2H_2^-, CN^-$	45	CO_2H^-	65		85		104	
27	$C_2H_3^-$	46		66		86		105	
28 }	Traces	47		67		87		106	
29		48	C_4^-	68		88		107	
30		49	C_4H^-	69		89		108	
31		50		70	Cl_2^-	90		109	
32	O_2^-	51	ClO^-	71		91		110	

associated with the target material and negative ions constituting different compounds of copper, principally CuH^-, $CuCl^-$, Cu_2H^-, etc., negative atomic and molecular copper ions also occur. The intensities of the peaks of the atomic copper ion are particularly high; after bombarding for some time and slightly heating, they become comparable with the Cl^-, O^-, and OH^- peaks, the strongest in the spectra. The absence of any link between the intensities of the copper and copper compound ions, as well as the monotonic increase in the peaks of the negative copper ions with energy, typical for sputtering processes, show that the Cu^- and Cu_2^- ions are sputtered from the main material and are not fragments of surface compounds. The proportion of nega-

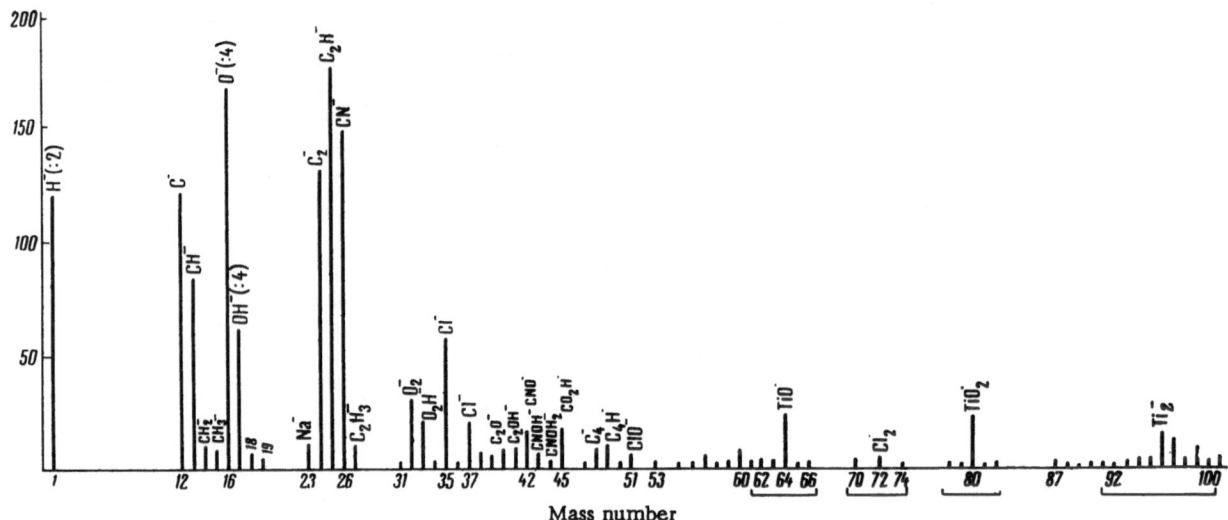

Fig. 7.54. Mass spectrogram of negative-ion sputtering on bombarding titanium with 1000-eV Cs^+ ions.

tive ions containing copper to the total number of negative ions is (judging by the spectrograms) quite considerable.

When working with a copper—beryllium alloy containing 2% Be, in addition to the peaks observed in the copper spectrum, there were a large number of negative beryllium ions, Be^- and BeH^-. On bombarding an aluminum target with 800- to 1000-eV Cs^+ ions, the ions associated with the nature of the metal itself included Al^-, AlO_2^-, and AlO^-. It should be remembered that these only appeared after many days of intensive bombardment of the aluminum with Cs^+ ions. The effect is presumably associated with the severe contamination of the aluminum surface.

On bombarding zirconium with Cs^+ ions, four groups of zirconium peaks: Zr^-, ZrO^-, ZrO_2^-, and ZrO_3^- may easily be identified from their characteristic isotopic structure.

On bombarding molybdenum foil, no negative atomic or molecular molybdenum ions appeared. A search for peaks associated with the target material showed that, in this case, these were represented by MoO_2^-, MoO_3^-, and MoO^- (in order of falling intensity).

Apart from the metallic targets just mentioned, we analyzed the composition of the negative-ion sputtering from silicon bombarded with Cs^+ ions. The composition of the silicon negative-ion sputtering spectrum was similar to that of the metallic targets. An important result in the case of silicon was the observation of negative silicon ions. Honig [440] bombarded silicon—germanium alloy targets with inert-gas ions but found no silicon or germanium negative ions. The observation of Si^- ions in our own case was apparently due to the creation of a cesium film on the surface.

Despite the fact that the electron affinity of silicon has a considerable value, the intensity of the Si^- peaks was low compared with the rest. This may have been because the silicon was not subjected to heat treatment, so that bombardment led mainly to the sputtering of the contaminated surface layer.

In order to obtain some information regarding the nature of the intensification of negative-ion sputtering, a titanium target 1 was sputtered in a chamber with two ion sources 2 and 3 (Fig. 7.53). In this chamber, the titanium target was bombarded successively with Na^+ and Cs^+

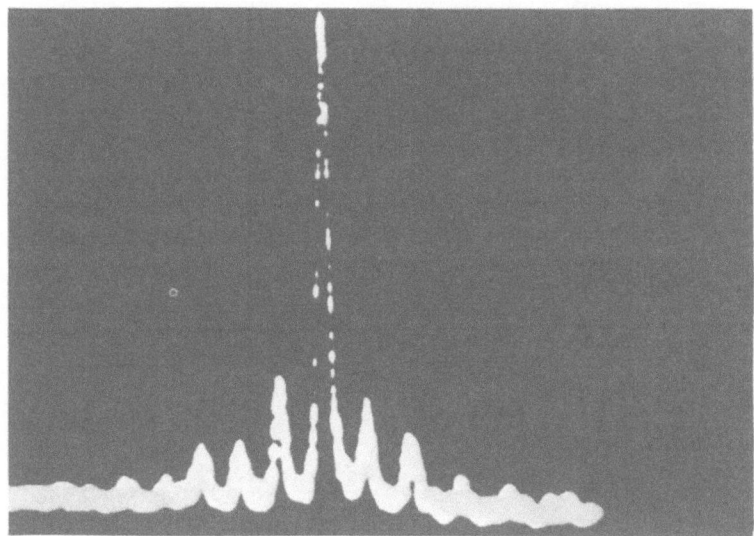

Fig. 7.55. Peaks of the negative TiO_2^- ion.

ions. The composition of the negative-ion spectrum obtained on bombarding titanium with Na^+ ions is shown in Table 7.4. We see from this table that negative ions of the target material and its chemical compounds with surface-contamination products are absent. However, there are peaks for $NaCl^-$, Na_2Cl^-, Cl_2O^-, and also a conversion peak for Na^-. If the titanium is bombarded with Cs^+ ions immediately after this, we first obtain a spectrum very similar in composition to that of Table 7.4. However, after prolonged bombardment with Cs^+ ions the negative-ion spectrum from the titanium changes in form and composition (Table 7.5), becoming richer and much stronger than that of Table 7.4. Whereas at first the peaks of chlorine and its compounds are the strongest, in the course of time these weaken and some peaks drop right out ($NaCl^-$, Na_2Cl^-, Cl_2O^-). The peaks of oxygen and its compounds increase and some new titanium-compound peaks appear: TiO^-, TiO_2^-, and apparently Ti_2^-. The intensity and shape of the spectrum change considerably in the neighborhood of these peaks. On disconnecting the Cs^+ ion source and reconnecting the Na^+ source, the negative ions given in Table 7.4 appear once more, and this is repeated every time. The peculiar isotopic structure of titanium also eases the unmistakable identification of the peaks corresponding to its compounds (Figs. 7.54 and 7.55).

The observation of the TiO^-, TiO_2^-, and Ti_2^- peaks on bombarding titanium with Cs^+ ions and their absence on bombarding with Na^+ indicates a great intensification of negative-ion sputtering on depositing Cs over the surface. The intensification arises not only in the sense of an increase in existing peaks but also in the appearance of new ones. The electron affinity of these compounds clearly has such a low value that only the presence of cesium on the surface, reducing the work function of the titanium more considerably than sodium, can activate the surface sufficiently to make the negative ions of these compounds detectable.

INTERACTION OF NEUTRAL ATOMS
WITH A SOLID SURFACE

§ 1. Brief Review of Investigations into the
Interaction of Neutral Atoms with a Solid Surface

The investigations described in earlier chapters showed that on bombarding targets with positive ions a considerable proportion of the primary particles were neutralized on colliding with the surface of the metal. The number of neutralized ions depended on the relation between the ionization energy of the bombarding particles and the work function of the metal. However, in setting up the balance of the particles (primary and secondary), the number of neutralized ions was not determined by direct measurement but by finding the difference between the primary and secondary ion currents. In addition to this, in order to set up the energy balance for the interaction of the particles with the solid surface, we must know the energy spectrum of the neutralized ions as well as their total number. In view of this, it is particularly interesting to study the various properties characterizing the emission of neutral particles on ion bombardment.

On the other hand, the study of secondary processes associated with the bombardment of metals by neutral atoms is very pressing both from the point of view of a deeper and more circumstantial penetration into the essence of the mechanisms underlying the interaction of atomic particles with a solid surface, and also from that of various practical problems.

Unfortunately, effects accompanying the bombardment of a surface by neutral particles have not so far been studied very fully. An obstacle to the development of such investigations has been the difficulty of obtaining a flow of monoenergetic neutral particles of controlled energy and intensity, and more particularly the recording of these.

In order to obtain beams of fast neutral particles, the following two methods are employed: resonance recharging (resonance charge exchange) and neutralization at a solid surface. The better way of obtaining neutral beams is that of resonance recharging, since this approach ensures a smaller energy spread than that of surface neutralization. The ions which capture electrons in the course of recharging and are thus neutralized experience hardly any change in energy and direction, and this offers the possibility of obtaining neutral atomic beams controlled in energy and intensity.

The picture is very different for surface ionization. When an ion falls on a target, it may eject a conduction electron from the latter and recombine with this. The ion is thus converted into an excited or metastable atom. However, the use of such particles for studying secondary processes is inconvenient owing to their large energy scatter. For this reason, in most investigations, the neutral-atomic beams are obtained by resonance charge exchange.

After obtaining such neutral beams, the possibility of studying their interactions with solid surfaces arises. Historically, experiments with neutral metastable atoms preceded experiments with ordinary neutral atoms. This was because the metastable atom carried an additional store of energy which, as experiments showed, was realized in electron emission.

The study of secondary processes on the surface of a metal bombarded with metastable atoms was first undertaken in 1924 by Webb [31]. In this investigation, nickel targets were bombarded with Hg atoms. Later, the collision of Hg atoms with the surface of a metal was studied by Franck and Einsporn [17], Messenger [41], Couliette [51], and Sonkin [81]. A number of investigations [52, 56, 64, 74] were devoted to secondary processes under the influence of metastable atoms of inert gases, Ne, He, and Ar. The most careful work in this line was carried out by Oliphant [55]. Studying the energy spectra of secondary electrons, Oliphant showed that the maximum energy of these electrons coincided with the difference between the excitation energy of the metastable state and the work function of the metal. This enabled the mechanism of the process to be understood. Oliphant also showed that some of the metastable atoms were scattered in an excited state; with increasing energy of the atoms, the scattering coefficient fell, and with increasing angle of incidence it rose.

Oliphant was unable to determine the absolute yield of electrons relative to one incident metastable atom γ, since the flux of metastable atoms remained unknown. Dorrestein [161] succeeded in doing this in experiments on the excitation of the metastable helium levels 2^3S and 2^1S by electron impact. Metastable atoms of thermal velocities were formed by collision of electrons with helium atoms. Dorrestein showed that, when one metastable He atom in the 2^3S state fell on a degassed platinum plate, on average 0.24 of an electron was emitted, while for atoms in the 2^1S state this was doubled.

In 1948, Chaudhri and Khan [185] studied secondary emission from nickel and molybdenum under the influence of fast Hg and K atoms, and also the scattering of these atoms. The targets were degassed by heating to 500°C in a vacuum of 10^{-6} mm Hg. A beam of Hg atoms was obtained by the charge exchange of an ion beam sliding along a wall, while the beam of K atoms was obtained by charge exchange in its own vapor.

The authors found that the K coefficient for the molybdenum target was larger than that of the nickel. Secondary electron emission in the energy range 1.5-2.5 keV increased almost linearly with increasing energy. However, the flux of neutral atoms to the target was not determined. The energy spectrum of the secondary electrons was similar to the Maxwell distribution.

In the same year, Berry [186] studied the ejection of electrons from tantalum by Ar, He, and N atoms and ions in the energy range 500-4000 eV. The atomic beams were obtained by charge exchange of a flow of ions in their own gas. The $\gamma(E_0)$ relation for both atoms and ions was roughly the same for each of the pairs Ar and Ar^+ and N and N^+; the γ_{He} curve rose more sharply than the γ_{He^+} curve.

In 1950, Greene [211] carried out some experiments on the secondary emission of electrons arising after bombarding a molybdenum target with Ne, He, Ar, and H atoms at energies of up to 1 keV. The vacuum in the apparatus before admitting the gases was 10^{-6} mm Hg; during the experiment it was limited by the necessity of igniting the arc discharge, being $8 \cdot 10^{-5}$ mm Hg in experiments with Ne and Ar, and $6 \cdot 10^{-4}$ mm Hg in experiments with He and H. The atomic beams were obtained by neutralization of the atoms in glancing collisions with walls. The author obtained the energy spectra of the secondary electrons and concluded from an analysis of these that on bombarding with metastable He, Ne, and Ar atoms, the main contribution to the secondary emission came from the potential mechanism associated with the transfer of the excitation energy of the metastable atom to the electrons; this was confirmed by the good agreement between the experimental upper limit to the energy of the electron and the excess of the excitation energy over the work function. As regards hydrogen, the author considered that the beam contained more molecules than atoms. Emission under the influence of the hydrogen beam was evidently associated with the kinetic mechanism of electron emission, which the author treated from the point of view of the local-heating hypothesis. It was also shown that the number of scattered ions fell with increasing energy E_0. Comparing his own results with those of Oliphant [55] and Chaudhri and Khan [195], Greene found that the conclusions of the other authors were in general supported.

Stier, Barnett, and Evans [296] studied secondary electron emission from an uncleaned nickel foil under the influence of H, He, Ne, Ar, and N atoms and ions with energies between 20 and 250 keV in a vacuum of $\sim 10^{-5}$ mm Hg. Considering that the negative emission comprised electrons only, the authors concluded that the coefficient of secondary emission for Ne, Ar, and N was independent of the charge of the incident particles. At the same time, the γ^0 associated with the normal incidence of H atoms exceeded the γ associated with H^+. The ratio of these coefficients γ^0/γ equalled 1.11 at 20 keV and 1.16 at 60 keV.

Stebbings [402] measured the γ^0 coefficient for gold attacked by metastable He atoms in the 2^3S state (obtained in a low-voltage arc with a heated cathode) more accurately than Dorrestein [161]. The flow of atoms passing through a collision chamber containing argon fell on a target with a grid in front of it to trap the secondary electrons. The intensity of the flow was determined from the number of positive argon ions formed as a result of the ionizing collisions of the metastable atoms with the argon atoms. The efficiency of extracting electrons from the nondegassed gold determined by Stebbings was 29%.

The problem as to the possibility of reflecting metastable atoms from the surface without relaxation was also studied. For this purpose, the detector was prepared in the form of a cone or sharp dihedral angle, which offered the prospect of multiple collisions. Experiment showed that the reflection of a metastable atom without the extraction of an electron was a comparatively rare process.

Berry [432] studied the secondary neutral-electron emission from tungsten bombarded with 300- to 3500-eV He atoms. The experiments were carried out in a vacuum of $7 \cdot 10^{-7}$ mm Hg (referred to the residual gases). In the presence of a monolayer of contamination on the tungsten surface, the emission threshold was ~ 300 eV, and the total electron yield increased almost linearly with increasing energy of the atoms, reaching 0.42 of an electron per He atom at 3500 eV. When several monolayers had been formed on the surface, the emission rose sharply.

Hasted [462] described the results of experiments on the collisions, at thermal velocities, of metastable He atoms in the 2^3S state and Ne atoms in the $3P$ state with metal surfaces in a vacuum of $2 \cdot 10^{-6}$ torr. As a result of the potential energy of the metastable atoms electrons were extracted from the metal by the Auger process. Experiments showed that the γ^0 due to the metastable atom was smaller than that due to the ion of the same element. The emission of electrons from molybdenum and tungsten caused by atoms of the same sort differed very little; this agreed with the results of Hagstrum [345] obtained on bombarding molybdenum and tungsten with inert-gas ions.

In recent years, a comparative study of electron emission under the impact of ions and neutral atoms in the normal state has been undertaken.

Medved, Mahadevan, and Layton [661] measured the coefficients of ion–electron emission γ and neutral-electron emission γ^0 on bombarding the cleaned surface of polycrystalline molybdenum with 500- to 2500-eV Ar atoms and ions. As in the previous cases treated in the present review, the fast neutral atoms were obtained by the resonance charge exchange of ions at gas atoms. The flux of neutral atoms was measured with a portable thermocouple probe calibrated by reference to the ion beam. In the calibration it was considered that the energy transfer coefficient for the ions and neutral atoms was the same for a specific energy. The coefficient of potential emission was 7.5% in the absence of any marked kinetic extraction of electrons. On further increasing E_0 there was an almost linear rise in γ and γ^0. The ratio of the slopes of the $\gamma(E_0)$ and $\gamma^0(E_0)$ straight lines was 1.5; in the opinion of the authors, this was due to the rise in the coefficient of potential emission with the velocity of the ion.

The transfer of energy to target particles on bombarding with 500- to 2500-eV Ne and Ar ions and neutral atoms at a pressure of $2 \cdot 10^{-7}$ mm Hg was studied in [667]. The detector bombarded was a thin platinum disc with a thermocouple (chromel—constantan) soldered to it. The scattered particles fell on a second detector of analogous construction. The energy-transfer coefficient was estimated from the ratio of the readings of the thermocouples on the first and second detectors. Putting the first detector at an angle of 45° to the direction of the beam and the second one in a plane parallel to the beam, the authors found an apparent difference in energy transfer between atoms and between ions, especially for low energies; they explained this difference as being due to the preferential reflection of fast neutral atoms in the specular direction as compared with ions.

Medved [679], comparing the result of Brunnee [383] on the reflection of K^+ ions with his own results for neutral atoms from clean molybdenum (obtained by the method described by Mahadevan and Layton [667]) found that the reflection coefficients were quite close in the two cases for primary particle energies between 500 and 2500 eV. At the same time the reflection coefficients for Ar^0 obtained from a platinum thermocouple probe were much larger than the corresponding coefficients for Ar^+. In the opinion of the author, this might indicate that the scattering process depended on the structure of the electron shell of the incident particle. It was suggested that the Auger process in some way affected the reflection mechanism. Since the transfer of energy differed for Ar^0 and Ar^+ beams, the author considered that the scattering coefficient associated with neutral atoms would be smaller than that associated with ions of the same energy.

Zscheile [693] bombarded electrolytically polished copper single crystals annealed at several hundred degrees Centigrade with beams of N_2 molecules and ions at a pressure of 10^{-6} mm Hg. The neutral particles were obtained by the charge exchange of ions drawn from a discharge and subjected to mass separation. A voltage of ±30 V was applied between the target and the collector during the measurements; this was insufficient to hold back all the secondary ions. The angular dependence of the secondary electron emission and the scattering of ions and neutral atoms was studied, as was the azimuthal distribution of the secondary particles for fixed incident angles φ. Also presented was a stereographic projection of the relation between γ and the electron escape angle (from 0 to 360°) for a constant φ on bombarding with 28-keV ions. The coefficient γ had a minimum when the direction of the ion beam coincided with the principal crystallographic axes of the target. The dependence of γ, γ^0 on φ for a constant azimuthal angle of 8° was represented for both ions and neutral particles ($E_0 = 28$ keV) by a single curve with different vertical scales.

Chambers [718] studied secondary electron and secondary positive-ion emission on bombarding a copper—beryllium alloy with H_1^+, H_2^+, H_3^+ ions and H_1^0, H_2^0, H_3^0 neutral particles in the energy range 2-55 keV. The intensity of the beam of neutral particles was measured calorimetrically, using a thermoresistance. In order to establish a constant temperature on the thermistor several minutes of continuous particle bombardment were required. The bombarded target was polished and then cleaned in acetone. Heat treatment and electropolishing in vacuum applied to another sample produced no changes in electron yield as compared with the former samples. Improving the vacuum from 10^{-4} to 10^{-6} mm Hg also produced no great changes in the results of the experiments. All this clearly gave grounds for considering that the target was covered with an equilibrium adsorbed film in all cases. The author found that for these conditions the electron emission on bombarding with neutral particles was greater than on bombarding with ions. The ratio γ^0/γ was slightly energy dependent. For an energy of 25 keV the value of γ^0/γ equalled 1.14 for H_1, 1.26 for H_2, and 1.41 for H_3. The values of γ^0 and γ increased with increasing φ roughly on a sec φ law. The energy spectra of the secondary electrons were very much the same for bombardment with ions and neutral particles. There was also a certain similarity between the spectra of the secondary positive ions in the two cases.

Devienne, Rouston, and Souquet [719] studied the emission of electrons from metals on bombarding with 500- to 3000-eV Ar atoms. Degassing was effected by bombarding with an Ar^+ ion beam at an intensity of 2.5 μA in a vacuum of ~10^{-6} mm Hg. The trace of the molecular beam on the target was ~1 cm^2 in section. The highest coefficient γ^0 in the series aluminum, gold, silver, copper, zinc, and stainless steel occurred in the case of aluminum and fell successively on passing to the steel. With increasing angle of incidence the value of γ^0 rose, this rise taking place most rapidly for a nondegassed surface. The effect of the adsorption of various gases (Ar, He, H_2, O_2) on the secondary electron emission was also studied.

The same authors studied the variation of γ^0 with the energy of an atomic Ar beam; they obtained a linear relationship for $\gamma^0(E_0)$, the value of γ^0 being fairly large: for $E_0 = 500$ eV, $\gamma^0 = 15\%$, and for $E_0 = 3000$ eV, $\gamma^0 = 91\%$. In this investigation the surfaces of the various metals were simply cleaned by ion bombardment, which proved to be insufficient for the purpose in hand.

After modifying the apparatus used in earlier work, Devienne [768] determined the velocity of the neutral Ar particles scattered from an aluminum surface. The velocity distribution of the scattered neutral particles was determined by time-of-flight measurements. By reference to the delay in the secondary-current pulses it was established that for a primary beam of neutral particles with an energy of 2500 eV the most probable velocity of the scattered particles was $5.2 \cdot 10^6$ cm/sec.

It follows from our review that the secondary emission of metals under the influence of atoms has been studied less than secondary processes due to the interaction of positive ions with metal surface. In the majority of cases the object of the investigation has been the emission of electrons by metals bombarded with atoms. However, the results are quite inadequate to establish fundamental relationships. This is largely because of the low accuracy of the method of measuring the intensities of neutral-atom beams and conducting experiments under relatively poor vacuum conditions.

In the following sections we shall set out our own investigations, in which we have attempted to study secondary processes arising from the interaction of atoms with a clean metal surface on the basis of a specially developed method of measuring the intensity of a flow of neutral particles.

§2. Study of the Properties of the Neutral Component of Secondary Emission Arising in the Ion Bombardment of a Solid

In the interaction of a beam of fast ions with a metal surface the scattering of ions is accompanied by scattering in the form of neutral particles. The graphical analysis of [495] established the proportion of primary ions neutralized on the surface of a metal. However, the laws governing the scattering in the form of atoms on bombarding the surface with a beam of ions were not studied. Several properties of the neutral component of secondary emission arising when a beam of alkali ions is scattered at a metal surface were studied by Arifov, Flyants, and Ayukhanov [550].

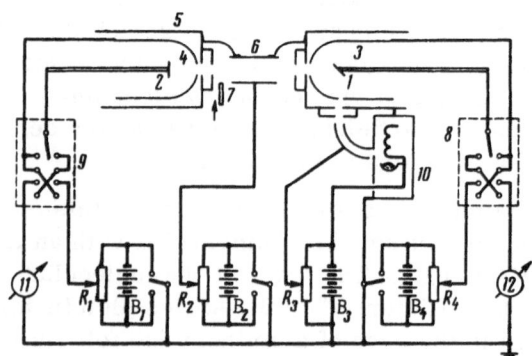

Fig. 8.1. Apparatus and main electric circuit.

In the experiments we used a glass apparatus with electrodes and other parts arranged as in Fig. 8.1. The apparatus consisted of a source of alkali ions 10 with a cylindrical condenser and two receiving parts with guard cylinders 5, collectors 3 and 4, and targets 1 and 2. The target 1 was bombarded with the primary ion beam, while target 2 served as a detector of secondary particles. In the space between the two targets the charged particles were duly deflected by the field of the flat condenser 6, so that only neutral particles leaving the surface of the first target could pass to the second target. If necessary, the access of particles to target 2 could be stopped by means of the shutter 7, controlled from outside with an electromagnet. The primary current to the target 1 and the secondary currents in the collector 3 were measured with the galvanometer 12 of sensitivity $1 \cdot 10^{-9}$ A/mm. By using the switch 8 we could measure the current to the target 1 and the collector 3, and also the primary current. The secondary currents to the collector 4 were measured with an electrometer 11 of sensitivity $1 \cdot 10^{-13}$ A/mm. Using the switch 9, the necessary voltages were applied to the target 2. For obtaining the individual points on the curves the targets were briefly heated to 2000°K and then their temperatures were reduced to 1200 and 950°K (1 and 2, respectively).

In order to study the neutral component of secondary emission arising as a result of bombarding the metal with positive ions, a detection method based on the ionization of atoms on the surface of heated metal was employed.

As indicated earlier, one of the important characteristics of secondary ion—ion emission is the elastic scattering of the ion from individual target atoms with the conservation of the charge. It may also be considered that some of the ions of the primary beam, after capturing an electron, are elastically scattered in the form of neutral atoms. Then the energy spectrum of such a neutral component of secondary emission should not differ greatly from the spectrum of elastically scattered ions. When such a flux of neutral atoms falls on the surface of the target 2 at high temperatures there may be surface ionization and secondary emission; these may be observed by measuring the currents in the collector 4 surrounding the target 2.

Figure 8.2 shows the retardation curves of the secondary ions appearing on bombarding the target 2 with secondary neutral atoms from target 1, the latter being bombarded with 100- and 200-eV Na$^+$ ions (curves 1 and 2, respectively). We see from the figure that the secondary ions from target 2 have considerable energies. The currents in collector 4 vanish on bombarding target 1 with 100- and 200-eV ions for positive potentials of 60 and 100 V, respectively, on collector 4. The existence of secondary ions with such large energies can only be explained by the fact that the neutral atoms bombarding target 2 have high energies, i.e., the neutral component of secondary emission arising as a result of the bombardment of target 1 with ions contains (in addition to secondary ions) scattered neutral atoms, which leave the surface, as it would appear, as a result of elastic collisions with individual atoms of target 1. The limiting energies of such neutral atoms may be found, of course, from relation (1.9). These neutral atoms, striking target 2, leave the latter in the form of positive ions with the conservation of a considerable proportion of the original energy. The limiting energies of these ions may be calculated from relation (1.9) by assuming that in the latter case also elastic collisions take place, and that the initial energy of the particles falling on target 2 equals the limiting energy of the particles leaving the surface of target 1. For example, if target 1 is composed of tantalum and

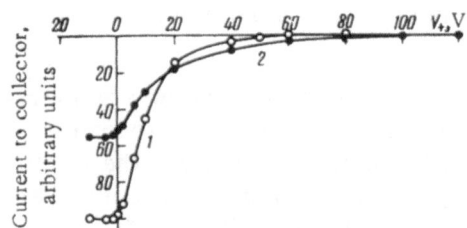

Fig. 8.2. V/A characteristics obtained on bombarding target 2 (Fig. 8.1) with the neutral component of secondary emission from target 1.

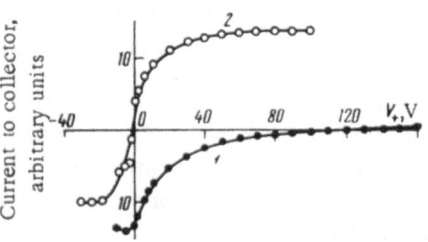

Fig. 8.3. V/A characteristics obtained on bombarding clean and contaminated targets (curves 1 and 2, respectively), with the neutral component.

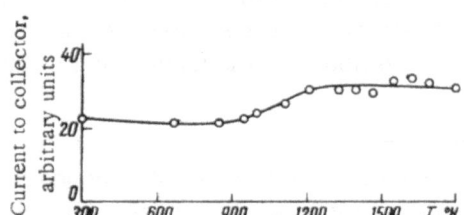

Fig. 8.4. Secondary neutral emission as a function of the temperature of the first target.

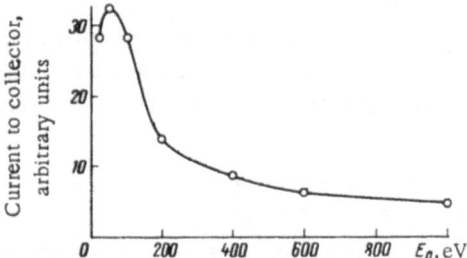

Fig. 8.5. Energy dependence of the current of secondary ions obtained under the influence of the neutral component.

is struck by Na^+ ions with an energy of $E_0 = 100$ eV, relation (1.9) gives the energies of the secondary ions leaving target 1 as $E = 77$ eV. Then, if we consider that target 2 is bombarded with neutral atoms having a maximum energy of 77 eV, we obtain the value $E = 59$ eV for the energy of the secondary ions leaving the target 2. We see from Fig. 8.2 that the calculated value almost coincides with that measured experimentally. A comparison was also made between the energy distributions of the secondary ions resulting from the bombardment of target 2 with the neutral and ionic components of secondary emission from target 1.

In the second case all the secondary particles from target 1 were directed to target 2 when there was no voltage on the plates of the plane condenser. This experiment showed that the energy distribution of the secondary ions on bombarding target 2 with secondary particles arriving from target 1 agreed closely with the distribution of secondary ions obtained on bombarding target 2 with the neutral component only.

As to the retention of a considerable proportion of energy by the scattered neutral atoms, we may also judge this by reference to the secondary emission of negative particles produced by the neutral atoms falling on the cold, contaminated surface of target 2. Figure 8.3 shows the V/A characteristics obtained on bombarding the clean and contaminated surface of target 2 (curves 1 and 2, respectively) with the neutral component (ion—neutral emission) from target 1. The curves were obtained by bombarding target 1 with 600-eV Na^+ ions. We see from the figure that on bombarding the clean target 2 there is no emission of negative particles; only the scattering of the neutral component in the form of positive ions (neutral atom—ion emission) take place. On bombarding the contaminated surface of target 2, however, the neutral component also produces the secondary emission of negative particles. We see from the V/A characteristic (curve 2) that the energies of the secondary negative particles extend to ~15 eV and the secondary positive ions to ~50 eV. Clearly, the secondary emission of negative particles can only arise from particles possessing considerable energies.

Elastic scattering in the form of fast neutral atoms is also observed for other temperatures of target 1. Figure 8.4 shows a curve characterizing the dependence of the secondary neutral emission on the temperature of the clean target 1. The vertical axis gives the total current of positive ions formed at target 2 under the influence of the neutral component in arbitrary units. The temperature of target 2 is in this case equal to 1300°K. We see from the form of the curve that the magnitude of the neutral component remains almost constant up to T ~ 1000°K. On further raising the temperature there is a rise in current. By applying a small potential between the target 2 and its collector to delay the ions, it was found that the rise in current took place as a result of ions with thermal velocities, the appearance of which could only be due to evaporated neutral atoms from target 1. Thus, the second neutral emission may include the same groups of secondary particles as the secondary ion emission.

Considering what has been said in respect to the curve shown in Fig. 8.4, and remembering that the coefficient of ion scattering for the secondary ion—ion emission is independent of target temperature [458], we may consider that the total number of scattered particles (ions and atoms) remains constant as the temperature of the target bombarded increases from 300 to 1800°K.

It is interesting to study the dependence of the neutral component on the energy of the primary ions. We see by comparing curves 1 and 2 of Fig. 8.2 that the current of secondary positive ions from target 2 due to the neutral component is different in the two cases. Figure 8.4 shows the current of the secondary positive ions arising under the influence of the neutral component as a function of the energy of the ions bombarding the first target. The form of this curve is reminiscent of that of the coefficient of secondary ion—ion emission expressed as a function of the energy of the primary ions [383, 494, 542]. However, we must remember in this case that on reducing the energy of the neutral atoms bombarding the metallic surface their scattering coefficient in the form of ions should increase (see Chapter 4, §3), the degree of detection at the second target also becoming greater. We should therefore expect that the real behavior of the coefficient of secondary ion—neutral emission for small values of energy E_0 would be characterized by a rise less steep than that found in the curve of Fig. 8.5.

Thus, on bombarding metals with atomic particles (ions or atoms) there is elastic scattering of the primary particles at individual target atoms, both in the form of positive ions and also in the form of neutral atoms, the corresponding energy spectra being very similar to one another, while the maximum energy is determined by the law of elastic collision.

§ 3. Comparative Study of the Secondary Emission
of Metals and Metallic Films under the
Influence of Alkali Atoms and Ions

In order to elucidate the way in which fast neutral atoms interacted with solid surfaces, it was interesting to make some comparative measurements of secondary emission on bombarding a surface with ions and neutral atoms under identical conditions. Investigations of this kind were carried out by Arifov, Flyants, Rakhimov, and Ayukhanov [678, 706, 707] for Na and K ions and neutral atoms impinging on metals both in the pure (clean) state and during the deposition of films on their surfaces. These experiments made it possible to compare data relating to secondary emission obtained under a wide variety of surface conditions.

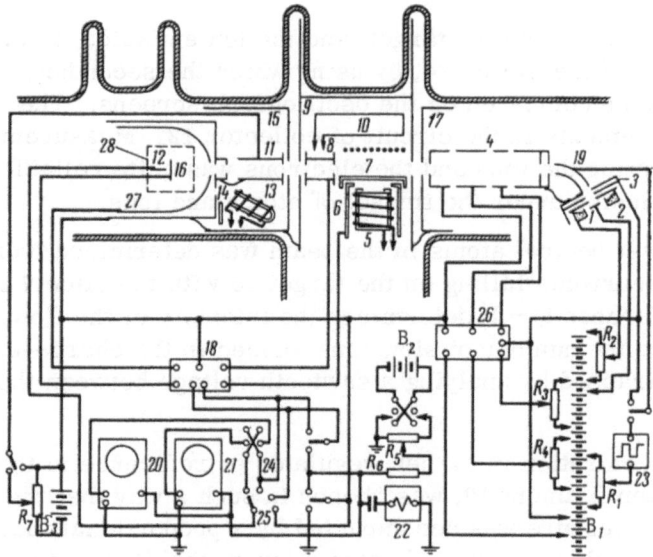

Fig. 8.6. Vacuum apparatus and electric circuit for studying secondary processes on bombarding solids with neutral alkali atoms.

Apparatus and Measuring Method

Beams of neutral alkali atoms were obtained by the resonance charge exchange of ions [678] in a stream of vapor from the same elements. The ion source shown in Fig. 8.6 constituted a combination of an immersion objective 1-3 and a single electrostatic lens 4, separated by a 64° spherical condenser 19. The surface emitting the ions was a tungsten strip 1 dusted with an alkali halide salt. The ion beam formed by the source, 2 mm in diameter, was intensity-modulated by generator 23 and passed through a stream of vapor created in furnace 5, which had a slit aperture. Freezing of the expended jet took place in the jacket 9. The ions which had not undergone charge exchange were deflected, after emerging from the charge-exchange chamber, by the field of the flat condenser 11, and neutral atoms modulated at the same frequency as the ion currents were directed toward the target 16. Bombardment of the target with ions was effected by shutting off the vapor jet with the magnetic slide 7. In this case, the upper plate of condenser 11 was connected to the guard cylinder. The target was a tantalum or molybdenum strip. Preliminary cleaning of the target was effected by prolonged baking at 2500°K, and immediately before each measurement by brief heating to 2200°K. The film of alkali metal was deposited on the cooling target by means of evaporator 13. The jet of metal vapor could be shut off at will by the magnetic shutter 14. The concentration of molecules in the jet ensured the deposition of the film at a rate of up to 20 monolayers/sec. The apparatus, which was divided into three parts by cylinders 15 and 17, was evacuated with three oil-vapor pumps. The working vacuum in the receiving part was $\sim 1 \cdot 10^{-6}$ mm Hg.

The measuring system was similar to that employed in the double-modulation method. The circuit included a switch 18 by means of which one of the oscillographs (20 or 21) could be connected either to the target—collector circuit (16, 27) or to the system of the charge-exchange chamber.

In experiments requiring the separation of the negative emission into electron and ion components, a transverse magnetic field was employed. Figure 8.6 shows a projection of the pole tips of the magnet on the plane of the sketch (28). The electron emission was taken to a plate collector 12 situated beyond the target, and the ion emission to the collector 27. The electrical circuit included the switch 24, by using which the secondary currents passing to collectors 12 and 27 could be compared on the oscillograph screens. The switch 25 served for control current measurements in the circuit of collector 12. Measurements showed that the method used for separating the ions and the electrons was quite reliable for a comparative study of secondary emission under the impact of atoms and ions.

The number of fast neutral atoms in the beam was determined from the difference between the primary ion currents falling on the target 16 with the slide 7 respectively closed (I_0) and open (I_0'). The difference $I_0 - I_0'$ determines the intensity of the flow of neutral atoms and is in magnitude similar to the number of slow ions formed in the charge-exchange chamber. These particles were trapped by applying a sawtooth voltage between the reticular hemicylinder 8 and the casing 6.

The working concentration in the jet, regulated by reference to the current of surface ionization in the tungsten filament 10, was chosen in such a way that the increase in the difference $I_0 - I_0'$ with rising pressure was accompanied by a proportional increase in the secondary currents from the target; this clearly indicated a low probability of twofold collisions between primary-beam particles and the atoms of the jet. In addition to this, control galvanometric measurements of the currents in a special electrode placed between the plates of condenser 11 and the entrance aperture of the collector 27 shows that the scattering of fast atoms and ions which had experienced collisions but not undergone charge exchange [188] could be neglected, since the influence of such scattering was outside the limits of measuring error.

Secondary Emission in the Bombardment of
Clean Targets with Neutral Atoms

Measurements of the intensity of the primary beam of neutral atoms by the method indicated in the foregoing enabled us to determine the coefficients of secondary emission of positively and negatively charged particles from clean targets [678, 706]. The results of the measurements showed that for neutral-atom energies up to 1000 eV the emission of negatively charged particles was only slight. A considerable number of the neutral atoms falling on the target were scattered from the clean surface in the form of positive ions. The use of a modulated beam of neutral atoms made it possible to compare the oscillograms of the V/A characteristics of secondary currents arising from the impact of ions and neutral atoms, respectively.

In both cases, the emission of secondary positive ions with almost the same energy distribution (see, for example, the first series in Fig. 8.9) from a clean target was observed. Special experiments showed that the saturation of the V/A characteristic obtained under the incidence of neutral atoms set in at values of the retarding voltage corresponding to the energy given by relation (1.9). Hence, as in the case of ion bombardment, some of the fast neutral atoms falling on the target were reflected as a result of elastic collisions with individual target atoms [220, 238, 383].

The coefficients of scattering in the form of positive ions on bombarding with ions (K_s) and neutral atoms (K_s^0) of equal energy also coincide. The coefficients K_s and K_s^0 were determined from the formulas

$$K_s = \frac{I}{I_0}, \quad K_s^0 = \frac{I^0}{I_0 - I_0'},$$

where I and I^0 are the saturation currents of the positive ions on the retardation curves for a target heated to 1300°K obtained on bombarding with beams of positive ions and neutral particles, respectively.

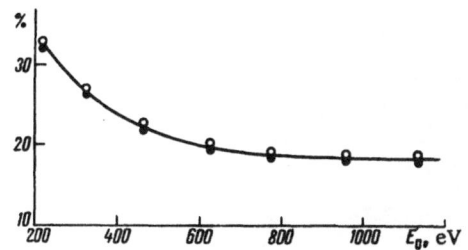

Fig. 8.7. Coefficients of scattering in the form of positive ions as functions of the energy of the bombarding Na atoms and ions.

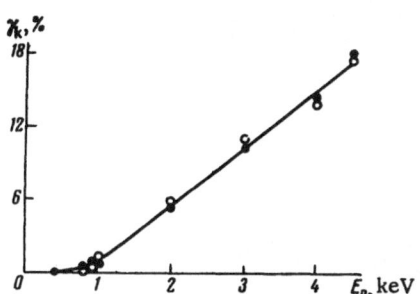

Fig. 8.8. Coefficients of kinetic electron emission γ_k as functions of the energy of the Na ions and atoms.

Figure 8.7 shows the $K_s(E_0)$ (black circles) and $K_s^0(E_0)$ (light circles) relationships for Na ions and neutral atoms bombarding a tantalum target at energies of 200–1200 eV. Individual points on the curve were obtained in both cases immediately after a brief heating of the target to 2000°K, which ensured satisfactory cleanliness. We see from the curves that these relationships are similar over the whole energy range studied. Thus, scattering in the form of positive ions on bombarding clean tantalum is independent of the charged state of the bombarding particle.

The effect of the charge of the bombarding particle on the electron kinetic-emission yield was studied for a clean molybdenum surface. In all investigations so far carried out, this problem has been solved by using ions of inert gases [656] or multiply charged ions [314]. In both cases, the kinetic electron emission had to be determined by subtracting the potential component from the total emission. Under the influence of alkali ions and neutral atoms, however, only kinetic emission should occur. Figure 8.8 shows the coefficient γ_k as a function of the kinetic energy E_0 of a beam of Na ions (black circles) and atoms (light circles). We see from the figure that at energies up to 1000 eV the value of γ_k is small for both ions and atoms. On raising the energy to 4500 eV there is an almost linear rise in γ_k. However, the most important fact is that over the energy range studied the curves relating γ_k to energy coincide for both atoms and ions (Na). The form of the $\gamma_k(E_0)$ curves and the absolute values agree satisfactorily with the data of [493]. An analogous result was obtained for a molybdenum target bombarded with K atoms and ions. Thus the results show that the charge on the ion has no serious effect on the kinetic emission of electrons.

Study of Secondary Negative Emission Obtained by Bombarding Potassium and Sodium Films Deposited on a Refractory Substrate with Ions and Neutral Atoms

It is well known that, on bombarding sodium films intensively deposited on the surface of pure tantalum with positive ions, the secondary emission of positive ions almost ceases after a time, while that of negative ions passes through a maximum and then falls to a constant value [422]. Further study of this phenomenon showed that changes taking place in secondary emission during the bombardment of sodium films on a tantalum surface with fast Na atoms were of an analogous character. Figure 8.9 shows oscillograms of the V/A characteristics of secondary emission obtained by bombarding targets in three different states with atoms and ions. The oscillograms in the first row correspond to a tungsten surface at 1200°K. The second row shows oscillograms of the V/A characteristics of the secondary currents corresponding to a thin film of sodium on the tungsten surface. In this case there is hardly any emission of positive ions, while the emission of negative particles reaches a maximum. The third row shows oscillograms for the case in which there is a thick film of sodium on the bombarded surface. By comparing the oscillograms obtained on bombarding the tungsten with atoms (a) and ions (b) during the

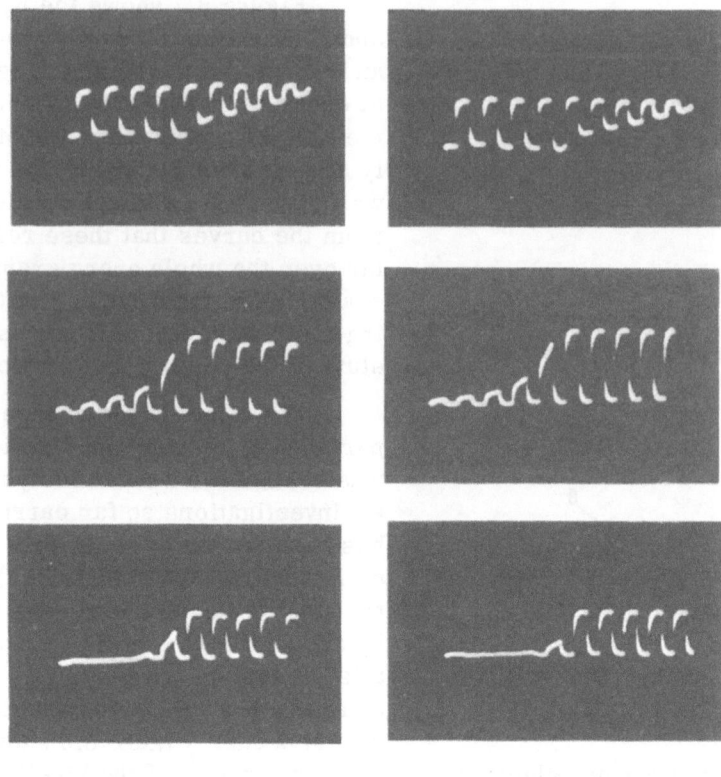

a   b

Fig. 8.9. Oscillograms of the V/A characteristics obtained
while depositing a sodium film on tungsten while bombarding
with Na atoms (a) and ions (b).

deposition of the sodium film we see that the character of the changes in secondary emission is
entirely the same in both cases. Measurements showed that the absolute values of the coefficients of secondary emission on bombarding tungsten with atoms and ions while depositing a
sodium film on its surface were also the same. An analogous result was obtained on bombarding the sodium film with K atoms and ions. A different result was obtained on bombarding a
potassium film with K atoms and ions while depositing the film on a molybdenum target [707].

The experiments were carried out in such a way as to separate the electron emission
from the negative ions. The oscillograms for the electron and negative-ion emission were
studied simultaneously on the screens of two oscillographs which had the inputs of their vertical amplifiers, respectively connected to collectors 12 and 27. The oscillograms were photographed with a motion-picture camera at a rate of 4 frames/sec. The results obtained from a
series of oscillograph pictures taken consecutively while depositing the film were analyzed; the
results of the analysis appear in Fig. 8.10 in the form of a graph representing the variation in
the coefficients of secondary electron emission γ and secondary negative-ion emission K^- on
bombarding a potassium film with 235-eV Na ions (black circles) and atoms (light circles). The
horizontal axis gives the time of depositing the film in seconds. The broken curve I_t characterizes the variation (on the same time scale) of the unmodulated thermionic (electron) current
appearing as a result of the fall in work function as potassium is adsorbed on the cooling target.
It is clear at the present time that the adsorption minimum of the work function occurs for an
optimum coverage of $\theta_{opt} = 0.67$ of a monolayer [23, 295], while the maximum of the thermionic
emission occurs still earlier, at $\theta = 0.55$ of a monolayer [23]. The abscissa of the thermionic-

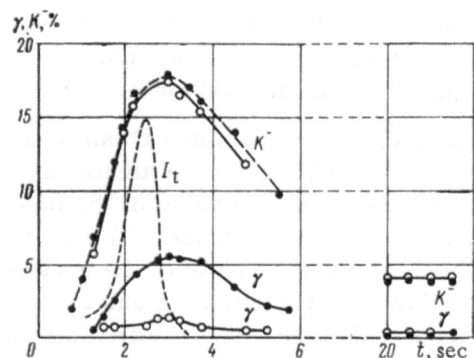

Fig. 8.10. Dependence of the coefficients γ and K^- on the time of depositing the film on bombarding with Na ions and atoms.

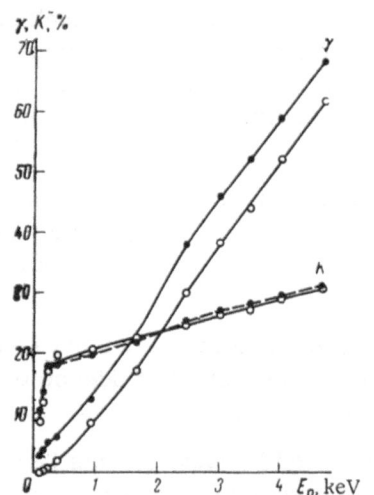

Fig. 8.11. Dependence of the coefficients γ and K^- on the energy of Na ions and atoms.

emission maximum may serve as a special kind of reference point for the curves of transitional processes under the influence of ions and neutral atoms, since the position of the maximum is stable for the same temperature and vacuum conditions and for the same rate of film deposition. These conditions may easily be kept constant during the measurements. Hence, for two transitional processes recorded one after the other only one I_t curve appears.

We see from Fig. 8.10 that the coefficients of secondary electron emission under the influence of ions and atoms, like the coefficients of negative-ion emission under analogous conditions, have a characteristic maximum. The positions of the maximum are almost the same for all the curves. The same applies to the absolute values of the coefficients of secondary negative-ion emission K^- for any states of the bombarded surface. At the same time, the coefficients of secondary electron emission γ differ considerably for intermediate states of the film. For large film thicknesses the coefficients γ and K^- fall and reach almost constant values. Analogous $\gamma(t)$ and $K^-(t)$ curves are obtained for other energies also. In all cases, as the coefficients γ and K^- vary, the thermionic-emission current rises; however, the I_t curve is displaced in the direction of thinner films and the emission of thermoelectrons vanishes soon after the onset of the fall in the K^- and γ coefficients. Nevertheless, the maxima of the γ and K^- curves lie in the same range of thicknesses as that corresponding to thermionic (electron) emission. Since the existence of thermionic emission at comparatively low temperatures is due to a reduction in the work function of the surface [22, 23, 29, 295], the increase in the emission of negative ions and secondary electrons should also be related to a fall in the work function of the surface. The presence of negative-ion emission in this case indicates the presence of foreign molecules with electronegative properties in the potassium film. Zingerman and Ishchuk [613] note a correlation between the film-thickness dependence of the coefficients γ and K^- and the way in which the work function varies with film thickness under analogous vacuum conditions.

Figure 8.11 shows the dependence of the values of γ and K^- corresponding to the maxima of the $\gamma(t)$ and $K^-(t)$ curves presented on the energy of the Na ions (black circles) and atoms (light circles) bombarding a potassium film. We see from the figure that the coefficients K^- agree closely for both ions and atoms over the whole energy range studied. The general form of the curves agrees with the behavior of the sputtering coefficient of alkali halide films sputtered in the form of negative ions on bombarding with alkali metal ions observed by Moroz and Ayukhanov [643]. The emission of electrons under the influence of atoms and ions increases almost linearly with energy. However, in the energy range studied, the electron emission for ions is much greater than that corresponding to atom bombardment. Of course, for any values φ in the case of Na atoms only kinetic electron emission can take place. Hence, if we consider

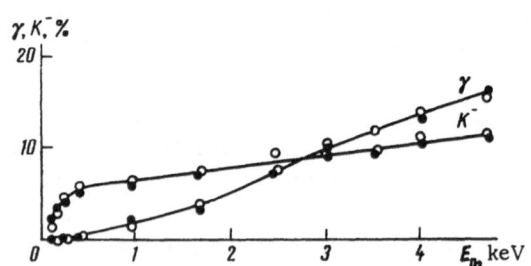

Fig. 8.12. Dependence of the coefficients γ and K^- on the energy of Na ions and atoms bombarding a thin potassium film.

that kinetic emission is almost the same for ions and atoms (see Fig. 8.8), the excess of the $\gamma(E_0)$ curves for ions over that for atoms must be attributed to the potential emission of electrons.

Accepting the excess of the $\gamma(t)$ curve obtained for ions (Fig. 8.10) over the analogous curve for atoms as giving the value of the potential emission, we consider that the $\gamma(t)$ curve for atoms describes the behavior of the coefficient of secondary kinetic emission due to neutral Na atoms and Na ions. The maximum of this emission, like the maximum of the potential emission, occurs at the same film thickness, corresponding to a low φ. Since the dependence of the potential emission on φ is regarded as established, the fact that the $\gamma(t)$ curves are the same for ions and atoms may indicate the existence of a specific relationship between the kinetic emission and the work function of the thin film.

Figure 8.12 shows the relations between γ and K^- and the energy of the ions (black circles) and atoms (light circles) for fairly thick films, for which these coefficients have attained steady values. In contrast to the case of thin films, the values of γ are almost the same for ions and atoms at any energy. For this thickness, the values of K^- corresponding to atoms and ions also come very close together.

Thus, on bombarding a potassium film on molybdenum with Na^+ ions, potential electron emission takes place for comparatively small thicknesses ($V_i > 2\varphi$) [22, 29, 613]. The value of this emission is almost independent of the energy of the bombarding ion within the energy range studied.

For all states of the film (i.e., states differing in work function and in the density of the adatoms on the substrate), and for all energies of the primary particles, the dependence of the coefficient K^- on the charge of the bombarding particles is barely noticeable.

The results which have just been presented were also supported by experiments in which K atoms and ions were used as bombarding particles. We should also note that in this case, in view of the fact that the difference between V_i and 2φ was very slight, the coefficient γ corresponding to ions exceeded the corresponding coefficients for atoms by no more than 1 or 2%.

§4. Secondary Electron Emission of Metals
Under the Influence of Inert-Gas Atom and
Ion Bombardment

It is well known that, on bombarding pure metals with inert-gas ions having energies up to 1 keV, potential extraction of electrons tends to predominate. For energies above 1 keV potential emission is accompanied by the kinetic emission of electrons [493, 645]. The separate study of potential and kinetic electron emission under the influence of ions only, is difficult owing to the simultaneous development of these two processes. Such a study becomes possible, however, if we use beams of ions and atoms derived from the same inert gas, since on bombarding a metal with neutral atoms only kinetic electron emission takes place. The use of inert-gas atomic and ionic beams is convenient because the number of particles scattered in the form of ions is very small ($\leq 1\%$). On the other hand, a comparative study of the emission of electrons on bombarding solids with ions and atoms of the same gas enables one to determine the effect

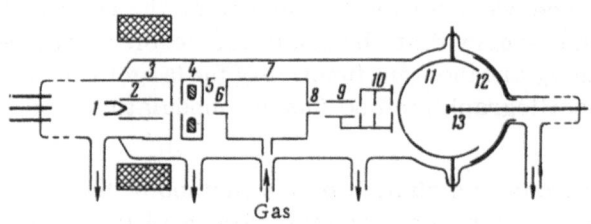

Fig. 8.13. Arrangement of the experimental apparatus.

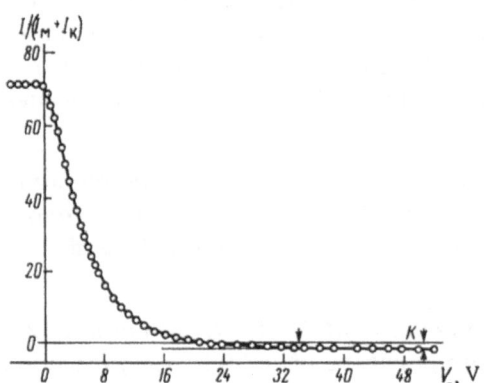

Fig. 8.14. Retardation curve for secondary electrons on bombarding molybdenum with 5-keV He$^+$ ions.

of the velocity of the ion on the potential emission of electrons. The kinetic and potential emission of electrons was studied by Rakhimov and Dzhurakulov [716] and Arifov, Dzhurakulov, and Rakhimov [775] by measuring the γ coefficients and the energy distribution of the electrons emitted on bombarding metals with ions and atoms of inert gases at energies up to 5 keV.

Apparatus and Measuring Method

The experiments were carried out in the apparatus shown schematically in Fig. 8.13. The apparatus consisted of five sections, containing an ion source, an electrostatic lens 4–5, a charge-exchange chamber 7, a plane condenser 9 with a guard cylinder 10, the electrodes of the measuring section (spherical collectors 11 and 12 with a diameter of 80 mm), and the target 13. The ions were obtained by electron impact in the gap between the heated filament 1 and the anode 2, situated in the longitudinal magnetic field of a solenoid. The flow consisted of singly charged ions in the unexcited state; this was ensured by applying an appropriate anode voltage (31 V for Ar, 56 V for Ne, and 70 V for He). The plane condenser served to remove ions which had not undergone charge exchange from the beam when working with neutral atoms. The sections of the apparatus separated by plane (3, 10) and tubular (6, 8) diaphragms were evacuated with individual mercury pumps. The charge-exchange chamber was not connected directly to the pump, but was evacuated through tubes 6 and 8; this made it possible to measure its gas pressure to a fair accuracy.

The beam of neutral atoms was obtained by the resonance charge-exchange of ions in their own gas. The intensity of the neutral atoms in the beam was determined from the difference in the intensity of the beam of ions passing to the target before and after admitting gas to the charge-exchange chamber. The operating conditions of the ion source were kept constant by having a pressure drop between the individual sections of the apparatus, of 2–2.5 orders for Ar and Ne and 1.5–2 orders for He.

The degree of scattering of the flow of atoms was checked in each case by studying the dependence of the secondary electron current on the gas pressure in the charge-exchange chamber on bombarding the surface of the degassed target with a flow of neutral atoms under conditions such that the ion flow at the entrance into the charge-exchange chamber was constant. The relationship was linear up to certain pressure values, after which there was a fall in the secondary electron current, evidently due to a change in the energy and intensity of the flow attributable to scattering. The working pressure was chosen on the linear part of this characteristic. Thus, for example, in the case of Ar, the experiments were carried out with a pressure of $2 \cdot 10^{-4}$ mm Hg in the ion source and $8 \cdot 10^{-5}$ mm Hg in the charge-exchange chamber.

The target was a molybdenum foil $12 \times 7 \times 0.02$ mm in size, subjected to careful heat treatment at 2000 and 2300°K for 30 and 5 h, respectively, in a vacuum of 10^{-7} mm Hg. In measuring the coefficients γ, the state of the target surface was checked by measuring the coefficient γ_p of potential electron emission, and in studying the energy distribution of the secondary electrons, was checked by simultaneously measuring both γ_p and the work function of the target φ [645, 775]. The measurements were made both by the oscillograph method (double modulation) and by static techniques.

The energy distribution of the secondary electrons was studied by the method of the retarding potential in the field of a spherical condenser. By way of example, Fig. 8.14 illustrates the retardation curve for secondary electrons on bombarding the target with He ions at energies of $E_0 = 5.0$ keV. In the figure the vertical axis gives the ratio $\rho = I/I_0$ (I = secondary particle current to the collector; I_0 = primary ion current), and the horizontal axis gives the retarding potential between the target and the collector. We see from the figure that for a positive potential on the collector ρ reaches saturation and in absolute magnitude equals the coefficient of ion—electron emission γ. For a negative and fairly high (over 40 V) potential on the collector, the value of ρ changes sign; in Fig. 8.14, K denotes the coefficient of secondary positive-ion emission, which evidently includes the true emission of positive ions from the target and the emission of tertiary electrons from the collector.

The differential curve of the secondary-electron energy distribution was found by numerical differentiation [245] of the retardation curve. Differentiation was carried out with respect to data averaged over several measurements, the maximum deviation of individual points from the mean being no greater than 5%.

Individual points on the retardation curve were taken every 0.5 V of the collector potential, 3-4 sec after briefly heating the target to the flash temperature of 2300°K, i.e., for a cooling target. Special experiments showed that the target temperature was over ~700°K under these conditions.

The contact potential difference between the target and the collector was determined by the method of thermionic (electron) and ion—electron emission and allowed for when plotting the curves of the electron energy distribution. The work function φ was measured under conditions similar to those governing the measurement of the secondary electron distribution. The work function was measured by the Richardson straight-line method, and each point on these lines was also determined (after flashing at T = 2300°K) over the temperature range 1370-1740°K. In determining φ, the current of the thermoelectrons was measured in a Faraday cylinder situated on the far side of the bombarded surface at a distance of 30 mm from the target. The target temperature was measured by comparing its brightness with that of a molybdenum filament, the temperature of which was determined from the heating current.

In the first stage of heat treatment, the work function of the molybdenum target was high (6-7 V), and then during degassing it fell continuously. After 30 h of degassing at T = 2000°K and subsequent heating in flashes at 2300°K the value of φ became more stable and equal to 4.29 ± 0.05 eV. The spread in the value of φ (± 0.05 eV) was probably due to the continuous rearrangement (recrystallization) of individual parts of the polycrystalline target [471].

Dependence of the Coefficient of Ion—Electron and Neutral—Electron Emission on the Energy of the Primary Particles

Figure 8.15 shows the behavior of the coefficient of electron emission $\gamma(E_0)$ under the influence of Ar, Ne, and He ions (light circles) and atoms (black circles) bombarding a carefully cleaned molybdenum target. We see from the graphs that for fairly low energies (up to 0.8 keV) the secondary electron emission under the influence of the neutral atoms is practically absent, while for the ions Ar^+, Ne^+, and He^+, $\gamma = 11$, 27, and 26%, respectively, in good agree-

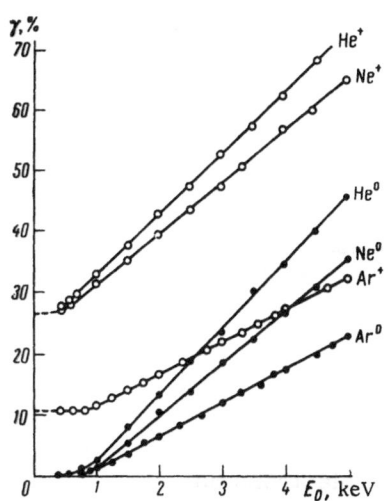

Fig. 8.15. Dependence of the coefficient of secondary emission on the kinetic energy of Ar, Ne, and He atoms and ions.

ment with the results of [493, 345]. On raising the energy to 5 keV in the case of both atoms and ions, there is a linear rise in γ_k. It is quite clear that the inclination of the $\gamma_k(E_0)$ line to the E_0 axis is the same for the same sort of ions and atoms. Allowing for the smallness of the electron emission at low energies of the atoms, we may consider that there is no appreciable quantity of excited particles in the flow of atoms. Thus, the electron emission observed at high energies of the atoms may be entirely ascribed to kinetic ejection.

These results show that the coefficient of potential electron emission is independent of the velocity of the ions and that the charge on the bombarding particle has no serious effect on the kinetic emission of electrons. The fact that the kinetic emission of electrons associated with ions and atoms of the same element is nearly the same may be explained by considering that the kinetic ejection of electrons takes place when the incident particle is much closer to the surface of the metal than the critical Auger-neutralization distance. In other words, the collision of an ion with the surface of the metal takes place after it has been neutralized and converted into an atom. On the other hand, the coincidence of the curves relating the coefficient of kinetic electron emission to the energy for ions and atoms of the same element indicates that the charge of the bombarding particle has no effect on the process of kinetic ejection. This conclusion agrees with the experimental data of Flaks [314] and Tel'kovskii [349] and the theoretical conclusions of Parilis and Kishinevskii [539, 618].

It should be noted that the secondary emission of positive ions from a degassed target was very small on bombarding with either ions or atoms. Secondary ion emission became considerable when the surface was contaminated by the adsorption of residual-gas molecules, evidently as a result of the sputtering of the adsorbed atoms in the form of positive ions.

Study of the Energy Distribution of Secondary Electrons by the Static Recording Method

Figure 8.16 shows the energy distributions of secondary electrons emitted from molybdenum under the influence of bombardment with Ar ions and neutral Ar atoms. The areas under the curves are proportional to the coefficient γ for the corresponding energy of the bombarding particles. Figures 8.17 and 8.18 show analogous curves for the bombardment of the target with Ne and He atoms and ions. The corresponding curves for atoms relate to purely kinetic emission. On bombarding with ions, both potential and kinetic electron emission takes place.

Let us consider the energy distribution curves of the kinetically emitted electrons. First we note that for a given E_0 the electron energy distribution is harder under bombardment by the lighter atoms. As the kinetic energy of the bombarding atoms increases we find: 1) an increase in emission mainly as a result of electrons with energies between 0 and 7 eV in the case of Ar, and 0–10 eV in the case of He; the number of electrons with energies outside the ranges indicated is not more than 15% of the total number; 2) a slight displacement in the position of the maximum of the energy distributions in the direction of higher electron energies $\mathscr{E}_k$ (for $E_0 = $ 5 keV this equals 2 eV for Ar and 3 eV for He). Calculation showed that on increasing the kinetic energy of the bombarding particles E_0 from 1.5 to 5.0 keV, the mean energy of the secondary electrons $\mathscr{E}_{av}$ rose from 3.6 to 5.5 eV in the case of He and from 2.7 to 3.5 eV in that of Ar.

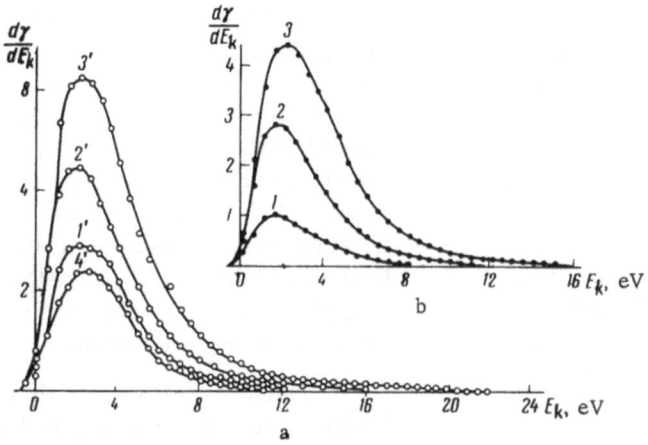

Fig. 8.16. Energy distributions of secondary electrons on bombarding the target with Ar^+ ions (a) at an energy of E_0 equal to: 1') 1.5; 2') 3; 3') 5; 4') 0.4 keV, and Ar atoms (b) at an energy of E_0 equal to: 1) 1.5; 2) 3; 3) 5 keV.

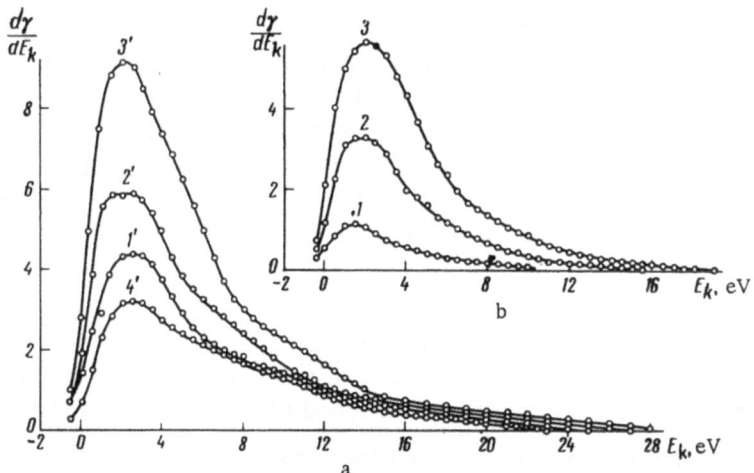

Fig. 8.17. Energy distributions of the secondary electrons on bombarding the target with Ne^+ ions (a) at an energy of E_0 equal to: 1') 1.5; 2') 3; 3') 5; 4') 0.4 keV, and Ne atoms (b) at an energy of E_0 equal to: 1) 1.5; 2) 3; 3) 5 keV.

By considering the energy distribution curves of the secondary electrons emitted from the target under the influence of ion bombardment, we see that for a given E_0 the energy spectra of the electrons ejected from the target by the ions differ from those associated with the atoms in that the electron energy spectrum is wider in the former case.

In order to understand the effect of the velocity of the ion on the electron energy spectrum for potential emission in more detail, let us consider Fig. 8.19. Curve 4' is taken from the results of Fig. 8.18a and constitutes the energy distribution of electrons emitted from the target on bombarding with 0.4-keV He^+ ions. Curve 3'-3 represents the difference between the corresponding curves in Fig. 8.18a,b, i.e., the energy spectrum of the electrons ejected by virtue of the potential energy of the ion at $E_0 = 5.0$ keV. Here we assume that the energy dis-

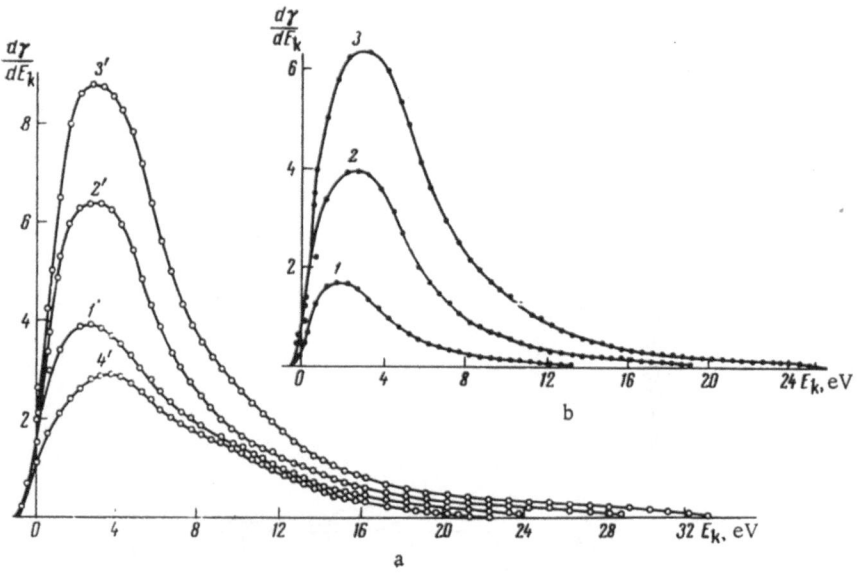

Fig. 8.18. Energy distributions of the secondary electrons on bombarding the target with He$^+$ ions (a) at an energy of E_0 equal to: 1') 1.5; 2') 3; 3') 5; 4') 0.4 keV, and He atoms (b) at an energy of E_0 equal to: 1) 1.5; 2) 3; 3) 5 keV.

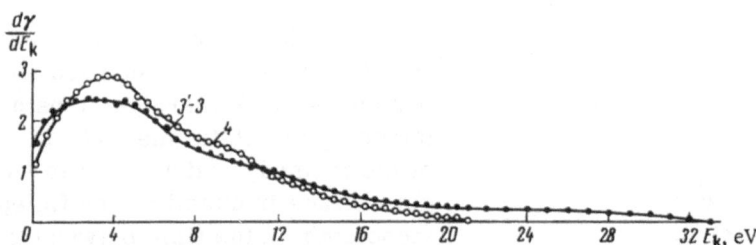

Fig. 8.19. Comparison of the energy distributions of potential-emission secondary electrons for $E_0 = 0.4$ and 5 keV.

tribution of the electrons emitted on bombarding the target with ions and atoms by virtue of their kinetic energy is the same in each case. We see from the figure that for an energy of $E_0 = 5$ keV, the electron spectrum becomes harder; this is expressed by an increase in the maximum energy and a fall in the number of slow electrons. Hence, the effect of the velocity of the ion on the energy spectrum of the potential-emission electrons reduces to an increase in the number of fast electrons and a reduction in the relative number of the slow ones. The relative yield of electrons emitted as a result of the potential energy of the ions remains constant, in agreement with [645].

Figure 8.20 shows area-normalized secondary-electron distribution curves illustrating the relative change in the energy spectrum of kinetic-emission electrons with changing energy of the bombarding He atoms. We see that, with increasing E_0, first the relative proportion of fast electrons in the total emission increases more rapidly than the relative proportion of the slow ones, and, secondly, the maximum energy of the electrons increases.

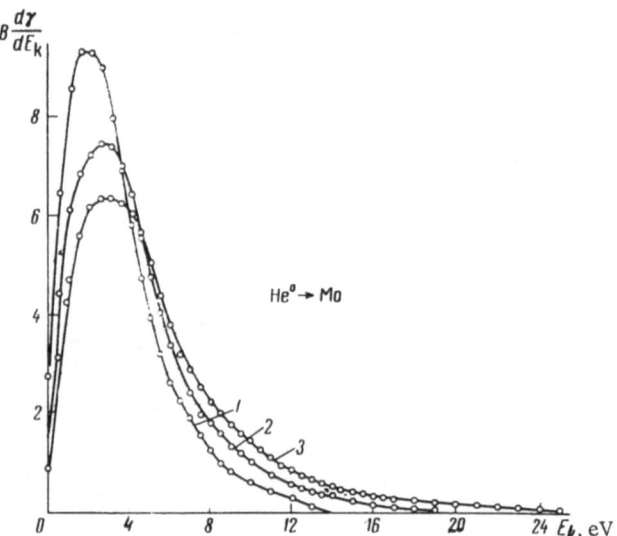

Fig. 8.20. Secondary electron distribution curves on bombarding the target with He atoms at energies of: 1) 1.5; 2) 3.0; 3)
5.0 keV.

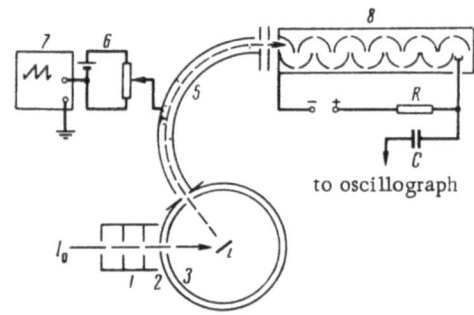

to oscillograph

Fig. 8.21. Arrangement of the measuring part of the apparatus with the cylindrical condenser.

Study of the Energy Distribution of Secondary Electrons by the Oscillograph Method of Recording

The foregoing data relating to the energy distribution of electrons emitted from molybdenum
bombarded with inert-gas ions and atoms were obtained by the static method. In view of the inertia
of the method, residual gas is adsorbed on the target surface in quantities sufficient to form a whole
monolayer in the time between the instant of cleaning the target by high-temperature heating and that
of effecting the measurement in a vacuum of $\sim 2 \cdot 10^{-7}$ mm Hg.

In order to conduct the investigations under
better conditions, subsequent measurements were
carried out by the oscillograph method [352, 743, 653] in which the electrons were energy-ana-
lyzed in the field of a cylindrical condenser.

The experiments were carried out in a vacuum system analogous to that described
earlier (Fig. 8.13). Only the construction of the receiving part of the apparatus was modified;
the construction of the electrodes is shown schematically in Fig. 8.21. Electrodes 1 and 2
served to provide the final shaping of the primary beams of atoms and ions and prevented them
from falling directly on the collector 3. For energy-analyzing the electrons, a cylindrical
Hughes–Rojansky condenser 5 with outer and inner plates of radius 60 and 54 mm, respectively,
was used. For entrance and exit slits of 0.8 and 0.4 mm, the analyzer 5 had a resolving power
of 0.5 eV. The beam of primary ions or atoms fell on the surface of the target 4 at an angle
$\varphi = 45°$ to the normal, the direction of which coincided with the axis of the electron beam cut
out by the entrance of the cylindrical condenser and the slits in electrodes 2 and 3. Thus, the
geometry of the analyzer and the other electrodes was similar to that used by Harrower [352]

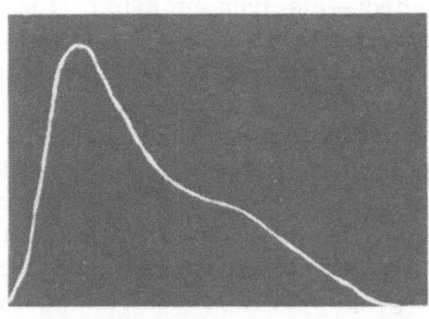 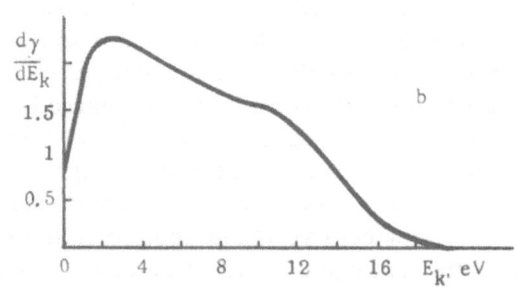

Fig. 8.22. Oscillogram (a) of the energy spectrum of electrons ejected from molybdenum by He$^+$ ions with $E_0 = 0.5$ keV, and electron energy-distribution curve (b) for the case of bombarding molybdenum with He$^+$ ions having $E_0 = 0.6$ keV.

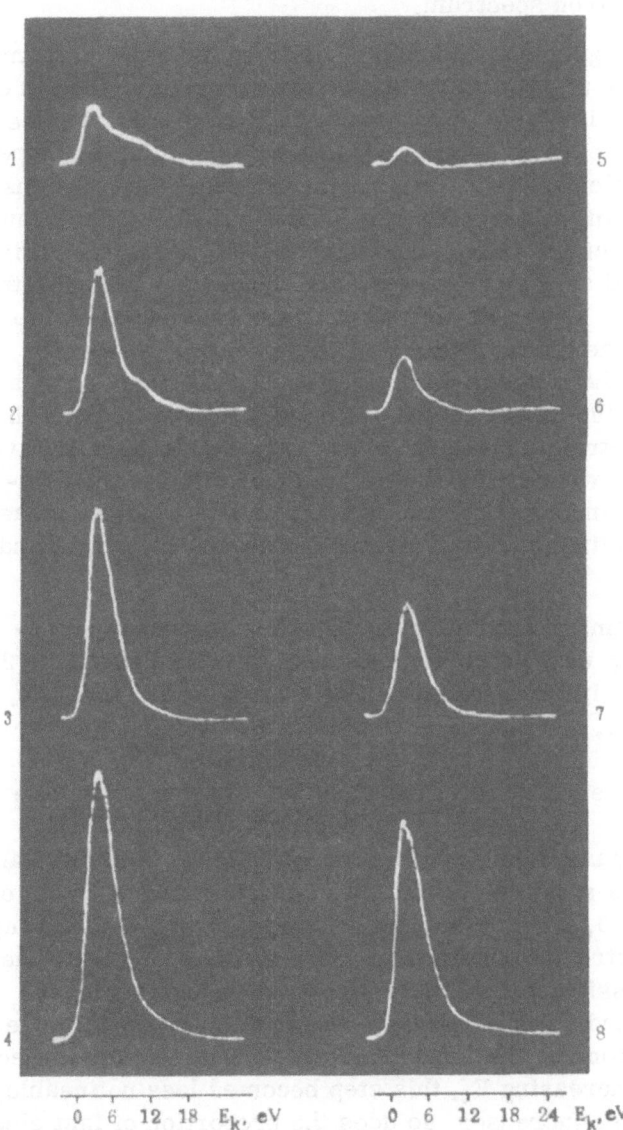

Fig. 8.23. Series of oscillograms of the energy spectra of electrons emitted from molybdenum on bombarding with He atoms and ions.

for studying the characteristic losses of energy in electron—electron emission. The fraction of secondary electrons emitted an an aperture angle of about 2.5° fell into the space between the plates of the cylindrical condenser. The radial field between the condenser plates was created by applying a sawtooth voltage from generator 7. The amplitude and frequency of the sawtooth voltage could be varied over the range desired. The separated electrons fell on the first dynode of electron multiplier 8 (of the open type) with an amplification factor of $2.5 \cdot 10^8$. The last dynode of multiplier 8 was connected through a condenser to the input of the vertical oscillograph amplifier. The frequency of the horizontal sweep of the oscillograph was synchronized with that of the sawtooth voltages applied to the outer plate of the cylindrical condenser from generator 7 and kept equal to about 25 cps. The oscillograph screen presented a stationary picture of the energy distribution of the electrons emitted from the target 4. The resultant pictures were photographed and subjected to further analysis.

The energy scale was calibrated by measuring the displacement of a selected point on the curve on further accelerating the electrons by a specific amount in the space between the target and the collector. The zero-energy position was determined by comparing the energy spectra of the secondary electrons with the thermoelectron spectrum.

Figure 8.22a shows an example of oscillograms of the energy distribution emitted from molybdenum on bombarding with 0.5-keV He^+ ions; Fig. 8.22b shows the energy distribution of electrons ejected from molybdenum by 0.6-keV He^+ ions, taken from Hagstrum [345]. We see from the figure that the distribution curve represented by the oscillogram is similar in shape to the curve obtained by Hagstrum for an atomically clean molybdenum surface; the positions of the curve maxima and the values of the maximum energies agree quite satisfactorily. However, on the curve shown in Fig. 8.22a, the spectrum is enriched with slower electrons. This difference is associated with the fact that in all analyzers (electric and magnetic) acting on the principle of deflecting the charged particles the sensitivity of the analyzer is proportional to the energy of the electrons analyzed. Hence, the energy analysis of electrons with energies close to zero is very difficult. The energy of the secondary electrons may be increased by a constant amount if we apply a voltage between the target and the collector. However, even then there is a certain distortion of the energy spectrum on account of the variation in the effective slit width and the preferential acceptance of slow electrons due to the application of the additional voltage. This distortion, however, remains constant and has no effect on relative measurements. For all measurements the accelerating potential difference between the target and the collector was kept equal to 35 V.

Figure 8.23 shows a series of oscillograms of the energy spectrum of secondary electrons emitted from molybdenum on bombarding with He atoms and ions. Oscillograms 1, 2, 3, and 4 were obtained by bombarding the target with He^+ ions at energies of $E_0 = 0.5, 1.5, 3.0,$ and 5.0 keV, respectively, while oscillograms 5, 6, 7, and 8 were obtained by bombarding the target with He atoms at energies of $E_0 = 0.5, 1.5, 3.0,$ and 5.0 keV, respectively. The horizontal axis in these oscillograms represents the energies of the electrons $\mathscr{E}_k$ in eV and the vertical axis a quantity proportional to $d\gamma/d\mathscr{E}_k$. We see from the oscillograms that all the curves have maxima at about 2-3 eV. An increase in the kinetic energy of the bombarding particles leads to a broadening of the curves and a simultaneous rise in the maximum for both ions and atoms. We also see from oscillograms 5, 6, 7, and 8 that for an energy of the bombarding He atoms equal to 0.5 keV the emission of electrons is negligible. This is associated with the fact that in the case of atoms only kinetic emission, which has a threshold value, can occur. The existence of steps on the oscillograms 1 and 2 for the energy range $\mathscr{E}_k = 8$-10 eV is due to the superposition of kinetic-emission electrons on the comparatively wide spectrum of electrons arising from potential emission. With increasing E_0, this step becomes less noticeable, since, as the energy of the bombarding particles increases, so does the proportion of fast electrons in the kinetic-emission spectrum.

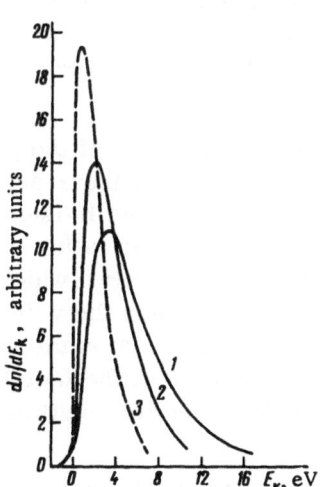

Fig. 8.24. Energy spectrum of electrons emitted from molybdenum on bombarding with He and Ar atoms.
1) He, $E_0 = 5.0$ keV; 2) Ar, $E_0 = 5.0$ keV; 3) He, $E_0 = 0.5$ keV.

Analogous oscillograms were obtained in the case of bombarding the target with Ne and Ar atoms and ions. The way in which the distribution of secondary electrons varied with the energy of the bombarding particles was, in general features, similar to the He case. However, we should note that for $E_0 = 0.5$ keV in the case of Ne and Ar, the potential-emission electrons constituted the whole of the emission, whereas, in the case of He (oscillogram 5, Fig. 8.23), kinetic emission also occurred.

For comparison, Fig. 8.24 shows the energy spectrum of electrons emitted on bombarding the target with He and Ar atoms (kinetic emission) at $E_0 = 5.0$ keV and for He atoms at $E_0 = 0.5$ keV. We see from the figure that for the specified E_0 the energy spectrum of the electrons is harder for the target bombarded by the lighter atoms, while for the case of He the maximum of the distribution curve is displaced in the direction of larger $\mathcal{E}_k$. This indicates that the energy spectrum of the kinetic-emission electrons depends not only on the energy but also on the nature of the bombarding particles. However, in order to explain the part played by the nature of the bombarding atom in the kinetic emission of electrons, a comparison of the data obtained for similar kinetic energies is insufficient. The kinetic emission of electrons depends on the velocity of the bombarding particle and on its individual properties, for example, on the structure of the electron shell. In order to elucidate the effect of the latter, we must compare data obtained for two types of particles at the same velocity. In Fig. 8.24, the broken curve characterizes the energy distribution of electrons emitted from molybdenum on bombarding with 0.5-keV He atoms; at this energy the velocity of the He atoms is similar to that of an Ar atom at $E_0 = 5.0$ keV. We see that the energy spectrum of the electrons in the case of the He atoms is much softer, i.e., it is poor in fast electrons, and the maximum is displaced in the direction of lower energies.

Thus, the results show that for the same velocity of the bombarding particles the energy spectrum of the kinetic-emission electrons is rather harder when bombarding with atoms having a more complex electron shell.

The energy spectrum of the potential-emission electrons changes considerably in the range $E_0 = 0.5$-5.0 keV. A certain number of fast electrons appear; we may associate the appearance of these with the broadening of the electron energy-distribution curve in accordance with the Heisenberg uncertainty principle. We suppose here that the energy spectrum of the kinetic-emission electrons is the same for ions and atoms of the same element; the validity of this assumption is justified by the fact that the process leading to the kinetic emission of electrons takes place after the neutralization of the ion close to the surface of the solid, i.e., after potential emission, as follows from Hagstrum's theory [291]. The energy spectrum of kinetic emission varies with the energy of the bombarding atoms in such a way that with increasing E_0 the proportion of fast electrons in the total emission increases and the electron distribution curve broadens.

The results show that with increasing velocity of the ion the energy spectrum of the electrons in the energy range studied undergoes certain changes, although it is known that the coefficient of potential emission of electrons changes very little with the velocity of the ion (see Chapter 6). This expresses itself in the fact that, with increasing velocity of the ion, the relative proportion of the fast electrons increases at the expense of a reduction in the number of slow electrons. The value of the maximum electron energies also increases. For example, in the case of He ions, the value of the maximum energy, equal to 21 eV at 0.4 keV, rises to 33 eV at $E_0 = 5.0$ keV.

According to Hagstrum's theory [291], the variation in the energy spectrum of the potential-emission electrons with the velocity of the ion is due to the displacement of the energy levels of the ion near the metal surface, and to the broadening associated with the Heisenberg uncertainty principle. In actual fact the probability function of the Auger neutralization of the ion $P_t(S, v_0)$ has a maximum value at a distance S_{max} from the surface of the metal, which equals $(1/a) \ln(A/v_0)$ (v_0 if the velocity of the ion; a and A are constants). When the velocity of the bombarding ion rises, S_{max} falls, but the shape of the $P_t(S, v_0)$ curve remains the same, i.e., with increasing E_0 the Auger transitions take place closer to the surface of the metal. This leads to a reduction in the effective ionization energy $E_i^!$, which displaces the distribution curve in the direction of lower $\mathscr{E}_k$.

The influence of the Heisenberg uncertainty principle is associated with the finite nature of the lifetime of the initial state. The energy broadening in this case, according to Hagstrum [291], equals $\Delta \mathscr{E}_k = \frac{\hbar}{\tau} = \hbar a v_0$, where h is Planck's constant, τ is the lifetime of the initial state, v_0 is the velocity of the ion, and a is a constant equal to 3 Å^{-1}. For He$^+$ ions at a velocity corresponding to 5.0 keV, $\mathscr{E}_k \sim 15$ eV. Allowing for this value, the maximum energy of the potential-emission electrons for $E_0 = 5.0$ keV is roughly equal to 31 eV, in good agreement with the experimental value.

According to the theory of Parilis and Kishinevskii [539, 621], the kinetic emission of electrons takes place, in the same way as the potential emission, partly as a result of the Auger process. The difference is simply that, whereas the potential emission of electrons arises as a result of the energy evolved in the Auger neutralization of the bombarding ion by the electrons of the conduction band, the kinetic emission occurs partly by the excitation of electrons in the conduction band resulting from the energy released in the Auger recombination of electrons with the hole formed in the filled band of the metal by the collision between the ion and a metal atom.

A number of special features in the experimental data (the increase in the average and maximum energies with increasing energy of the bombarding atom, etc.) may be satisfactorily explained on the grounds of this theory. The energy spectrum of the kinetic-emission electrons is determined (according to the theory) by the hole distribution in the filled band of the metal. With increasing kinetic energy of the bombarding particle and correspondingly increasing activation energy of the target atoms, holes are formed at a greater depth in the filled band, and on Auger recombination of these with electrons of the conduction band electrons with higher energies are excited. This should lead to an increase in the maximum and average electron energies, and hence to a rise in the total electron yield.

The maximum electron energy, according to theory, is $\mathscr{E}_{max} = \delta - 2\varphi$, where δ is the greatest possible depth of the position of the hole on the filled band of the metal. Hence, the value of $\mathscr{E}_{max}$ is closely related to the structure of the energy bands in the metal. In the neighborhood of the threshold energy for the kinetic emission of electrons, δ corresponds to the upper edge of the filled band of the metal. In the case of bombarding a molybdenum target with Ar^0 atoms at $E_0 = 1.5$ keV (threshold energy) the results of Fig. 8.18 (allowing for the effect of the Heisenberg uncertainty principle) give $\mathscr{E}_{max} = 6.6$ eV and $\delta = 15.1$ eV, which agrees with experimental data relating to the position of the maximum of the density of states in the filled band of molybdenum ($\delta = 16$ eV) observed by Harrower [352] in studying the characteristic losses of energy in the case of secondary electron—electron emission.

It should be mentioned that the experimentally observed electron energy distributions cannot be fully explained by the distribution of holes in the filled band of the metal. In actual fact, the excited electrons emerging from the metal lose energy on colliding with electrons from the conduction band and with atomic cores in the lattice. There may be a certain retardation of the primary electrons and excitation of the secondary electrons. In a later development of the theory [670], this part of the kinetic emission was taken into account, and it was shown that the Auger recombination supplied only about half the electrons emitted (see Chapter 11). In addition to this, there may be excitation of the electrons in the metal as a result of the energy released in the Auger relaxation of excited atoms, which are formed on collision with the surface [743].

However, on the basis of the experimental data obtained in the present treatment, it is very difficult to estimate the respective contributions of the electrons excited by these individual mechanisms to the total emission of slow electrons.

CHAPTER 9

SPUTTERING OF A SOLID WITH ION BEAMS

§1. Brief Review of Investigations into Cathodic Sputtering

We have already noted that beams of ions or neutral atoms striking a solid surface engender collisions between the incident particles and target atoms accompanied by the scattering and penetration of the incident particles and the emission of electrons. At the same time, some of the energy and momentum is transferred to the target atoms, and as a result of this the atoms may leave the target in the form of atoms, ions, or molecules [1]. This phenomenon, taking place in parallel with the scattering of the primary particles, was originally observed on the cathodes of ionic devices and was accordingly called cathodic (or cathode) sputtering. In the presence of adsorbed particles on the target surface (arising from the primary ions and residual-gas atoms), these may also be subject to sputtering. As a result of this, in addition to neutral atoms sputtered from the film, the secondary ions may include positive and negative ions arising as a result of film sputtering; these are quite indistinguishable from the rest. Thus, the phenomena of the scattering of primary ions and cathodic sputtering are closely related and sometimes experimentally indistinguishable. In view of this these phenomena cannot be considered in isolation, either theoretically or experimentally, although, until now, in view of experimental difficulties, secondary emission and cathodic sputtering have been considered separately.

The study of cathodic sputtering is of great importance in connection with explaining the mechanisms of processes taking place on the surface of structural elements of ion—electron equipment and fast-moving aircraft, in the technology of producing thin coatings, in plasma work, in ionic motors, etc. Cathode sputtering has formed the subject of many dozens of articles and detailed reviews [66, 89, 107, 155, 156, 174, 269, 283, 324, 515, 634, 686, 730]. These have considered the coefficients of cathodic sputtering, together with the thresholds of the phenomenon and the dependence of the coefficients on the mass, energy, and incident angle of the bombarding ion, the gas pressure, and the mass of the atoms comprising the solid target, as well as the temperature, thermal conductivity, specific heat, electrical conductivity, latent heat of vaporization, and surface state of the latter. The structure, properties, and spectral composition of the products of cathodic sputtering have also been studied.

The amount of sputtering target material Δm is of course proportional to the primary ion current I_0 and the sputtering time t:

$$\Delta m = \frac{A}{e} N I_0 t,$$

where N is the coefficient of cathodic sputtering, numerically equal to the number of sputtered target atoms associated with one primary ion, A is the atomic weight of the sputtered particles, and e is the charge of the electron. The coefficient of cathodic sputtering N is an important parameter characterizing the process of sputtering from a solid target.

Table 9.1

Ion	Material	Energy of the ion, keV	Sputtering coefficients, atoms/ion	Authors
Hg⁺	Fe	1—30	0.5—4.5	Eremeev and
	Steel		0.5—4.5	Estrinov [246]
	Graphite		0.1—2	
Cu⁺	Cu	0.5—6	3.4—8.7	Dobretsov and
Mn⁺	Mn	1.6—6	2.6—7.7	Karnaukhova [240]
Fe⁺	Fe	0—6	1.3—3.5	
Pb⁺	Pb	0.5—5	8.4—36.2	
Sn⁺	Sn	1.0—4	5.6—12.2	
He⁺	Ag	0.4—6	0.9—4.5	Keywell [325]
Ne⁺	Ag	0.4—5	1.2—5.8	
Ar⁺	Ag	0.4—6	4.5—9.2	
Ke⁺	Ag	1—5	7.0—11.5	
Ar⁺	Cu	0.4—4	1.2—4.2	
Ke⁺	Cu	0.4—4	2.8—5.4	
Ar⁺	Pb	0.4—4	1.5—5.4	
He⁺	Pb	0.4—3,5	0.5—2.5	
Hg⁺	W	0.5—2	0.5—1.9	Günterschulze [326]
	Mo	0.5—3	0.58—2.4	
	Ni	0.5—3	0.8—3.5	
	Fe	0,5—3	1.1—3.0	
	Al	0,5—3	1.0—2.1	
Hg⁺	Ni + 1% Co⁶⁰	0.01—0.1	$1 \cdot 10^{-4} - 40 \cdot 10^{-4}$	Morgulis and Tishchenko [358]
Hg⁺	Fe	2—13	3.0—6.0	Strachan and
	U	2—13	0.8—5.2	Harris [356]
	Si	2—6	4.0—8.0	
	Re	Up to 7	Up to 6	
	C	0,06—20	0.044—2.2	
Ar⁺	Fe	2—13	1.3—2.3	
	U	2—13	0.8—1.2	
	C	0.03—13	0.44—0.8	
Ar⁺	Ag	0.04—0,240	0.01—1.6	Koedam [439]
Hg⁺	Pd	0.0—0.3	0.2—0.9	Wehner [399]
	Co	0.12—0.35	0.02—0.48	
	U	0.05—0.35	0.01—0.8	
	Ta	0.1—0.35	0.02—0.4	
	Ir	0.1—0.3	0.07—0.8	
	Nb	0.2—0.4	0.1—0.3	
	Rh	0.07—0.3	0.02—0.65	
	Zr	0.1—0.4	0.01—0.35	
	Ag	0.05—0.25	0.01—1.3	
	Ti	0.15—0.4	0.05—0.45	
	Au	0.05—0.25	0.05—1.3	
	Al	0.12—0.35	0.02—0.35	
	Cu	0.06—0.3	0.01—0.85	
	Fe	0.1—0.4	0.01—0.5	
	Ge	0.12—0.4	0.05—0.62	
Hg⁺	C	0.05—0.5	0.01—0.18	Wehner [399]
	Ni	0.07—0.4	0.01—0.7	
	H	0.12—0.4	0.05—0.5	
	Re	0.12—0.35	0.07—0.6	

Table 9.1 (continued)

Ion	Material	Energy of the ion, keV	Sputtering coefficients, atoms/ion	Authors
Hg⁺	W	0.06—0.4	0.001—0.55	
	Th	0.1—0.42	0.02—0.65	
	Mo	0.15—0.4	0.03—0.43	
	Cr	0.1—0.3	0.02—0.42	
	Si	0.12—0.3	0.01—0.1	
Hg⁺	Cu	4—15	8.0—13.0	Wehner and Rosenberg [562]
	Ni	4—15	5.0—8.0	
	Co	4—15	4.0—7.0	
	Fe	4—15	3.5—5.5	
	V	4—15	2.0—4.0	
	Ti	4—15	2.0—3.5	
	W	4—15	2.5—4.5	
	Ta	4—15	2.5—4.5	
	Pb	4—15	6.0—11.0	
	Au	4—13	12.0—25.0	
	Ag	4—13	12.0—27.0	
	Pb	4—15	5.0—10.0	
	Rh	4—15	4.5—7.0	
	Mo	5—15	3.0—6.5	
	Cu	0.1—1	0.3—3.2	
	Ni	0.1—1	0.1—1.7	
Xe⁺	Fe	0.1—1	0.5—1.4	
	Mo	0.1—1	0.02—1.4	
Xe⁺	Cu	0.1—1	0.3—3.7	
	Ni	0.1—1	0.2—2.2	
He⁺	Au	0.02—0.04—0.1	0—0.008	Scott [592]
Ne⁺		0.02—0.04—0.1	0—0.14	
Ar⁺		0.02—0.04—0.1	0—0.16	
Ne⁺	W	0.4—0.6	0.34—0.32	Laegreid and Wehner [556]
	Mo	0.40—0.6	0.44—0.54	
	Pt	0.6	0.7	
	Al	0.4—0.6	0.68—0.83	
	Ag	0.4—0.6	1.6—1.98	
	Ni	0.6	1.34	
	Cu	0.4	1.53	
	Au	0.4	1.0	
Hg⁺	W	0.2—0.8; 2—10	0.3—0.54; 1.3—2.0; 2.75—3.0	Colligon [588]
	Mo	0.2—0.8; 2—10	0.1—0.44; 1.0—2.0; 2.75—5.0	
	Fe	0.2—0.8; 2—10	0.13—0.5; 1.1—2.0; 2.75—4.5	
Ar⁺	Cu (100)	0.02—0.1	0.07—0.42	Kaminskii [730]
	Cu (111)	0.03—0.1	0.07—0.2	
	Al (111)	0.02—0.12	0.006—0.17	
Ar⁺	Au	0.02—0.07	0.06—0.1	McKlown [564]
Ar⁺	Si	0.05—0.3	0.0025—0.12	Vol'skii [593]
Ar²⁺	Si	0.05—0.3	0.01—0.5	Vol'skii and Zhdanyuk [558]
Xe⁺	Si	20—60	0.7—1.3	Hines and Wallor [560]

Table 9.1 (continued)

Ion	Material	Energy of the ion, keV	Sputtering coefficients, atoms/ion	Authors
Ar^+	LiF	2—10	1.3—2.2	Navinsek [732]
	KBr	2—10	0.3—0.6	
	KCl	2—10	0.9—2.0	
	NaCl	2—10	0.35—1.0	
Ar^+	Cu	5—25	4.2—5.5	Guseva [472]
Kr^+		5—25	6.5—9.0	
Xe^+		5—25	6.2—10.5	
Cu^+		5—25	7.2—11.5	
Ar^+	Ag	3—25	7.0—12.0	
Kr^+		3—25	8.0—19.0	
Ag^+		3—80	11.0—19.0	
Xe^+		3—18	9.0—21.0	
Kr^+	Ta	5—30	1.2—2.7	Guseva [594]
	Stainless steel	5—30	2.0—5.0	
D^+	Ta	10—30	0.0045—0.006	
	Stainless steel	6—30	0.05—0.03	
Cl^+	Cu	5—20	4.0—5.0	Rol, Fluit, and Kistemaker [516]
N^+		5—20	1.5—2.1	
Na^+		5—20	1.7—3.0	
K^+		5—20	3.5—5.0	
Cd^+		5—20	9.8—11.5	
Tl^+		5—20	6.3—14.0	
Hg^+		5—20	3.0—13.0	
Zn^+		5—20	4.8—7.4	
Si^+		5—20	8.0—4.3	
KJ^+		5—20	12.0—17.0	
N_2^+		5—20	3.8—4.0	
Ar^+		5—25	5.5—6.2	
Ne^+		5—25	2.8—3.0	
Cu^+		4—25	4.5—8.0	
Pb^+		5—15—25	2.8—4.65—4.2	
Ar^+	Cu	5—87,5	6.48—9.25	Yonts, Normand, and Harrison [559]
D^+		10—44	0.048—0.023	
He^+		15—40	0.23—0.025	
Ne^+	Cu	0.1—1	0.3—1.7	Weijsenfeld, Hoogendoorn, and Koedam [561]
	Ni	0.1—1	0.2—1.1	
	Fe	0.1—1	0.05—0.8	
Ar^+	Mo	0.1—1	0.03—0.45	
	Cu	0.1—1	0.3—2.6	
	Ni	0.1—1	0.2—2.1	
Kr^+	Fe	0.1—1	0.1—1.3	
	Mo	0.1—1	0.02—1.1	

At the present time, two fundamentally differing methods are used for studying cathodic sputtering: the sputtering of the sample in the atmosphere of a gas discharge [151, 179, 269, 324, 355-358, 399, 555, 592, 593] and sputtering in a high vacuum by an ion or neutral-particle beam shaped in a special source [240, 246, 283, 516, 557]. When working with a discharge it is easy to obtain large ion currents, and this greatly reduces the effect of contamination on the test surface by the adsorption of residual gases owing to the cleaning of the surface by the sputtering. When studying cathodic sputtering in a high vacuum control may be exerted over the state of the sputtered surface and the composition, energy, incident angle, and current of the primary ions.

One of the important parameters of the sputtering process is the sputtering threshold: the minimum energy of an incident ion for which a target atom may be ejected. The determination of threshold energies and the establishment of a relation between these and the parameters of the target material and incident ion have formed the subject of a large number of investigations [326, 355-358, 287, 564, 442, 519, 298, 443, 477, 729, 730]. However, until now the majority of authors have only indicated the region of the threshold. For example, Morgulis and Tishchenko [357, 358] showed that the sputtering thresholds of cobalt, zinc, silver, zirconium, antimony, tantalum, tungsten, iridium, and thallium for Hg^+ ions lay in the range 3-20 eV. The sputtering thresholds of the alkali metals by inert-gas ions were estimated by Bradley [297] as 2-5 eV. Some authors, on the other hand, gave the exact values of the threshold energies; it was asserted by Günterschulze [326] that on bombardment with Hg^+ ions tungsten, tantalum, molybdenum, nickel, iron, aluminum, zinc, manganese, and silicon started sputtering at the same energy, 190 eV. Stuart and Wehner [591] determined the exact values of the threshold energies for a large number of metals bombarded with Hg^+ and Ar^+ ions: The "thresholds" lay between 20 and 50 eV, and their values approximately equalled four times the heat of sublimation of the metal in question, being roughly equal to the energy of radiation-defect formation.

Using mass analysis, a number of authors studied the chemical composition of sputtering products [440, 459, 503, 512, 566, 597, 612]. It was found that on bombarding a solid surface with positive ions three sorts of particles were ejected: atoms of the target material and atoms and molecules of the impurity surface film formed by the adsorption of complexes of atoms corresponding to chemical compounds of the target atoms with sorbed particles. All these particles may leave the surface either in the neutral state or in the form of positive or negative ions. Unfortunately, the investigations based on the mass-spectrometric method give no quantitative estimates of the intensities of these three groups of particles. Only a small number of investigations are devoted to a quantitative analysis of the sputtering of target material in the form of positive ions [360, 459, 512, 566, 597, 612]. By studying the energy spectrum of the sputtered particles the majority of authors [166, 473, 498, 508, 517, 567, 568, 598-600, 637, 638, 640, 688, 781] have determined the average kinetic energies of the sputtering components. The results of these investigations are very disconnected and often contradictory. For example, Wehner [473] asserts that the average energies of the sputtered particles equal 9-16 eV for various ion—metal pairs and are independent of the primary-ion energy. Akishin [634] showed that the ejected particles have energies no greater than 0.15 eV. Thompson et al. [599] asserted that the energy distribution constitutes the superposition of two Maxwell distributions corresponding to temperatures of 300-1000 and 60,000°K. Kopizky and Stier [568] found a periodic dependence of the average velocities of the sputtered particles on the atomic number of the target element.

Wehner [324] observed that on sputtering single crystals with ions of energy up to 500 eV the atoms escaped not isotropically but along the closely packed chains of atoms in the crystal lattice. Work of Yurasova et al. [441, 474, 475] enabled this conclusion to be extended to bombarding-ion energies of 50 keV. Nelson and Thompson [476, 569] observed anisotropy in the escape of sputtered particles on irradiating gold foil "in transmission" with protons, these having an average energy of 300 keV after passing through the film. In all the experiments it was shown that on raising the particle energy the deposition pattern of the sputtered material on the collector became more complex owing to the escape of particles in the directions of less close packing. The identification of these second-order escape directions of the sputtered particles is even now the subject of discussion [400, 475, 518]. Yurasova [601] showed that changing the temperature of a sputtered copper single crystal from −170 to +200°C produced no marked changes in the arrangement of the deposited spots. The directional escape of the particles was preserved up to T = 950°C (onset of evaporation), although the general background between the spots increased sharply and the spots themselves were distorted.

It was shown in [557, 602, 603, 689] that the sputtering coefficient depended nonmonotonically on the angle of incidence of the ions on the sputtered face of a single crystal. On increasing the angle of incidence the coefficient of cathodic sputtering rose in inverse proportion to the cosine of the angle. However, when the direction of the ion beam approached the direction of the principal crystallographic axes of the target, the sputtering coefficient fell sharply.

Published experimental values of the coefficient of cathodic sputtering are quite disconnected and disagree sharply with each other. For example, the data relating to the sputtering of iron by mercury ions presented in [246, 326, 358, 399, 562] differ from one another by 100-300% (the sputtering coefficients obtained by various authors are given in Table 9.1). It may nevertheless be regarded as established that the coefficient of cathodic sputtering depends on a whole series of parameters: the energy, mass [515], and atomic number of the incident ion, the angle of incidence on the target surface [557, 565], the mass, atomic number [562], and crystal-lattice binding energy of the target atom [39, 199, 356, 636], and the state and temperature of the bombarded surface [355, 359, 687]. However, when one is considering ways of describing the dependence of the sputtering coefficient on various parameters, and in particular seeking analytical expressions for these, the opinions of different authors disagree. For example, Dobretsov and Karnaukhova [240] found a parabolic dependence of the sputtering coefficient on the energy of the bombarding ions, while the sputtering-coefficient−energy relationship appeared to saturate in [246]; the coefficient varied linearly with $E_0^{3/2}$ in [326], while in [596] there were two distinct regions, in one of which the sputtering coefficient was proportional to the square root of the energy of the ion while in the other a relationship of the form $N \sim E_0^{3/2}$ held.

The temperature dependence of the sputtering process has been very little studied. Wehner [399] presented some results according to which the coefficient of cathodic sputtering of metals by mercury ions in the range T > 3000°C was independent of target temperature. On the other hand, Penning and Moubis [151] found under analogous experimental conditions that as the target temperature increased from 400 to 1000°C, the rate of cathodic sputtering almost doubled.

Thus, the accumulation of reliable experimental results characterizing the true phenomenon of cathodic sputtering is only in a rudimentary state. Broader investigations are urgently required.

Theories of Cathodic Sputtering

Several theories of cathodic sputtering have been proposed at various times; these have differed not only in the strictness of their mathematical apparatus, but also in the essence of the proposed mechanism of the phenomena. The chemical theory (Morgulis [174]) considers cathodic sputtering as the result of a certain chemical reaction between the bombarding ions and the atoms of the sputtered surface; the radiation theory (Buck and Smith [19]) associated cathodic sputtering with the absorption of radiation arising when the ions are sharply retarded at the cathode; the explosion theory (Berliner [3]) explains cathodic sputtering by the explosion of occluded gas in the cathode; the theory of cumulative sputtering reduces the elementary sputtering act to cumulation under the influence of the bombarding ion [19]. The thermal theory (Kapitsa [24], Hippel [39]) and the momentum theory (Stark [11], Langmuir [25]) explain sputtering in terms of certain well-known models. The two latter theories deserve further attention at the present time.

The thermal theory of cathodic sputtering was formulated in 1908 by Stark [11] and revived in 1926 by Hippel [39]. This theory considers cathodic sputtering as the evaporation of atoms during local heating to an extremely high temperature in a microscopic section of the sputtered surface as a result of the impact of the ion. Attempts at improving these principles were made in [92, 132, 133, 169].

A slightly modified form of the thermal theory, the mechanism of "thermal wedges," was proposed by Thompson and Nelson [605, 731]. These authors considered that the sputtering process could not be described on the basis of focused collisions only (see later). The process of momentum transfer along a closely packed row is executed in a period of ~10^{-13} sec. However, after this there remains a severely heated region or "wedge" in which a cascade of collisions has taken place. It is considered that the "wedge" is in fact a sphere, the radius of which depends on the energy of the incident ion. Within the "wedge" the lattice has a constant temperature. The "wedge lifetime" is determined by the thermal conductivity of the lattice; there is hardly any transfer of lattice energy by the electrons. If such a "thermal wedge" intersects the surface and remains hot for a fairly long time, the atoms are freed from the surface by evaporation. The number of atoms evaporated from the surface of one "wedge" (the sputtering coefficient) is calculated in accordance with Maxwell—Boltzmann statistics. On the basis of their own special experiments, the authors of the papers cited estimated the temperature and size of the "wedge" and suggested that the part played by this mechanism in the sputtering process increased with increasing mass and energy of the ion. However, the results of Kaminsky's experiments [604, 641] at high energies failed to confirm these conclusions. The distribution of sputtered material in a spot on the collector was better described by a Gauss function rather than a cosine law.

The momentum theory was created by Stark [11], Langmuir and Kingdon [25], Holst [30], Eliseev (see [174]), and Lamar and Compton [99]. At the present time there are several forms of the momentum theory, differing in their choice of the potential function used as a basis for the interaction between the incident ion and the target atom.

For high energies of the incident ion, the Coulomb interaction between the particles is used [99, 444, 520, 521, 522]. Comparison of the majority of the theoretical calculations with experiment is made difficult because the final conclusions contain a certain number of parameters which cannot be determined experimentally. Only in the theory of Pease [520] is an expression derived (after a number of simplifications) for the energy dependence of the coefficient of cathodic sputtering in such a way as to be capable of experimental test:

$$N = \frac{1}{4}\,\sigma n^{\eta_s}\left[1 + \sqrt{\frac{\ln \bar{E}/E_s}{\ln 2}}\right]\frac{\bar{E}}{E_d}\,,$$

where $\bar{E}$ is the average energy given by the ion to the target atom, E_s is the energy required to move a target atom into an interstice, E_d is the energy required to remove a target atom from the surface (heat of vaporization), n is the number of target atoms per unit volume, and σ is the cross section corresponding to the displacement of target atoms into the interstices.

An experimental check applied to the Pease formula shows that in some cases it agrees with Kaminsky's results [604, 641] while in others it gives a value of N double that obtained experimentally [523].

Some authors [520, 559], considering sputtering in the energy range $E_A < E_0 < E_B$, where

$$E_A = 2E_R Z_1 Z_2 (Z_1^{1/3} + Z_2^{1/3})^{1/2} \frac{m_1 + m_2}{m_1},$$

$$E_B = 4E_R^2 Z_1^2 Z_2^2 (Z_1^{1/3} + Z_2^{1/3}) \frac{m_2}{m_1} \frac{1}{E_s},$$

use the screened Coulomb potential for the calculation; here, E_R is the Rydberg energy (13.68 eV; Z_1, Z_2 are the atomic numbers of the colliding particles; m_1 is the mass of a target atom; m_2 is the mass of the ion; and E_s is the energy required to displace a target atom into an interstice. Comparing the analytical expressions with the results of their own experimental investigations, the authors obtain satisfactory agreement.

In the range $E_0 < E_A$, the collision of an incident ion with a target atom is considered by Henschke as the paired collision of elastic spheres [400, 401]. In the case of sputtering with a single collision, in order to produce sputtering the ion must have a certain minimum energy ("threshold"). The calculated values of E_g agree closely with the threshold values obtained in Wehner's experiments [399].

Keywell [247, 325] used an analogy with the retardation of fast neutrons in order to explain the sputtering process [327]. According to this theory, sputtering occurs by diffusion-type collisions of the sputtering ion with atoms in the crystal lattice. The coefficient of cathodic sputtering is proportional to $\sqrt{E_0}$. Keywell supports these conclusions by his own experimental data. However, the experiments recorded in [442, 445] disagree with this theory. The mathematical methods of neutron-diffusion calculations were also used by Harrison [348] to construct a theory of cathodic sputtering; however, no conclusions capable of being tested experimentally were obtained.

Rol, Fluit, and Kistemaker [516] consider that sputtering occurs if a target atom receiving energy from an ion lies no further from the surface than a distance equal to the free path of the ion; this distance is in turn dependent on the energy of the ion. Only the first collisions of the ions with the target atoms are considered here. The coefficient of cathodic sputtering is proportional to E_0. These authors' theory agrees with their own experiments [516].

The momentum theory has been used as a basis for a large number of attempts to explain the effects of cathodic sputtering associated with the structure of the sputtered single crystals [476, 478, 524, 525]. All these are based on the mechanism of the focused propagation of energy in the single crystals established theoretically by Silsbee [403]. According to these principles, the energy received by an atom in the single crystal is propagated mainly in directions corresponding to the most densely packed rows in the lattice. Focusing of this type is possible if $D_{hkl}/d < 2$, where D_{hkl} is the distance between the lattice points and d is the effective radius of a lattice atom. The angle between the direction of the momentum and the direction of the densely packed row falls monotonically with increasing number of collisions. This process is possible when the energy carried along the crystal axis is no greater than a certain specific amount (usually of the order of 100 eV). Numerical calculation on the basis of Silsbee's principles is extremely complex, since the problem reduces to a many-body problem. A successful attempt at solving this problem was made by Vineyard (see [554]). The basis of the model is paired interaction of the Born-Mayer type. Using an electronic computer, Vineyard obtained solutions for a finite block of copper crystal. The energy in the crystal lattice was transferred mainly by the "relay" mechanism in the [110] and [100] directions. The [110] chain focuses for $E \leq 30$ eV, the [100] chain for $E \leq 40$ eV. In a well-focused chain the energy losses in the collision are 2-3 eV for the [110] and 7-8 eV for the [100].

We see from the foregoing review that the theories proposed for the mechanism of cathodic sputtering cannot explain all the experimental facts observed. Apart from errors in the actual theories this is associated both with the inadequate knowledge of the experimental conditions corresponding to particular experimental results, and also with the difficulties involved in determining microscopic quantities of sputtered material. In view of this, it became essential to develop a method of determining microscopic quantities of sputtered material more accurately, and also to conduct the experiments in relatively pure conditions.

§2. Method of Studying Metal Sputtering

An analysis of possible reasons for the disagreement between the experimental results of various authors led to the conclusion that these might lie primarily in the imprecise methods of measuring the amounts of material sputtered. From this point of view special attention should be directed on the method of determining microscopic quantities of sputtered material by weighing the sputtered target or the deposit in the collector. Both forms of this method involve the same difficulty. This lies in the necessity of allowing for the weight of foreign inclusions. In determining the amount of sputtered material by weighing either in air or in vacuum, we must be sure that it is really the target material which is being weighed and not products of its chemical interaction with sorbed gases. Other methods of determining microscopic quantities of sputtered material (by reference to the change in the resistance of a filament cathode, by photometering a transparent screen, by deactivation of the cathode, etc.) are also imperfect and provide no absolute measurements. The second cause of disagreement lies in the influence of the contaminated state of the surface subjected to ion bombardment. Here it must be remembered that "contamination" may be formed on the target surface as a result of the adsorption of gas from the main volume of the apparatus, by the diffusion of gases earlier occluded by the target, or by adsorption and penetration of the bombarding ions. Depending on the experimental conditions, the influence of each of these factors will differ. Clearly, the results of a reproducible experiment correspond to a state of equilibrium between the contamination of the surface and its cleaning by ion bombardment.

In accordance with the foregoing, it is clear that two conditions must be satisfied in studying the process of cathodic sputtering: 1) a method of determining microscopic amounts of the sputtered material as accurately as possible must be developed; 2) surface conditions must be kept in a constant state of cleanliness throughout the experiment. We considered that it was most important to obtain cathodic-sputtering data for the surfaces which constituted the main targets in the experiments described in earlier chapters.

In studying cathodic sputtering it is very important, in order to preserve the clean state of the target, to work with small primary-ion currents. This made it necessary to develop special methods of investigation based on measuring microscopic quantities of material by means of radioactive indicators. The limit of sensitivity of this method is mainly determined by external interference (the background of the recording counter); quantities of sputtered material down to 10^{-15} g may be measured in this way.

In order to study the process of cathodic sputtering we first employed the apparatus shown in Fig. 9.1 [283]. This included an ion source 1 with a cylindrical condenser 5 (see Chapter 1, §2) enabling the ions to be separated from the neutral atoms. The receiving part of the apparatus consisted of a collector 3 and a small cylinder with a base of aluminum foil; these were placed as a kind of internal "plate" for the collector 3. The false collector was intended to collect the sputtered target atoms. After each experiment the "collector" could be removed and the amount of material deposited upon it could be measured by reference to the activity of the deposit. The target 2 consisted of two S-shaped nonradioactive tantalum holders; one end of each was fixed to the heater pin and a flat radioactive tantalum plate was fixed to the other. The guard

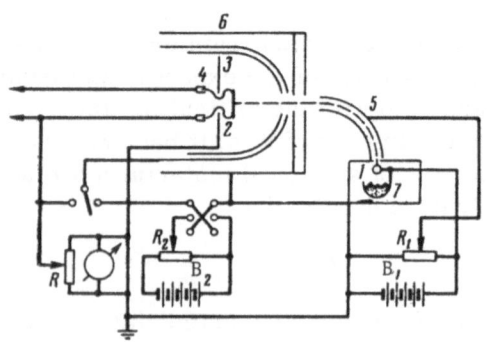

Fig. 9.1. Apparatus for studying the cathodic-sputtering process.

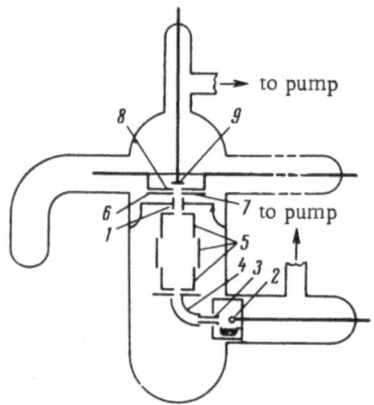

Fig. 9.2. Experimental apparatus for studying cathodic sputtering by the radioactive-isotope method.

disc 4 made of soot-blackened nickel trapped and grounded any stray ions. The vacuum in the apparatus was kept at a level of $\sim 10^{-6}$-10^{-7} mm Hg the whole time. The electrical circuit of the apparatus and the method of measuring the currents were the same as those used for the other secondary processes.

The measuring apparatus for determining the number of sputtered target atoms consisted of an end-window counter, a lead container screening the counter, a measuring cylinder, and a counting circuit.

In practice, one is always concerned with specimens in which a certain proportion of the total number of atoms are radioactive. In the experiments about to be described, tantalum containing a trace of $_{73}Ta^{182}$ and tungsten containing the isotope $_{74}W^{185}$ were used. The specific activity of the tungsten used in our experiments enabled amounts of material down to $1 \cdot 10^{-9}$ g to be measured.

The absolute values of the induced activity could not be directly determined in all the experiments. The ratio between the active and inactive atoms was therefore determined by an indirect method. For this purpose, special standards were prepared from the irradiated material. For example, a piece of activated tungsten foil was burned in a muffle furnace to W_2O_5. From the resultant mass, 10-20 mg portions were taken and mixed with ~1000 mg of quartz powder. To ensure uniform mixing, the mixture was crushed in an agate mortar with water for several hours. The resultant suspension was dried in air. The amount of radioactive tungsten in each milligram of mixture was determined. Ten to fifteen standards were prepared from the mixture. The amounts of powder mixture thus determined were marked on graduated discs. The specific quantities of W_2O_5 standards thus prepared were checked in a counter and the number of pulses per minute n_0 corresponding to 1 mg of the radiometric mixture, averaged over 10-15 standards, was determined. Then the microscopic quantity of material sputtered from the target and caught by the collector was calculated quite simply by reference to the activity (n pulses per minute) recorded from the collectors.

The experiments were carried out in the following way. The radioactive target was subjected to ion bombardment with fixed parameters of the ion beam, vacuum conditions, and state of the target surface. A specific fraction of the sputtered material was caught by the collector. After this, the apparatus was opened and the collector replaced. In this way, one point of the relationship under examination was obtained in each experiment. In seeking the next point the whole process was repeated. This method involved considerable difficulties associated with the repeated reproduction of the same experimental conditions in order to determine the desired relationship. In addition to this, it was impossible to study the kinetics of the process. The radioactive-isotope method thus called for considerable improvement.

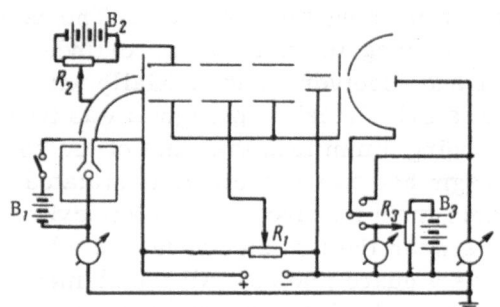

Fig. 9.3. Electrical measuring circuit.

A new experimental apparatus devised by Shustrov, Khasanov, and Ayukhanov [683, 727] is shown in Fig. 9.2. The apparatus is free from glass joints and the measuring part is separated from the source part by a vacuum diaphragm 1. Experiments were made in this with alkali-metal ions. This group of ions was chosen mainly because of the practical importance of the W(Ta) alkali-ion systems. In addition to this, experiments on the sputtering of metals by other than gas ions were interesting because of the special features created by the adsorption of ion-beam material on the target surface. The alkali ions were obtained by a method based on surface ionization (see Chapter 1, § 2).

The ion beam formed by the source 2 and electrical section 3-5 passed through the vacuum diaphragm 1, was shaped by the guard plate 6, and after passing through the aperture 7 in the collector 8 fell normally on the target 9. An ion source of this construction yielded an ion current density of up to 10 μA/cm^2. In order to obtain ion currents of high density, the vacuum diaphragm and cylindrical condenser had to be dispensed with. In this form of the apparatus the ion source was arranged so that the extracting funnel was coaxial with the lens. These changes enabled the density of the ion current to be raised to 800 μA/cm^2.

The sputtering products were collected by collectors made of nickel strip 0.05 mm thick having the shape of a hemicylinder with a generator of 40 mm and a diameter 20 mm. Up to 20 such collectors were arranged on molybdenum guide rods in the plane of the target. During the experiments, these interchangeable collectors were moved along the guide rods with an electromagnet and placed under the target one by one. After finishing a measurement, the interchangeable collector was directed into a special side tube of the apparatus to be kept until the end of the experiments. The use of interchangeable collectors made it possible to obtain 15-20 points in a single experiment under the same vacuum conditions, i.e., to plot one or two curves of the relationship being investigated. An apparatus of this construction also enabled us to solve problems associated with the kinetic aspects of sputtering.

Figure 9.3 shows the electrical measuring circuit. The operating principle is obvious from the diagram.

The method of radioactive indicators in its usual form cannot provide information regarding the chemical nature of the sputtering products. However, this method enables us to distinguish the sputtered particles in respect to their charged state. For this purpose potentials V_k of different signs relative to the target are applied to the collector.

Differential pumping of the vacuum apparatus was effected by means of three mercury-vapor pumps, one of which was a booster for the other two. The mercury vapor was frozen by a system of traps containing liquid nitrogen. In the measuring part of the apparatus the residual-gas pressure was less than $1 \cdot 10^{-7}$ mm Hg. In the working ion source the pressure rose to $3 \cdot 10^{-7}$ mm Hg.

During the work it became necessary to study the sputtering process at temperatures such that the target started evaporating. The difficulty was then that the earlier-observed [683] evaporation of the material in question was of a plainly nonequilibrium character. Hence, the evaporation could only be precisely regulated at the same time as the particular sputtering experiment. The necessity of solving this problem led to the creation of a new form of experimental apparatus. The receiving part of the apparatus was altered (Fig. 9.4). As receiver for the sputtered material we used a nickel cylinder 1, the side surface of which was divided by the generators into 12 equal parts, each constituting a collector. In the center of each collec-

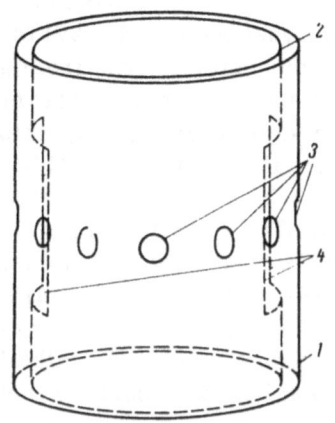

Fig. 9.4. Arrangement of the receiver for sputtered material.

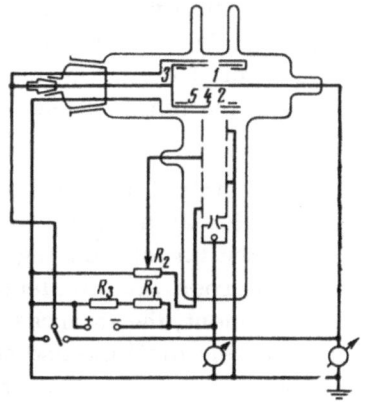

Fig. 9.5. Second experimental apparatus and electrical measuring circuit.

was an aperture 3 for passing the ion beam. The target (Fig. 9.5) was set along the axis of the cylinder. Between the target 1 and receiver 5 and coaxially with respect to the latter, a cylindrical diaphragm 2 was fixed, the diameter of this being 2 mm less than that of the receiver. This diaphragm had two diametrically situated cuts 4. The ions came from a source without any cylindrical condenser. Between the last electrode of the lens and the cylinder 5 was a guard plate 3. All the elements of the receiving part were fixed in a ground-glass joint. The cylinder 2 and the guard plate 3 were fixed rigidly and the cylinder 5 was on a small ground-glass joint placed in the middle of the large one. By using this joint, the cylinder 5 could be rotated around its own axis and the collectors could thus be interchanged. An apparatus of this construction [728] enabled us to measure two processes taking place at the same time. The collector on the side of the beam entrance collected the n' atoms evaporated and sputtered from the target; the collector on the opposite side collected just the n" atoms evaporated from the target. In this case, the difference n = n' − n" characterizes the sputtering of the material for a fixed target temperature. In practice, the use of this apparatus enabled us to obtain six points of the cathodic-sputtering relationship of present interest in a single experiment.

The target temperature was determined with a pyrometer and later by reference to the target heating current. By extrapolating the $T = f(I_h)$ curve thus obtained, temperatures under 800°C could be read directly.

The electrical measuring circuit is shown in Fig. 9.5. The shape and size of the ion beam at the target was studied visually by reference to the fluorescence of an Na^+ ion beam. This showed that the fluorescence was roughly circular, 2-2.5 mm in diameter, and situated in the center of the target. The density of the current in these experiments could be extended to $\sim 10^{-3}$ A/cm^2. In sputtering experiments at high temperatures the thermionic current from the target was one or two orders higher than the primary ion current. This made it impossible to measure the ion current with a heated target, since the thermocurrent losses through the apertures in the collectors became comparable with the primary ion current. At every change of collector the target heating was therefore disconnected and the ion current was recorded. Since this procedure took no more than 10 sec, while the exposure of each collector in the sputtering process was usually at least 3 min, it was considered permissible to neglect the error thus introduced.

In the form of apparatus without the vacuum diaphragm the experimental conditions were rather worse. In this case, the residual pressure was usually (1-2) · 10^{-6} mm Hg. The pressure during the operation of the ion source fluctuated between 3 · 10^{-6} and 7 · 10^{-6} mm Hg, depending on the salt used in the experiment and the magnitude of the ion current.

§3. Influence of the Contamination of a
Solid Surface on the Sputtering Process

Some experiments were made to establish the time dependence of the coefficient of cathodic sputtering N from an uncleaned, cold (~300°K) target. In these experiments, described by Shustrov [727], the target was bombarded with 1600-eV K$^+$ ions for several hours continuously. The sputtering products were collected by interchangeable collectors exposed under the target for 1 h each. The density of the ion current in these experiments was ~10^{-5} A/cm^2. The results of the experiments showed that the sputtering coefficient N was independent of the time of bombardment. We therefore proceeded to study the energy dependence of the cathodic-sputtering coefficients [727]. Experiments with cold targets not subjected to any preliminary treatment in which tungsten was bombarded with alkali ions showed poor reproducibility. For example, in the case of Li$^+$ ions, with constant experimental parameters, two successive measurements might differ from one another by over 100%. This might well have been due to the serious effect of surface contamination on the sputtering process.

In order to verify this supposition experiments were made in which a cold tungsten target previously degassed by vacuum heating was subjected to sputtering. Degassing was effected by gradually raising the target temperature to T ~ 2000°K, leaving it at this temperature until the pressure in the apparatus reached ~10^{-6} mm Hg. After this the target heating was switched off and the experiments proceeded at room temperature.

Under these conditions reproducible energy curves of the coefficients of cathodic sputtering were obtained for tungsten targets attacked by Cs$^+$, Rb$^+$, K$^+$, and Na$^+$ ions (Fig. 9.6). As before, the sputtering of tungsten by Li$^+$ ions remained irreproducible. The curves representing this relationship for a gas-free target were qualitatively similar. The sputtering coefficient for all these ions varied by a factor of two or more between 1000 and 2500 eV.

In the experiments just described, the surface subjected to ion bombardment could be regarded as fairly clean at the beginning of the experiment. However, in order to obtain a single point on the desired curve the previously cleaned target had to be bombarded for several minutes. It is well known that at a pressure of ~10^{-6} mm Hg a monolayer of adsorbed atoms is formed on the surface in a few seconds. The fact that this layer of "contamination" took part in the sputtering of the main target material is supported by the results of a number of authors who observed, among the ejected particles, not only positive and negative ions of the contaminants but also products of chemical combination between residual gases and the target material. Owing to the fundamental difficulties in quantitatively determining the neutral component of the sputtering, none of these investigations (carried out with the help of mass spectrometers) give any quantitative estimate of the contributions made to the total sputtering by individual particles of various kinds. If this contribution is large, then in analyzing the experimental data one must allow for the simultaneous occurrence of two processes: the direct ejection of atoms belonging to the bombarded target, and the ejection of target material in the form of chemical compounds with adsorbed gases. Clearly we cannot assert in advance that these two processes obey the same laws. Neglect of the second process demands justification. Thus, the question reduces to estimating the proportion of total sputtering associated with the ejection of material in the form of chemical compounds. It was impossible to analyze the sputtering products with respect to chemical composition during the experiments. As already noted, however, the method employed enabled the sputtered particles to be distinguished by reference to their charged state [633, 704, 727].

In the experiments a tungsten target held at room temperature after degassing by the method mentioned was bombarded with Cs$^+$ ions. Curve 2 in Fig. 9.7 corresponds to the experimental condition in which there was no potential difference between the target and the col-

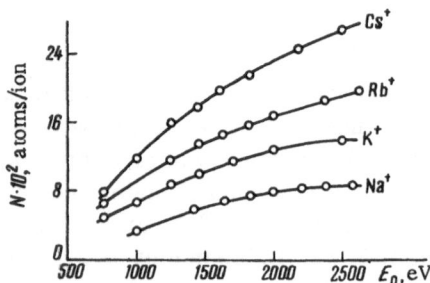

Fig. 9.6. Coefficient of cathodic sputtering of a cold, clean tungsten surface as a function of the energy of the alkali ions.

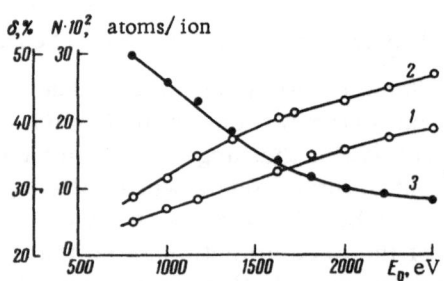

Fig. 9.7. Results of experiments on the sputtering of a cold, clean tungsten surface by Cs^+ ions.

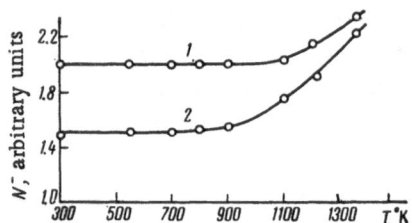

Fig. 9.8. Sputtering of tungsten in the form of negative ions as a function of the temperature of the target bombarded.

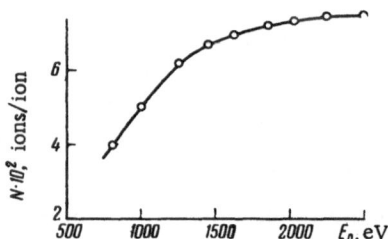

Fig. 9.9. Energy dependence of the coefficient of cathodic sputtering of tungsten in the form of negative ions of tungsten-containing compounds on bombarding with Cs^+ ions.

lector, i.e., V_k = 0 (these data have already been given in Fig. 9.6). In the case of V_k > 0, no change in the amount of radioactive deposit on the collector was observed. This meant that if, indeed, the sputtering products contained any positive tungsten-containing ions, then the amount of these was below the sensitivity of the method employed. Curve 1 was obtained for the condition V_k = −300 V, i.e., no negatively charged sputtered particles with an energy of under 300 eV could fall on the collector. Special experiments showed that for a voltage of V_k = −300 V, practically all the negative ions containing tungsten were stopped. Hence, curve 1 describes the neutral component of the cathodic sputtering of tungsten. The difference between the ordinates of curves 1 and 2 corresponds to the negative component of sputtering. Clearly this number of negative tungsten ions can only be associated with the sputtering of tungsten-containing compounds. This conclusion agrees with mass-spectrometric investigations of Ayukhanov and Iskhakov [512] and Krohn [612]. Additional experiments were carried out in order to verify this conclusion. If the target temperature is raised, then the alkali film formed by adsorption of the beam evaporates and the work function of the target increases. This should be accompanied by a fall in the proportion of negative tungsten-containing ions in the sputtering as a whole. The results of this experiment, shown in Fig. 9.8, support the foregoing discussion. Curve 1 corresponds to the experimental condition V_k = 0, curve 2 to V_k = −300 V. We see that the difference in the ordinates of these curves falls with increasing temperature, i.e., the proportion of negative tungsten-containing ions in the total sputtering diminishes.

Let us return to the results presented in Fig. 9.7. Curve 3 reflects the energy dependence of the proportion δ of negative tungsten ions among the total sputtering. It was shown in [612] that, on bombarding tungsten with Cs^+ ions, in general only compounds of tungsten with

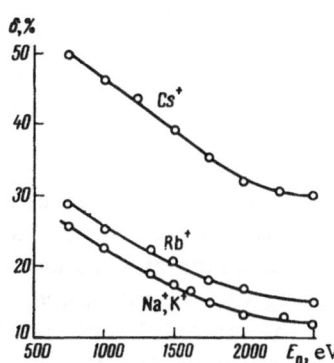

Fig. 9.10. Proportion of negative ions of tungsten-containing compounds in the total sputtering as a function of the energy of the bombarding ions.

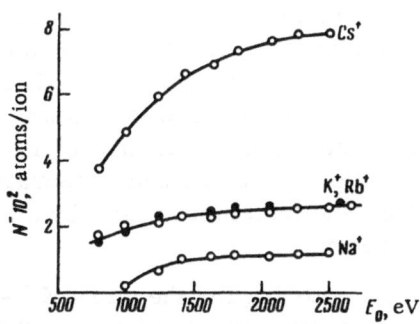

Fig. 9.11. Coefficient of negative-ion sputtering of tungsten as a function of the energy of the bombarding ions.

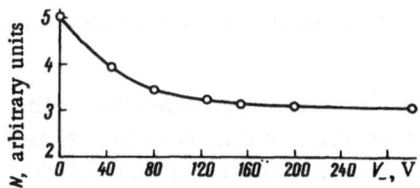

Fig. 9.12. Amount of tungsten settling in the collector as a function of the negative potential of the latter (primary Cs^+ ion energy 2500 eV).

oxygen were ejected in the form of negative ions. From the data presented in Fig. 9.7, we may determine the coefficient of sputtering in the form of negative ions N^-. For this we must plot a curve based on the difference between the ordinates of curves 1 and 2. The energy dependence of N^- on bombarding tungsten with Cs^+ ions is shown in Fig. 9.9.

Figures 9.9 and 9.10 give the results of analogous experiments in which tungsten was bombarded with Rb^+, K^+, and Na^+ ions. We see from these figures that, for vacuum conditions between 10^{-6} and 10^{-7} mm Hg, on bombarding a clean, cold tungsten surface with alkali ions a considerable proportion of the sputtered material contains negative ions of chemical compounds formed between the tungsten and target-surface contaminants (Fig. 9.10). For example, in the Cs^+-W system, $\delta \sim 50\%$ for 800-eV primary ions. On increasing E_0, the value of δ falls for all the ions (Fig. 9.10). In the energy range 800–2000 eV, δ varies by a factor of about two in each case.

The value of N^- rises slightly with increasing E_0 for $E_0 < 1600$ eV. For $E_0 > 1600$ eV, the value of N^- ceases depending on the primary-ion energy except in the case of Cs^+. The $N^- = f(E_0)$ curve for Cs^+ still fails to saturate. In the energy range considered for the primary Cs^+ ions, N^- varies by a factor of about two.

If the retarding potential V_k is varied between 0 and -300 V, we obtain a special kind of "stopping curve" for the negative tungsten-containing ions [727]. The results of an experiment with 2500-eV Cs^+ ions are shown in Fig. 9.12. Of course, the accuracy of the experiment ($\sim 10^{-9}$ g) will not allow us to establish the limiting energies of the negative ions ejected. However, we may conclude that the number of these ions with energies of $E_0 > 100$ eV is negligibly small.

The results presented in this section indicate that under the conditions described the sputtering of pure, cold tungsten cannot be studied. This assertion may clearly be extended to bombardment with any ions. Experiments with alkali ions are only convenient in this respect in that the presence of an alkali coating created by the beam itself reduces the work function of tungsten, and hence a large fraction of the sputtered tungsten-containing compounds leave the surface in the form of negative ions. Bombardment of tungsten with any other ions will also be accompanied by the ejection of tungsten-containing compounds, but in these cases the compounds will leave the surface mainly as neutral particles.

In sputtering with alkali ions the presence of a negative-ion sputtering component may serve as a convenient criterion for estimating the cleanness of the surface bombarded.

§4. Sputtering of Solids under

Pure Surface Conditions

The difficulties arising in the study of cathodic sputtering as a result of the ejection of contaminants may clearly be to a large degree eliminated in the sputtering of a target heated to a high temperature. Accordingly it seemed desirable to study sputtering at various target temperatures. The variation in the coefficient of cathodic sputtering with temperature was studied in some of our papers [283]. However, the evaporation of the material could not be taken into account when using the apparatus shown in Fig. 9.1. Special experiments by Shustrov, Khasanov, and Ayukhanov [683, 727] showed that at temperatures up to 1200°K for tungsten and 1450°K for tantalum there was no transfer of target material to the collector as a result of evaporation. When studying cathodic sputtering at target temperatures of 1200-1500°K for tungsten and 1450-1850° K for tantalum, evaporation of metal-containing compounds did occur, but this could easily be taken into account by successively carrying out experiments with evaporation only and with evaporation accompanied by sputtering. Since the evaporation was of an equilibrium nature in these temperature ranges, the amount of sputtered material was found as the difference between the amounts of material received by the collector in these two experiments. For higher temperatures the rate of evaporation fell with time at constant temperature. The nonequilibrium processes observed may be explained by the evaporation of metal-containing coatings from the surface of the metal [683, 727]. At such temperatures cathodic sputtering can only be studied by simultaneously observing evaporation by itself and combined evaporation and sputtering. Experiments on the sputtering of heated tungsten and tantalum targets by all the alkali ions were carried out in apparatus of two types of construction.

The temperature dependence of the coefficient of cathodic sputtering for the pairs K^+-W, Na^+-W, and Li^+-Ta is shown in Figs. 9.13-9.15. Starting from a temperature of over 900°K, the coefficient of cathodic sputtering increases with rising temperature for all the systems studied and passes through a maximum lying between 1100 and 1300°K. The behavior of the sputtering coefficient at higher temperatures will be described subsequently. A comparison of the $N = f(T)$ curves obtained for the same primary-ion energy shows that this process certainly depends on the nature of the ion. For example, in the case of 1600-eV K^+ ions the ratio N_{max}/N_0 of the maximum sputtering coefficient to the coefficient at room temperature equalled 1.7, while for Na^+ ions it equalled 3.3.

It is natural to expect that, owing to the sputtering of metal-containing compounds, the energy dependence of the coefficient of cathodic sputtering will differ for different target temperatures. Experiments show that in the energy range studied not only the magnitude of the sputtering coefficient, but also the shape of the $N = f(E_0)$ curve change with changing temperature. The corresponding curves are shown for the K^+-Ta system in Fig. 9.16. We see from the figure that the curves are distorted for low values of energy E_0. For values of $E_0 > 600$ eV the relationship becomes linear and the quantity $\Delta N/\Delta E_0$ depends little on temperature.

These curves illustrate the result of two simultaneous processes: the ejection of metal atoms and the sputtering of the metal in the form of chemical compounds with sorbed gases. In order to explain the laws governing the sputtering of the clean metal surface we must eliminate the possibility of the metal atoms being lost by the second process. This process will only fail to appear when there are no metal-containing compounds on the surface. This requirement is satisfied for a fairly high target temperature if the vapor tension of these compounds is greater than that of the metal itself. Experiments on sputtering at high temperatures showed that, starting from some temperature T_H, the sputtering coefficient ceases depending on temperature

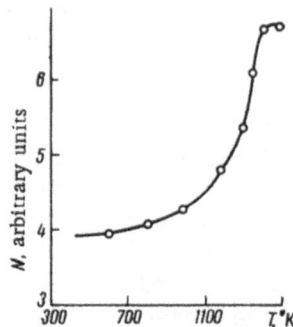

Fig. 9.13. Sputtering of tungsten as a function of temperature on bombarding with 1600-eV K⁺ ions.

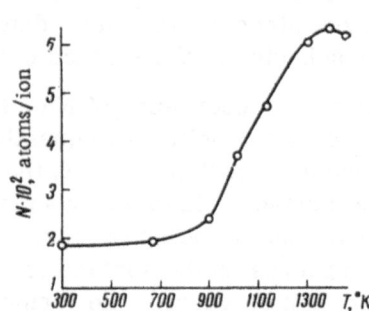

Fig. 9.14. Sputtering of tungsten as a function of temperature on bombarding with 1600-eV Na⁺ ions.

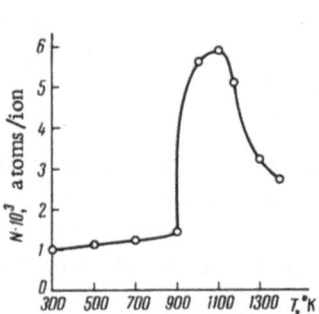

Fig. 9.15. Sputtering of tantalum as a function of temperature on bombarding with 500-eV Li⁺ ions.

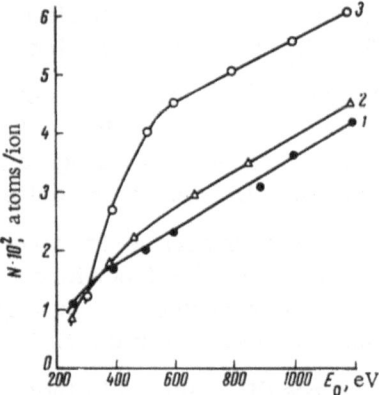

Fig. 9.16. Energy dependence of the coefficient of cathodic sputtering of tantalum by K⁺ ions at temperatures of: 1) 300; 2) 1000; 3) 1250°K.

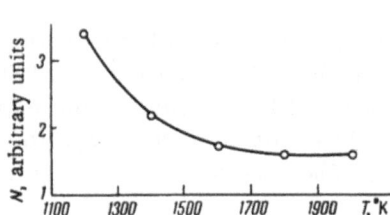

Fig. 9.17. Coefficient of cathodic sputtering of tungsten by 2600-eV Na⁺ ions as a function of target temperature.

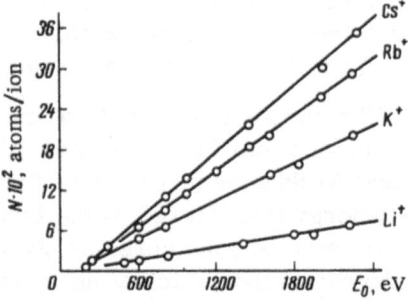

Fig. 9.18. Energy dependence of the coefficients of cathodic sputtering of a clean tungsten surface on bombarding with ions of the alkali elements.

(Fig. 9.17). The temperature T_H is almost the same for all the alkali ions. This would really be expected, since T_H is mainly determined by the nature of the metal-containing compounds and their adhesion to the surface of the metal.

From the dependence of N on target temperatures for $T > T_H$ we may draw two conclusions. First, under the conditions in question, it is a clean metal surface which is being sputtered. The results enable assertion, in fact, that for $T > T_H$ there are no metal-containing compounds on the target surface. Chemically active gases, for example oxygen, falling on the surface from the main volume of the apparatus, enter into reaction with the target material, but the lifetime of the compounds on the surface is very short. For $T > T_H$, we shall record such compounds as evaporation products. The formation of these and their leaving the surface will not affect the sputtering process. On the other hand, for the high temperatures considered, the surface is also free from coatings which would develop on the cold target as a result of the adsorption of alkali atoms from the ion beam itself.

Secondly, in the temperature range above T_H, the coefficient of cathodic sputtering of the clean metal surface will not depend on target temperature within the limits of experimental accuracy. This assertion, experimentally verified for $T > T_H$, may evidently be extended to the range $T < T_H$ if the target surface remains clean under this condition.

Using the information so obtained, we carried out an experiment on the sputtering of a hot tungsten target by Li^+ ions. As already indicated, no reproducible results could be obtained on sputtering a cold tungsten surface with Li^+ ions. Under the new conditions the experiments with Li^+ ions were quite reproducible.

Analogous experiments were also carried out for the Cs^+—W, Rb^+—W, and K^+—W systems. In all cases the dependence of the coefficient of cathodic sputtering on the primary-ion energy was represented by a straight line between 300 and 2500 eV. The straight line $N = f(E_0)$ had a maximum slope for the Cs^+—W system and a minimum slope for Li^+—W (Fig. 9.18).

The results obtained in these experiments characterize the sputtering of a clean tungsten surface. Comparison of these results with those relating to the sputtering of a cold degassed surface lead to the conclusion that the cleaning of the bombarded surface has the following two results: 1) the coefficients of cathodic sputtering N are increased; for example, in the case of Rb^+ at an energy of 2200 eV the coefficient N increased by 60%; 2) there is a considerable change in the way in which the coefficient of cathodic sputtering varies with the energy of the bombarding ions.

In the old form of apparatus (Fig. 9.1) the energy dependence of the sputtering coefficient was obtained on bombarding clean tantalum with Cs^+ ions. For primary ions with energies between 0.2 and 1 keV the relationship was also linear.

The absolute values of the coefficients of cathodic sputtering obtained on bombarding a clean surface may be compared with the fundamental theoretical view of Arifov [283] and Shustrov and Ayukhanov [684], according to which the ejection of material takes place at the expense of the energy transferred to the target by the incident ion. The interaction of the ion with a target atom may be considered as the collision of elastic spheres. Then the average energy $\overline{E}_1$ transferred to the lattice atom by the ion equals

$$\overline{E}_{1,\,\text{rec}} = \int\limits_0^\pi E_\vartheta \, dw\,(\vartheta) = \frac{2m_1 m_2}{(m_1 + m_2)^2} E_0,$$

where m_1 and m_2 are the masses of the target atom and the ion E_0 is the energy of the incident ion, ϑ is the scattering angle in the center-of-mass system, E_ϑ is the energy given to the target atom by an ion scattered through an angle ϑ, and $dw(\vartheta)$ is the probability of scattering at an angle ϑ. If we consider that sputtering takes place as a result of the energy received by the lattice in the first collision [516], then the coefficients of sputtering should be proportional to the quantities $m_1 m_2 / (m_1 + m_2)^2$. This conclusion does not agree with the experimental results. One possible reason for the disagreement between theory and experiment is the coarseness of the theoretical approximation, according to which it is considered that sputtering takes place as a result of the energy transferred to the lattice in the first collision. This error should be revealed particularly sharply in the case of light ions.

Let us calculate the energy received by the target with due allowance for twofold collisions between the ions and lattice points. Since the depth of target into which the ion transfers energy is considerable, secondary collisions should be considered in the case in which the ion is reflected backward in the first collision. The average energy given to a target atom by the ion in a second collision equals

$$\bar{E}_{2,\text{rec}} = \frac{a(1-a)}{(1+a)^2} \left[1 - \frac{a}{(1+a)^2} (2-a)^2 \right] E_0,$$

where we have introduced the notation $a = m_2 / m_1$. The average energy transferred to the lattice by the ion in the two collisions under consideration equals

$$\bar{E}_{\text{rec}} = \bar{E}_{1,\text{rec}} + \bar{E}_{2,\text{rec}} = x E_0,$$

where

$$x = \frac{2a}{(1+a)^2} \left\{ 1 + \frac{1-a}{2} \left[1 - \frac{a}{(1+a)^2} (2-a)^2 \right] \right\}.$$

In this approximation the sputtering coefficients should be proportional to x. Experiment confirms this conclusion.

The correlation established between the coefficient of cathodic sputtering and the energy $\bar{E}_{\text{rec}}$ received by the target from the incident ion in two successive collisions with lattice atoms enables us to draw qualitative comparisons with the momentum mechanism of sputtering. For normal incidence of the ion, sputtering may take place as a result of at least two collisions. In the first collision of the ion two fast particles moving through the lattice arise: the scattered ion and the recoil atom. If they have fairly high energies these two particles may lead to sputtering in their subsequent collisions with lattice atoms. From this point of view it is reasonable to look for a relation between the coefficient of cathodic sputtering and the total energy transferred to the lattice by an ion in two successive collisions.

CHAPTER 10

SECONDARY ELECTRON—ELECTRON EMISSION

§1. Brief Review of Experimental and Theoretical Investigations into Secondary Electron—Electron Emission

The secondary electron emission of metals under the influence of electrons, first discovered by Austin and Starke [5] in 1902, has taken on considerable practical importance in recent years.

Secondary electrons are usually divided into three groups: 1) electrons belonging to the emitter and leaving it as a result of excitation by the primary electrons; this is true secondary emission (n_1); 2) primary electrons elastically scattered by the surface of the solid (n_2); 3) primary electrons inelastically scattered within the solid (n_3). Corresponding to this we have the following quantitative characteristics of secondary electron emission: the coefficient of true secondary emission

$$\delta = \frac{n_1}{n_0}, \tag{10.1}$$

the coefficient of elastically scattered electrons

$$\eta = \frac{n_2}{n_0} \tag{10.2}$$

and the coefficient of inelastically scattered electrons

$$r = \frac{n_3}{n_0}. \tag{10.3}$$

The sum of these coefficients

$$\sigma = \frac{n}{n_0} = \delta + \eta + r \tag{10.4}$$

we call the coefficient of secondary electron emission. Here n_0 is the number of primary electrons causing the appearance of electrons n_1, n_2, and n_3.

However, until now there has been no strict criterion for this classification. It is considered that for fairly high energies of the primary electrons ($E_0 > 2000$ eV) all the secondary electrons with energies lower than 50 eV are truly secondary, while all the secondary electrons with energies higher than 50 eV belong to the elastically and inelastically scattered class. In recent years the greatest attention of research workers has been devoted to studying the energy distribution of secondary electrons, particularly the characteristic electron energy losses [361, 404, 479]. The special interest in this is due to the fact that an explanation of the nature of the characteristic energy losses will offer the possibility of making a more specific judgment on the electron structure of solids. Studies in this field provide useful information for electron microscopy, cathodoluminescence, and semiconductor electronics. However, these investigations are still insufficient for a complete understanding of the mechanism of inelastic scattering, which is an indivisible part of secondary electron emission.

A number of theoretical and experimental investigations have been made in the low-energy primary-electron field in order to elucidate the mechanism of secondary electron emission [21, 34, 125, 140, 168, 216, 248, 270, 362-364, 414-416, 446]. However, the results of the early experiments relate to contaminated surfaces.

It was shown by Shul'man and Myakinin [248, 270, 362] and also Myers [216] that for energies E_0 not exceeding the work function of the metal only elastic scattering of the electrons occurs, while for larger energies truly secondary and inelastically scattered electrons start appearing. With increasing E_0 the number of elastically scattered electrons first falls sharply and then the fall becomes slower, in accordance with the theoretical calculations of McColl [140].

For primary-electron energies up to a few tens of electron volts the reflection coefficient varies between 0.05 and 0.35 for pure metals and between 0.3 and 0.5 for metals covered with films reducing the work function. This is important for thermionic emission.

The energy spectra of secondary electrons are given in [248, 270, 362, 417]. These spectra have two maxima corresponding to truly secondary and elastically scattered electrons. As E_0 rises in the range below 15 eV, the low-energy maximum moves in the direction of higher energies. For $E_0 > 15$ eV, the position is independent of E_0, in accordance with theory [140].

Energy Distribution of Secondary Electrons

A detailed study of the secondary-electron energy distribution offers the prospect of understanding the manner in which primary electrons interact with the electrons of the metal more deeply.

Sternglass [299] used the retardation method to study the secondary-electron energy distribution for carbon, platinum, tantalum, and iron in the range $E_0 = 200$-2000 eV. Inelastically scattered electrons tended to dominate. The energy spectrum of these electrons was independent of E_0 between 500 and 2000 eV. For the targets studied the energy of the inelastically scattered electrons rose sharply with increasing atomic number of the target.

Shul'man [365] studied metals and dielectrics by the single-pulse method and showed that the energy distribution for metal targets was independent of E_0 over a wide range of E_0 values. The width of the maximum of truly secondary electrons for dielectrics was much narrower than for a metal. For $E_0 > 200$ eV, the most probable energy of the secondary electrons from metals was of the order of a few electron volts, while for dielectric targets it was close to zero.

Bronshtein and Kovalenko [609] studied the secondary electron emission for beryllium, aluminum, nickel, mercury, and lead films over the range $E_0 = 200$-5000 eV under good vacuum conditions and showed that the energy of the primary electrons had no effect on the shape of the energy distribution of inelastically scattered electrons.

Bronshtein and Fraiman [563] studied the relation between the secondary emission properties of many substances and the atomic number of the element Z. In the first half of each period of the Mendeleev Table, δ rose with increasing Z, and in the second it fell. The rise in δ with increasing Z and the similarity between the δ values for neighboring elements (metals and semiconductors) indicated that not only conduction electrons but also electrons of the lower filled bands took part in the formation of slow secondary electrons.

The inelastic scattering of electrons and the secondary emission of barium, bismuth, silver, and platinum deposited on various substrates in a vacuum better than 10^{-7} mm Hg were studied by Bronshtein and Segal' [480, 481]. In the range $E_0 = 0.1\text{-}3.6$ keV the inelastically scattered electrons played a considerable part in the secondary electron emission of metals. The depth of penetration of the primary electrons varied on an E^n law, where $n \approx 1.2\text{-}1.5$ for different metals, while the depth of λ from which secondary electrons with energies ≥ 15 eV escaped was no more than 10 atomic layers.

The effect of the work function of the surface on the secondary electron emission was studied in [482] for the case of barium adsorbed on tantalum. As φ fell from 4.2 to 2.6 eV the maximum of the slow secondary electrons moved in the low-energy direction, in agreement with the result of Kadyshevich [172].

Bronshtein and Shuchinskii [694a] used the spherical-condenser method to study the energy distribution of secondary electrons from films of elements belonging to the fourth period of the Mendeleev system (potassium, calcium, titanium, chromium, iron, nickel, copper, zinc, germanium, and selenium). Films of these metals were obtained on previously-cleaned metallic substrates. The secondary-electron energy distribution curves were independent of E_0 in the range $E_0 = 500\text{-}2000$ eV for all the substances. In the first half of the period the most probable energy of the slow secondary electrons moved in the high-energy direction and the half width of the curve maximum increased, while in the second half the most probable energy moved in the low-energy direction and the half width of the maximum diminished.

Morozov and Shul'man [365a] studied the component of elastically scattered electrons with a well-degassed tungsten target in the range $E_0 = 100\text{-}2000$ eV in sealed vacuum systems at about $(2\text{-}5) \cdot 10^{-9}$ mm Hg. The measurements showed that for a tungsten target at a temperature of 1275°K the coefficient r reached 2-3% and the $r = f(E_0)$ curve had a maximum at $E_0 = 600$ eV. On cooling the target to room temperature, this curve exhibited some excellently reproducible additional maxima, vanishing on heating the target. The position of the main maximum depended on the atomic number of the metal.

Nakhodkin and Mel'nik [642] used a cylindrical condenser to make a comparative study of the secondary and photoelectron spectra excited in sputtered silver, germanium, and copper films by primary electrons and soft x-ray quanta with energies of about 1.5 keV. In contrast to the secondary-electron spectra, the photoelectron spectrum had a considerably smaller number of fast electrons with an energy close to the energy of the quantum, and a considerably larger number of Auger electrons.

A large number of investigations [272, 273, 302, 330, 331, 390, 405, 447, 483] have been devoted to a study of the secondary electron emission of metals and their compounds for high primary-electron energies (from 20 keV to 235 MeV).

Characteristic Energy Losses of Electrons in Solids

The question of the characteristic energy losses in solids first arose in 1924 in the investigations of Becker [28]. Studying the energy distribution of reflected electrons with $E_0 = 200$ eV, Becker observed a group of electrons which had lost a considerable proportion of their original energy. These losses Becker associated with the transition of the metal electrons from the deeper bands in the quasicontinuous spectrum of the conduction band under the influence of the primary electrons.

Further investigations into characteristic losses were carried out by Rudberg [111] when studying the secondary-electron emission distribution for silver, gold, copper, calcium, barium, calcium oxide, and barium oxide. Rudberg showed that the losses depended on the nature of the solid and were independent of the energy of the bombarding electrons up to $E_0 = 1000$ eV.

Ruthemann [160, 173] studied the penetration of electrons through thin films of silver, gold, beryllium, and Al_2O_3 and found that electrons with energies from 2-8 keV underwent discrete energy losses (from 10 to 20 eV), the value of which depended on the material of the target. The values of the later losses were multiples of the initial, lowest energy loss.

The energy spectra of secondary electrons from beryllium, aluminum, nickel, copper, barium, and platinum and their oxides were studied by Lander [271], using a cylindrical condenser with an aperture angle of 90°. In the low-energy part of the spectrum there were some steps which, according to the author, were due to Auger electrons. In the high-energy part of the secondary-electron spectrum of the metals under consideration, characteristic loss peaks peculiar to the metal in question appeared. The results for the characteristic losses of barium were qualitatively similar to those of Ruthemann [160]. However, Lander pointed out that the plasma-transition mechanism which Ruthemann used to explain his results was incapable of explaining his own. In addition to this, characteristic-loss peaks qualitatively agreeing with the results of Rudberg [111] for copper, barium, and barium oxide were obtained.

The analysis of characteristic losses at high primary-electron energies, up to 100 keV, became possible as a result of the creation of an electrostatic velocity analyzer with a high resolving power by Möllenstedt [200, 201, 251]. Marton and Leder [303, 304] used this analyzer to study the characteristic losses of a number of elements (gold, aluminum, silver, beryllium, magnesium, zinc, nickel, cadmium, indium, antimony, lead, silicon, and tellurium) and also certain compounds (SiO_2, TeO_2, MgO, PbS, Sb_2S_3); they found that the secondary-electron energy spectra of a metal and its compounds were very much alike. The energy losses of the electrons in the metals were slightly lower than in their chemical compounds.

Gornyi [300, 301] developed a method of electrical differentiation for the V/A characteristic of the secondary currents obtained in the field of a spherical condenser and established the existence of fine structure in the low-energy part of the secondary-electron spectrum from a single-crystal copper surface. This method was also used [366] to observe characteristic losses in the inelastic reflection of primary electrons, which the author ascribed to interband transitions.

The spherical-condenser method was used by Shul'man and Farbshtein [328] with nickel and molybdenum targets to study the secondary-electron energy spectra, which were obtained directly on an oscillograph screen by electrical differentiation. Loss peaks with energies of 5.5, 11.5, and 17 eV were observed for molybdenum.

Shul'man and Fridrikhov [329] studied inelastic scattering of electrons in NaCl and KCl single crystals. The primary beam was directed in periodic pulses perpendicular to the plane of the NaCl and KCl crystals (heated to 300°C); this eliminated the possibility of the surface becoming charged. The spherical-condenser method was employed to analyze the electron velocities. The resultant electron spectra consisted of a peak of elastically scattered electrons, a peak of truly secondary electrons, and several intermediate maxima belonging to inelastically scattered electrons. The energies corresponding to these maxima agreed satisfactorily with the calculated energy values obtained from the Vyatskin theory [217].

Watanabe [332] studied the characteristic losses as a function of the film thickness and showed that with increasing thickness the number of characteristic losses also became greater.

Harrower [352, 367] studied the energy distributions of secondary electrons from molybdenum and tungsten for a primary-electron energy range up to 2000 eV, using a 127° electrostatic velocity analyzer of the Hughes-Rojansky type [54]. During the measurements the pressure remained under $1 \cdot 10^{-10}$ mm Hg. The author was able to observe fine structure in almost all regions of the secondary-electron energy spectrum. In Harrower's opinion, the characteristic-loss peaks were associated with the transfer of energy to the electrons in the outer shells of the target atoms, while the maxima in the low-energy part of the spectrum were evidently due to Auger processes.

Losses equal to multiples of 14.9 and 10.7 eV, respectively, were found for aluminum and magnesium by Blackstock et al. [333]; for copper there were single losses of 22.6 eV. The experiments failed to establish with any certainty whether energy were transferred to free or individual bound electrons, since the energy losses were of the same order in both cases.

Good agreement was observed by Lederer et al. [370] between the characteristic energy losses of electrons in solids and the fine structure in the short-wave part of the x-ray absorption spectrum. In the majority of cases the value of the characteristic loss was inversely proportional to the square of the lattice constant.

Sternglass [371] considers all the observed characteristic losses as the result of a combination of three main types of independent atomic processes: ionization by the removal of electrons from the shell following the valence level, and the ionization and excitation of valence electrons.

Robins [527, 573, 608], studying the characteristic energy losses of scattered electrons from many films deposited on solid substrates, came to the conclusion that the main mechanism producing characteristic energy losses for metals not belonging to the transition and noble classes was the excitation of plasma oscillations. In the spectra of transition and noble metals interband transitions acquired a considerable significance.

Characteristic losses were also studied in [368, 369, 372-376, 406-408, 449, 450, 484, 485, 572].

Angular Dependence of the Secondary-Electron Energy Spectrum

Ternbull and Farnsworth [126], and also Reichertz and Farnsworth [202] studied the secondary-electron energy spectrum as a function of the escape angle θ for a copper single crystal, using the Hughes—Rojansky cylindrical-condenser method. In these investigations the primary beam, up to 200 eV in energy, fell normally on the (100) plane. The angles corresponding to the maximum intensities of elastic scattering and characteristic losses coincided.

Kushnir and Krumin [162] studied the energy distribution of truly secondary electrons as a function of θ for silver and molybdenum at $E_0 = 200-600$ eV. With increasing θ, the maximum corresponding to slow secondary electrons moved in the direction of high energies; the angular distribution of secondary electrons had sharp maxima and minima.

Jonker [227] studied the angular dependence of secondary-electron energy distribution for electrons ejected from polycrystalline nickel on bombarding with 25-, 100-, and 450-eV primary electrons, using the retardation method at a pressure of 10^{-7}-10^{-10} mm Hg. The incident angles φ of the primary electrons were 0, 30, and 45°, respectively. The distributions of slow secondary electrons and electrons of medium energies obeyed a cosine law; elastic scattering occurred mainly to the rear (large maximum) and at the specular angle (small maximum). With increasing E_0 the small maximum diminished and the large one increased.

In another paper [275], Jonker used soot as a target. The angular distribution of secondary electrons with medium velocities arising from the soot was considerably drawn out in the direction of incidence of the primary electrons, while the angular distribution of slow electrons tended toward the cosine law with increasing E_0 and was practically independent of the direction of the primary beam.

Fine structure of the angular distribution was observed by Timofeev and Afanas'eva [120] in studying the angular distribution of secondary electrons from the (100) face of copper and nickel single crystals.

Bronshtein and Segal' [480] used a spherical condenser to study the dependence of the coefficients δ and η on the incident angle φ for platinum, tungsten, bismuth, silver, and beryllium vacuum-deposited at about 10^{-7} mm Hg. Both coefficients rose with increasing angle of incidence. The $\delta(E_0)$ relation passed through a maximum, while $\eta(E_0)$ rose first rapidly and then more slowly with increasing E_0.

Bronshtein and Denisov [694] studied the relative coefficient of inelastically scattered electrons $\eta_{rel} = \eta_\alpha / \eta_0$ (where η_α and η_0 are the coefficients of inelastically scattered electrons for incident angles of α and 0) and the effective depth of emergence of inelastically scattered electrons as functions of E_0 and the angle of incidence for a number of metals. The work was carried out over the energy range $E_0 \approx 0.1$-5 keV. The incident angle was varied in steps of 15° from 0-60°. The beryllium, aluminum, silicon, titanium, nickel, and lead targets used were obtained by vacuum evaporation at about 10^{-7} mm Hg. With increasing angle of incidence and an energy of 100-200 eV, there was a considerable rise in the relative inelastic-scattering coefficient η_{rel}, the $\eta_{rel}(E_0)$ curves for light and heavy materials being of a different character. This difference was explained by the authors as being due to different inelastic-electron-scattering mechanisms in the case of light and heavy elements.

Lenard [307] showed that for beryllium, aluminum, and gold films deposited on collodion the maximum number of electrons experiencing the first characteristic loss and the interference maximum of elastically scattered electrons corresponded to the same scattering angle. In addition to this, it was shown that the ratio of the intensity of the elastically scattered electrons to that of the inelastically scattered electrons remained constant for all scattering angles.

It was found by Marton et al. [309, 575] and Watanabe [335] that up to a scattering angle of about $0.5 \cdot 10^{-2}$ rad, the intensity of the elastically scattered electrons was greater than that of the electrons which had experienced characteristic losses. For scattering angles greater than the value indicated, the dominant role was taken by electrons of the characteristic-loss type. This relationship was preserved until the position of the diffraction maximum had been reached.

A study of angular distribution based on severely diaphragmed electron beams ($E_0 = 6$ keV) was undertaken by Fridman [308, 377] for selenium layers. In order to eliminate the influence of multiple scattering, very thin layers were employed. Experiment showed that the angular distribution of the inelastically scattered electrons with energy losses fell 100 times faster with increasing θ (in the range 10^{-3} to $2 \cdot 10^{-3}$ rad) than the theoretically calculated distribution of elastically scattered electrons.

Geiger [610] studied the spectrum of characteristic energy losses for electrons scattered at various angles from a beryllium layer 300 Å thick. The curves obtained for scattering angles of $2.1 \cdot 10^{-3}$ to $1.5 \cdot 10^{-2}$ rad had two characteristic-loss maxima at energies of 15.3 and 30.6 eV.

Secondary Electron Emission of a Metal in Various States of Aggregation

It is interesting to study the secondary-electron distribution of metals in different states of aggregation in order to throw some light on the mechanism of secondary-electron excitation.

Trelor and London [127] measured the secondary-emission coefficient σ for nickel and cobalt targets near the Curie point and showed that the transformation from the hexagonal to the face-centered cubic lattice taking place at this point had no effect on the variation in σ.

No substantial changes in $\sigma(E_0)$ were found either by Brophy [164] or by Kushmir et al. [143] as gallium, lead, and mercury passed from the solid to the liquid state.

According to Morozov [163], σ changed sharply at the melting point for lead and tin. On passing into the liquid state the σ of lead and tin rose, while that of bismuth remained almost constant. The changes in σ were no greater than 10%. However, Brunning's results [448] were almost the reverse.

The dependence of the secondary electron emission on the state of aggregation of tin, copper, and germanium was studied by Bol'shov and Seleznev in [378] by the spherical-condenser method. It was found that on heating from room temperature to 232°C for tin and to 1033°C for copper, the maximum secondary emission coefficient σ_{max} changed by no more than 1%. In the case of germanium, on heating from room temperature to 959°C, σ_{max} fell by about 5–6%. It was also found that on melting σ experienced a jump, the direction of this being the same for all values of E_0 between 100 and 1500 eV. The values of σ_{max} for the solid and liquid substance roughly corresponded to the same primary-electron energy E_0. On melting, the value of σ for copper fell by 5%, while in the case of tin and germanium σ increased by 14 and 9%, respectively.

A study of the secondary-electron energy distribution for tin showed that the rise in σ on passing into the liquid state was mainly due to an increase in the yield of truly secondary electrons. The character of the secondary-electron distribution remained constant on melting.

Bol'shov and Zar ibin [487] measured the coefficient δ and η separately for indium and lead in the solid and liquid states and showed that as E_0 rose from 100 to 2000 eV, δ was greater for the solid sample than for the liquid, while the opposite was the case for the coefficient η (29% for lead and 13% for indium).

The characteristic losses of primary electrons with initial energies of 25 keV on passing through solid, liquid, and gaseous mercury and also through solid and gaseous silver were measured by Beorch et al. [611]. There was a great similarity in the dispositions of the characteristic losses and in the relative peak intensities for the various phases.

Theory of Secondary Electron Emission

Existing theories of secondary electron emission developed on the basis of various models of the solid state [448] explain individual aspects of the phenomenon. However, up to the present time there have been no theories satisfactorily explaining secondary electron emission as a whole.

In the theory of Kadyshevich [136, 152, 172] and Baroody [218] the Sommerfeld model was used to calculate the transition energy of conduction electrons on the basis of classical laws. Baroody [218] and Jonker [249] showed that it was possible to obtain a certain universal curve for the number of secondary electrons as a function of the primary-electron evergy.

Kadyshevich [136, 152, 172] started from the idea that the secondary electrons were formed in the course of collisions between primary and free electrons. A certain proportion of the secondary electrons, after suffering a fair number of elastic collisions might ultimately leave the surface. According to Kadyshevich's calculations, when the primary electrons fell perpendicularly to the surface, the coefficient of secondary electron emission was expressed by the relation

$$\sigma = \frac{0.13\,\lambda_1\lambda_2^2}{IE_0\,(\lambda_1 + 0.56\,\lambda_2)}, \tag{10.5}$$

where λ_1 and λ_2 were the average ranges of the primary and secondary electrons for all kinds of collisions within the metal, and l was the mean free path relative to elastic collisions between the secondary electron and the ionic lattice.

Vyatskin [142] also started from the Sommerfeld model of a metal; for the energy dependence of the coefficient of secondary electron emission he obtained a saturation law given by the expression

$$\sigma = AW\mu^{1/2}\varphi^{-3/2}\arctan(2E_0/\varphi)^{1/2},\tag{10.6}$$

where W is the height of the barrier, φ is the work function, and μ is the Fermi energy. This theory explained the reason for the small values of the emission coefficients in the case of pure metals.

Frölich [75], Frölich and Wooldridge [141], and Dekker and Van der Ziel [250] made a quantum-mechanical study of the collision between a primary electron and a single metal electron. In order to explain the emission of electrons it was supposed that the field of the crystal lattice had an influence on the electrons situated in the metal. The effect of the lattice was such that the electron was able to move in a direction opposite to that of the primary electrons.

In Izmailov's theory of secondary electron emission [581], the contribution of primary electrons scattered at various depths (losing their energy while returning to the surface) to the emission is considered. After a number of assumptions, Izmailov considers the dependence of the coefficient η on the energy E_0 for normal incidence and determines the energy distribution of the inelastically scattered electrons. The effect of inelastic scattering on the emission of slow secondary electrons is then considered. The way in which the quantities $\sigma(E_0)$ and $\eta(E_0)$ vary with energy depends greatly on the power of the energy term in the laws giving the energy losses and the scattering of the primary electrons.

Bethe and Bloch [65, 76, 85–87] and also Massey [252] considered the retardation of electrons in the medium on the Born approximation. Massey [252] considered the effect of the interatomic bonds on the scattering of electrons in solids in a simple manner.

Bohm and Pines [274, 361] studied the interaction of fast electrons with a Fermi gas and showed that for small changes of momentum the transfer of energy could be considered as the result of the excitation of collective plasma oscillations. The amount of energy transferred was determined by the sum of the elementary energy losses $h\omega_p$, where ω_p was the frequency of the plasma oscillations

$$\omega_p^2 = \frac{4\pi e^2}{m}N'.\tag{10.7}$$

The influence of the material was expressed in terms of the density of the electron gas N'.

Vyatskin [217, 253, 306, 451, 452] developed a theory of characteristic energy losses for electrons in solids, based on interband and intraband transitions.

For paired Coulomb interaction between the incident electrons and the electrons of the lattice, the latter being considered on the single-electron weak-coupling approximation, the transitions are divided into two forms (free transitions, interband p-transitions). For an interband p-transition the characteristic mean energy transferred to a metal electron is given by the equation

$$\mathscr{E}_n = \frac{\hbar^2}{2m}(n^2 + 2qn + q^2),\tag{10.8}$$

where **q** is the difference in the wave vectors and $\hbar$ is Planck's constant.

For a cubic lattice,

$$\mathscr{E}_n = \frac{\hbar^2}{2m}\left(\frac{2\pi n}{a}\right)^2 \approx \frac{150n^2}{a^2 A^2}\,\text{eV}, \tag{10.8a}$$

where $n^2 = n_1^2 + n_2^2 + n_3^2$, n_i are the indices of the crystal, and a is the constant of the cubic lattice. It is also shown that the losses of interband p-transitions are of the same order of magnitude as the losses in collisions with free electrons.

Vyatskin [452] offered a number of arguments in favor of the fact that interband transitions had a greater probability than plasma transitions [274, 361]. In addition to this it was shown that interband transitions took place over the whole range of scattering angles and plasma transitions only over that of small scattering angles (long-range forces).

We see from the foregoing review that at the present time the most interesting yet contraversial question is that of the characteristic losses of energy by electrons in matter. A number of authors [299, 367, 609] make an attempt to associate the secondary-electron energy spectrum with the atomic number Z of the substance under consideration. Sternglass [299] indicates a weak dependence of the number of inelastically scattered electrons on Z. Bronshtein and Kovalenko [609] analyze the energy spectrum obtained by differentiating stopping curves and conclude that there is no such dependence on Z. However, Harrower [367], starts from the fact that elements with a similar electron configuration should have a secondary-electron spectrum of the same form. We thus see that there is no universal opinion regarding the dependence of the secondary electron spectrum on Z.

Some papers by Farnsworth [126, 202] are the only ones in the field of low primary-electron energies in which the development of the argument enables us to understand the interaction of electrons with solids more deeply.

We also see from the review that there are two views as to which electrons in a solid produce characteristic losses. A number of research workers [161, 217, 253, 274, 301, 306, 328, 335, 361, 363, 366, 451, 452] consider that the losses are due to interaction with free electrons; others [111, 180, 271, 370, 126, 202, 371] hold to the view that the energy losses occur in transitions of electrons from low-lying bands or levels to higher allowed unoccupied states.

In some cases the Coulomb couplings between the lattice electrons are so considerable that the portion of energy transferred is distributed among a multitude of lattice electrons, each of which increases its energy by a small amount. As a result of this, the electrons in the solid acquire the capacity to execute collective oscillations at a high frequency, and this leads to plasma transitions.

It was shown by Watanabe [368] and Geiger [610] that the angular dependence of certain losses coincides with the theoretical predictions of Bohm [274] and Pines [361] of collective oscillations. However, the results of [307-309] are not in favor of this view. Furthermore, no difference was found by Marton et al. [575] between the results obtained for metals and their compounds in a study of the angular distribution of characteristic losses. This leads to the idea that the characteristic losses are not the result of the collective interaction of an incident electron with free electrons of the metal, leading to plasma oscillations. Of course, this does not exclude the possibility that for a few metals certain characteristic losses may be due to oscillations of the collective type [361]. At the same time, the characteristic losses observed for many semiconductors such as Sb, Te, Ge, Si, PbS, PbTe, PbSe, MgS_2, TeO_2, and dielectrics NaCl and KCl are hard to explain from the point of view of plasma transitions.

Thus, up to the present time there has been no single view as to the mechanism of characteristic losses, and there is no theory capable of explaining many aspects of this phenomenon. In order to understand the mechanism of characteristic losses it is interesting to study these in all parts of the secondary-electron spectrum, and this cannot be done by firing through thin films. A study of the complete secondary-electron energy spectrum and the associated characteristic losses may provide information on the relation between these and the atomic number of the sample material. In addition to this, the thin-film method only provides results for low scattering angles. It is therefore essential to study the dependence of the characteristic losses on the incident and escape angles. The method of electron reflection is also very promising when studying the dependence of the characteristic losses on the state of aggregation of the material.

An interesting question is that of the relation between the secondary electron emission and the state of aggregation of the metal. The results obtained by various research workers for the secondary-electron emission coefficient and the secondary-electron energy distribution of the same material fail to agree with each other; sometimes they are quite contradictory. This is apparently the result of the imperfection of the methods used in such investigations, or of insufficiently clean conditions on the surface of the sample material.

§2. Dynamic Method of Studying Secondary Electron—Electron Emission

We have seen from the review that there are several methods of studying secondary electron emission. These methods each have advantages and disadvantages. Analysis shows that for studying the secondary-emission properties of various materials the most convenient method is that based on a combination of spherical and cylindrical condensers, taken together with the method of double modulation employed by Arifov and Kasymov [644, 655]. This combination offers the possibility of combining the high resolving power of the cylindrical condenser and the high transmission of the spherical condenser in a single apparatus on the basis of the high-speed oscillograph recording of the processes.

Figure 10.1 illustrates our own apparatus for studying secondary electron emission. The apparatus consists of an electron source 1-3, a receiving system 4-6, a target 7, a cylindrical condenser 10-11, and an electron multiplier 12. All the components of the apparatus were made of copper and tantalum. In all cases the primary beam was directed onto the target surface at an angle of $\varphi = 45°$. The electron collector was a sphere 5, 80 mm in diameter. In order to protect it from external electric fields, the collector was placed within the conducting screen 4. In order to reduce the effect of tertiary currents a spherical antidynatron grid 6 with a transparency of 90-95% was employed. The target was a plate of sample material 1 × 1 × 0.001 cm in size. It was shown in [57] that the classical Lukirskii—Prilezhaev condition was valid for a target of any form if the maximum transverse dimensions were no greater than the diameter of the corresponding spherical target. The secondary-electron analyzer was a cylindrical condenser 10-11 (Fig. 10.1) with an aperture angle of 127° and a 50-mm radius of the equilibrium trajectory. In order to screen it from external electric fields, the condenser was enclosed in a metal housing 9. The entrance and exit slits (8, 8') had dimensions 0.9 × 10 and 0.5 × 10 mm. Energy analysis was applied to the secondary electrons escaping from the target at an angle of $\varphi = 0°$. The high resolution of the analyzer was achieved at the expense of a reduction in transmission. The currents were amplified by four or five orders in an electron multiplier of the open type. As in all the previous experiments, special attention was paid to obtaining a clean target surface.

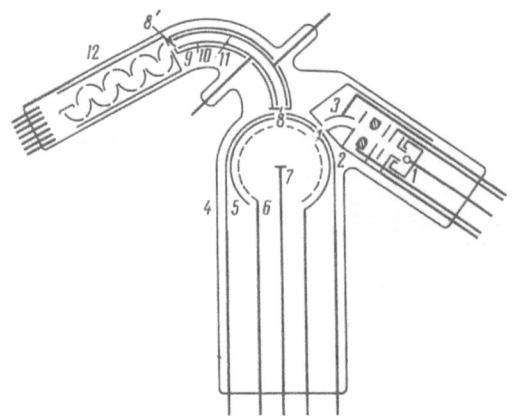

Fig. 10.1. Apparatus used for studying the secondary-electron energy distribution.

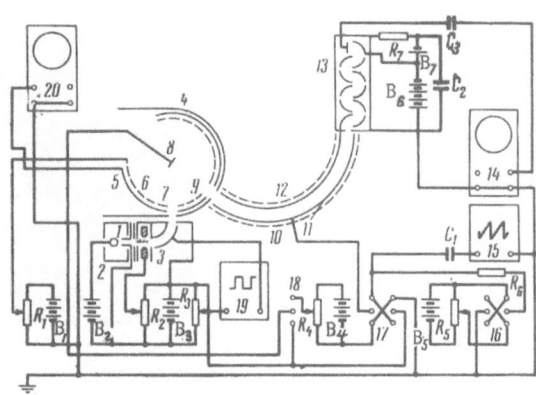

Fig. 10.2. Principal electric circuit of the method.

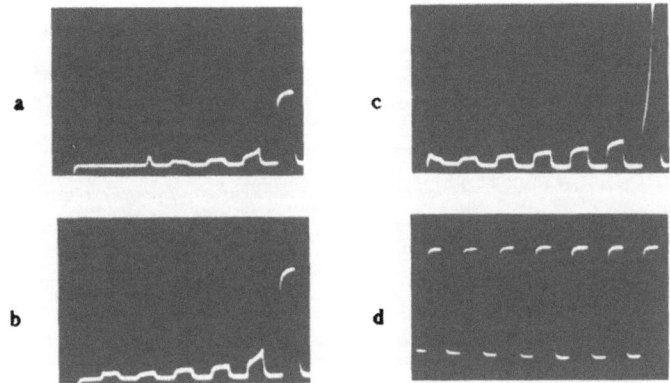

Fig. 10.3. Oscillograms of the V/A characteristics of the secondary electrons.

A considerable advantage of the method thus developed is the fact that the integral and differential pictures of the secondary-electron energy distribution and the total secondary electron current are recorded simultaneously on the screens of two oscillographs.

Measurements of secondary electron emission by the oscillograph method were carried out by means of the electrical circuit shown in Fig. 10.2. The primary electrons emitted from the heated spiral 1 in the source 2 are accelerated by batteries B_2 and B_3. Modulation of the primary electron beam is effected by applying a rectangular voltage from generator 19 to the plates of the cylindrical condenser 3. The intensity-modulated electrons fall onto the target 8. The secondary electron currents are measured by the collector 7 with the guard cylinder 4 and 5, using the antidynatron grid 6. Some of the secondary electrons pass through the diaphragm 9 to the cylindrical analyzer 12 with the screen 10.

We see from Fig. 10.2 that for one position of the switches 17 and 18 the integral energy distribution of the secondary electrons (V/A characteristic) is taken from collector 6 and recorded on the screen of oscillograph 20. For another position of switches 17 and 18 some of the electrons, passing through the cylindrical condensers 11 and 12, are subjected to energy analysis. Then the current of these electrons is amplified by three or four orders in the photo-

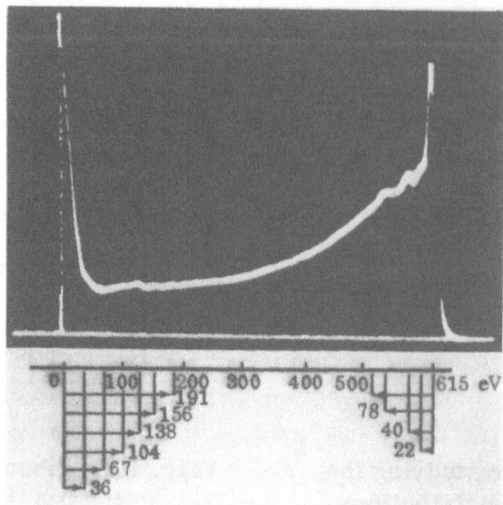

Fig. 10.4. Oscillograms of the secondary-electron energy
spectrum for a molybdenum target with $E_0 = 615$ eV.

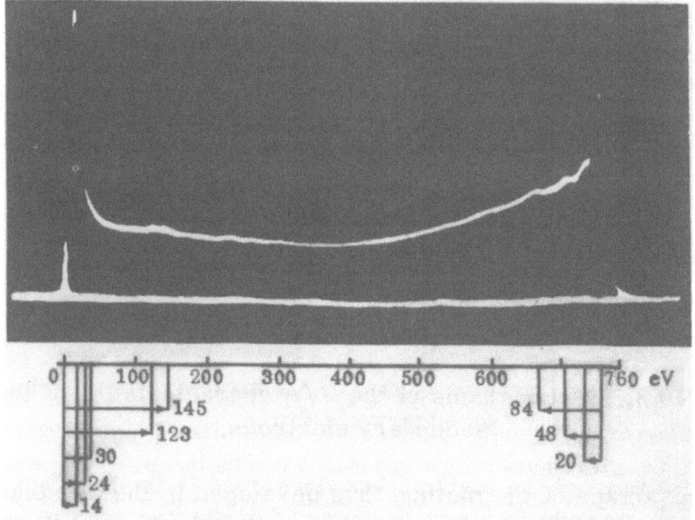

Fig. 10.5. Oscillogram of the secondary-electron energy
distribution for a clean tantalum target with $E_0 = 760$ eV.

multiplier 13. Oscillograph 14 records the energy spectrum of the secondary electrons (differential distribution). At the same time oscillograph 20 records the total current of secondary electrons (15 is a sawtooth-voltage generator and 16 a switch).

It should be noted that in such investigations the linearity of the sawtooth voltages plays an essential part. The amplitude of the generator may be regulated from 0–400 V so as to expand individual parts of the spectrum in order to reveal the fine structure.

The oscillograph method offers the possibility of obtaining stable pictures of the V/A characteristics, the total secondary electron current from the spherical condenser, and the energy distribution of the secondary electrons from the cylindrical condenser.

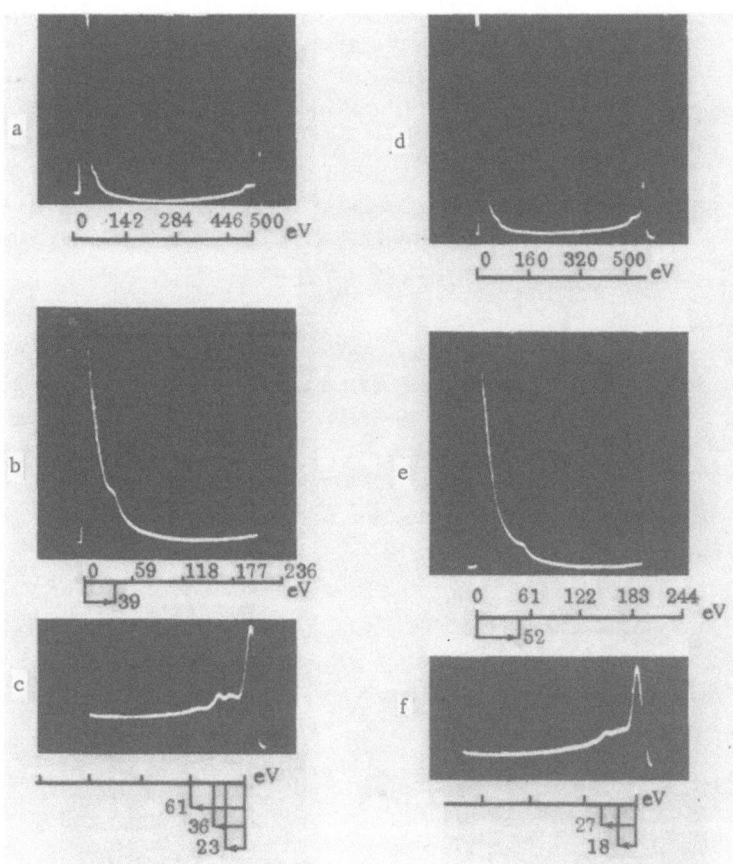

Fig. 10.6. Oscillograms of the secondary-electron energy
spectrum for E_0 = 500 eV in the case of iron and nickel.

Figure 10.3 shows some oscillograms of secondary electron V/A characteristics in which the primary electrons had energies of 180 (a), 272 (b), and 360 (c) eV; Fig. 10.3d gives the total current of the secondary electrons for a primary-electron energy of 272 eV. The envelope of the peaks of the rectangular pulses on these oscillograms gives the integral secondary-electron energy distribution. These curves clearly illustrate the changes in the V/A characteristic of the secondary electrons on increasing E_0.

An oscillogram of the secondary-electron energy spectrum for a molybdenum target at E_0 = 615 eV is shown in Fig. 10.4. This figure shows maxima corresponding to the truly secondary electrons (maximum to the left) and the elastically reflected electrons (large maximum to the right), and also the fine structure in the intermediate parts of the spectrum. This oscillogram indicates the complexity of the secondary-electron spectrum; it consists of a number of maxima which cannot be observed on the secondary current V/A characteristics (see Fig. 10.3).

§3. Study of the Complete Energy
Spectrum of Secondary Electrons

In this section we present the results of a study of secondary electron emission from a number of metals of the transition groups (molybdenum, tungsten, tantalum, iron, nickel, titanium, and zirconium) carried out by Arifov and Kasymov [644, 654]. These metals are interesting because they contain unfilled electron shells, leading to a great variety in their physicochemical properties.

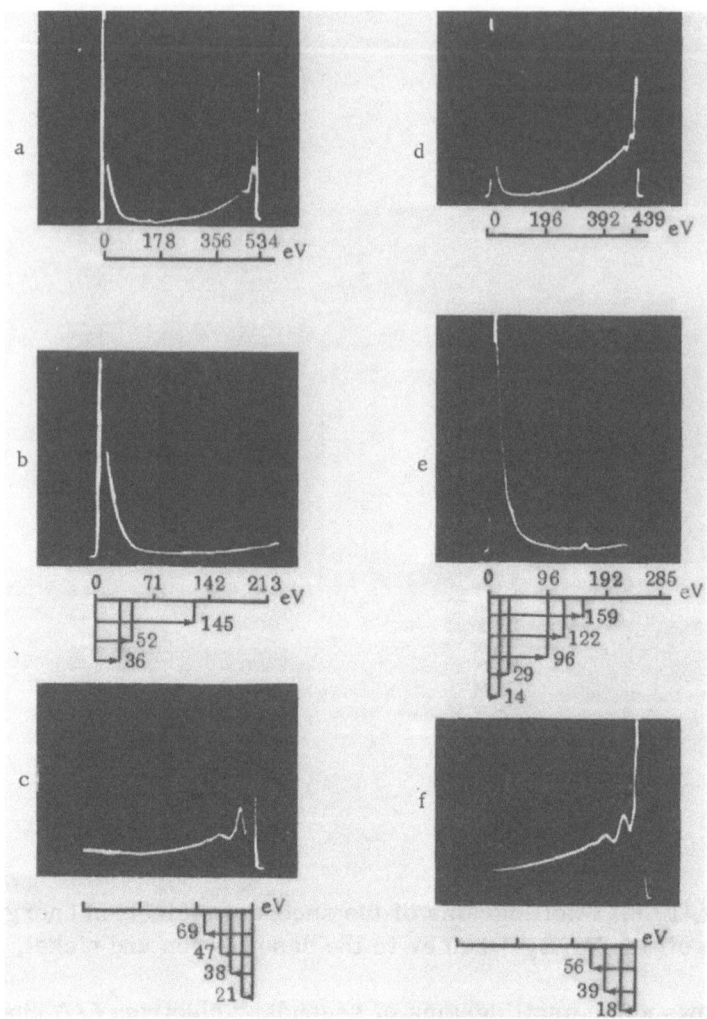

Fig. 10.7. Oscillograms of the secondary-electron energy spectrum for titanium at $E_0 = 490$ eV and for zirconium at $E_0 = 475$ eV.

Tantalum, Tungsten, and Molybdenum. An oscillogram of the secondary-electron energy distribution for a clean tantalum target with primary-electron energies of 760 eV is shown in Fig. 10.5. Apart from the peaks of the slow and elastically reflected electrons, we notice characteristic-loss peaks at 20, 48, and 84 eV. In the other part of the spectrum there are maxima at energies of 123 and 145 eV. Analogous oscillograms were obtained for tungsten and molybdenum targets. For the tungsten target the characteristic energy losses are 27 and 57 eV and for the molybdenum target 22, 40, and 78 eV; in addition to this, the spectrum shows maxima corresponding to energies of 104, 138, 156, and 191 eV, and also breaks (unresolved maxima) at energies of 36 and 67 eV.

Iron and Nickel. The total energy spectra of secondary electrons for iron and nickel with $E_0 = 500$ eV are shown in Fig. 10.6 (oscillograms a and d). Oscillograms b, e and c, f, respectively, give the low- and high-energy parts of the spectrum on an enlarged scale; this was achieved by reducing the amplitude of the modulating voltage on the secondary-electron analyzer. Figure 10.6a,c shows the characteristic-loss peaks corresponding to electron energies of 22, 36, and 61 eV for iron.

Figure 10.6a,b clearly shows a "step" around 39 eV; this appears after brief heating of the iron targets and is independent of the primary-electron energy in the range studied (100-2000 eV). Figure 10.6d,e shows a smoothly falling part of the spectrum and a step at energy 52 eV for nickel, as in the case of iron. However, the height of the step in the nickel spectrum is always lower than that in the iron spectrum. In Fig. 10.6d,f we notice the presence of two energy-loss peaks corresponding to 18 and 27 eV. It is interesting to note that, if there is a trace of nickel (over 2%) in the iron, then the low-energy part of the spectrum will show two steps, instead of one of pure iron. These steps correspond to 38 and 52 eV.

Titanium and Zirconium. Figures 10.7a,b,c show oscillograms of the total secondary-electron energy spectrum and the low- and high-energy parts of this separately for titanium at $E_0 = 490$ eV. Figures 10.7a,b show unresolved maxima (bends) at energies of 36 and 52 eV and a slight maximum around 145 eV. Figures 10.7a,c also show a large characteristic-loss peak with an energy of 21 eV and two neighboring peaks with energy losses of 38 and 47 eV, and also a fourth, barely perceptible but always reproduced flat maximum relating to energy losses of 63 eV. Figures 10.7d,e,f show the analogous oscillograms of energy spectra for zirconium bombarded with 475-eV electrons. Figures 10.7d,f show maxima at energies of 159, 122, and 96 eV. In addition to this, in Figs. 10.7e,f we also see characteristic-loss peaks with energies of 18, 39, and 56 eV.

It should be noted that for all the targets studied the maxima in the low-energy range were very sensitive to surface conditions; the cleaner the surface, the stronger were the peaks, and the better was their separation from their neighbors.

Analyzing the results obtained, we come to the conclusion that the secondary-electron energy spectra depend on the electron structure of the sample metal. This law is particularly clear in the low-energy part of the spectrum. The maxima in these parts of the curves evidently characterize the distribution of the density of levels in the energy bands of the metal. One notices the step in the secondary-electron energy spectrum for iron and nickel. Owing to the typical electron structure of the ferrous metals, the energy spectrum of the secondary electrons should in fact have the singularity actually observed in the form of a step for iron and nickel. In addition to this, the difference between the ordinates of the steps observed in the iron and nickel spectra indicates that the width of the unfilled part of the d band is different in these two elements.

As regards the high-energy parts of the secondary-electron spectra, no special difference was observed between iron containing traces (over 2%) of nickel and Armco iron, which is free from such impurities. In this connection it should be noted that Leder and Marton [304], studying differences in the energy spectra of metals and their compounds, in fact found no great difference between the two cases. These results led to the idea that the fine structure of the low-energy part of the secondary-electron energy distribution was more sensitive to various impurities than the high-energy part in the spectra of metals and their compounds.

If we suppose that the characteristic energy losses of electrons in solids are the result of interaction in which electrons of the upper bands take part, then we should obtain similar secondary-electron spectra for metals with similar outer electron shells. Comparison of the oscillograms for iron and nickel (Fig. 10.6) in fact shows a great similarity between the spectra of these metals. In addition to this, agreement is quite satisfactory for both the low- and high-energy parts of the secondary-electron spectra of titanium and zirconium (Fig. 10.7).

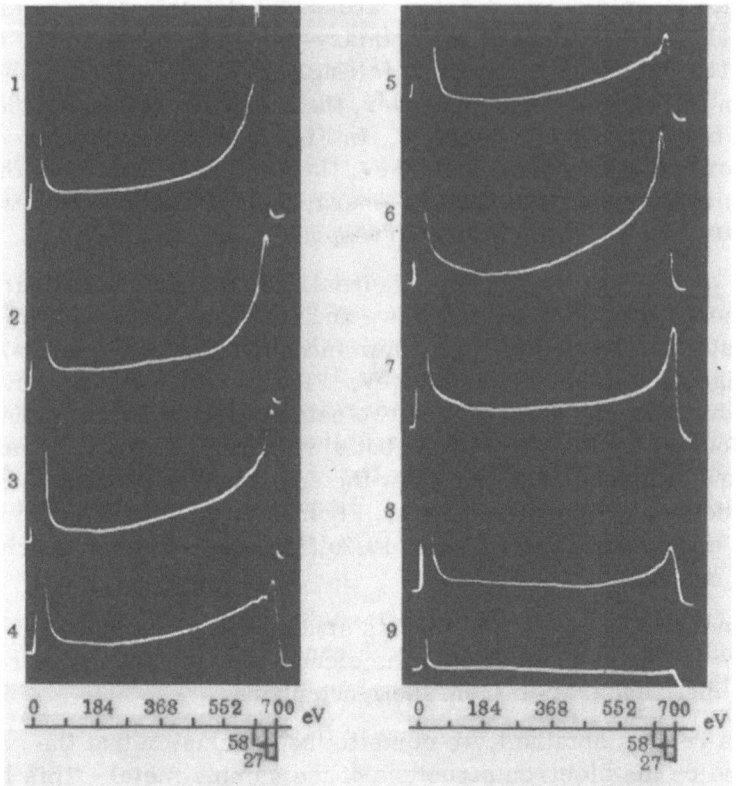

Fig. 10.8. Series of oscillograms of the complete secondary-
electron spectrum for a clean tungsten target at various
angles of incidence of the primary electrons (E_0 = 700 eV).

§4. Angular Dependence of the Energy Spectrum

of Secondary Electrons

A study by Arifov and Kasymov [695, 735] of the angular dependence of secondary electrons offers the prospect of obtaining information regarding the mechanism of interaction between the primary electrons and the electrons in the metal.

In order to study the relation between the total secondary electron spectrum (and individual parts of this) and the incident angle of the primary electrons, we used the apparatus described in Chapter 5 (Fig. 5.9). The apparatus enabled us to observe and photograph the complete energy spectrum of secondary electrons as a function of the primary-electron incident
angle for a given escape angle without breaking the vacuum and other conditions of the experiment. The relation between the intensity of individual parts of the energy spectrum of the secondary electrons and the primary-electron incident angle was obtained by measuring the heights
of various maxima for small sawtooth-voltage amplitudes. In this way various individual parts
of the complete secondary-electron energy spectrum were recorded across the whole oscillograph screen.

A series of oscillograms of the complete secondary-electron spectrum for angles of incidence between 0 and 70° obtained for E_0 = 700 eV with a clean tungsten target is shown in Fig.
10.8. Characteristic-loss peaks at energies of 27 and 58 eV may clearly be seen in the figure.
On varying the angle of incidence from 0 to 70° the position of the maximum of slow secondary

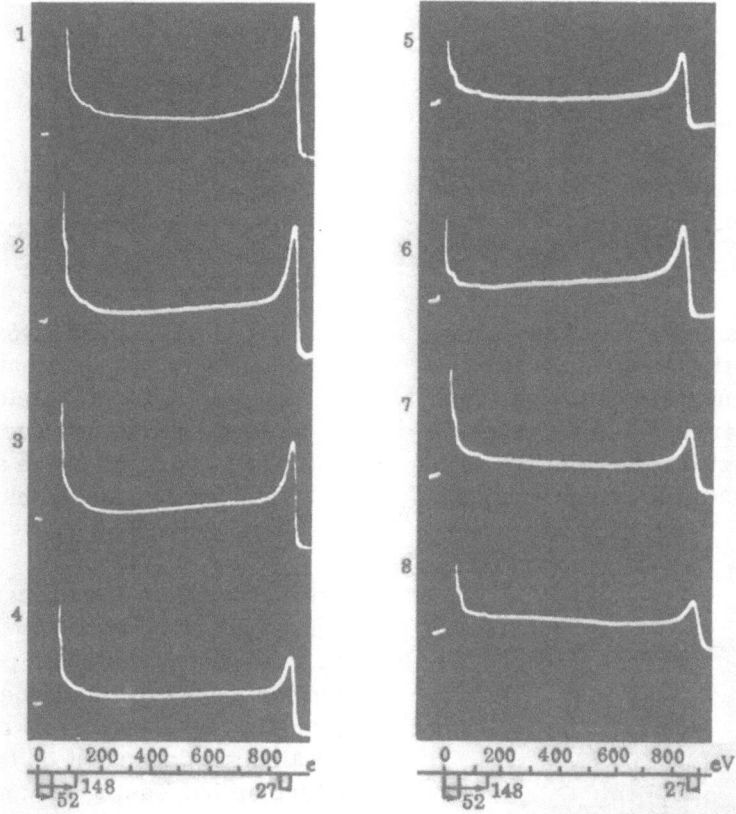

Fig. 10.9. Series of oscillograms of the complete secondary-electron spectrum for a clean nickel target at various angles of incidence of the primary electrons $(E_0 = 900$ eV$)$.

electrons from nickel and tungsten remains constant to within the accuracy of the measurements (about 1 eV); for nickel it lies at about 4 eV.

The relation between the height of the step corresponding to an energy of 52 eV for nickel and the incident angle of the primary electrons is very similar to that holding for slow secondary electrons (Fig. 10.9).

We also measured the angular dependence of the heights of the maxima in the low-energy part of the secondary-electron spectrum obtained from tungsten for $E_0 = 158$ and 266 eV and from nickel for $E_0 = 148$ eV. The height of these maxima depends greatly on the energy of the primary electrons and is almost independent of the angle of incidence.

In addition to this, we measured the half width of the slow secondary-electron maximum and the positions of the maxima in the low-energy region for tungsten at $E_0 = 158$ and 266 eV and for nickel at $E_0 = 148$ eV as functions of the angle of incidence. We found that all these quantities were independent of the angle of incidence between 0 and 70° within the limits of experimental error (about 1 eV).

Figures 10.10 and 10.11 show the relative intensity of the currents of slow, elastically reflected, and inelastically reflected electrons (the latter being the characteristic-loss peaks) as a function of the angle of incidence of the primary electrons for both metals of present interest (nickel and tungsten). We see from the figures that the number of slow secondary electrons

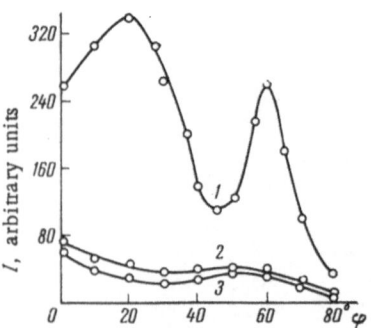

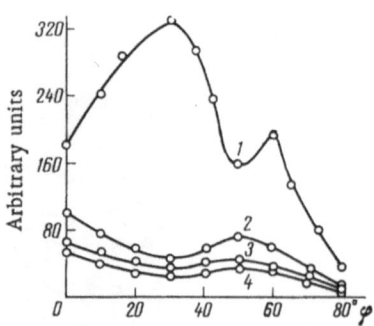

Fig. 10.10. Intensities of slow secondary (curve 1) and elastically reflected (curve 2) electrons and characteristic-loss peak (curve 3) at an energy of 27 eV as functions of the angle of incidence of the primary electrons for a nickel target with E_0 = 900 eV.

Fig. 10.11. Intensities of slow secondary (curve 1) and elastically reflected (curve 2) electrons and characteristic-loss peaks at energies of 27 eV (curve 3) and 57 eV (curve 4) as functions of the angle of incidence of the primary electrons for a tungsten target with E_0 = 900 eV.

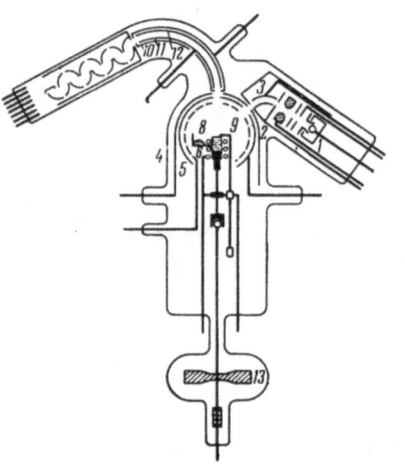

Fig. 10.12. Apparatus for measuring the spectra and coefficients of secondary electron emissions.

first increases with increasing angle of incidence of the primary electrons until it reaches a certain value; then, falling, it passes through a minimum for angles of incidence close to the reflecting angle, rises again, and reaches a maximum with a height dependent on the nature of the metal. On further increasing the angle of incidence, the intensity falls. With increasing primary-electron incident angle the intensity of the emission of elastically and inelastically reflected electrons diminishes, and after passing through a minimum reaches a maximum close to the reflection angle. In addition to this, the intensity of these groups of electrons at the maximum depends on the energy of the bombarding electrons, being the greater, the higher E_0. This indicates that with increasing E_0 the reflected electrons group themselves in greater and greater numbers around a preferential direction approximately coinciding with the angle of specular reflection.

These relationships indicate that as the number of elastically and inelastically reflected electrons falls the intensity of the slow-electron emission increases. This latter fact may clearly be explained as being due to the substantial participation of inelastically scattered electrons in creating electrons of the slow secondary type. For example, the existence of a maximum in the angular dependence of slow secondary-electron emission may easily be explained if we consider the sharp rise in the number of elastically and inelastically reflected electrons for angles of incidence close to the reflection angle. This may clearly be seen from Figs. 10.10 and 10.11. Figure 10.10 shows the angular dependence of the relative intensity of slow secondary (curve 1), elastically reflected (curve 2), and inelastically reflected (curve 3) electrons for a clean nickel target at E_0 = 900 eV. Figure 10.11 shows the angular dependence of the relative intensity of slow secondary inelastically reflected electrons with energy losses of 27 eV (curve 3) and 57 eV (curve 4) for a clean tungsten target at E_0 = 900 eV.

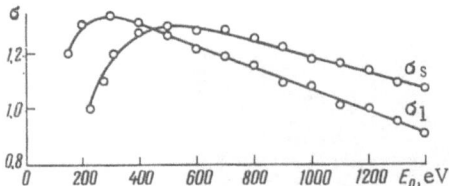

Fig. 10.13. Coefficient of secondary electron emission as a function of the primary-electron energy for uncleaned indium in the solid (σ_s) and liquid (σ_l) states.

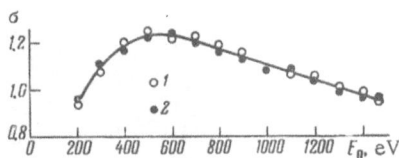

Fig. 10.14. Coefficient of secondary electron emission as a function of the primary-electron energy for a pure, repeatedly cleaned indium target in the liquid (1) and solid states (2).

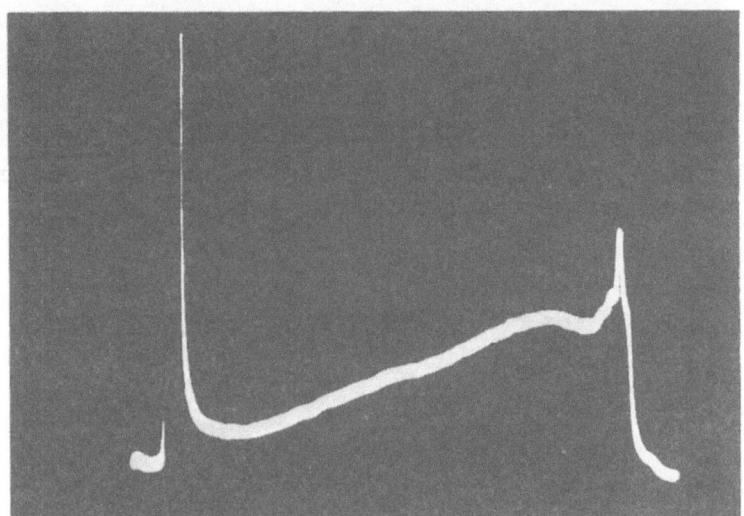

Fig. 10.15. Oscillogram of the energy distribution of secondary electrons for indium taken immediately after cleaning the surface with a scraper at E_0 = 500 eV.

§ 5.   Study of Secondary Electron—Electron

Emission from Metals in Various

States of Aggregation

The characteristic losses have been studied as a function of the state of aggregation of the metal in only a single case (Beorch et al. [611), based on the penetration method, for high energies of the bombarding electrons. However, this method is imperfect, since the spectrum of the substrate is superimposed on that of the characteristic losses for the metal under consideration.

The apparatus in which the integral and differential secondary-electron spectra were obtained and the coefficient of secondary electron emission was measured for indium and tin is shown in Fig. 10.12 [410, 655]. This consisted of an electron source 1-3, a spherical collector 5 surrounded by a guard cylinder 4, a secondary electron analyzer 11, 12, an electron multiplier 13, and a target 8. The rest of the apparatus is described in detail in § 2 of this chapter.

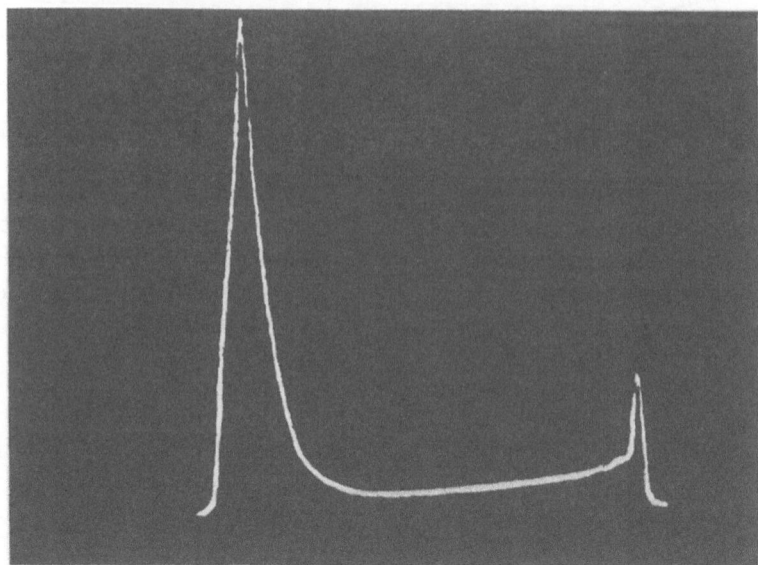

Fig. 10.16. Oscillogram of the complete secondary-electron energy spectrum for cleaned tin at $E_0 = 250$ eV.

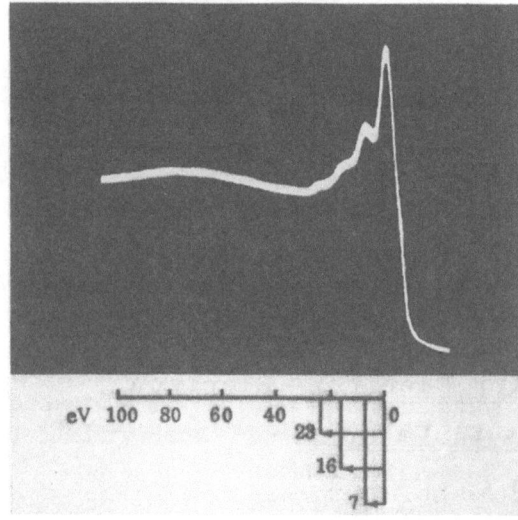

Fig. 10.17. Oscillogram of the high-energy part of the secondary-electron distribution of clean indium at $E_0 = 500$ eV.

The target consisted of a porcelain cylinder with an internal diameter of 10 mm holding the sample material, and a piston machined from stainless steel with a concave spherical tip. The piston was controlled from outside with a magnet. The metallic target was melted by passing a steady current through a spiral wound on the porcelain cylinder. All the parts of the target were arranged so as to disrupt the configuration of the field in the space between the target and collector, and hence between the target and analyzer, as little as possible.

In such investigations the vacuum conditions are particularly important, as is also the cleaning from the target surface of various kinds of contaminant. It must be remembered that

Table 10.1

Elements	Maxima of low-energy region, eV					Characteristic-loss peaks, eV				Author
Mo	37.2	112	140	152	172		25.7	40.7	76.9	Harrower [367]
	36	104	138	156	191		22	40	78	Our data
	40									Vyatskin [452]
Ta						21			47	Richardson
	30	58	123	145		20			48.84	Our data
W			160	264			26.8		58	Harrower [367]
			158	266			27		57	Our data
							30		60	Vyatskin [452]
Fe	39						23		62	Ruthemann [173]
							22	36	61	Our data
							18.3	36.6	54.9	Vyatskin [452]
Ni						17.6	23.4			Marton and Leder [303]
		52	148			18	27			Our data
						12.2	24.4			Vyatskin [452]
Ti						17.6	36.6	48.3	61.8	Robins and Swann [527]
	36	52	145			21	38	47	63	Our data
						23.2	40.6	50		Vyatskin [452]
Zr						18.2	41.6	51.1		Robins [573]
		96	122	159		18	39	56		Our data
In						12.4	18			Marton and Leder [303]
						10	18	32		Our data
						15.9	19.4			Vyatskin [452]
						15				
Sn						5.9	12.2	22.3		Klein [369]
						7	16	28		Our data
						6.1	132	34.4		Vyatskin [452]

these metals cannot be heated to too high a temperature without starting to evaporate. On the other hand, some of the easily melting metals, rapidly absorbing oxygen from the residual gas, become coated with a thick oxide film, which is hard to remove even at temperatures close to the melting point of the metals in question. This demands care when working with such metals. The choice of indium and tin as targets was made because these targets have a low vapor tension; the indium tested was 99.9999% and the tin 99.999% pure.

In order to obtain a clean target surface, we used a scraper (Fig. 10.12) controlled by a magnet so as to remove the contaminated film from the surface of the liquid metal.

The use of the oscillograph method offered the possibility of observing the changes in the secondary-electron spectra visually during the melting and solidifying of the material under the same experimental conditions.

We measured the coefficient of secondary electron emission σ as a function of E_0 for various states of aggregation of indium by analyzing the oscillograms simultaneously recording the current pulses of the primary and secondary electrons. Figure 10.13 shows the dependence of on E_0 for uncleaned indium in various states of aggregation. Figure 10.14 shows the $\sigma(E_0)$ relationship for pure indium in the liquid and solid states. We see from the figure that for pure

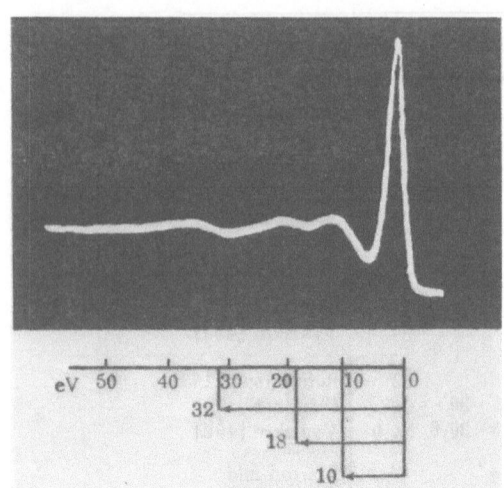

Fig. 10.18. Oscillogram of the high-energy part of the secondary-electron distribution of clean tin on an enlarged scale.

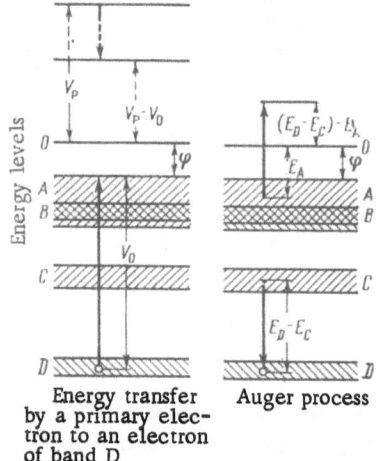

Fig. 10.19. Typical scheme of the Auger process.

indium the difference between the σ and E_0 curves for the solid and liquid material lies within the limits of experimental error (2-3%). The oscillogram of the complete secondary-electron spectrum for the clean, liquid surface of indium is shown in Fig. 10.15.

The oscillogram of the complete secondary-electron energy spectrum of clean liquid tin at E_0 = 250 eV is shown in Fig. 10.16. It should be noted that the fine structure in the high-energy part of the spectrum of tin appears in the same way as in the case of indium after cleaning the surface of the tin with a scraper; it remains constant for a long time (10-15 min) after passing into the solid state. These results show that until the film of surface contamination has been removed the molten surface remains insufficiently clean, in spite of preliminary heat treatment.

Figure 10.17 shows an oscillogram of the high-energy part of the secondary-electron spectrum for pure indium on a different scale; we notice that the secondary-electron energy spectrum of indium contains characteristic-loss peaks at energies of 7, 16, and 23 eV.

A similar oscillogram of the high-energy part of the secondary-electron spectrum of pure tin is shown in Fig. 10.18. In this figure we also see three loss peaks at energies of 10, 18, and 32 eV.

The position of the characteristic-loss peaks and the coefficient σ remain unchanged on passing from one state of aggregation to another with the limits of measuring error. The rise or fall in the coefficient of secondary-electron energy sion on melting observed by many authors may have been due to changes in the conditions on the target surface.

Our results for the coefficients σ, the complete secondary-electron energy spectrum, and the characteristic-loss peaks indicate (within the limits of experimental error) that the melting of metals has no effect on the secondary-emission properties of the substance under consideration or on the positions of the characteristic losses.

On considering the energy distribution of the secondary electrons, we encounter the problem: To which group of electrons in the metal do the individual maxima owe their origin?

The maxima and bends (inflections) in the low-energy part of the spectrum observed in the present investigation (Table 10.1) for a number of metals (molybdenum, tungsten, tantalum, nickel, titanium, and zirconium) relate to electrons which have left the target as a result of the absorption of a certain amount of energy transferred from the bombarding electrons. Among these groups of electrons there are probably those which have passed out into the vacuum directly after receiving energy from a primary electron.

The maxima in the low-energy part of the spectrum corresponding to energies of 104, 138, 156, and 191 eV for molybdenum, 158 and 266 for tungsten; 30, 58, 123, and 145 eV for tantalum; 148 eV for nickel; 36, 52, and 145 eV for titanium; and also 96, 122, and 159 eV for zirconium apparently correspond to the Auger process. The principle of the Auger process is illustrated in Fig. 10.19. In this process two electrons take part. The first electron, which initially lies, for example, in band C, fills a vacant place in band D which has been formed as a result of the interaction of a primary electron with one of the electrons in this band. Thereupon an energy $E_D - E_C$ is released and this is used in ejecting a second electron originally situated, for example, in band A. This second electron leaves the metal with a kinetic energy of $E_D - E_C - E_A$, which corresponds to a maximum in the low-energy part of the secondary-electron spectrum.

If we explain the peaks of characteristic energy losses by the inelastic scattering of primary electrons in solids, then groups of electrons possessing specific energies should exist in the secondary electron emission.

If both inelastically reflected and secondary electrons take part in the formation of intermediate maxima, then in general we should observe two systems of maxima, each of which, on varying the primary-electron energy, will retain a constant position relative to either the origin of coordinates or the peak of the elastically reflected electrons.

Table 10.1 gives the energies corresponding to the characteristic-loss peaks for the metals studied in our experiments. For comparison, the table also gives data obtained by other methods and calculated values for the characteristic-loss peaks obtained by the Vyatskin procedure [452] for interband transitions. Our own results agree satisfactorily with the values of the characteristic-loss peaks found by other methods, in particular by the method of passing fast electrons through thin films. The table also shows experimental values of the maxima and inflections in the low-energy part of the secondary-electron spectrum. These features cannot be observed by the method of passing the electrons through thin films in the liquid or solid state: the target has to be supported by some substrate. The result is that the spectra of energy losses obtained from the sample material are overlapped by the loss spectrum of the substrate. This shows the advantage of the dynamic method of recording complete secondary-electron spectra.

CHAPTER 11

THEORY OF CERTAIN PHENOMENA TAKING
PLACE IN THE COLLISION OF
ATOMIC PARTICLES WITH A SOLID SURFACE

Of the phenomena listed on page 7, only electron—electron emission, and to a certain extent cathode sputtering, have been reasonably well studied from the theoretical point of view. These theories are still far from complete and are still continuing to be developed. As regards the remaining phenomena, these have until recently received no satisfactory theoretical treatment, although they have been discussed on the basis of various models.

§1. Theory of Ion—Electron Emission

The emission of electrons from a metal under the impact of various ions takes place in two qualitatively different ways, associated with two possible mechanisms of electron excitation. The difference lies not only in the mechanism of energy transfer but also in the category of electrons participating in the emission. Accordingly, the two possible forms of ion—electron emission are usually observed in different energy ranges of the bombarding ions and in association with different types of the latter.

"Potential" Emission of Electrons

For this mechanism of electron extraction the source of energy is the potential energy of the metal electrons in the field of a positive ion. Emission of this kind has therefore become known as "potential." The kinetic energy of the bombarding ion plays no part (or hardly any) in this process, and the phenomenon takes place with roughly the same efficiency for fairly high energies of the incident ions (1 keV and above) and for almost zero energies. The experimentally observed slight dependence of the emission on the kinetic energy of the ions is indirect. The phenomenon of potential emission is in this respect closely associated with such processes as surface ionization, neutralization, and charge exchange. In essence the phenomenon constitutes a special kind of "collision of the second kind" [6, 312].

When a positive ion approaches the surface of a metal, the energy corresponding to the neutralization of the ion by a metal electron is released, and this makes it possible for an electron to be emitted. The possibility of such an emission was first indicated by Holst and Oosterhuis [20]; it was observed by Penning [46, 60] and Oliphant [61]. Oliphant and Moon [62] developed a theory of this phenomenon on the basis of semiclassical considerations. In this theory the problem of the neutralization of the ion reduced to the calculation of the field emission of an electron from the metal under the influence of the field of the ion, which reached $2 \cdot 10^8$ V/cm when the ion approached to within 10^{-7} cm of the surface of the metal. In order to calculate the number of electrons emitted for one incident ion, the authors used the Fowler—Nordheim formula for electron field emission.

For a more accurate description of neutralization phenomena, Oliphant and Moon considered the transition of an electron from the metal through the potential barrier to a discrete level of the ion; they assumed that electrons passed from the metal to an excited level of the atom when the latter was still at a certain distance from the surface of the metal, and after this the excited atom released its energy either by the emission of a photon or by the transfer of energy to another electron. The critical distance at which neutralization could take place was, according to Oliphant and Moon, $4 \cdot 10^{-8}$ cm. The calculations showed that during the time taken to travel from the place of neutralization to the metal, the excited atom could not emit a photon. The excitation energy was therefore transferred to a second electron in the metal. If this energy were greater than the work function, the electron would pass out into the vacuum.

On the basis of this type of consideration, Oliphant and Moon obtained accurate values for the minimum and maximum energies acquired by the electron:

$$E_{\max} = V_i - \varphi, \quad E_{\min} = V_i - \mathscr{E}_F - \varphi,$$

where $\mathscr{E}_F$ is the energy of the Fermi level; they showed that the necessary condition for potential emission was $V_i > 2\varphi$.

A quantum-mechanical treatment of the problem based on the ideas of Oliphant and Moon was developed by Massey [68], who considered a simplified picture of the two-stage neutralization of an ion close to the metal on the assumption that levels of equal energy necessarily existed in the metal and the ion. This assumption is not justified in any of the cases of practical importance. The problem was also simplified by a number of other approximations: no allowance was made for the existence of the exponentially falling part of the wave function of the electron, then no allowance was made for the presence of a surface potential barrier, and only the case of spherical symmetry was considered. This led to a calculated 100% yield of electrons from the metal, whereas experiment gives a value of the order of 10% (referred to one incident ion).

Cobas and Lamb [171] represented the ejection of electrons by He$^+$ ions colliding with the metal surface as a consequence of two distinct processes: neutralization in the 2^3S state with the formation of a metastable He atom, and the relaxation of this atom with the ejection of an electron from the metal. Using wave functions differing from those of Massey and also considering the dependence of the exciting potential on the distance from the metal, these authors obtained the following expression for the probability of the process as a function of the distance of the ion from the surface of the metal:

$$W = \frac{A_0}{v_0} \exp\left[-\frac{A_0}{v_0}\exp\left(-\alpha l\right) - \alpha l\right].$$

For $v_0 = 10^5$ cm/sec the quantity W has a maximum at $l = 2a_0$ (where a_0 is the Bohr radius), while for $v_0 = 3 \cdot 10^7$ cm/sec the maximum is at $l = 0.1a_0$, whereas Massey obtained a value of the order to $3a_0$ for this velocity.

The correct presentation and solution of the problem were given by Shekhter [117]. In this treatment the ion plays the part of a new energy level arising in the system, lower than the occupied levels of the metal. Thus the metal—ion system is an excited one. The relaxation of this system, leading to the release of the energy of neutralization of the ion, is considered as a phenomenon analogous to the Auger effect [33, 243], not requiring the existence of a resonance level. On the basis of a two-electron problem, the author considers the transfer of the energy thus released to another electron in the metal, leading to its possible emission into the surrounding vacuum. The probability of a process of this kind is given by the expression

$$W = \frac{2\pi}{\hbar} |V|^2 \delta(\mathscr{E}),$$

where

$$|V| = \int \psi_l^* (1, 2) \frac{e^2}{r_{1,2}} \psi_{\max} (1, 2) \, d\tau_1 d\tau_2$$

is the matrix element of the transition between the initial state described by the wave function ψ_l and the final state described by the function $\psi_{\max}$; $\delta(\mathscr{E})$ is the Dirac delta function needed to satisfy the law of energy conservation.

After some cumbersome calculations on the Sommerfeld model, Shekhter determined the probability W:

$$W = 5.8 \cdot 10^{23} l^{-4.8 \cdot 10^8 a} [166a^2 + 2.7 \cdot 10^{-8} a + 8 \cdot 10^{-19}],$$

where l is the distance of the ion from the surface of the metal, and a is a dimensionless parameter. The total probability may be derived by integrating with respect to l:

$$\gamma = \int\limits_0^\infty W \frac{dl}{v_0} = 1 \cdot 10^{-1},$$

where v_0 is the velocity of the ion. Thus Shekhter obtained the correct order of magnitude of the emission.

Shekhter considered the competing process of the neutralization of the ion near the surface of the metal with the emission of light and showed that the probability of this taking place was several orders below that of the Auger process. In addition to this, Shekhter considered the case described by Massey and found the probability of neutralization associated with an excited energy level of the ion (if this exists) with subsequent transfer of the energy by a collision of the second kind.

Shekhter's treatment lays the foundation of the modern theory of the potential emission of electrons from a metal. However, owing to the cumbersome nature of the calculations, the theory was never brought to a state of final comparison with experiment. This was partly because, at the time, there had been no careful and reliable experimental investigations into potential emission. The phenomenon had simply been observed, not studied in detail.

Hagstrum's careful experimental study of potential emission [262, 290, 345] enabled him to develop Shekhter's theory.

Hagstrum [291] considered the two Auger processes possible when an inert-gas ion approaches the surface of a metal: Auger neutralization and Auger relaxation. Without directly calculating the matrix element of the transition, Hagstrum attempted, by comparison with experiment, to allow for the relation between the matrix element and the distance of the ion from the surface of the metal and also the angle between the electron velocity and the normal to the surface.

A number of other factors not considered by Shekhter were taken into account in Hagstrum's work: the displacement of the energy levels of the ion near the surface of the metal, the finite lifetime of the initial state, the part played by resonance processes in determining the relative probabilities of Auger neutralization and Auger relaxation, etc.

Hagstrum was able to give a logical explanation for a whole series of experimental data relating to potential emission obtained in his experimental work. The weak point of the theory is the over-large number of "fitting" parameters. The form taken for the angular distribution of the excited electrons within the metal, in particular, is controversial and requires further investigation.

In the first form of the theory, Hagstrum's treatment referred only to a pure (clean) metal surface. Later, Hagstrum also considered Auger neutralization of ions on the surface of a semiconductor of the silicon type [579]. The general scheme of the theory remained the same as for emission from metals; however, the special properties of the energy-band structure of such semiconductors were taken into account, and surface effects were considered.

Propst [664] expressed the view that some of the emitted electrons constituted secondary electrons emitted as a result of the collision of Auger electrons with electrons from the metal. According to Propst, such electrons might constitute up to 50% of the whole emission. Hagstrum and Takeishi [741] did not agree with Propst's estimate, and showed by analyzing the experimentally observed energy spectra of the electrons that the secondary electrons made up no more than 10% of the total emission.

Hagstrum and colleagues [742, 743] also studied the nonadiabatic broadening of the energy distribution of Auger electrons observed on increasing the velocity of the ion.

Using Hagstrum's method, Takeishi considered the Auger extraction of electrons by inert-gas ions from BaO [351] and nickel [433].

Izmailov and Furman [212] and Furman [395] showed that neutral electronegative gas molecules bombarding the surface of a metal could also produce the potential emission of electrons. In calculating this effect, the authors confined attention to the semiclassical method of Oliphant and Moon which reduced the problem to the field emission of electrons in the field of the molecules.

On the whole, we may consider that the theory of the potential emission of electrons has been brought to such a satisfactory state by the investigations of Oliphant and Moon, Shekhter, and Hagstrum, that the mechanism and principle aspects of the phenomenon are broadly understood, and further development of the theory will tend to involve refinement of the parameters and the consideration of more complicated cases.

"Kinetic" Emission of Electrons

There is yet another source of energy for the emitted electrons: the kinetic energy of the ion. Despite the fact that the mechanism of the transfer of kinetic energy from the ion to the metal electrons has remained obscure for a long time, the source of this energy has never raised any doubts. The corresponding electron emission has earned the title of "kinetic" emission.

In contrast to potential emission, which is only observed for certain relations between the ionization potential of the ion and the work function of the metal surface $V_i > 2\varphi$, kinetic emission is evidently universal and occurs when solids are bombarded by ions of very different properties.

The first attempt to develop a method of theoretically calculating this phenomenon was made by Kapitsa [24], who considered that the ion (in the first experiments, an α particle), striking the surface of the target, gave up some of its energy to a very small volume of metal, producing brief but severe heating ("local heating"). The temperature thus developing, calculated from the classical heat-conduction equation, was such as to make thermionic emission possible. Thus, Richardson's formula could be used to calculate the emission. The formula was derived for α particles, for which the loss of energy could be considered as uniform along the path, thus ensuring cylindrical symmetry. For heavy ions one must consider that the energy is evolved at a point on the surface and that spherical symmetry holds. In this case an analogous but rather complex formula is obtained. In order to calculate the leakage of heat from the microscopic region of local heating, Kapitsa used the classical macroscopic heat-conduction equation.

It should be noted that Kapitsa specifically acknowledged the inapplicability of the Richardson equation derived from the equilibrium Maxwell distribution to this case. Kapitsa was not proposing to create a final theory but only trying to estimate the order of magnitude of the emission correctly; he considered that the overwhelming effect of surface conditions on the value of the emission would make more detailed calculations useless.

The main stimulus for the creation of such a theory, which has earned the title "thermal," was evidently the character of the energy spectrum of the electrons emitted; this constituted a Maxwell distribution with a maximum in the region of 2-4 eV, very like the spectrum of thermionic electrons.

Hippel [38, 39] took the idea of local heating in an attempt to create a thermal theory of cathodic sputtering. Morgulis [90, 92], developing Hippel's theory, considered the cooling of the heated microvolume by heat conduction and radiation, using the distribution of temperature in space and time from a momentary heat source. Not being acquainted with Kapitsa's work, Hippel also considered electron emission from the region of local heating and obtained similar results quite independently.

A decisive factor for the thermal theory of kinetic electron emission was Sommermeyer's indication [106] of the very poor energy exchange between the lattice of the metal and its conduction electrons in the nonstationary case. Since the energy of the ion is transferred directly to the metal atoms, it can only be transferred to the electrons through these. Owing to the vast difference between the masses of the atoms and the electrons, the transfer of energy to the electrons is so small that the electrons in the microvolume simply cannot heat up during the period of existence of the local heating. This decided the fate of the thermal theory, demonstrating its fundamental unsuitability (Morgulis [133] also soon attached himself to Sommermeyer's point of view and abandoned the thermal theory).

Izmailov [132], continuing the discussion of the thermal theory, sought a more accurate expression for the energy spectrum of the electrons, criticizing Kapitsa and Morgulis for inaccurate handling of the calculations. However, Izmailov later [434] noted that his own calculations also led to the conclusion that the thermal theory was untenable unless a very low value was taken for the effective range of the electron in the heated metal. The impossibility of heating the electrons is ultimately expressed by the fact that the dimensions of the region of local heating are smaller than the free path of the electrons in the metal.

Gurtovoi [144] directed attention to the fact that the main argument of the thermal theory (the Maxwellian character of the energy spectrum of the electrons emitted) was also untenable, since, first, this form was not necessarily a result of thermal motion (for example, the spectrum of secondary electron emission due to the action of fast electrons indicates a nearly Maxwellian form); secondly, a properly developed theory of local heating, allowing for the rapid outflow of heat, should in fact not lead to a Maxwell distribution of the velocities of the emitted electrons, so that the existence of such a distribution constitutes an argument against the theory rather than for it.

In view of the untenability of the thermal theory, Morgulis [133] and later Gurtovoi [144] again turned to the Kapitsa and Sommermeyer view that the correct way of regarding the emission of electrons under the action of ions lay in considering the ionization produced by the ions on colliding with the metal atoms.*

———————
*Disturbed by the necessity of bringing the concept of ionization to bear on an atom in a solid

Morgulis indicated two possible sources of electron ejection when an ion collided with the surface of a metal: 1) an atmosphere of electrons bound by a surface potential barrier (analogously to the photoeffect); 2) deeper bound levels of the surface atoms.

Gurtovoi [144], analyzing his experiments, favored the second variant; he also considered that the main part in secondary emission was played by the atoms in the first surface layer of the metal, while the atoms of succeeding layers provided no direct supply of electrons but simply influenced the conditions of ionization of the surface atoms. The contribution of succeeding layers increased with increasing energy of the ion. Gurtovoi emphasized that the more strongly the electrons were bound in the metal (i.e., the greater the work function), the more easily they accepted the energy of the ion, and hence the higher was the emission.

Unfortunately, the "impact-ionization" hypothesis was not at that time developed into a theory, possibly owing to the absence of a quantum-mechanical theory of the ionization of atoms by ion impact.

Frenkel' [154] proposed a "mechanical" theory of kinetic ion—electron emission based on the work of Migdal [134], who considered the ionization of an atom under the impact of a neutron as the result of the "shaking" of its electron shell when there was a sudden change in the motion of the nucleus. Frenkel' applied this method to the collision of an ion with the surface of the metal and found that the effective emission of electrons as a result of this mechanism should only take place for an ion energy of 0.1 to 1 MeV. The probability of ionization as a result of shaking for a central collision equals

$$\gamma = 0.28 \frac{m_e E_0}{m_2 V_i} ,$$

where m_e is the mass of an electron, m_2 being the mass of the ion. For an energy of $E_0 = 10$ keV, γ equals $3 \cdot 10^{-3}$. Even without considering the probability of a central collision and the escape of the electron from the metal, this is some two orders below the observed value. Thus, the proposed mechanism cannot explain the value of the emission in the energy range in which the motion of the electrons in the colliding atoms has an adiabatic character.

Izmailov [132] indicated that the true picture of the collision of two atoms differs considerably from that described by Frenkel', since it is the electron shells of the colliding systems which first come into contact, while the nuclei are retarded later.

Avak'yants [222, 223, 232] developed Frenkel's idea, considering the shaking as the internal conversion of brems quasiquanta emitted by the nucleus in the electron shells of the colliding atomic systems. Avak'yants found that the probability of ionization when a proton collided with an atom was given by the expression

$$\gamma \sim \frac{Z_1^2}{30} \left(\frac{e^2}{hc}\right)^2 \frac{E_0}{m_2 c^2} \left(\frac{m_e c^2}{V_i}\right) \operatorname{arcsec} \frac{E_0}{V_i} .$$

Izmailov [428], criticizing Avak'yants' work, repeated the objection raised against the "shaking" theory; he also pointed out the illegitimacy of describing the motion of electrons in colliding systems by hydrogen-like functions.

---

target, Sommermeyer proposed that ionization took place in an atom lying within a microscopic region brought into a gaseous state as a result of the impact of the ion; this region had a high temperature but a density close to that of the solid. Gurtovoi correctly noted that in this case it was inconceivable how an ion which had just given up its energy in forming the gaseous region could immediately afterwards produce ionization.

Roos [389] made an attempt to calculate the coefficient of ion—electron emission by the quantum-mechanical method of perturbed stationary states, but the method was not applied correctly enough and it is difficult to make any judgment on the validity of the theory.

Sternglass [390] developed an "impact-ionization" theory to describe the emission of electrons from metals under the action of high-energy ions (of the order of MeV). In order to calculate the ionization produced by the ion on penetrating into the metal, Sternglass used the Bohr—Bethe theory of the penetration of fast particles through matter. Sternglass obtained the following expression for γ:

$$\gamma = \frac{4AP\pi e^4 Z_i^2}{\overline{V} \alpha' \sigma_g} \left(\frac{Z_1}{E_R E_{eq}} \right)^{1/2},$$

where P is the probability of the escape of an electron from the metal, the constant A is determined by the distribution of the initial secondary-electron velocities (for a symmetrical distribution it equals roughly $1/2$), $\overline{V}$ is the mean energy expended in forming an ion in the solid, Z_i is the effective charge on the incident particle, E_{eq} is the energy of an electron traveling at the same velocity as the ion, E_R is the Rydberg energy, Z_1 is the atomic number of the metal, σ_g is the cross-sectional area of an atom of the metal, determined by the covalent radius, and α' is a constant characterizing the secondary-electron scattering cross section.

By an analogous method Ghosh and Khare [619] calculated the emission of electrons from aluminum after impact by hydrogen ions and atoms.

The Sternglass theory correctly explains the presence of a maximum on the $\gamma(E_0)$ curve and the subsequent fall at very high energies. The yield of electrons from the metal is considered as a consequence of their diffusion to the surface. However, the principal emission of electrons occurs in the range 1-1000 keV, where the Sternglass theory is inapplicable. In this range the motion of the electrons in the colliding atoms has a quasi-adiabatic character, and special methods are required to deal with this.

Izmailov [338, 428, 464, 465, 581] proposed a "radiation" theory of electron emission under the action of ions and atoms; in this the extraction of electrons is considered as a consequence of the absorption of quanta and quasiquanta (emitted during the instantaneous arrest of the ion as a result of a central collision with a surface atom) by free electrons of the metal. In a later version of the theory [581], the effect of the recoil of target particles was also taken into account. The difficulty of this theory is the impossibility of explaining the experimentally observed independence of the emission in relation to the charge of the ion. It is well known that the bremsstrahlung of a charged particle increases rapidly with its charge. However, experiment shows that kinetic ion—electron emission is independent of the charge on the ion. Fast inert-gas ions reach the surface of metals after already having been almost completely neutralized, but nevertheless produce intense kinetic emission. The emission of electrons under the influence of neutral atoms has the same value as ion—electron emission. In order to explain this fact, Izmailov [465] extended the theory by allowing for the bremsstrahlung derived from the (also instantaneous) stopping of the nucleus and the stopping of the electron shell of the atom separately. Since this mechanism was universal, it should be just as effective for the ion of a multielectron atom. Hence, it follows from the theory that, as a result of the simultaneous operation of both mechanisms, the ion—electron emission should be twice as great as the atom—electron emission. Experiment refutes this conclusion. It also follows from the theory [581] that there should be no emission resulting from the impact of atoms having the same nature as those in the target; here again, experiment disagrees. In the experiments of [349] the emission from molybdenum under the influence of molybdenum ions was greater than in the case of other ions. Experiments of ionization in gases confirm this [492].

The method itself also draws a number of objections. Only central collisions of the ion with the metal atom are considered, since only these lead to the sudden stopping of the ion and, hence, according to the theory, only these are effective. The relative probability of these collisions is not estimated; it is assumed that each ion on colliding with the surface experiences instantaneous stopping. However, the proportion of central collisions in the thin surface layer of the metal supplying the secondary electrons is small. The real nature of the motion of the ion, even for a central collision, is far from an instantaneous halt. The intensity of the field of bremsstrahlung is extremely sensitive to the acceleration, and the replacement of a finite acceleration by an infinite one cannot be regarded as justified. This prevents the extension of the Frölich theory [339] of the brems mechanism of secondary electron emission (valid for fast electrons) to slow ions.

The Frenkel'—Avak'yants—Izmailov mechanism is evidently responsible for a certain proportion of the emission caused by central collisions with a very high energy of the ions; however, the whole of the emission corresponding to the principal energy range can hardly be explained in this way.

Thus the large number of different theories proposed at various times in order to explain the phenomenon of kinetic ion—electron emission not only resulted in no final theory, but (in contrast to the case of potential emission) yielded no common point of view regarding the mechanism of electron excitation, since the theories in question were based on models disagreeing with each other as well as with experiment.

§2. Mechanism of Electron Emission from
Solids Under the Action of Atomic Particles

There are three possible sources of electron emission. The thermal and radiation theories consider the free electrons as the source of ion—electron emission. This means the electrons forming a cloud on the surface of the metal and the conduction electrons within the metal. The mechanical theory considers the source as being the electrons of the bombarding particle or electrons from the valence and deeper bands, and also electrons bound to the surface. It will be shown later that neither of these extreme points of view is correct. The process of electron emission breaks down into a number of stages, and in each of these electrons of a particular kind play the predominant part.

Transfer of the Kinetic Energy of the Ion
to the Electrons in the Solid

The energy which may be directly transferred to a free electron on collision with an ion is negligibly small. In fact, when an ion of mass m_2 and energy E_0 collides elastically with a free electron an energy $E_k = (m_e/m_2)E_0$ is transferred to the latter. Even for the light ion Ne^+ this energy only exceeds the work function of tungsten ($\varphi \approx 4.5$ eV) for an energy of $E_0 \approx$ 150 keV, i.e., in the energy range corresponding to the fall in emission. For the same reason it is impossible to heat the electron gas with lattice atoms set in intensive motion after collision with fast ions. Before a free electron can collect enough energy as a result of multiple collisions, the local "temperature" falls below the thermionic-emission threshold.

There remain bound electrons. The question as to which of the bound electrons (surface or volume) play the decisive role has been discussed. Ploch [221], Gurtovoi [144], and others inclined to the view that only surface electrons were excited. However, this is not so. The observation of an isotopic effect bears witness to the volume nature of the phenomenon. The isotopic effect can only be explained by the fact that ions of different mass lose energy and come to a halt on penetrating into a solid in different ways. The magnitude of the effect indicates that a considerable proportion of the electrons are excited in the internal layers of the lattice. The screening of the low-lying atomic layers recently observed in ion—electron emission from single crystals also indicates this.

It is well known from the theory of atomic collisions that the mechanism of electron excitation varies according to the relation between the forward velocity of the ion v_0 and the mean velocity of the bound electrons v_{max}. For $v_0 \gg v_{max}$ there is a sudden change in the potential acting on the bound electron and this produces excitation or ionization of the atom. For $v_0 \lesssim v_{max}$ (and this is the region most important for ion—electron emission) the motion of the electrons adiabatically follows the motion of the nuclei, and the probability of the excitation of an electron calculated by means of the adiabatic theory falls exponentially with increasing parameter $a\Delta E/\hbar v_0$. In this region the Born approximation is inapplicable, although special quantum-mechanical methods, for example, the method of excited stationary states, find a limited application in special cases (for one- and two-electron systems). The character of the collision between many-electron atoms is such that practically all the electron shells surrounding the nucleus experience deformation, mutual penetration, and perturbation (for $E_0 \approx 1$-100 keV the distance of closest approach is $R_0 \approx 10^{-9}$-10^{-10} cm).

The transfer of energy to the electrons is effected not by the impact of the ion as a whole but by the interaction of its electrons during mutual penetration of the electron shells, electrons from both the colliding atoms taking part in the energy-transfer process. Under these conditions, a statistical consideration of the excitation of the gas of bound electrons surrounding the "quasimolecule" formed by the colliding atoms is in order [454, 466]. On the one hand, the statistical method is necessitated because the process is a many-electron one, and the energy levels of the excited states are arranged very densely. On the other hand, the statistical method of Thomas and Fermi enables us to obtain fairly universal results quite simply, independently of the individual characteristics of electron-shell structure in the colliding atoms.

An analogous consideration was applied to the problem of ion—electron emission by Parilis and Kishinevskii [539], since the electrons of the inner shells of heavy metal atoms are weakly collectivized and form an ionic group similar in properties to an isolated atom.

In the course of collisions between an ion and a solid atom considerable electron exchange takes place. In view of the fact that the motion of the electrons is almost adiabatic, the exchange is accompanied by the transfer of the average momentum of forward motion acquired by the electron in the other atom, $m_e(\dot{R}_a - \dot{R}_b)$, through the surface S separating the regions in which the potentials of the two atoms act.* Certain dissipative effective forces act between the atoms:

$$F = \pm m_e(\dot{R}_a - \dot{R}_b) \int_S \frac{nv}{4} dS, \qquad (11.1)$$

analogous to frictional forces. Here, $\dot{R}_a$ and $\dot{R}_b$ are the velocities of the atomic nuclei, n is the concentration of the electrons in the Thomas—Fermi model,

$$n = 2^{3/2} (m_e\varphi)^{3/2}/3\pi^2\hbar^3, \qquad (11.2)$$

v is the average velocity of the electrons according to Thomas and Fermi,

$$v = \frac{3(3\pi^2)^{1/3}\hbar n^{1/3}}{4m_e} \qquad (11.3)$$

---

*When the charges differ by no more than a factor of 4, the surface may be replaced by a plane equidistant from the nuclei and perpendicular to the line joining them.

and φ is the potential at the surface S, which may be expressed in terms of the Thomas–Fermi function χ, the nuclear charges Z_1 and Z_2, the distance of the point on the surface from one of the nuclei r, and the Bohr radius $a_0 = \hbar^2/m_e e^2$:

$$\varphi = \frac{(Z_1 + Z_2)\,e}{r}\,\chi\left[1.13\,(Z_1 + Z_2)^{1/3}\,\frac{r}{a_0}\right]. \tag{11.4}$$

The work of the dissipative force on moving the atoms by $dR = dR_a - dR_b$ gives the electron excitation energy:

$$\mathscr{E} = m_e \int \left(\int_S \frac{nv}{4}\,dS\right)\dot{R}\,dR. \tag{11.5}$$

Or, on allowing for (11.2) and (11.3),

$$\mathscr{E} = \frac{m_e^2 e^2}{4\pi^2\hbar^3}\int\left(\int_S \varphi^2\,dS\right)\dot{R}\,dR. \tag{11.6}$$

If we confine attention to considering cases in which the nuclei of the colliding atoms move rectilinearly, i.e., collisions with large collision parameters, leading to scattering through small angles, we obtain the following simple formula for the amount of energy transferred [466]:

$$\mathscr{E} = \frac{0.35\,(Z_1 + Z_2)^{5/3}\,\dfrac{\hbar v_0}{a_0}}{\left[1 + 0.16\,(Z_1 + Z_2)^{1/3}\,\dfrac{R_0}{a_0}\right]^5}. \tag{11.7}$$

In order to extend the calculations, we need an expression for the amount of energy transferred in the case of any arbitrary collision parameters, without the simplification associated with rectilinear, uniform motion of the nuclei.

Classical consideration of the motion of the atoms (in this range of energy such a consideration is permissible for heavy particles) leads to the following expression for the energy transferred:

$$\mathscr{E}(P) = \frac{\hbar v_0}{\pi a_0^2}\,(Z_1 + Z_2)^2 \int_{R_0}^{\infty} \frac{\left[1 - \dfrac{V(R)}{E}\right]dR}{\sqrt{1 - \dfrac{V(R)}{E} - \dfrac{P^2}{R^2}}}\int_{R/2}^{\infty}\frac{\chi(\rho)\,d\rho}{\rho}, \tag{11.8}$$

where E is the energy of relative atomic motion, P is the collision parameter, and V(R) is the repulsive potential acting between the atoms at close quarters.

In the limiting case of small scattering angles (11.8) leads to (11.7).

Excitation of Electrons Belonging to the Metal

We now have the question as to the way in which the energy of excitation transferred to the electrons as a result of the kinetic energy of the ion is realized.

The first attempt was to associate this energy with the heating of the small volume of electron gas to which it was transferred, to ascribe a "temperature" to the electrons, determined by the average amount of energy transferred per electron, and, using the Fermi distribution for electrons in a metal, to calculate the emission of the electrons in the same way as thermionic emission.

However, the corresponding calculation showed that the energy spectrum of the emitted electrons obtained in this way varied continuously with varying velocity of the ion (and hence with varying transferred energy $\mathscr{E}$). As v_0 increased, there was a sharp rise in the maximum electron energy in this spectrum and the maximum moved sharply in the high-energy direction. At the same time experiment showed that the form of the kinetic ion—electron-emission energy spectrum was independent of the energy of the ion over a very wide range. We shall shortly consider this question in more detail; here we simply note that the fact in question indicates that the energy is communicated to electrons which are only capable of receiving and transferring a quite specific, large amount of energy. Hence, the excitation mechanism just discussed is only effective for bound electrons and cannot be applied to the free electrons of the metal. Excitation comprises the transition of a bound electron into the conduction band. This excitation may be considered as a kind of "ionization" of the atomic core of the metal.

We may estimate the upper limit of the cross section representing the single ionization of an atomic core by this mechanism; this equals

$$\sigma_1 = \pi P_1^2, \tag{11.9}$$

where P_1 is the collision parameter, in which $\mathscr{E}(P) = \delta_1 - \varphi$ (δ_1 is the depth of the hole in the filled band and φ is the work function of the metal).

Collisions with a large parameter are accompanied by the transfer of energy less than $\delta_1 - \varphi$, and hence insufficient to excite a bound electron.

The cross section calculated from (11.9) is insufficient to explain the value of the ion—electron emission and the principal laws governing the variations in this, particularly at high energies.

In order to obtain the correct behavior of the emission curve, it is desirable to estimate the cross section for the excitation of more than one electron into the conduction band. It is natural to base an approximate estimate of the cross section for this kind of excitation on the formula

$$\sigma = 2\pi \int_0^{P_1} \frac{\mathscr{E}(P)}{I} P \, dP, \tag{11.10}$$

where I is the average ionization potential for the outer shells of the atom. This quantity, which is known from the theory of the transmission of particles through matter, equals between 20 and 30 eV for metals.

For small collision parameters and an ion energy in the region of a few keV, the ratio $\mathscr{E}(P)/I$ is much greater than unity. Using (11.8) and (11.10) with the condition $\frac{1}{4} < Z_1/Z_2 < 4$, we may obtain the following expression for the cross section:

$$\sigma(v_0) = \frac{1.39 a_0 \hbar}{I} \left(\frac{Z_1 + Z_2}{\sqrt{Z_1} + \sqrt{Z_2}} \right)^2 S(v_0). \tag{11.11}$$

Two main regions of variation of $S(v_0)$ may be distinguished. These are due to different relations between the closest approach of the nuclei R_0 and the velocity v_0. For high velocities, R_0 is close to P, and for low velocities very different from this. Corresponding to these regions we have two regions for $\sigma(v_0)$: one of low velocities ($v_0 < 3 \cdot 10^7$ cm/sec) in which the relation $\sigma(v_0)$ is almost quadratic, and one of high velocities in which this relation is almost linear. The curve of $\sigma(v_0)$ has a very sharp threshold. By definition, the threshold velocity corresponds to $P_1 = 0$. For collisions with a lower velocity the amount of energy transferred fails to reach $\delta_1 - \varphi$, even for a central collision.

The integral

$$S(v_0) = \frac{1.28}{a_0^2} v_0 (\sqrt{Z_1} + \sqrt{Z_2})^2 \int_0^{P_1} P\,dP \int_{R_0}^{\infty} \frac{1 - \frac{V(R)}{E}}{\sqrt{1 - \frac{V(R)}{E} - \frac{P^2}{R^2}}}\,dR \int_{R/2}^{\infty} \frac{\chi^2 \left[1.13(Z_1 + Z_2)^{1/3} \frac{\rho}{a_0} \right]}{\rho}\,d\rho \qquad (11.12)$$

has to be found numerically. Despite the considerable differences in the masses of the ions and metal atoms, the $S(v_0)$ curves are very similar to one another; within the limits of the accuracy of the theory, they may be approximated by the function

$$S(v_0) = 5.25 v_0 \arctan [0.6 \cdot 10^{-7} (v_0 - v_{min})], \qquad (11.13)$$

where v_{min} is the threshold velocity. This differs for different pairs and equals $(0.6-0.7) \cdot 10^7$ cm/sec.

In the calculations we used the screened repulsive potential acting between the atoms [435]:

$$V(R) = \frac{Z_1 Z_2 e^2}{R} \chi \left[(\sqrt{Z_1} + \sqrt{Z_2})^{2/3} \frac{1.13\,R}{a_0} \right]. \qquad (11.14)$$

Coefficient of Ion—Electron Emission

If an ionizing collision between an ion and a lattice atom takes place inside the metal at a depth x, the secondary electrons will experience a number of collisions before they come out at the surface. It is usually considered that the flow of secondary electrons falls off in accordance with an exponential law $e^{-x/\lambda}$ with a range λ. Then the coefficient of secondary-ion—electron emission, i.e., the number of electrons emitted for one ion falling upon the surface, equals

$$\gamma = \int_0^{x_n} \sigma(v) W N e^{-\frac{x}{\lambda}}\,dx, \qquad (11.15)$$

where W is the probability that an excited electron will emerge from the metal, N is the number of metal atoms per cm³, x_n is the depth to which the ion retains its capacity to ionize, and v is the velocity of the ion within the metal.

The fall in the velocity v with depth must be taken into account. The mean loss in the velocity of the ion as a result of elastic collisions with metal atoms in a path dx equals [196]

$$\overline{dv} = - \frac{m_1 m_2}{(m_1 + m_2)^2} v N \sigma_d\,dx, \qquad (11.16)$$

where σ_d is the diffusion-scattering cross section, which for the potential of (11.14) reduces to the following form in the energy range considered:

$$\sigma_d = \frac{1.24 \pi a_0 e^2 Z_1 Z_2 (m_1 + m_2)}{v^2 (\sqrt{Z_1} + \sqrt{Z_2})^{2/3} m_1 m_2}, \qquad (11.17)$$

from which we obtain a simple law for the fall in velocity with depth of penetration

$$v_0^2 - v^2 = Kx; \qquad (11.18)$$

here, v_0 is the initial velocity of the ion and the quantity

$$K = \frac{2.48 \, \pi N a_0 e^2 Z_1 Z_2}{(\sqrt{Z_1} + \sqrt{Z_2})^{1/2}(m_1 + m_2)} \tag{11.19}$$

has the sense of a retardation. By definition, x_n is the depth in which the velocity falls to v_{min}.

Relation (11.18) enables us to convert to an integration with respect to velocities

$$\gamma = \frac{2NW}{K} \int_{v_{min}}^{v_0} v\sigma(v) \exp\left(-\frac{v_0^2 - v^2}{K\lambda}\right) dv = NW\lambda \, [\sigma(v_0) - \Delta\sigma(v_0)], \tag{11.20}$$

where

$$\Delta\sigma(v_0) = \exp\left(-\frac{v_0^2}{K\lambda}\right) \int_{v_{min}}^{v_0} \exp\left(\frac{v^2}{K\lambda}\right) \frac{d\sigma(v)}{dv} \, dv \tag{11.21}$$

allows for the reduction in the ionization cross section associated with the retardation of the ion in the metal.

According to (11.11) and (11.13),

$$\frac{d\sigma(v)}{dv} \sim \frac{dS(v)}{dv} \sim \frac{d}{dv} \, [v \ \arctan \ 0.6 \cdot 10^{-7} (v - v_{min})]. \tag{11.22}$$

The last derivative is closely approximated by the function

$$\frac{\pi}{2} - \frac{v_{min}}{v}\left[\frac{\pi}{2} - 0.6 \cdot 10^{-7} v_{min}\right]. \tag{11.23}$$

Whence

$$\Delta\sigma(v_0) = \exp\left(-\frac{v_0}{K\lambda}\right)\left\{\frac{\pi}{2}\sqrt{K\lambda}\left[\Phi\left(\frac{v_0}{\sqrt{K\lambda}}\right) - \Phi\left(\frac{v_{min}}{\sqrt{K\lambda}}\right)\right] - \right.$$
$$\left. - \frac{1}{2} v_{min}\left(\frac{\pi}{2} - 0.6 \cdot 10^{-7} v_{min}\right)\left[E_i\left(\frac{v_0^2}{K\lambda}\right) - E_i\left(\frac{v_{min}^2}{K\lambda}\right)\right]\right\}, \tag{11.24}$$

where

$$\Phi(x) = \int_0^x e^{-t^2} dt, \quad E_i(x) = \int_{-\infty}^x e^t \, \frac{dt}{t}.$$

The effect of the term $\Delta\sigma(v_0)$ is greatest at low velocities, when the path in which the incident particle ionizes the metal atoms is no longer than the distance from which the electrons escape from the metal ($x_n < \lambda$). For higher velocities, $x_n \gg \lambda$, so that the velocity of the ion within the layer supplying the secondary electrons remains almost constant. Here, $\Delta\sigma(v_0) \ll \sigma(v_0)$.

If we introduce the effective ionization cross section

$$\sigma^*(v_0) = \sigma(v_0) - \Delta\sigma(v_0), \tag{11.25}$$

the formula for the coefficient of ion—electron emission will take the simple form

$$\gamma = N\sigma^*(v_0) \lambda W. \tag{11.26}$$

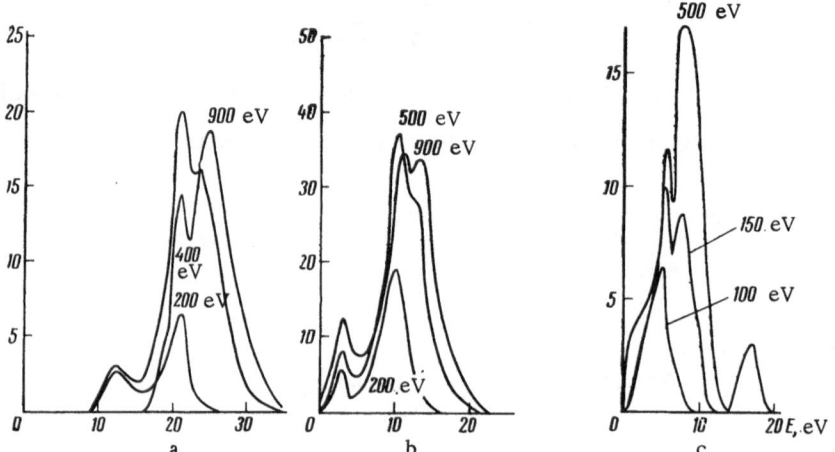

Fig. 11.1. Energy spectra of electrons emitted during the collision
of K^+ ions with: a) neon; b) argon; c) krypton [455].

Escape of Electrons from the Metal

In order to calculate W, i.e., the proportion of excited electrons leaving the metal, we
have to know their energy and angular distributions. The foregoing mechanism of energy transfer to the bound electrons of the metal certainly tends to favor a spherically symmetrical angular distribution of the excited electrons.

The matter is more complicated with the energy distribution. Unfortunately, the theoretical calculation of the energy spectrum of the electrons emitted in the course of ion—atom
ionizing collisions in the range of present interest is complicated by the fact that at the moment there are hardly any reliable experimental data on this subject. Only two investigations
[455, 580] have been carried out as a first attempt to measure the energy distribution of the
ionization electrons. We are forced to base our discussion on the laws which may be derived
from these sparse data and attempt the estimation of W in this manner [670].

Moe and Petsh [455] measured the energy distribution of electrons arising from the ionization of neon, argon, and krypton by the impact of K^+ ions for various energy values (Fig.
11.1). It may be noted that the curves contain either one or two dominant maxima close together; the positions of these and the shape of the whole spectrum are independent of the energy of the ion. The maximum energy rises slightly with increasing initial energy. If we suppose that the distribution $\bar{N}_i(\mathscr{E}_k)$ of excited electrons in the metal above the Fermi level (Fig.
11.2) has an analogous form, then, allowing for the geometrical escape factor

$$P_e(\mathscr{E}_k) = \begin{cases} \dfrac{1}{2}\left[1 - \sqrt{\dfrac{\mathscr{E}_0}{\mathscr{E}_k}}\right] & \text{for } \mathscr{E}_k > \mathscr{E}_0, \\ 0 & \text{for } \mathscr{E}_k < \mathscr{E}_0, \end{cases} \tag{11.27}$$

corresponding to a spherically symmetrical distribution, we obtain the energy spectra of the
electrons outside the metal (Fig. 11.3):

$$N_0(E_k) = N_i(\mathscr{E}_k)P_e.$$

Here the energy of the excited electron $\mathscr{E}_k$, the energy of the Fermi level $\mathscr{E}_F$, and the vacuum
energy level $\mathscr{E}_0$ are reckoned from the bottom of the conduction band, while E_k is the energy
of the electron outside the metal ($E_k = \mathscr{E}_k - \mathscr{E}_0$).

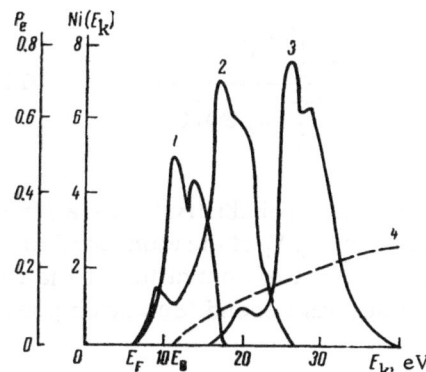

Fig. 11.2. Spectra of Fig. 11.1 inside the metal. 1) K^+—Kr, $E_0 = 150$ eV; 2) K^+—Ar, $E_0 = 500$ eV; 3) K^+—Ne, $E_0 = 400$ eV; 4) the function P_e.

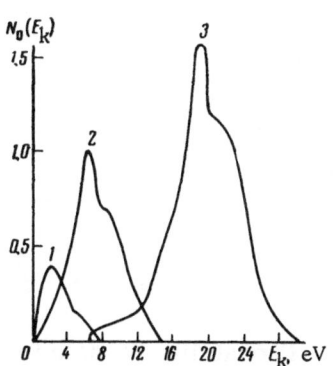

Fig. 11.3. Spectra of $N_0(E_k)$ obtained from the spectra of $N_i(E_k)$ (Fig. 11.2) for P_e in accordance with (11.27).

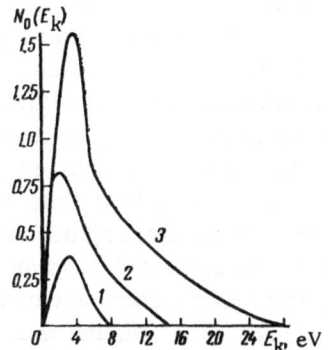

Fig. 11.4. The same spectra (as in Fig. 11.3) but after allowing for the cascade process of Wolff [313].

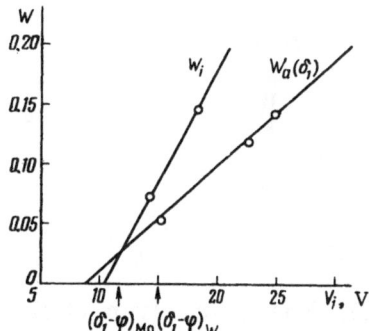

Fig. 11.5. Probability W_i as a function of V_i or $\delta_1 - \varphi$ and probability W_a as a function of δ_1. The arrows indicate the values of $\delta_1 - \varphi$ for molybdenum and tungsten.

A characteristic feature of these spectra is the independence of the position of the maximum and the shape of the spectra relative to the energy of the ion. These are only dependent upon the nature of the colliding atoms. The electrons arising from the depths of the material are not only scattered but also retarded, generating secondary electrons; as a result of this, there is an additional displacement of the maximum in the direction of lower energies and a slight leveling of the spectra for different initial energies (Fig. 11.4). Allowance for this cascade process [313] leads to a slightly different form for formula (11.27):

$$P_e = \frac{1}{2}\left[1 - \sqrt{\frac{\mathscr{E}_0}{\mathscr{E}_k}}\right]\left(\frac{\mathscr{E}_{init.}}{\mathscr{E}_k}\right)^x,$$

where x varies from 4 to 2 as $\mathscr{E}_k \, \mathscr{E}_F$ varies from 1 to 5. The ratio of the areas under the corresponding curves gives the probability of a direct ejection of an electron into the surrounding vacuum:

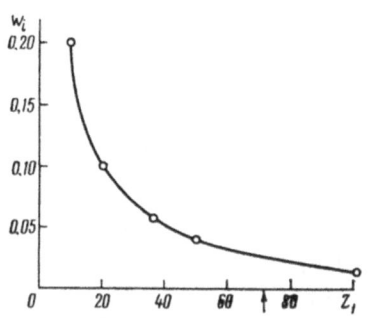

Fig. 11.6. Probability W_i as a function of Z_1. The arrow shows the Z_1 of tungsten.

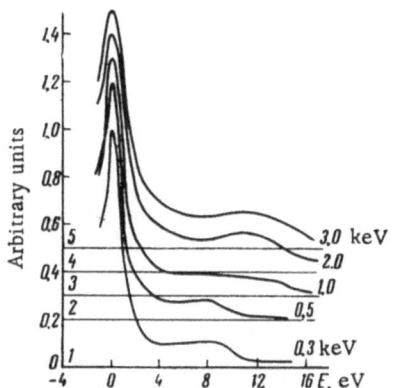

Fig. 11.7. Energy spectrum of electrons emitted on the collision of neutral Ar^0 atoms [580]. For clarity the curves relating to different energies are displaced in the vertical direction.

$$W_i = \frac{\int_0^{E_{max}} N_0(E_k)\, dE_k}{\int_{E_F}^{E_{max}} N_i(\mathscr{E}_k)\, d\mathscr{E}_k} . \qquad (11.28)$$

Figure 11.5 shows the probability W_i as a function of the ionization potential V_i of the atom bombarded (the probability of repeated ionization of the ion is very small, since its second ionization potential is high).

The relationship is linear and may be used to estimate the probability $W_i(\delta_i - \varphi)$. For tungsten, $W_i \approx 0.08$; for molybdenum, $W_i \approx 0.03$. These values are rather small. However, if we plot the relation for $W_i(Z_1)$ (this is quite simple: $W_i = 2/Z_1$, Fig. 11.6) and extrapolate to tungsten ($Z_1 = 74$) we find that there should be hardly any emission in heavy metals. This kind of extrapolation is probably inaccurate, since the mean energy of the electrons (and hence W_i) correlates with V_i and not Z_1. The relation between these latter existing for inert gases is not universal. On the other hand, according to Berry [580], the greater proportion of electrons ejected from Ar^+ by ions or atoms of the same element with energies between 0.3 and 3.0 keV have an energy far less than 4 eV (the spectrum being entirely independent of the energy of the ion, Fig. 11.7). This means that the greater proportion of the electrons should remain inside the metal. However, as we shall see later, this does not mean that no electrons will pass into the surrounding vacuum.

Auger Effect in Solids

Even in those extreme cases in which an electron ejected from the filled band into the conduction band lacks the energy required to overcome the potential barrier at the surface of the metal, the emission of an electron may still take place. The point is that the ionization of the atomic core of the metal is accompanied by the formation of a hole in the filled band. The ionization cross section of the atomic core $\sigma(v_0)$ is in the nature of things the cross section for the formation of an electron—hole pair.

Under the conditions existing in a metal, there is a high probability of the radiationless recombination of an electron from the conduction band with the hole so formed, and this is accompanied by the transfer of energy to another electron (the so-called Auger recombination, Fig. 11.8).

Let us estimate the probability of this effect. The problem is a two-electron one. In the initial state both electrons are in the conduction band and in the final state one is in the filled band and the second above the Fermi level.

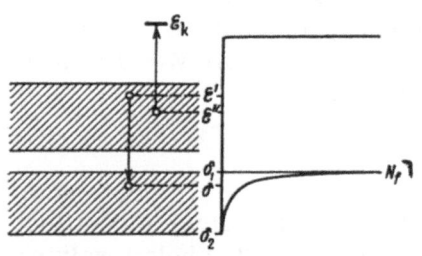

Fig. 11.8. Scheme of the Auger re-combination of an electron and a hole in the metal.

As the perturbation leading to the transition we may take the screened Coulomb interaction of the electrons

$$V(r_{1,2}) = \frac{e^2}{r_{1,2}} e^{-\frac{r_{1,2}}{\lambda_e}}, \tag{11.29}$$

where λ_e is the screening radius in the metal and $r_{1,2} = |r_1 - r_2|$. In accordance with perturbation theory (first approximation) the transition probability in unit time) equals

$$W = \frac{2\pi}{\hbar} N(\mathcal{E}_k) |V_{i,f}(r_{1,2})|^2, \tag{11.30}$$

where N is the density of the final states,

$$\mathcal{E}_k = \mathcal{E}' + \mathcal{E}'' + \delta - \mathcal{E}_0 \tag{11.31}$$

is the energy of the excited electron; $\mathcal{E}'$, $\mathcal{E}''$ and $\mathcal{E}_0$ are the initial energies of the electrons and the energy of the vacuum level, reckoned from the bottom of the conduction band, and δ is the depth of the hole, reckoned from the vacuum level.

The matrix element of the Auger transition has the form

$$|V_{i,f}| = \left| \int \psi_i^*(r_1) \psi_i^*(r_2) V(r_{1,2}) \psi_f(r_1) \psi_k(r_2) d\tau_1 d\tau_2 \right|, \tag{11.32}$$

where $\psi_i(r_i) = U(r_i) e^{i(k_i, r_i)}$ is the Bloch wave function of the initial state, ψ_f is the wave function of the final state of the electron in the ionized atomic core, and ψ_k is a function which for $\mathcal{E}_k > \mathcal{E}_0$ is the wave function of the free electron and for $\mathcal{E}_F < \mathcal{E}_k < \mathcal{E}_0$ (i.e., outside the metal) falls off exponentially. The density of the final states equals

$$N(\mathcal{E}_k) = \frac{4\pi \left(\frac{4}{3} \pi R^3\right)(2m_e)^{3/2}}{h^3} \mathcal{E}_k^{1/2} \tag{11.33}$$

(since $\mathcal{E}_k \gg \mathcal{E}_F$); here R is the radius of the range of interaction between the electrons in the field of the hole, which may be taken as equal to the screening radius

$$R \approx \lambda_e. \tag{11.34}$$

An exact calculation of the probability W is not required in our problem. We simply wish to estimate the order of magnitude of W in order to compare it with the probability of the competing form of recombination involving the emission of a photon. For this purpose we put the matrix element $|V_{i,f}|$ in the form [337]

$$|V_{i,f}| = N_e \frac{e^2}{R} \left(\frac{4}{3} \pi R^3\right), \tag{11.35}$$

where N_e is the density of conduction electrons in the metal. Then from (11.30), (11.33), (11.34), and (11.35) we obtain

$$W = \frac{64 \sqrt{2} \pi^3}{3} \frac{m_e^{3/2} e^4 \mathcal{E}_k^{1/2} \lambda_e^7 N_e^2}{h^4}. \tag{11.36}$$

In contrast to other recombination processes, the probability of Auger recombination is proportional to the square of the electron concentration in the conduction band N_e. In the case of tungsten, $N_e = 3 \cdot 10^{23}$ cm^{-3}, while for $\mathscr{E}_k \approx 10$ eV and $\lambda_e = 1$ Å, we obtain $W = 10^{15}$ sec^{-1}. Hence the lifetime of a hole in relation to Auger recombination equals

$$\tau_a = \frac{1}{W} \approx 10^{-15} \text{ sec.}$$

Another possibility of estimating the probability arises from the principle of detailed equilibrium. The process opposite to the recombination of an electron with a hole involving the transfer of energy of the order of 10–15 eV to another electron is the process in which the atomic core is ionized by impact with an electron of this kind of energy. The probabilities of the forward and reverse processes should be equal [231]. Hence,

$$W \approx \sigma_i v N_e, \tag{11.37}$$

where σ_i is the ionization (hole-formation) cross section corresponding to electron impact and v is the velocity of the electron. With $N_e = 3 \cdot 10^{23}$ cm^{-3} for $\mathscr{E}_k = 12$ eV (v $= 2 \cdot 10^8$ cm/sec) and $\sigma_i \approx 1 \cdot 10^{-17}$ cm^2 we obtain $W = 6 \cdot 10^{14}$ sec^{-1}, whence $\tau_a = 1.7 \cdot 10^{-15}$ sec^{-1}, which agrees with the previous estimate.

These estimates agree with the results of Pincherle [337], Sosnovskii [411], and Bess [412], who showed that even in semiconductors the radiationless impact recombination of electrons in the conduction band was far less than in the metal.

It is well known that the probability of the radiative capture of an electron by a hole under the same conditions is many orders lower. The lifetime in relation to recombination with the emission of a photon equals $\tau_p \approx 10^{-8}$ sec. Furthermore, in the metal a photon would immediately be absorbed as a result of the photoeffect, which would lead to the same result as Auger recombination. Hence, radiative recombination cannot compete with Auger recombination.

It is interesting that the time τ_a is comparable with the time of the atomic collision causing the formation of the hole, so that the emission takes place during the actual collision or soon after it.

Essentially speaking, the Auger recombination of an electron from the conduction band with a hole in the filled band considered in the foregoing is only the final stage in a series of Auger transitions accompanying the redistribution of energy among the electrons in the atom excited by the collision with an ion. The process of transferring the kinetic energy of the ion to the electrons just described is characterized by the fact that, in the course of the collision, the energy is transferred in small portions of the order of $m_e v_0^2/2$, comprising fractions of an electron volt at $v_0 \sim 10^7$ cm/sec. Hence, in the course of energy accumulation, an individual electron cannot acquire a considerable amount of energy at one stroke, and the energy transferred is in the first stage more or less uniformly distributed among all the electrons, but in such a way that none of them should have energy sufficient for ionization. Here we have an analogy with the state of a compound nucleus, with the essential difference, however, that the excitation energy of the nucleus is initially concentrated in one captured neutron and is only afterwards distributed among the neutrons, whereas, in atomic collisions, the relatively uniform initial distribution is determined by the actual process of "heating" the electron shells by "friction."

A state with a large number of excited electrons is unstable, and even after a time of the order of 10^{-15} sec the energy is concentrated on individual electrons by way of a series of Auger relaxations, the result of which is clearly represented by the spectra shown in Fig. 11.1. In these spectra we particularly notice the sharply discrete form, the independence of the position of the maxima relative to the energy of the ion, and the correlation between the system of maxima and the scheme of energy levels of the colliding atoms; in particular, the average energy of the electrons approximately equals the ionization potential of the atom [739, 787].

A knowledge of the probability of Auger recombination at every instant is still insufficient for calculating the probability of the emission of an Auger electron into the surrounding vacuum. This probability is associated with the angular and energy distributions of the excited electrons within the metal; these determine the capacity of the electrons to overcome the potential barrier at the surface.

Hagstrum [291], after calculating the coefficient of potential ion—electron emission for the Auger neutralization of the ion outside the metal, was forced to introduce an artificial form of angular distribution in order to match the theory with experimental results. In this distribution the probability P_Ω of emission into the solid angle $d\Omega$ in the direction of the polar angle θ had two discrete values:

$$P_\Omega = P_{\Omega_1} \text{ for } \theta < \theta_k,$$
$$P_\Omega = P_{\Omega_2} \text{ for } \theta > \theta_k \tag{11.38}$$

and symmetrically for the left-hand half space. The value of $f = P_{\Omega_1}/P_{\Omega_2}$, constituting an additional parameter of the theory, was chosen in such a way that the greater proportion of the electrons should be directed into precisely the solid angle for which escape from the metal was possible, i.e., θ_k was the angle in the direction of which the normal component of the kinetic energy of the electron was sufficient to overcome the barrier:

$$\cos \theta_k = \sqrt{\frac{\mathscr{E}_0}{\mathscr{E}_k}}. \tag{11.39}$$

According to Hagstrum's theory, f was independent of the distance from the ion to the surface of the metal, and also independent of the ion velocity.

Parilis showed that there is a possibility of describing the angular distribution of the excited Auger electrons quite naturally without introducing any additional arbitrary parameters apart from those already contained in the theory [620]. The probability of excited electrons escaping from the metal is here given by the expression

$$P_e(S, \mathscr{E}_k) = \frac{1}{2} \left\{ 1 - \sqrt{\frac{\mathscr{E}_0}{\mathscr{E}_k}} \exp \left[-aS \left(\sqrt{\frac{\mathscr{E}_k}{\mathscr{E}_0}} - 1 \right) \right] \right\}, \tag{11.40}$$

the factor $\frac{1}{2}$ being associated with the symmetry properties of the matrix element. The function P_e depends on S; as $S \to \infty$, the value of P_e tends to $\frac{1}{2}$ and as $S \to 0$, the value of P_e tends to $\frac{1}{2} \left[1 - \sqrt{\frac{\mathscr{E}_0}{\mathscr{E}_k}} \right]$ i.e., to the value obtained for a spherically symmetrical angular distribution of the excited electrons. Thus, the way in which P_e varies with S appears quite clearly.

We notice that, if the argument devised by Hagstrum to justify the form chosen for the angular distribution is developed consistently, then the quantity f, equal to the ratio of the matrix element of (11.32) for the wave function of the emitted electron to the value of (11.32) for the exponentially falling function of an electron unable to overcome the barrier, should also increase sharply with S, and this will lead to analogous behavior of $P_e(S, \mathscr{E}_k)$.

The formation of an electron—hole pair takes place in the direct collision of an ion with a metal atom (i.e., for $S \leq 0$). For Auger recombination we therefore take a spherically symmetrical distribution of excited electrons.

Let us now consider the energy spectrum of the Auger-recombination electrons. Figure 11.8 contains an energy diagram illustrating the scheme of Auger recombination. An electron with energy $\mathscr{E}'$ recombines with a hole occupying a level δ. The energy $\mathscr{E}_0 - \mathscr{E}_1 - \delta$ thus released is transferred to another electron lying at a level $\mathscr{E}''$. As a result of this, the electron is excited to a level with energy

$$\mathscr{E}_k = \mathscr{E}' + \mathscr{E}'' - \mathscr{E}_0 - \delta. \tag{11.41}$$

The minimum energy is established by Pauli's principle:

$$\mathscr{E}_k \geqslant \mathscr{E}_F, \tag{11.42}$$

the maximum is determined by the fact that $\mathscr{E}'$, $\mathscr{E}'' < \mathscr{E}_F$:

$$\mathscr{E}_{max} = 2\mathscr{E}_F - \mathscr{E}_0 - \delta. \tag{11.43}$$

The condition for emission into the vacuum takes the form

$$\mathscr{E}_{max} > \mathscr{E}_0, \tag{11.44}$$

whence

$$|\delta| > 2(\mathscr{E}_0 - \mathscr{E}_F), \quad \text{i.e.,} \quad \delta > 2\varphi. \tag{11.45}$$

This necessary condition for emission determines the boundary level:

$$\delta_b = 2\varphi. \tag{11.46}$$

The Auger recombination of holes lying above δ_b introduces no contribution into the emission of electrons into the vacuum. This enables us to determine the energy threshold of kinetic emission due to the recombination of holes more accurately. This corresponds to the transfer of energy $|\delta_b| - \varphi = \varphi$ in a central collision. This conclusion is interesting because, in a two-stage emission mechanism the minimum energy required for the escape of an electron from the metal is also no greater than φ; the free electrons of the metal are ultimately emitted.

In order to calculate the energy spectrum of an Auger electron and then the probability of the emission of such particles, we may put W in the form [291]

$$W(\mathscr{E}', \mathscr{E}'', \mathscr{E}_k) = F(\mathscr{E}', \mathscr{E}'') N(\mathscr{E}_k), \tag{11.47}$$

where $F(\mathscr{E}', \mathscr{E}'')$ is the probability that electrons with energies $\mathscr{E}'$ and $\mathscr{E}''$ will take part in the process, while $N(\mathscr{E}_k)$ is the density of the final states. Clearly,

$$F(\mathscr{E}', \mathscr{E}'') = \text{const } N_e(\mathscr{E}') N_e(\mathscr{E}'') \delta(\mathscr{E}' + \mathscr{E}'' - \mathscr{E}_0 - \mathscr{E}_k - \delta), \tag{11.48}$$

where $N_e(\mathscr{E}') = N_e(\mathscr{E}'')$ is the density of the electrons in the conduction band, and $\delta(x)$ is a delta function ensuring conservation of energy.

The total probability of the recombination of a hole at a level δ in unit time equals

$$R_t(\delta) = \iiint W \, d\mathscr{E}' \, d\mathscr{E}'' \, d\mathscr{E}_k. \tag{11.49}$$

In order to integrate with respect to $\mathscr{E}'$ and $\mathscr{E}''$ we introduce the variables $\mathscr{E}$ and Δ (Fig. 11.8):

$$\mathscr{E}' = \mathscr{E} + \Delta, \quad \mathscr{E}'' = \mathscr{E} - \Delta. \tag{11.50}$$

Then integration with respect to $\mathscr{E}$ and Δ gives

$$\int_0^{\mathscr{E}_F/2} \int_0^{\mathscr{E}} N_e(\mathscr{E}+\Delta)\,N_e(\mathscr{E}-\Delta)\,\delta\,(2\mathscr{E}-\delta-\mathscr{E}_0-\mathscr{E}_k)\,d\Delta\,\Delta\mathscr{E}\,+$$

$$+\int_{\mathscr{E}_F/2}^{\mathscr{E}_F} \int_0^{\mathscr{E}_F-\mathscr{E}} N_e(\mathscr{E}+\Delta)\,N_e(\mathscr{E}-\Delta)\,\delta\,(2\mathscr{E}-\delta-\mathscr{E}_0-\mathscr{E}_k)\,d\Delta\,d\mathscr{E}\,=$$

$$=\int_0^{\mathscr{E}_F} T(\mathscr{E})\,\delta\,(2\mathscr{E}-\delta-\mathscr{E}_0-\mathscr{E}_k)\,d\mathscr{E}=T\,[(\mathscr{E}_k+\mathscr{E}_0+\delta)/2], \qquad (11.51)$$

where

$$T(\mathscr{E}) = \begin{cases} \int_0^{\mathscr{E}} N_e(\mathscr{E}+\Delta)\,N_e(\mathscr{E}-\Delta)\,d\Delta & \text{for} \quad 0<\mathscr{E}<\dfrac{\mathscr{E}_F}{2}, \\ \int_0^{\mathscr{E}_F-\mathscr{E}} N_e(\mathscr{E}+\Delta)\,N_e(\mathscr{E}-\Delta)\,d\Delta & \text{for} \quad \dfrac{\mathscr{E}_F}{2}<\mathscr{E}<\mathscr{E}_F, \\ 0 & \text{for} \quad \mathscr{E}<0,\ \mathscr{E}>\mathscr{E}_F. \end{cases} \qquad (11.52)$$

The transformation $T(\mathscr{E})$, first introduced by Lander [271], is known as the Auger transformation.

If we put

$$N_e(\mathscr{E}) = \begin{cases} K & \text{for} \quad 0<\mathscr{E}<\mathscr{E}_F, \\ 0 & \text{for} \quad \mathscr{E}>\mathscr{E}_F, \end{cases} \qquad (11.53)$$

then $T(\mathscr{E})$ is a simple linear function with a break at the point

$$T(\mathscr{E}) = \begin{cases} K^2\mathscr{E} & \text{for} \quad 0<\mathscr{E}<\dfrac{\mathscr{E}_F}{2}, \\ K^2(\mathscr{E}_F-\mathscr{E}) & \text{for} \quad \dfrac{\mathscr{E}_F}{2}<\mathscr{E}<\mathscr{E}_F, \\ 0 & \text{for} \quad \mathscr{E}<0,\ \mathscr{E}>\mathscr{E}_F. \end{cases} \qquad (11.54)$$

The real distribution should be smooth owing to the natural broadening of the spectrum; it is due to the broadening of both the initial and final states. The former is broadened as a result of the very existence of a probability that an Auger recombination may take place; the latter contains holes in the filled part of the conduction band, and these in turn may suffer a secondary Auger process as a result of electrons lying high up in the band. Although the recombination of these holes, which as a rule lie above δ_b, cannot give any contribution to the emission, its probability broadens the final state. Landsberg [204] showed that this broadening was a maximum when the secondary hole lay at the bottom of the conduction band ($\mathscr{E}'$, $\mathscr{E}''=0$), and equal to zero when $\mathscr{E}'$, $\mathscr{E}''=\mathscr{E}_F$. Hence, the displacement of the lower boundary of $T(\mathscr{E})$ is due to the finite lifetimes of both the initial and the final states, while the upper boundary is only associated with the initial state. We are interested in the upper boundary. The displacement of this as a result of broadening may be taken into account by means of the factor

$$I(\Delta\mathscr{E}) = \frac{1}{(\Delta\mathscr{E})^2+\left(\dfrac{\hbar R_t}{2}\right)^2}. \qquad (11.55)$$

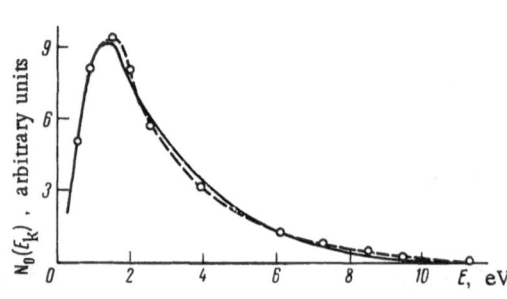

Fig. 11.9. Energy spectrum of kinetic ion—electron emission. Continuous curve, theoretical; broken curve, experimental [424].

then

$$T_1(\mathscr{E}) = T(\mathscr{E}) I. \qquad (11.56)$$

Returning to $R_t(\delta)$, we obtain

$$R_t(\delta) = \text{const} \int_{\mathscr{E}_F}^{\infty} N(\mathscr{E}_k) T_1 [(\mathscr{E}_k + \mathscr{E}_0 + \delta)/2] \, d\mathscr{E}_k. \quad (11.57)$$

The probability that an electron will be excited to the level $\mathscr{E}_k$, i.e., the energy distribution of the excited electrons $N_i(\mathscr{E}_k)$, equals

$$N_i(\mathscr{E}_k) = \int_{-\infty}^{\delta_b} \frac{N(\mathscr{E}_k) T_1 [(\mathscr{E}_k + \mathscr{E}_0 + \delta)/2] N_f(\delta)}{\int_{\mathscr{E}_F}^{\infty} N(\mathscr{E}_k) T_1 [(\mathscr{E}_k + \mathscr{E}_0 + \delta)/2] \, d\mathscr{E}_k} \, d\delta, \quad (11.58)$$

where $N_f(\delta)$ is the probability of the formation of a hole at a depth δ. Integration with respect to δ is carried out between $-\infty$ and δ_b. However, as already indicated, the mechanism of the transfer of the kinetic energy of the ion to the electrons of the atomic core is such that the probability of the formation of a very deep hole is low. In accordance with this, the depth distribution of the holes is expressed in the form

$$N_f(\delta) = e^{-f(\delta - \delta_1)}, \qquad (11.59)$$

where δ_1 is the threshold of the filled band.

The energy distribution outside the metal has the form

$$N_0(E_k, \delta_1) = N_i(\mathscr{E}_k, \delta_1) P_e(\mathscr{E}_k, \mathscr{E}_0), \qquad (11.60)$$

$$E_k = \mathscr{E}_k - \mathscr{E}_0. \qquad (11.61)$$

Figure 11.9 shows the theoretical and experimental $N_0(E_k)$ distribution curves for tungsten. The agreement of these curves indicates that the Auger-recombination electrons have the same spectrum as the other electrons participating in the emission. The probability of the emission of an electron into the vacuum as a result of Auger recombination is given by the expression

$$W_a(\delta_1) = \frac{\int_0^{E_{max}} N_0(E_k) \, dE_k}{\int_{\mathscr{E}_F}^{\mathscr{E}_{max}} N_i(\mathscr{E}_k) \, d\mathscr{E}_k}. \qquad (11.62)$$

Since $N_i(\mathscr{E}_k)$ is normalized to one hole, we have

$$\int_{\mathscr{E}_F}^{\mathscr{E}_{max}} N_i(\mathscr{E}_k) \, d\mathscr{E}_k = 1 \qquad (11.63)$$

and

$$W_a(\delta_1) = \int_0^{E_{max}} N_0(E_k) \, dE_k. \qquad (11.64)$$

This probability is illustrated in Fig. 11.5.

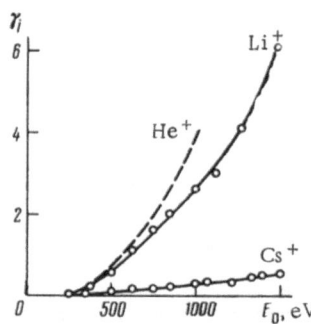

Fig. 11.10. Threshold of kinetic ion—electron emission according to [424]. The broken line shows the excess over the potential part of the emission for He$^+$ according to [262].

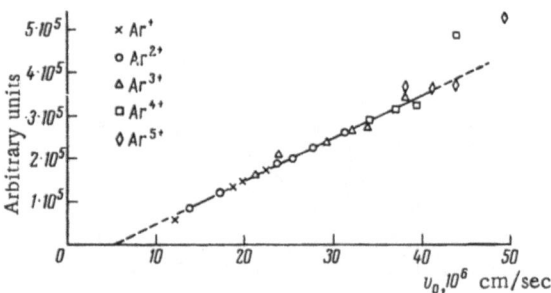

Fig. 11.11. Ion—electron emission from a copper—beryllium alloy (98% Cu) [746] under the action of Ar^{n+} ions (n from 1-6).

The total probability W to be substituted into formula (11.26) equals

$$W = W_i(\delta_1 - \varphi) + W_a(\delta_1). \qquad (11.65)$$

We shall now give a comparison of the theoretical conclusions with experiment.

Let us first consider the dependence of the ion—electron emission on the velocity of the ion. This comparison is very important, since the energy dependence of γ enables us to judge the mechanism of the process. The $\gamma(v_0)$ relation is entirely contained in the cross section $\sigma^*(v_0)$. We may distinguish three ranges of variation of $\sigma^*(v_0)$ due to the dependence of R_0 on P.

In the region near the threshold, the coefficient of ion—electron emission increases slowly as a result of the influence of $\Delta\sigma(v_0)$ (as a polynomial of the fourth degree in v_0), then in the range $(1-3) \cdot 10^7$ cm/sec, $\sigma^*(v_0) \sim v_0^2 - (\tfrac{3}{2}v_{min})^2$, which corresponds to a linear increase with energy, and finally in the high-velocity range, where $\Delta\sigma(v_0)$ is negligibly small, $\sigma^*(v_0)$ behaves in the same way as

$$\sigma(v_0) \sim v_0 \arctan 0.6 \cdot 10^{-7}(v_0 - v_{min}),$$

asymptotically tending toward the straight line $(v_0 - v_1)$, the continuation of which intersects the axis of velocities at the same point for a variety of ions, namely, at $v_1 = 1.05 \cdot 10^7$ cm/sec. It is exactly in this way that $\gamma(v_0)$ behaves experimentally.

We may consider that the existence of a threshold of kinetic emission for pure metals has been proved experimentally. Although this threshold is not expressed very sharply, we may reliably indicate a range of energies for which γ is practically zero. The position of the threshold on the experimental curves depends on the accuracy of the measurements. In the nature of things, in every experiment we consider the threshold to be that value of E_0 for which γ begins to exceed the minimum measurable value. Extrapolation from the region near the threshold is not very precise, since on approaching the threshold the function $\gamma(E_0)$ continuously changes its behavior. Watters [424] studied this region most carefully (down to $\gamma \sim 10^{-6}$) and found that the curve $\gamma(E_0)$ approached the E_0 axis more slowly than would follow from the coarser measurements (Fig. 11.10). From the point of view of theory, the relation $\gamma(E_0)$ might conveniently be replaced by the relation $\gamma(v_0)$, where v_0 is the velocity of the ion.

It is important to emphasize that the position of the threshold depends sharply on the cleanness of the surface. In fact, the observation of a region in which $\gamma = 0$ only became possible after a clean metal surface had been achieved [536]. For a surface covered with gas films kinetic emission is observed down to very low energies indeed [548].

In general, the dependence of kinetic emission on the presence of contaminations in the neighborhood of the threshold is so great that the measurement of γ is one of the most accurate methods of checking the cleanness of a surface.

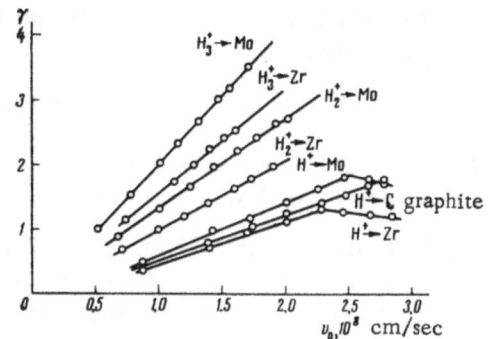

Fig. 11.12. The $\gamma(v_0)$ relationship in the linear region [349].

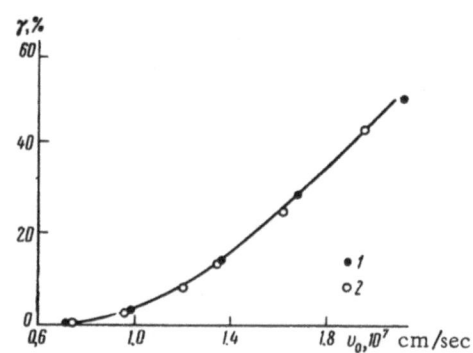

Fig. 11.13. The $\gamma(v_0)$ relationship for Ar^+ ions bombarding a tungsten surface. Continuous curve, theoretical. 1) [536]; 2) [504-506].

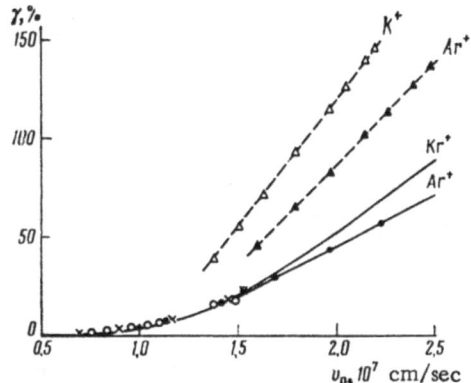

Fig. 11.14. The $\gamma(v_0)$ relationship for various ions bombarding molybdenum. Continuous curves, theoretical. Experimental points: Ar^+ [536]; Kr^+ [536]; K^+ [383]; Ar^+ [349].

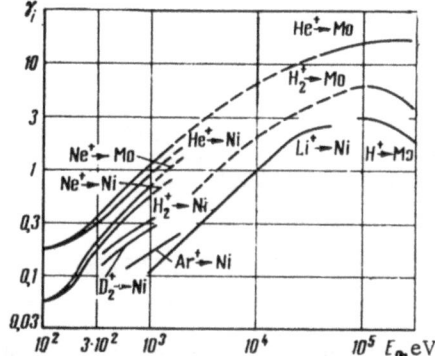

Fig. 11.15. Emission of electrons from molybdenum and nickel under the influence of light ions with energies between 100 and 10,000 eV [491].

According to the most reliable data, the threshold occurs at $v_{min} = (0.6-0.7) \cdot 10^7$ cm/sec, in good agreement with the theoretical value.

Medved, Magnuson, Carlston, and Layton, in a number of investigations, compare their experimental data with the theoretical conclusions and indicate excellent agreement in relation to the dependence of the emission on the velocity and the position of the threshold [661, 744]. The same is indicated by Klein [763] when comparing the experimental emission threshold with the calculated value.

Kistemaker and colleagues [746] also directed attention to the good agreement between the calculated and observed values of the threshold of kinetic emission and the manner in which this varied with the velocity of the ion (Fig. 11.11).

After the region near the threshold we may distinguish a region in which $\gamma \sim Kv_0$ is constant. This region on the $\gamma(E_0)$ curves corresponds to a linear rise in emission and is therefore sometimes called the linear range, but it is really more accurate to call it quadratic. The quadratic range extends up to about $v_0 = (1.6-2) \cdot 10^7$ cm/sec.

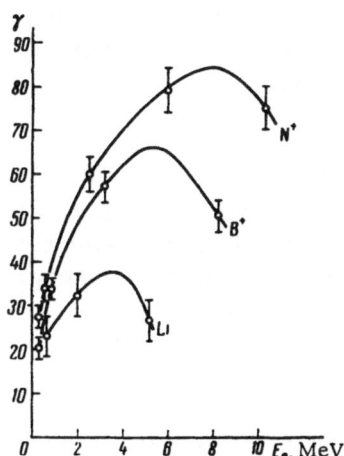

Fig. 11.16. Electron energy on bombarding a copper—beryllium alloy with Li^+, B^+, and N^+ ions in the range 1-10 MeV [461].

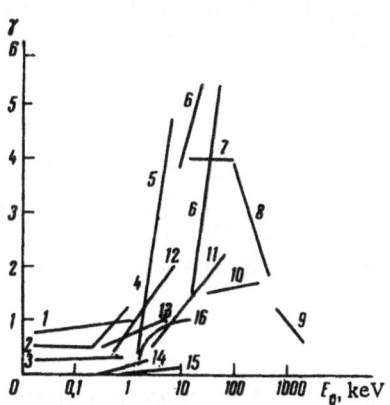

Fig. 11.17. Composite graph of the $\gamma(E_0)$ relationship according to the results of various authors. 1) [262]; 2) [61]; 3) [262]; 4) [279]; 5) [289]; 6) [310]; 7) [69]; 8) [135]; 9) [288]; 10) [138]; 11) [255]; 12) [221]; 13) [147]; 14) [109]; 15) [311]; 16) [256].

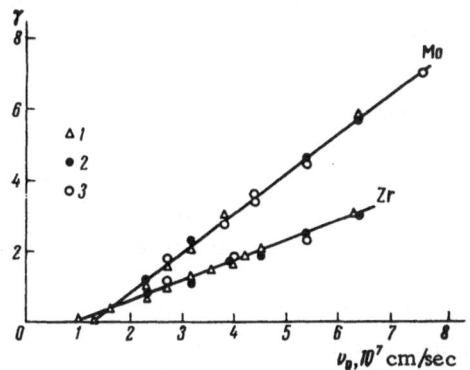

Fig. 11.18. Emission from molybdenum and zirconium under the action of the following ions: 1) Ar^+; 2) Ar^{2+}; 3) Ar^{3+} [349].

The linear range starts immediately after the quadratic and reaches up to $v_0 \sim (2.5\text{-}3) \cdot 10^8$ cm/sec. On the graphs we now have a slow rise (of the $\sqrt{E_0}$ type) with a gradual transition to a maximum. Both in the quadratic and in the linear regions, the inclination of the curves to the horizontal axis depends sharply on the cleanness of the surface. The dirtier the surface, the greater is the slope (i.e., the greater is γ). The slope also depends on the kind of ion (see Figs. 6.16 and 11.12; it is interesting to note that in Fig. 11.12 the continuation of all the straight sections intersects the horizontal axis at the same point $v_1 \approx 1 \cdot 10^7$ cm/sec).

Figures 11.13 and 11.14 show the theoretical and experimental curves of $\gamma(v_0)$ for a number of ions on tungsten and molybdenum. The theoretical curves are calculated from formula (11.26) with the following values of the parameters, taken from [352, 454]: $\delta_1 = 19$ eV and $\lambda = 11$ Å for tungsten; $\delta_1 = 16$ eV and $\lambda = 13.3$ A for molybdenum.

The average ionization potential I which has to be substituted in $\sigma^*(v_0)$ equals 39 eV for tungsten and 28 eV for molybdenum, in good agreement with the theoretical estimates of a number of authors [65, 176].

As the velocity of the ion increases further, a maximum appears on the experimental curves (Figs. 11.15-11.17) at about $2.5 \cdot 10^8$ cm/sec, which corresponds to the range between 100 keV and 10 MeV for various ions. This is usually explained by the penetration of the ions into the lower parts of the metal. Actually, this is in fact the ordinary maximum of the cross section for ion—atom inelastic collisions, associated with the disruption of the adiabatic motion of the atomic electrons. The presence of a maximum and the subsequent fall in the curve are satisfactorily described on the Born approximation [454], which is applicable in this range of velocities [224, 225].

The existence of a maximum may be demonstrated on the basis of the foregoing theory also, using the mechanisms of O. B. Firsov. In actual fact, the transferred energy $\mathscr{E}$, determining the degree of excitation of the electrons is roughly equal to

$$\mathscr{E} = \left(\frac{m_e v_0^2}{2} \right) n \frac{a}{v_0} , \tag{11.66}$$

where $m_e v_0^2/2$ is the mean excitation energy transferred from one atom to another by the electron; n is the number of exchanges per second; and a/v_0 is the time of the collision between the atoms. If we consider that as the velocity v_0 increases the degree of adiabatic adaptation of the electron and hence the energy transferred by the electron diminish, then

$$\overline{\mathscr{E}} = \frac{m_e v_0^2}{2} e^{-\frac{v_0}{v_{av}}} ; \tag{11.67}$$

the quantity $e^{-v_0/v_{av}}$ may be called the accommodation factor or coefficient of an electron belonging to one atom in the shell of the other, while $\mathscr{E}$ and γ with it will pass through a maximum at $v_0 = v_{av}$, the average velocity of orbital motion in the shell. After the maximum, in the range of high ion velocities, a fall will take place and this will lead to extremely low values of γ.

The general character of the energy dependence of the coefficient of ion—electron emission appears clearly on the composite graph of Fig. 11.17; the logarithmic scale of this enables us to follow the changes in γ over the whole range, starting from potential emission and ending on the falling kinetic branch.

Another factor requiring comparison with experiment is the independence of the kinetic emission relative to the charge on the ion. In contrast to potential emission, which increases sharply with increasing multiplicity of the charge on the ion, kinetic emission is almost independent of this. This independence can at any rate be traced up to eight- to ninefold charged positive ions. This is shown both by special direct experiments [267, 349, 746] (Figs. 11.18 and 11.19) and also by indirect data obtained in [536]. In the latter experiments it was shown that the kinetic emission under the action of inert-gas ions, determined by subtracting the potential contribution from the total emission, coincided with the kinetic emission associated with the neighboring alkali ions (within the limits of experimental error). This also constitutes an indirect confirmation of the fact that the emission is independent of the charge, since on approaching the metal surface the inert-gas ion is neutralized with a probability of almost unity [291] and collides with the surface as a neutral atom, whereas, in the case of the alkali ion, there is no such neutralization.

The additivity of the potential and kinetic emission is confirmed by the experiments of [536] (Fig. 8.17).

Independence of the charge on the ion is not characteristic of the radiation theory, but it is characteristic of the mechanism proposed (the charge on the ion is entirely absent from the theoretical equations).

Let us analyze the data relating to the dependence of γ on the nature of the ion. For kinetic emission the dependence on the nature of the ion is complicated, since it includes the dependence on the mass of the ion m_2 and on its atomic number Z_2. The difference between the masses of the ions, in particular, makes the relationship which may be derived from a comparison of the $\gamma(E_0)$ curves untypical, since ions of different mass have different velocities for the same energy. In order to eliminate this effect one must compare γ for different ions at the same velocity (Figs. 11.20 and 6.32).

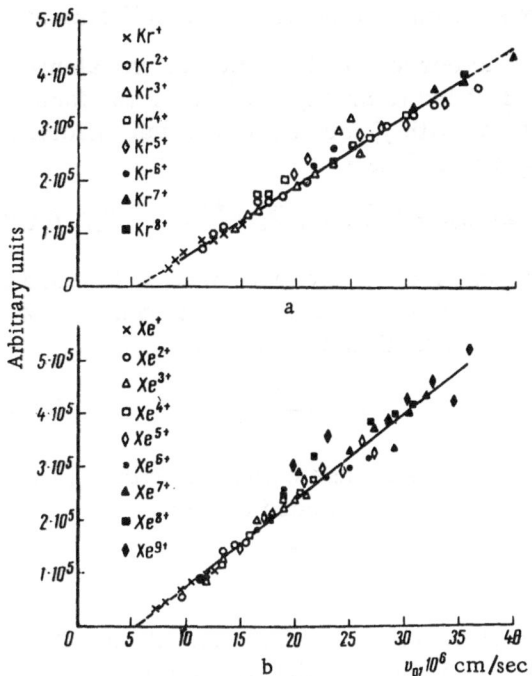

Fig. 11.19. Ion—electron emission from a copper—2% beryllium alloy [746] under the influence of Kr^{n+} and Xe^{n+} ions (a and b, respectively).

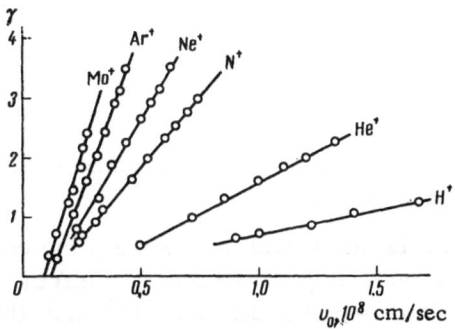

Fig. 11.20. Kinetic emission from molybdenum on bombarding with various ions [349].

In theory the dependence of γ on the nature of the ion is included in $\sigma^*(v_0)$. The cross section $\sigma^*(v_0)$ consists of two terms depending in different ways on the nuclear charges and masses of the colliding particles. The two terms have a common factor

$$f(Z_1, Z_2) = \left[\frac{(Z_1 + Z_2)}{\sqrt{Z_1} + \sqrt{Z_2}}\right]^2 . \qquad (11.68)$$

In addition to this, $\sigma(v_0)$ contains the function $S(v_0)$, which, as indicated earlier, in practice depends very little on the parameters of the particles. On the other hand, the term $\Delta\sigma(v_0)$ associated with the energy lost by the ion in elastic collisions depends greatly on the relation between the masses of the ion and the atom, by way of the retarding acceleration (retardation)

$$K \sim \frac{Z_1 Z_2}{(\sqrt{Z_1} + \sqrt{Z_2})^{1/2}(m_1 + m_2)}. \qquad (11.69)$$

The character of this relationship is such that $\Delta\sigma(v_0)$ is a maximum for $m_1 = m_2$ and falls for $m_1 \lesssim m_2$. For low velocities, at which the effect of $\Delta\sigma(v_0)$ is considerable, this relationship compensates the effect of the factor $f(Z_1, Z_2)$, so that γ almost coincides for different pairs of particles, as indicated by Figs. 11.14 and 11.15.

For high velocities, at which the term $\Delta\sigma(v_0)$ may be neglected,

$$\gamma \sim \sigma(v_0) \sim f(Z_1, Z_2) = \left[\frac{Z_1 + Z_2}{\sqrt{Z_1} + \sqrt{Z_2}}\right]^2 . \quad (11.70)$$

Thus, in Fig. 11.14, the ratio of the slopes for $Mo^+ \rightarrow Mo$ and $Ar^+ \rightarrow Mo$ is 1.32. The theoretical value of this ratio is 1.34. It should be noted that in the same investigation the slope of the straight lines for the lighter atoms apparently favored a different form of the function $f(Z_1, Z_2)$, which depended considerably on the electron-density distribution in the system formed by the colliding atoms (quasimolecule).

In the paper by Kistemaker and colleagues already mentioned [746], there is also good agreement between the experimental results and formula (11.70). This relationship is obeyed to an accuracy of 10% for Ne, Ar, Kr, and Xe ions.

It has been suggested by Parilis and Kishinevskii that when one of the two quantities Z_1 and Z_2 is small, the energy transfer is limited by the number of electrons in the light atom [539]. In this connection the following problem arose: to establish the amount of transferred energy calculated in [539] more accurately, allowing for the fact that the electron density is lower near the light atom than in the middle of the axis joining the nuclei of the colliding atoms.

The corresponding calculation (Kishinevskii [617]) led to an expression for the factor f_2 coinciding with f_1 for heavy ions but differing for light ions [618]:

$$f_2(Z_1, Z_2) = (\sqrt{Z_1} + \sqrt{Z_2})(\sqrt[6]{Z_1} + \sqrt[6]{Z_2})^3. \quad (11.71)$$

For comparison with experiment, it is more convenient to refer all values of γ to the case $Z_1 = Z_2$, i.e., to calculate the value of

$$F = \frac{\gamma(Z_1, Z_2)}{\gamma(Z_1, Z_1)}, \quad (11.72)$$

which in the first case coincides with f_1, i.e., $F_1 = f_1$, and in the second equals

$$F_2 = \frac{1}{16}\left(1 + \sqrt{\frac{Z_2}{Z_1}}\right)\left(1 + \sqrt[6]{\frac{Z_2}{Z_1}}\right)^3. \quad (11.73)$$

Figure 11.21 shows the functions F_1 and F_2, the broken line indicating the function F_1 in the range of Z_2/Z_1 values for which its use is illegitimate. For comparison, the curve of $F_3(Z_1, Z_2)$ plotted on the basis of experimental data from [349, 607] is presented. Thus, the proposed mechanism of kinetic ion—electron emission may be extended to light ions as well.

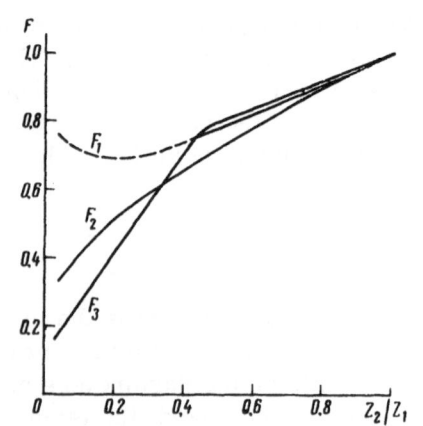

Fig. 11.21. Dependence of F_1, F_2, and F_3 on Z_1/Z_2.

Table 11.1

	$v_0 = 1.1 \cdot 10^7$ cm/sec	$v_0 = 0.75 \cdot 10^7$ cm/sec
Experimental	$+0.4 \pm 0.2$	$+0.5 \pm 0.2$
Theoretical	$+0.36$	$+0.45$

In order to separate out the dependence of γ on the mass of the ion in pure form there is only one possibility: to study the emission under the action of different isotopes of the same element. The "isotopic effect" is the name given to the quantity β, equal to

$$\beta = \left(\frac{\Delta\gamma}{\gamma}\right)_{v_0 = \text{const}} \left(\frac{\Delta m_2}{m_2}\right)^{-1}.$$

Here, $(\Delta\gamma)_{v_0 = \text{const}}$ is the difference in the values for the two isotopes for the same velocity v_0, while Δm_2 is the difference between the masses of the isotopes.

An attempt was made by Ploch [221] to measure the isotopic effect; however, the contamination of the surface prevented this from being observed. A positive isotopic effect was found by Brunnee [383] for ions of the isotopes K^{39} and K^{41} ($\beta = 0.4 \pm 0.2$) and Rb^{85} and Rb^{87} ($\beta = 0.5 \pm 0.2$). The potential emission was independent of the mass of the ion.

From formula (11.26) the isotopic effect may be obtained in the following way. We have

$$\left(\frac{d\gamma}{dm_2}\right)_{v_0 = \text{const}} = \frac{d\gamma}{dK}\frac{dK}{dm_2}. \quad (11.74)$$

Hence,

$$\beta = \left(\frac{\Delta\gamma}{\gamma}\right)_{v_0 = \text{const}}\left(\frac{\Delta m_2}{m_2}\right)^{-1} = -\frac{m_2}{\sigma^o}\frac{d}{dK}\Delta\sigma\frac{dK}{dm_2}. \quad (11.75)$$

Table 11.1 shows the experimental and theoretical values of the isotopic effect calculated from formulas (11.24) and (11.75).

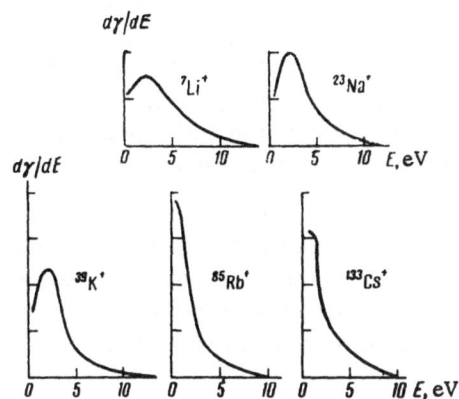

Fig. 11.22. Energy distribution of electrons for pure molybdenum ($E_0 = 0.5$ keV) [383].

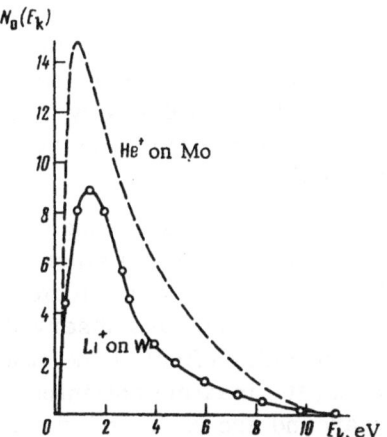

Fig. 11.23. Energy distribution of electrons for pure tungsten [424]. The broken line shows the kinetic-emission spectrum for molybdenum according to [262], $E_0 = 1000$ eV.

In addition to the coefficient of ion—electron emission γ the energy spectrum in usually measured in experimental work. Information regarding the energy spectrum of the electrons is no less important than information regarding the total yield of the emission, since it enables us to discover which electrons take part and what is the mechanism of their emission. Unfortunately, a systematic experimental study of the energy distribution of the secondary electrons and the dependence of this distribution on the various parameters characterizing kinetic ion—electron emission has never yet been carried out by anyone. Only disjointed measurements have been made for certain ions and metals. A careful measurement of the potential-emission spectrum was carried out by Hagstrum [262]. Kinetic-emission spectra were studied by Brunnee [383] (Fig. 11.22), Watters [424] (Fig. 11.23), Tel'kovskii [349] (Fig. 11.24), Pradal and Simon [427] (Fig. 11.25), and Arifov, Rakhimov, and Dzhurakulov [645].

It follows from the data presented in Figs. 11.22-11.25 that the kinetic-emission energy spectrum has a maximum at $E_k \approx$ 1-3 eV, the position of this (like the shape of the whole spectrum) not depending (within very wide limits) on the energy of the bombarding ions. The maximum energy of the electrons from pure metals is not greater than 10-20 eV. The maximum electron energy increases slightly with increasing energy of the ion; however, this increase is certainly not commensurable with the rise in E_0. As E_0 varies by several orders, the maximum energy increases by a few eV.

The energy distribution of the electrons for kinetic emission has a certain similarity to a Maxwell distribution. For this reason electrons are sometimes given an equivalent "temperature" (30,000-80,000°K); this, however, is independent of the ion energy [349, 427].

In the theory just presented, $W = W_i + W_a$, the two terms being roughly equal. Hence, the energy spectrum should be a combination of spectra of the type shown in Figs. 11.14 and 11.10. As a result of the first, the maximum energy of the electrons may exceed $\delta_1 - 2\varphi$. The two spectra are independent of the energy of the ion and correspond to the general character of the experimentally observed energy distributions of the electrons emitted (Figs. 11.22-11.25 and 8.18-8.22).

Abroyan [577], studying the kinetic emission from dielectrics, found that γ varied regularly with the width of the forbidden band (Fig. 11.26).

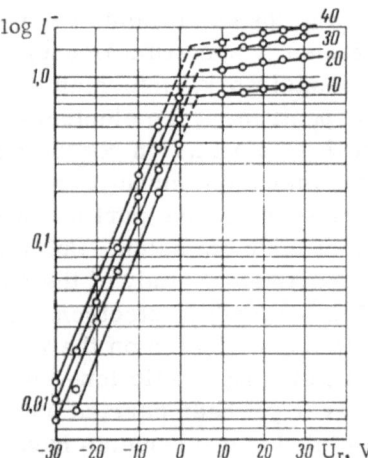

Fig. 11.24. The log I⁻ = $f(U_r)$ relationship according to [349]. I⁻) Current; U_r) retarding voltage. Figures on the curves give the energies of the ions in keV.

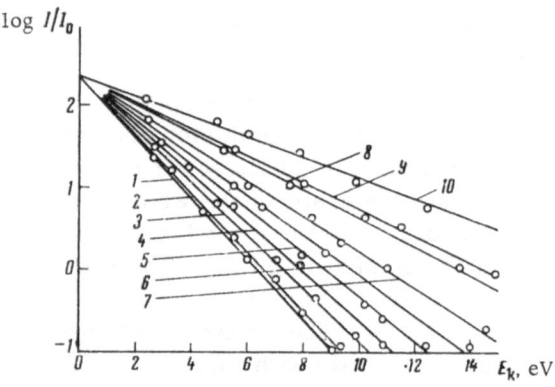

Fig. 11.25. The log (I/I_0) = $f(E_k)$ relation obtained on bombarding a number of metals with Ar^+ ions (E_0 = 40 keV). 1) Al, 31,500°K; 2) N, 33,000°K; 3) Cu, 36,000°K; 4) Co, 39,450°K; 5) Ta, 43,600°K; 6) Zn, 48,300°K; 7) Au, 55,200°K; 8) Mo, 68,200°K; 9) Ni, 72,500°K; 10) Pt, 89,000°K.

We note that the kinetic emission from dielectrics and metals coated with dielectric films exceeds the emission from a pure metal by a factor of several times [427, 578]. On the other hand, the potential emission falls sharply on forming a dielectric coating and may even vanish altogether [349, 536].

In some old investigations [61, 138] the relation between γ and the angle of incidence of the ion on the surface of a polycrystalline metal was considered; the relation was of the $\gamma \sim$ sec φ type.

Without going into the experimental reliability of the results, we may note that a relation of this kind might be derived naturally from the theory, since the emission is proportional to the path d sec φ traveled by the ion in the surface layer of thickness d which supplies the secondary electrons. Such a relation should occur in a polycrystalline aggregate, when the energy of the ion falls very little along the path and the electron yield is not very dependent on the depth.

The dependence of γ on the angle of incidence of an ion on the face of a single crystal was studied in [724, 790]. A complicated, non-monotonic $\gamma(\varphi)$ relationship was observed; this was due to the screening effect of the atomic layers. However, the maxima and minima of this curve lay around sec φ. This indicated that the law in question was valid for a polycrystalline surface.

In general the mechanism underlying the emission of electrons from a single crystal is no different from that relating to a polycrystalline aggregate. The difference in the angular relationships is only associated with the different "transparency" of the atomic layers for different directions of the bombarding beam. The change in this transparency with increasing temperature (associated with the vibrations of the atoms in the lattice) produces corresponding thermal effects [771, 790].

On the basis of the mechanism set out in the present chapter, Martynenko [771] calculated the ion—electron emission from the face of a single crystal, allowing for the screening of the deeper atomic layers by those lying nearer the surface; he obtained the correct relation between the emission and the angle of incidence of the ion on the single crystal and calculated the effect of the thermal vibrations of the atoms on the structural effects of electron emission.

Harrison and others [745] also used the mechanism in question to explain the laws governing the emission of electrons from a single crystal. These authors obtained the proportionality

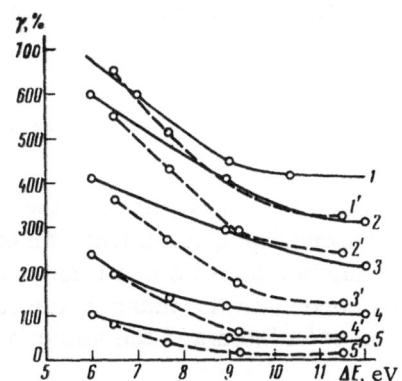

Fig. 11.26. Dependence of γ on the width of the forbidden band ΔE for E_0 equal to: 1) 6; 2) 4; 3) 2; 4) 1; 5) 0.5 keV. Continuous curve, theoretical; broken curve, experimental [577].

between the emission and the packing density of the face naturally arising from formula (11.75) and considered the limitations imposed on the value of the collision parameter by the dimensions of the crystal cell. The latter is unnecessary, as the parameter (11.8) and the cross section (11.25) are small compared with the cell dimensions. The calculation of the potential of elastic interaction between the ion and the lattice atom from electron-emission data seems a little strange. This method of calculating the potential cannot be regarded as either simple or accurate. It is far more natural to judge the potential from the angular distribution of the ions scattered by the lattice atoms.

§3. Theory of the Reflection of Atomic Particles from a Solid Surface

After Compton and Lamar's 1933 attempt [82] to calculate the accommodation factor and the associated coefficient of ion reflection, Avak'yants, Gurvich, and Umarov [266, 320] turned their attention to the theory of ion reflection in 1952, seeking the probability of the angular and energy distributions of the scattered ions in general form:

$$F(\theta, E) = \sum_{n=1}^{\infty} \varphi_n(\theta)\, \psi_n(\theta, E),$$

where $\varphi_n(\theta)$ is the probability density of the escape of an ion from the target at an angle θ to the normal after n collisions, and $\psi_n(\theta, E)dE$ is the probability that after n collisions an ion escaping at an angle θ will have an energy between E and E + dE.

The attempt to find specific expressions for $\varphi_n(\theta)$ and $\psi(\theta, E)$ encounter certain mathematical difficulties. Even for a scattering cross section spherically symmetrical in the laboratory system, the free range of the ion in the metal being independent of the energy, the second-order functions $\psi_2(\theta, E)$ contain integrals which cannot be expressed in elementary functions. Starting from the third-order term of the series, i.e., for $n \geq 3$, none of the functions $\varphi_n(\theta)$ and $\psi(\theta, E)$ can be represented in elementary terms. The general method has therefore not found any application in the creation of a theory of ion—ion emission.

More specific methods for calculating the reflection of ions and atoms from a solid surface were developed by Roos [392], who considered the process of secondary ion—ion emission as the collision of a flow of ions with a "gas" of atoms composing the solid. This kind of consideration is reasonable when all the collisions between the ions and the atoms may be regarded as paired, the ions losing energy only as a result of such paired collisions. With these assumptions the process of ion—ion emission may be described by methods developed for calculating the diffusion of neutrons in a solid, i.e., we may solve the kinetic Boltzmann equation:

$$\frac{\partial N(\mathbf{r}, \mathbf{v}, t)}{\partial t} = N\sigma(\mathbf{v})\,\mathbf{v} N(\mathbf{r}, \mathbf{v}, t) + N \int f(\mathbf{v}, \mathbf{v}')\, \mathbf{v}'\sigma(\mathbf{v}')\, N(\mathbf{r}, \mathbf{v}', t)\, d\mathbf{v}',$$

in which $N(\mathbf{r}, \mathbf{v}, t)$ is the distribution function of the ions moving in the metal; N is the number of metal atoms per unit volume; $\sigma(\mathbf{v})$ is the effective scattering cross section; and $f(\mathbf{v}, \mathbf{v}')$ is a function characterizing the probability that the velocity of the ion will change from $\mathbf{v}'$ to $\mathbf{v}$ on collision.

Roos [392] assumed that the binary interaction between the colliding ion and atom is described by a potential of the form

$$V(R) = \frac{A}{R} e^{-\frac{R}{r_0}},$$

(11.76)

and calculated the scattering cross section on the Born approximation. As a result of rather heavy calculations, using the double Laplace—Mellin integral transformation, Roos finds an expression for the scattering of ions in the form of the ratio of the number of reflected ions to the number of ions incident on the target. The resultant expressions depend on the mass of the ion, which enables the isotopic effect and the energies of the ions to be calculated and the velocity distribution considered.

The absolute coefficients of ion—ion emission calculated by Roos for an energy of 2000 eV were several times greater than those measured experimentally by Brunnee [383]. The reason for this discrepancy lies in a number of errors in Roos's calculations. It is well known that the calculation of the scattering cross section on the Born approximation is inadmissible for the motion of an ion or atom in a solid. Expression (11.76) for the interaction potential of the atoms is inaccurate, and Roos's expansion of the cross sections in series with restriction to the first term is inadmissible, since it involves discarding large terms. The true meaning of this error is that the Born-approximation cross section was in effect replaced by the spherical-symmetry cross section, which failed to reflect the original interaction potential between the atom and the ion. This accumulation of errors made the Roos calculations deviate by a factor of several times from experiment. If Roos had consistently used the cross section obtained on the Born approximation, the calculated values of the coefficients of ion—ion emission would have been many times smaller than the experimental.

In addition to this, the experimental results themselves were not really complete, since it was the total ion emission which was usually measured, not the reflection proper of ions from the surface of the solid in the course of collision.

In recent years it has been possible to separate the directly reflected particles reliably from the total emission accompanying the bombardment of solid surfaces with ion beams. Certain properties of this component have already been studied experimentally [383, 504, 536, 622, 624, 698]; however, there has not yet been any systematic study of this phenomenon as a function of all its defining parameters.

Certain laws of reflection observed experimentally or expected on the basis of simple considerations of the elementary act of interaction between the ions and the atoms in the solid were described and estimated quantitatively by Parilis and Turaev [699, 747, 749, 750, 779, 780, 785, 788, 792]. These concern medium and high energies (over 1 keV) and the high-energy part of the reflected-particle spectrum.

We shall call the bombarding and scattered particles "ions" and the atoms in the solid "atoms," although the nature of the scattering is evidently independent of the charged state of the particle.

The following model forms the basis of our considerations. We suppose that the solid has either a polycrystalline structure with a random distribution of single crystals or else constitutes an ideal single crystal, the smooth surface of which coincides with one of the faces. The effects of microrelief and the thermal vibrations of the atoms on the character of the reflection are considered separately. Only paired collisions of ions with lattice ions (considered free) are treated.

It is well known that in the range of energies under consideration the motion of the ions may be considered classical. For a correct description of the reflection of an ion it is important to choose the repulsive potential actually acting between the ion and the atom. In the theory of sputtering and radiation effects one usually employs several types of potentials: the solid-sphere potential [715], the Bohr [530], Born—Meyer [525], Firsov [380], and Abrahamson [582] potentials, and the Lehmann and Robinson potential [713].

The solid-sphere model proved to be completely inapplicable. The same applies to its modified version in which the radius of the sphere is taken as equal to the distance of closest approach in a head-on collision, i.e., it depends on the energy.

Since back scattering at high energies was being considered, the relatively simple screened Coulomb potential of Firsov [380] was selected:

$$V(r) = \frac{Z_1 Z_2 e^2}{r} \chi \left[(Z_1^{1/2} + Z_2^{1/2})^{2/3} \frac{r}{a_{\mathrm{TF}}} \right], \tag{11.77}$$

where χ is the Thomas—Fermi function, Z_1 and Z_2 are the nuclear charges of the ion and atom, r is the distance between them, and $a_{\mathrm{TF}} = 0.468$ Å.

The corresponding differential-scattering cross section in the energy range of present interest (1-30 keV) in the laboratory system of coordinates equals [435]

$$\sigma(\theta) = \frac{\pi^2 (m_1 + m_2) a_{\mathrm{TF}}^2 Z_1 Z_2}{m_1 (Z_1^{1/2} + Z_2^{1/2})^{2/3}} \frac{13.68}{E_0} \frac{\pi - \theta}{(2\pi - \theta)^2 \theta^2 \sin\theta}, \tag{11.78}$$

$$\theta = \beta + \arcsin(\sin\beta/\mu),$$

where β is the scattering angle in the laboratory system of coordinates; θ is the angle in the center-of-inertia system; E_0 is the initial energy of the ion; $\mu = m_1/m_2$ is the ratio of the mass of the atom m_1 to the mass of the ion m_2; corresponding to the angle θ, the collision parameter $P(E_0, \theta)$ equals

$$P(E_0, \theta) = \frac{a_{\mathrm{TF}} (13.68 Z_1 Z_2)^{1/2} (m_1 + m_2)^{1/2}}{m_1^{1/2} (Z_1^{1/2} + Z_2^{1/2})^{1/3} E_0^{1/2}} (\pi - \theta) \left[\frac{1}{(2\pi - \theta)\theta} \right]^{1/2}. \tag{11.79}$$

The energy $E_1(E_0, \beta)$ retained by the ion after a single elastic collision into an angle β, equals [436]

$$E_1(E_0, \beta) = \frac{E_0}{(1 + \mu)^2} [\cos\beta + \sqrt{\mu^2 - \sin^2\beta}]^2 = \frac{E_0}{(1 + \mu)^2} F^2(\beta, \mu) \tag{11.80}$$

and is determined by the conservation laws, independently of the type of potential. It is interesting that, strictly speaking, the latter assertion is invalid for scattering through a specified angle β as a result of two successive collisions with two atoms fixed in position. In this case the energy has the form

$$E_2(E_0, \beta_1, \beta) = \frac{E_0}{(1 + \mu)^4} F^2(\mu, \beta_1) F^2(\mu, \beta_2) \tag{11.81}$$

and depends on the interaction potential, since the value of the angles β_1 and β_2 is determined (in the plane case) by the equations

$$\sin(\psi - \beta_1) = \frac{P_1(E_0, \beta_1) - P_2[E(\beta_1), \beta_2]}{d}, \tag{11.82}$$

$$\beta_1 + \beta_2 = \beta,$$

ψ is the angle between the initial direction of the ion and the line-of-centers of the atoms (glancing angle), and d is the distance between them (Fig. 11.27).

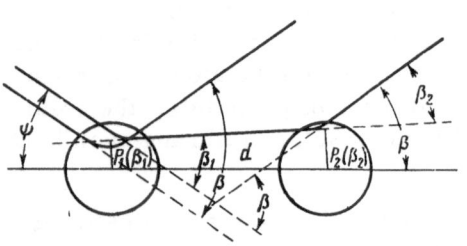

Fig. 11.27. Scheme of collisions.

In the general case, Eq. (11.82) also involves the azimuthal angle φ, while β_2 is determined by the equation

$$\cos \beta_2 = \cos \beta_1 \cos \beta + \sin \beta_1 \sin \beta \cos \varphi.$$

For calculating the inelastic energy loss $\mathscr{E}(\beta, E_0)$ there are no such exact formulas. Firsov's estimate [466] gives a value of $\mathscr{E}(E_0, \beta)$ approximately 50% lower than the experimentally observed value. In addition to this, experimental work [413, 492] showed a power dependence of $\mathscr{E}(E_0, \beta)$ on β. However, Firsov's formula

$$\mathscr{E}(E_0, \beta) = \frac{(Z_1 + Z_2)^{7/3} 4.3 \cdot 10^{-8} v_0}{[1 + 3.1(Z_1 + Z_2)^{1/3} \cdot 10^7 r_0(\beta)]^5} \tag{11.83}$$

remains as yet the only general formula usable in the absence of experimental data. Here, v_0 is the velocity of the ion, $r_0(\beta)$ is the distance of closest approach corresponding to scattering through an angle β. Following Filippenko [623], it was considered that the proportion of energy $\mathscr{E}_1(E_0, \beta)$, belonging to the ion equalled

$$\mathscr{E}_1(E_0, \beta) = \frac{Z_2}{Z_1 + Z_2} \mathscr{E}(E_0, \beta). \tag{11.84}$$

The method of analyzing the reflection of ions from a solid lay in distinguishing the ions reflected in a particular direction as a result of one, two, etc., collisions with the lattice atoms and discovering the special features characterizing each of these components individually as a function of the initial energy, the angles of incidence and escape, the orientation of the single crystal, and the temperature and surface relief of the latter. The energy and angular distributions of the scattered ions and the total reflection coefficient were calculated.

First of all, let us consider the energy distribution of the scattered ions. The most probable occurrence is the reflection of ions in a direction β as a result of a single collision with a lattice atom. The probability of this equals

$$K_1(E_1, \beta) = \frac{N}{\sin \psi} \int_0^\infty \sigma(E_0, \beta) \exp\left(-\frac{x}{\lambda(E_0)\sin\psi}\right) \exp\left(-\frac{x}{\lambda(E_1)\sin\alpha}\right) dx = \sigma(E_0, \beta) C(\beta, \psi, \alpha) \lambda(E_0) N, \tag{11.85}$$

where

$$C(\beta, \psi, \alpha) = \frac{\sin\alpha}{\sin\alpha + \dfrac{\lambda(E_0)}{\lambda(E_1)}\sin\psi} = \frac{\sin\psi}{\sin\alpha + \dfrac{(1+\mu)^2}{F^2(\beta)}\sin\psi}, \tag{11.86}$$

$\lambda(E_0)$ and $\lambda(E_1)$ are the ranges of the incident and reflected atoms; N is the number of atoms in 1 cm^3 of the solid. The factor $C(\beta, \psi, \alpha)$ gives the correct dependence of the probability of single scattering on the glancing and escape angles ψ and α, agreeing with experimental data [622, 698] (Fig. 11.28).

Experiment [536, 598, 624, 698, 708, 737] shows that the energy distribution of the ions reflected in a specific direction has a fairly sharp maximum at an energy of $E = E_1(E_0, \beta)$. The exception is the case $m_2 > m_1$, when single scattering is impossible for $\beta > \arcsin \mu$.

For $\beta < \arcsin \mu$ there should be two maxima corresponding to the two signs in formula (11.80). The intensities of these, according to elementary scattering theory, are proportional to $\sigma(\beta_1)$ and $\sigma(\beta_2)$, where β_1 and β_2 correspond to the two signs in the formula

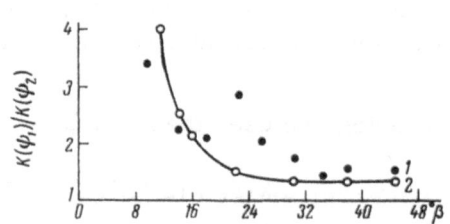

Fig. 11.28. Ratio $K_1(\psi_1) / K_2(\psi_2)$ for Ar^+—Cu at $E_0 = 25$ keV and glancing angles $\psi_1 = 4°$, $\psi_2 = 10°$. Points represent the experimental data of [708]; the continuous curve is theoretical.

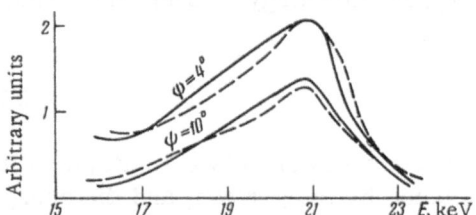

Fig. 11.29. Energy distribution of reflected Ar^+ ions from a copper surface at $E_0 = 25$ keV, $\beta = 30°$. Broken curves give the experimental results [708]; continuous curves, the theoretical.

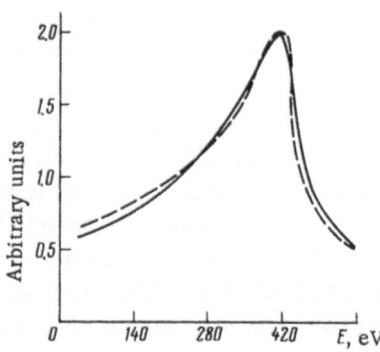

Fig. 11.30. Energy distribution of reflected Rb^+ atoms from a tantalum surface at $E_0 = 0.7$ keV, $\beta = 60°$, $\psi = 20°$. Broken curve gives the experimental results; continuous curve, the theoretical.

$$\cos \beta_{1,2} = -\frac{1}{\mu} (\sin^2 \beta \pm \cos \beta \sqrt{\mu^2 - \sin^2 \beta}). \quad (11.87)$$

For $\beta \rightarrow \arcsin \mu$ the maxima approach one another.

There is yet another exception. For glancing angles $\psi \leq \psi_e$, where

$$\psi_e \approx \arcsin \frac{P_1(E_0, \beta_1) - P_2(E_1, \beta_2)}{d}, \quad (11.88)$$

a single scattering through an angle β is impossible owing to the screening effect of the neighboring ions. Here d is the distance between the closest atoms on the surface in the reflection plane, and β_1 is the scattering angle at the screening atom. If this tangential collision leads to an energy loss

$$E_0 - E(E_0, \beta_1),$$

not exceeding the apparatus width of the principal maximum ΔE, we may speak of a single collision. The angle ψ_e varies with the energy of the ion and the plane of reflection. For Ar^+ ions traveling in the (110) plane and striking the (100) face of a copper target, for $E_0 = 30$ keV, $\Delta E = 250$ eV, $\beta = 70°$, $\psi_e = 8°$, and for $E_0 = 5$ keV, $\Delta E = 30$ keV, $\beta = 70°$, $\psi_e = 20°$. It should be noted that these estimates are based on an allowance for the mutual screening action of the two closest atoms on the surface. A more accurate result is obtained on allowing for scattering at a chain of atoms.

The part of the energy distribution of scattered ions lying outside the maximum is due to multiple scattering [598, 622]. It is interesting that, as shown by Parilis and Turaev [662] and Datz and Snoek [698], this relates not only to the low-energy but also to the high-energy part of the distribution, since the energy of an ion scattered through an angle β may be not only less than but also greater than $E_1(E_0, \beta)$. For twofold elastic collisions we find from (11.80) and (11.81) [747] that $E_2(E_0, \beta, \beta_1, \varphi_1) = E_1(E_0, \beta)$ in a cone

$$\cos \varphi_1 = \frac{1}{2 \sin \beta_1 \sin \beta} \left[(1 + \mu) \frac{F(\beta)}{F(\beta_1)} - 2 \cos \beta_1 \cos \beta \right], \quad (11.89)$$

where φ_1 is the azimuthal angle of the direction of motion of the ion after the first scattering. Outside the cone $E_2 < E_1(E_0, \beta)$ and inside $E_2 > E_1(E_0, \beta)$. This means that twofold collisions also introduce a contribution into reflection with energy $E_1(E_0, \beta)$.

The probability of a reflection with energy $E_2(E_0, \beta)$ at an angle β as a result of a twofold collision, together with $K_1(E_1, \beta)$ composes the main part of the total probability of scattering through an angle β.

In order to calculate the probability of a twofold scattering one uses the formula

$$K_2(E_2, \beta) = \sum \sigma(E_0, \beta_1) \sigma(E_1, \beta_2) C(\beta_1, \psi, \beta_1 - \psi) C(\beta_2, \psi - \beta_1, \alpha) \lambda(E_0) \lambda(E_1) N^2. \qquad (11.90)$$

The summation has to be carried out over cones of equal energies $E_2(E_0, \beta) = \varkappa E_0$,

$$\cos \varphi_1 = \frac{1}{2 \sin \beta_1 \sin \beta} \left[\frac{\varkappa(1 + \mu)^2}{F(\beta_1)} + \frac{(1 - \mu) F(\beta_1)}{(1 + \mu) \varkappa} - 2 \cos \beta_1 \cos \beta \right], \qquad (11.91)$$

relating β_1 and φ_1.

For multiple collisions the summation of the probabilities $K_i(E, \beta)$ must also be extended over all combinations of intermediate angles giving the specified energy as their final result. Finally,

$$K(E, \beta) = \sum K_i(E, \beta). \qquad (11.92)$$

For a polycrystalline aggregate we may consider that any intermediate scattering direction is realized with equal probability; summation is replaced by integration, and the energy distribution has a smooth character. Examples of such a distribution are shown in Figs. 11.29 and 11.30 [747].

In the case of a single crystal the situation is different, since the intermediate scattering angles β_1 and φ_1 can only take certain discrete values. For twofold collisions this means a discrete structure of the spectrum; this should evidently be observed experimentally on a general background due to multiple scattering.

The calculated probabilities are shown in relative units on a semilogarithmic scale in Fig. 11.31 as a function of energy for Ar^+ ions on copper and K^+ ions on tungsten. The (000) indices correspond to a single collision; the remaining indices denote the atom with which the second collision took place after a first collision with atom (000).

An approximate calculation by Parilis and Turaev [699] for Ar^+ ions striking copper at $E_0 = 25$ keV on the (100) face and K ions striking tungsten at $E_0 = 3$ keV also on the (100) face with various ψ and α, carried out with due allowance for elastic losses only, showed that the peaks of the twofold collisions had a considerable intensity, while the distance between them was fairly large. An attempt at experimental observation [779] led to the reliable separation of at least one peak, corresponding to a twofold collision along the [110] axis. The position of this peak agreed closely with its calculated position (allowing for inelastic energy losses), as in Fig. 11.32.

In order to carry out further calculations, a program was prepared for the M-20 electronic computer, introducing the crystallographic coordinates of the atoms in a crystallite of cubic symmetry containing about 60 atoms surrounding the atom (Fig. 11.33) with which the first collision took place, which was regarded as resting at the origin of coordinates. The computer calculation of scattering in a block of atoms forming part of a lattice of cubic symmetry revealed the fine structure of the spectrum as a function of the size and composition of the crystal lattice.

The calculated ratio of the probability of a twofold scattering to that of a single scattering in a specified direction [780] equals

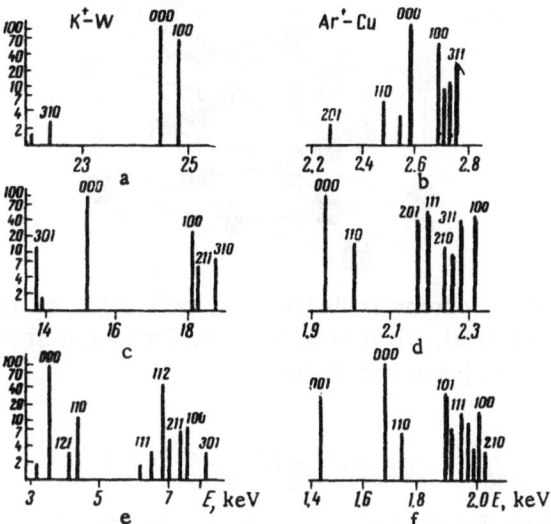

Fig. 11.31. Energy spectra of reflected Ar^+ ions from a copper single crystal ($E_0 = 25$ keV) for: a) $\psi = 4°$ and $\beta = 10°$, b) $\psi = 10°$ and $\beta = 50°$, c) $\psi = 70°$ and $\beta = 110°$; and of K^+ ions from a tungsten single crystal ($E_0 = 3$ keV) for: d) $\psi = 10°$ and $\beta = 50°$; e) $\psi = 45°$ and $\beta = 90°$; f) $\psi = 70°$ and $\beta = 110°$.

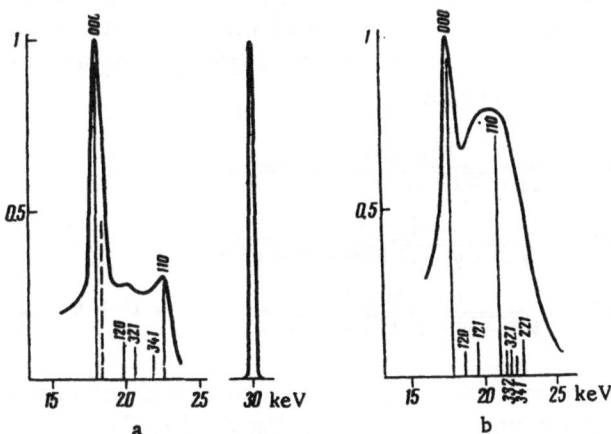

Fig. 11.32. Energy distribution of Ar^+ ions reflected from a copper single crystal [$E_0 = 30$ keV, reflection plane (110)]. a) Face (100), $\psi = 25°$, $\beta = 50°$; b) face (114), $\psi = 25°$, $\beta = 50°$. Peaks, theoretical; continuous curves, experimental; broken line indicates an isotopic peak for the single scattering of Ar^+ at the atom (000) Cu^{65} [749].

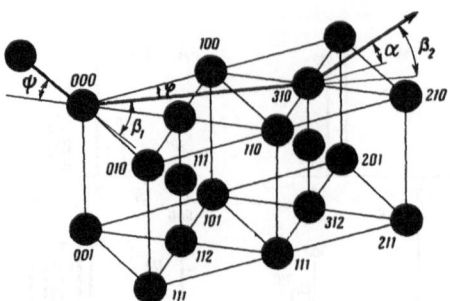

Fig. 11.33. Scheme indicating the reflection of an ion from a single crystal. Atoms in the side faces not shown; central atoms shown for a bcc lattice.

$$K_{1,2} = \frac{\sigma(\beta_1)\,\sigma(\beta_2)}{\sigma(\beta)}\;\frac{C(\beta_1,\,\psi,\,\beta_1-\psi)\,C(\beta_2,\,\psi-\beta_1,\,\alpha)}{C(\beta,\,\psi,\,\alpha)} =$$

$$= c(Z_1)\,\frac{R(Z_1,\,Z_2)\,\Phi(\beta_1,\,\mu_1)\,\Phi(\beta_2,\,\mu_2)}{d^2 E_0 F^2(\beta_1,\,\mu_1)\,\Phi(\beta,\,\mu_1)}\;\frac{C(\beta_1,\,\psi,\,\beta_1-\psi)\,C(\beta_2,\,\psi-\beta_1,\,\alpha)}{C(\beta,\,\psi,\,\alpha)}\,,$$

(11.93)

where

$$\Phi(\beta,\,\mu) = \frac{\left[\pi-\beta-\arcsin\dfrac{\sin\beta}{\mu}\right]\operatorname{cosec}\left[\beta+\arcsin\dfrac{\sin\beta}{\mu}\right]}{\left[2\pi-\beta-\arcsin\dfrac{\sin\beta}{\mu}\right]^2\left[\beta+\arcsin\dfrac{\sin\beta}{\mu}\right]^2}\,,$$

$$R(Z_1,\,Z_2) = \frac{Z_1 Z_2}{(Z_1^{1/2}+Z_2^{1/2})^{2/3}}\,, \quad c(Z_1) = 0.45\pi^2\,30.4\,\frac{m_1+m_2}{m_1}\left(\frac{a_{\mathrm{TF}}}{a_0}\right)^2,$$

$$a_{\mathrm{TF}} = 4\,68\cdot 10^{-9}\ \mathrm{cm};$$

here, Z_2, m_2, and Z_1, m_1 are the atomic numbers and masses of the incident ion and target atom, respectively; a_0 is the lattice constant; and $\mu = m_1/m_2$. The direction of incidence [hk$l$] and escape [mnp] and the coordinates (x, y, z) of the atoms of the single crystal were given in the form of Miller indices (Fig. 11.33); the scattering angles β_1, β_2 and total reflection angle β were also expressed in terms of these:

$$\cos\beta_1 = \frac{hx+ky+lz}{\sqrt{h^2+k^2+l^2}\,\sqrt{x^2+y^2+z^2}}\,,$$

$$\cos\beta_2 = \frac{xm+yn+zp}{\sqrt{x^2+y^2+z^2}\,\sqrt{m^2+n^2+p^2}}\,,$$

(11.94)

$$\cos\beta = \frac{hm+kn+lp}{\sqrt{h^2+k^2+l^2}\,\sqrt{m^2+n^2+p^2}}\,, \quad d^2 = x^2+y^2+z^2.$$

In setting up the program particular attention was paid not to increasing the size of the crystallite, but to ensuring the possibility of varying Z_1, Z_2, E_0, and μ, the type of crystal lattice, the lattice parameters, and the directions of incidence and reflection.

The program provided for possible differences in μ and Z due to the isotopic and chemical composition of the solid.

Figures 11.34 and 11.35 show some typical energy spectra calculated in this way. The indices (000) indicate the principal maxima; the secondary peaks are indicated by the indices of the direction of first scattering.

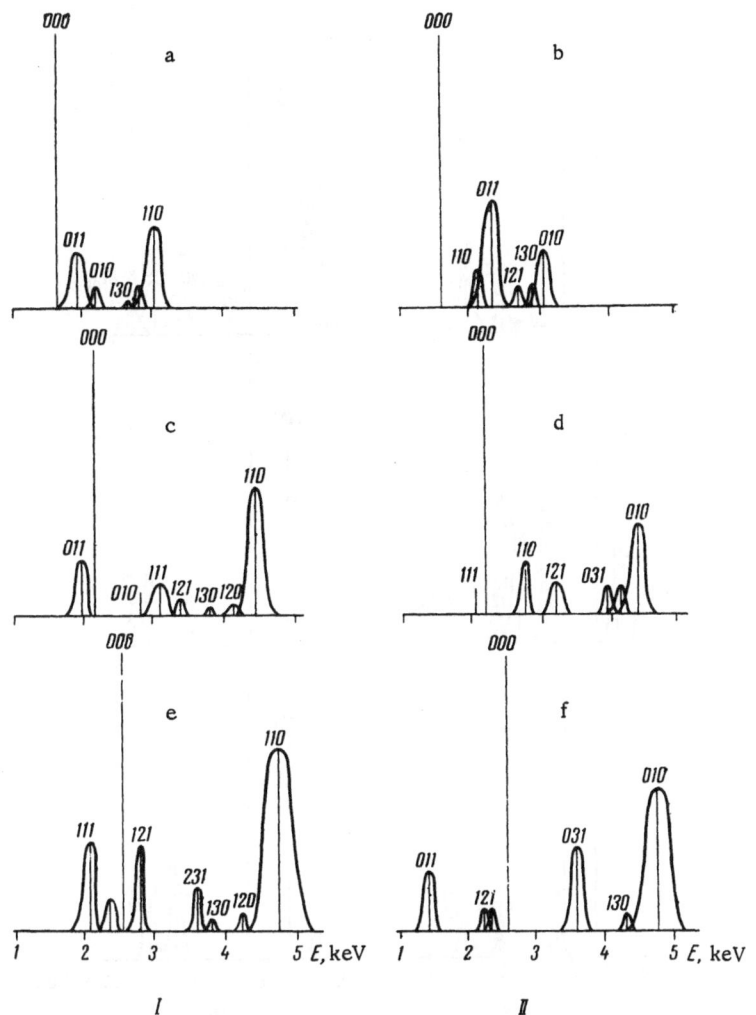

Fig. 11.34. Energy spectra of reflected K^+ ions on the (100) face of a nickel single crystal for $E_0 = 10$ keV and $\alpha = 50°$ in the case of reflection planes I (110) and II (100) for: a,b) $\psi = 70°$; c,d) 50°; e,f) 30°.

It is an interesting fact that in the case of a compound such as KCl there may be secondary peaks corresponding to the combinations K—K, K—Cl, Cl—K, Cl—Cl. The analogous quadruplication of the peaks corresponding to two isotopes (for example, Cu^{63} and Cu^{65}) could hardly be resolved experimentally. It should be mentioned, however, that Datz and Snoek [698] were able to resolve even the isotopic splitting of the principal peak.

The possibility of experimentally observing any particular peak is determined by its half width. The spread of the secondary peaks due to the thermal vibrations of the two atoms perpendicular to the line joining them has a half width

$$\Delta E_2 (\beta_1 \beta_2) = \frac{4E_2}{d} \sqrt{\frac{\ln 4}{\gamma}} \left[\frac{d}{d\beta_1} \ln F(\beta_1) - \frac{d}{d\beta_2} \ln F(\beta_2) \right], \qquad (11.95)$$

where γ is the parameter of the Gaussian distribution of the isotropic vibrations of the atoms at the lattice point. According to [340], $1/\gamma \approx 0.010 a_0^2 (T/T_{max})$; a_0 is the distance between the closest atoms in the lattice; T is the absolute temperature; and T_{max} is the melting point.

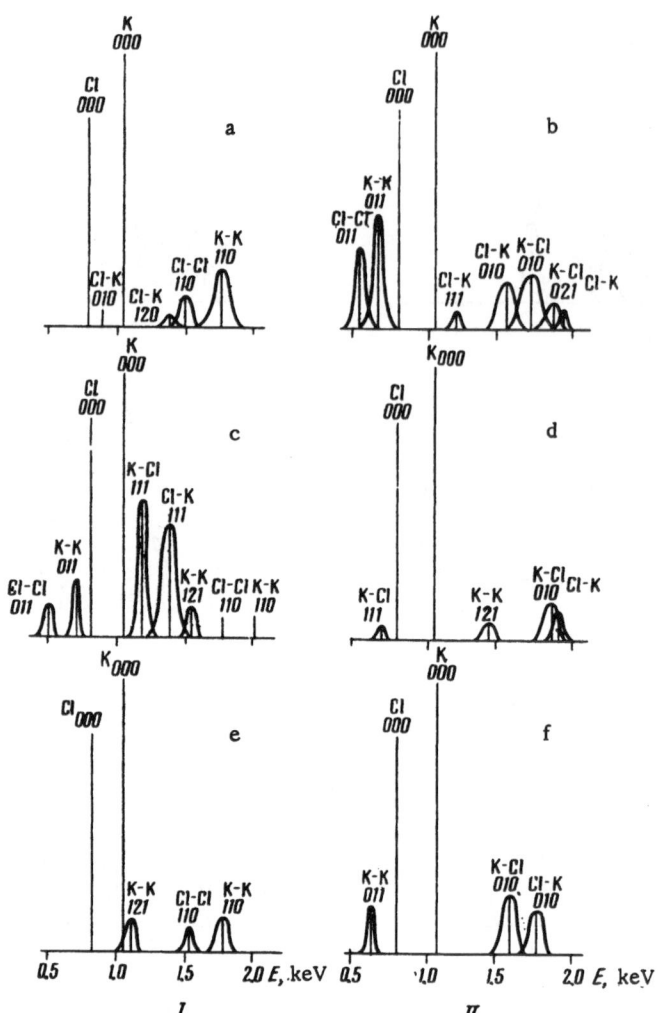

Fig. 11.35. Energy spectra of reflected Ne$^+$ ions on the (100)
face of a KCl single crystal for E$_0$ = 5 keV and β = 100° in the
case of reflection planes I (110) and II (100) for: a,b) ψ = 70°;
c,d) 50°; e,f) 30°.

Owing to the opposite changes in β_1 and β_2 during the vibration of the atoms the half width
of the peaks (11.95) is fairly small, constituting about 5–15% of their energy. The narrowest
peaks correspond to similar values of β_1 and β_2. The peak width includes a contribution from
crystal-structural defects, dislocations, etc. The increase in the intensity of a peak due to the
vibration of the atoms along the axis joining them, $\Delta K_2 = K_2(3/\gamma d^2) = 0.03\,(a_0^2/d^2)(T/T_{max})K_2$
is no greater than 1–2% at room temperature. Thus the fine structure of the energy spectrum
just described is best sought in the high-energy part of the spectrum for similar values of β_1
and β_2, high values of μ, and reasonably low temperatures.

It follows from experimental work [504, 536, 614, 710] that the maximum observed ion
energy E$_f$ is not proportional to E$_0$. The ratio E$_f$/E$_0$ falls with increasing E$_0$. Usually, this
fact is regarded as a contradiction of formula (11.81) and as a manifestation of coupling between the target atoms or the simultaneous collision of an incident ion with several surface
atoms [504, 540, 614].

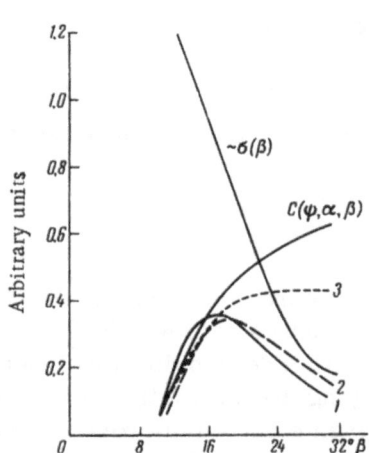

Fig. 11.36. Angular distribution of reflected Ar^+ ions from a copper surface at $E_0 = 25$ keV. 1) Theoretical curve; 2) experimental [622]; 3) solid-sphere model.

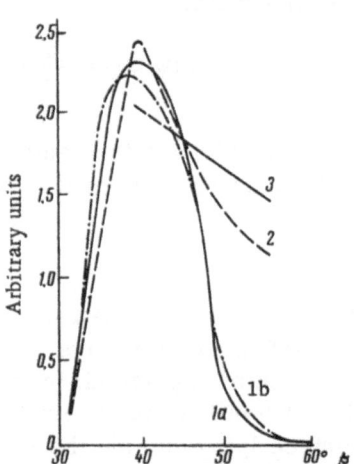

Fig. 11.37. Angular distribution of reflected Cs^+ ions from a molybdenum surface at $E_0 = 0.7$ keV. 1a) Theoretical curve; 1b) experimental [624]; 3) solid-sphere model.

It should be remembered, however, that the maximum observable energy of the ions depends on the sensitivity of the recording apparatus. We can only speak of an energy E_f corresponding to a specified low probability $K(E_f, \beta) = K_f$.

The latter equation relates the combination of the intermediate angles to E_0. For example, in the case of twofold scattering we obtain a falling function $\beta_1(E_0)$. In accordance with (11.81), E_f / E_0 is a rising function of β_1 (up to $\beta_1 = \beta/2$). Hence, even in the scheme of paired collisions E_f / E_0 falls with E_0. This must be taken into account when attempting to calculate the effective mass of the scatterer from the relation between E_f / E_0 and E_0 [614].

The angular distribution of the ions scattered from a solid polycrystalline surface constitutes a curve falling away sharply both in the high- and low-angle direction [622, 624, 697].

This may be attributed to the superimposition on the monotonically falling function

$$K(\beta, \psi) = \int K(\beta, \psi, E) \, dE$$

of a series of factors, the influence of which may be more conveniently examined for the case of single scattering, which supplies the majority of the reflected ions.

The results of the numerical integration are shown in Figs. 11.36 and 11.37. For comparison, we also show curves (curves 3 in both figures) corresponding to the solid-sphere model and a curve corresponding to an effective scattering-atom mass of $m_{eff} = 2m_1$ (curve 2 in Fig. 11.37). These curves differ sharply from the experimental.

In Fig. 11.37, $m_2 > m_1$. We know that in this case there is a limiting angle of single scattering β_{lim}. For Cs^+ ions striking molybdenum, $\beta_{lim} = 46°$. The falling branch for $\beta > \beta$ is naturally obtained as a result of multiple scattering, while the rising branch for $\beta \geq \psi$ is due to the factor $C(\beta, \psi, \alpha)$.

Microrelief of the surface reduces the intensity of reflection at small angles; its effect may be roughly modeled by means of a periodic relief of rectangular form, reducing to the inclusion of the factor

$$f(\psi, \alpha) = \frac{a}{a+c} + \begin{cases} \dfrac{c - h\,(\cot\psi + \cot\alpha)}{a+c} & \text{for} \quad \cot\psi + \cot\alpha < \dfrac{c}{h}, \\ 0 & \text{for} \quad \cot\psi + \cot\alpha \geqslant \dfrac{c}{h}. \end{cases} \tag{11.96}$$

Here, a is the width of the rectangular tooth; h is its height; and c is the distance between the teeth. The quantity $f(\psi, \alpha)$ lies between the limits

$$\frac{a}{a+c} < f(\psi, \alpha) < 1,$$

when $0 < \psi$, $\alpha < \pi/2$. The real profile of the relief is of course more complicated. By modeling the relief in various ways, we may obtain a family of $f(\psi, \alpha)$ curves with which to compare the experimental angular distributions. The use of ion beams with a de Broglie wavelength of the order of 0.001 Å may provide a unique potentiality for studying the microstructure of a solid surface lying outside the resolving power of optical and even electron microscopes [788]. This relates especially to microrelief caused by the ion bombardment itself.

For a smooth polycrystalline surface, the fall in the intensity of the beam reflected at small angles as a result of the increased length of the path in the surface layer corresponds to $C(\beta, \psi, \alpha) \rightarrow 0$ as $\alpha \rightarrow 0$. Even on the absolutely smooth surface of the face of a single crystal, however, the screening action of the atomic chains arranged in the reflection direction should have an appreciable effect. The first atom in this direction produces an additional scattering at an angle corresponding to the parameter

$$P_2\,[E_1(\beta_1), \beta_2] = P_1(E_0, \beta_1) + d\sin(\beta_1 - \psi), \tag{11.97}$$

as a result of which the escape angle $\alpha = \beta_1 + \beta_2 - \psi$ not only does not tend to zero when $\beta_1 \rightarrow 0$ but, starting from a certain $\alpha_{\min}$, rises again. The angle $\alpha_{\min}$ is found by simultaneous solution of Eq. (11.97) and the condition for the minimum of α:

$$\frac{dP_1}{d\beta_1} + d\cos(\beta_1 - \psi) = -\frac{dP_2}{d\beta_2}. \tag{11.98}$$

The minimum angle falls with energy and distance d; it is interesting to follow this experimentally for various orientations of the reflection plane with respect to the crystallographic axes.

Scattering at the forward-lying atoms leads to a fall in intensity even near the limiting angle for $\alpha \geq \alpha_{\min}$; it is this which causes the falling branch of the angular distribution on a smooth surface.

This does not only relate to the surface of the crystal. Reflection from deeper layers is screened in a similar manner by the atoms lying at higher levels, and we should expect a fall in the reflection in directions constituting a continuation of the low-index axes. This kind of special volume effect should be observed for high ion energies when the effective scattering cross sections are small, the crystal is fairly transparent, and the low-lying layers introduce a considerable contribution into the reflection. The effect was observed by Tulinov [772] when firing protons accelerated to 3 MeV in a cyclotron through a tungsten single crystal.

The limitation of the reflection at large angles is also associated with the screening action of the surface atoms. On continuously reducing ψ to small values, the surface atoms gradually come into the shade cast by the neighboring atoms. First of all, the parameters corresponding to scattering within the solid become unattainable, then this happens for small reflection parameters (large angles). The angular distribution, as it were, "cuts off" at an angle

β_{max} determined by the equation

$$P_2\,[E_1(\beta_1),\,\beta_{max}] = (R - d\sin\psi)\cos\psi,\qquad(11.99)$$

where $R = 2.2[l\,b^2/(1+\mu)]^{1/3}$ is the radius of the shadow at a distance l from the atom (this radius was calculated by Martynenko when solving the problem of the sputtering of a single crystal [714]); β_1 is the scattering angle corresponding to R, determined by the parameter

$$P_1(E_0,\,\beta_1) = \Big(\frac{\pi d\cos\psi b^2}{1+\mu}\Big)^{1/2};$$

b is the distance of closest approach for a head-on collision with energy E_0.

Thus, the reflected beam is limited by escape angles $\alpha_{min} < \alpha < \beta_{max} - \psi$. On reducing ψ the upper limit falls rapidly and the reflected beam contracts. It is evidently here that we should seek the explanation of the "mirror reflection" observed for low energies and small angles of incidence [48, 748], for which the effect in question intensifies. Firsov [789] showed that mirror reflection should be observed from a polycrystalline surface also for glancing angles of incidence and a fairly high energy of the ions.

Both the existence of a minimum scattering angle and the high-angle scattering cutoff apply to the deeper layers of the solid and to other crystallographic directions as well as to the surface. As a result of the shading of the atoms in the lower layers by those in the upper, the contribution of the former to the scattering at an angle β vanishes when the direction of incidence of the ions approaches the axis of close packing at an angle less than τ. The equations for τ are analogous to equations (11.99):

$$\cos\tau\,(R - d\sin\tau) = P_2\,[E_1(\beta_1),\,\beta],$$
$$P_1(E_0,\,\beta_1) = \Big(\frac{\pi d\cos\tau b^2}{1+\mu}\Big)^{1/2}.\qquad(11.100)$$

A similar effect was observed by Fluit, Kistemaker, and Snoek [697] on bombarding the (100) face of copper with Ar^+ ions in the [110] direction. Figure 11.38 shows the curves of [697] slightly refined by allowing for the dependence of the coefficient of electron emission of the detector on the energy of the reflected ions [539]. We see from Fig. 11.38 that the angular width of the minimum coincides with that calculated from formula (11.100).

However, it is clear that the estimate just presented can only be an approximate one, cince each of the two atoms shown in Fig. 11.39 is screened by the previous one and in turn screens the following atoms in the reflection plane. If the experiment is set up in such a way that the incident and reflected beams lie in a plane passing through one of the principal axes of the crystal, then, for glancing incidence, before falling into the detector, the ions undergo a number of deviations through small angles in this plane as a result of successive collisions with atoms of the surface chains parallel to the axis in question.

Below we present the results of a calculation by Kivilis, Parilis, and Turaev [785, 792], based on a plane model consisting of a single chain; these enable us to study certain interesting features in the reflection of ions from the face of a single crystal at glancing angles of incidence.

We chose an infinite surface chain (110) of copper atoms bombarded with a parallel beam of Ar^+ ions at energies $E_0 = 5$, 10, and 30 keV. The distance between the atoms d = 2.56 Å. Since this distance is comparatively large, the interaction of the chain with the ion for the energies in question reduces to successive collisions with individual chain atoms (Fig. 11.39).

It was considered that the first collision occurred with an atom for which the collision parameter was no greater than a certain limiting value P_b corresponding to scattering through a specified small angle $\beta_b \sim 30'$. For $E_0 = 30$, 10, and 5 keV, $P_b = 1.5$, 2.0, and 2.5 Å, respec-

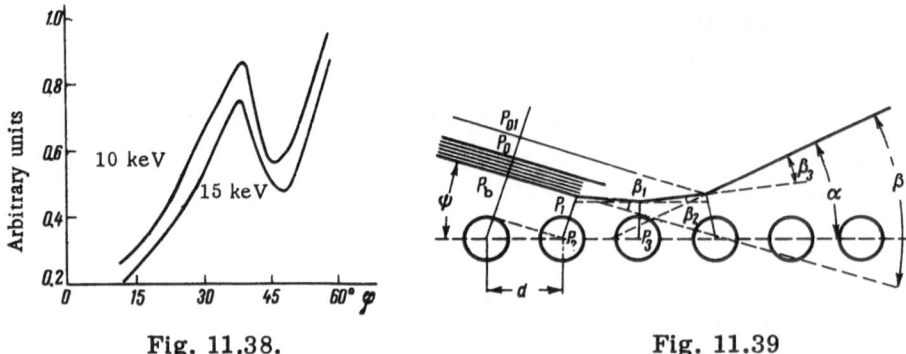

Fig. 11.38.   Fig. 11.39

Fig. 11.38. Angular distribution of Ar$^+$ ions reflected from a copper single crystal. Face (100), reflection plane (001) [697].

Fig. 11.39. Scattering of particles at a chain of atoms.

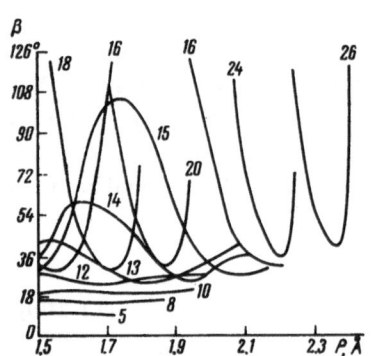

Fig. 11.40. The β(P) relationship for various ψ (figures on curves).

tively. The collision parameter of each successive collision is determined by the parameter P_i and scattering angles β_i. The $P_i(\beta_i)$ relationship is determined by the potential (Firsov [380]) allowing for all previous elastic and inelastic energy losses [466]. Interaction with the chain ceases when $P_i > P_b$, and the ion leaves the chain, scattered at an angle $\beta = \Sigma \beta_i$. Usually this happens after 5-10 collisions. For a given glancing angle ψ, the escape angle α and scattering angle $\beta = \psi + \alpha$ are determined unambiguously by the first collision parameter P, reckoned in a straight line perpendicular to the incident beam. It is clear that β(P) is a periodic function of P with a period d sin ψ (Fig. 11.40). We see from Fig. 11.39 that for all ψ the scattering angle has a lower limit of β_{min} (the escape angle is correspondingly limited by an angle α_{min}). The minimum escape angle α_{min} falls with E_0 and increases with ψ (Fig. 11.41). The scattering angle also has an upper limit of β_{max}. This rises with ψ and E_0 and only reaches π for fairly high values of ψ when the effect of screening vanishes and the scattering takes place independently at individual atoms. For small ψ the reflection is nearly specular (broken straight line in Fig. 11.41), this effect being expressed more strongly for low energies. Thus, Parilis' conclusions [750] regarding the limitation of the reflected beam obtained on a two-atom model are qualitatively confirmed.

It is interesting to consider the energy spectrum of the reflected ions.

In the two-atom model, scattering at a specified angle β may occur as a result of one or two collisions, giving different energies of the reflected ions E. In Fig. 11.42, this is represented as a function of β (broken curves denote single and dotted and dashed curves double scattering). It would appear that for multiple scattering in a chain this ambiguity should not be preserved. However, this is not so.

One of the unexpected results of the calculation was that the ambiguity of the function E(β) was in fact preserved. We see from Fig. 11.40 that each β(P) curve, within the limits of its own period, passes twice through a particular value of β at two different values of P. Since a different value of E corresponds to each P, the energy of the ions reflected in a specified direction is double-valued.

Fig. 11.42 shows the energy of ions reflected from a chain as a function of β for various characteristic angles of incidence (continuous lines). The left-hand edge of each oval corresponds to β_{min} and the right-hand edge to β_{max}. The oval only passes between the single- and

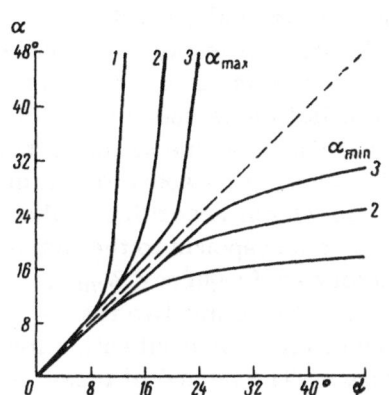

Fig. 11.41. Limiting escape angles for:
1) $E_0 = 30$; 2) 10; 3) 5 keV.

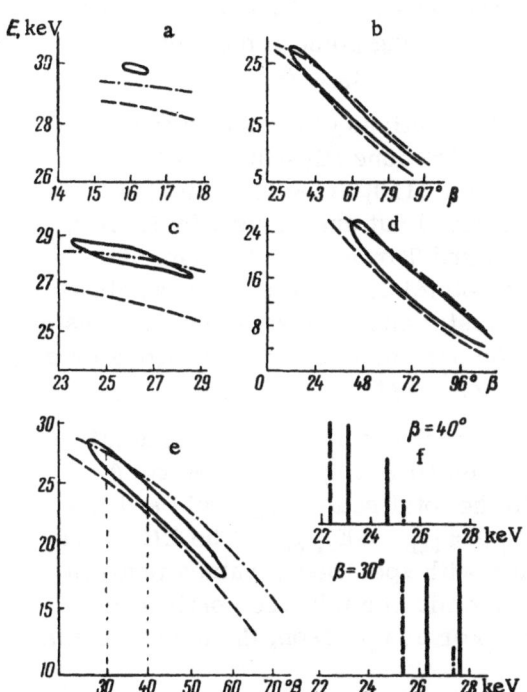

Fig. 11.42. The $E(\beta)$ relationship for reflected ions with glancing angles of: a)
$\psi = 8°$; b) 16°; c) 12°; d) 26°; e,f) 14°.

double-scattering curves for high glancing angles. For lower angles the width of the oval is smaller than the distance between the broken lines. This means that for glancing collisions the distance between the two peaks in the energy distribution of reflected ions becomes shorter than the distance calculated from the model of reflection from two surface atoms. In addition to this, with falling ψ the whole oval is lifted upward, and its lower arc may be at or above the level of the upper curve. This makes us wary of identifying experimentally observed peaks as "single" or "twofold," according to their position on the energy scale. Sometimes the apparent vanishing of the "single" peak may simply be a consequence of this kind of shift, while the vanishing of a "twofold" peak observed by Mashkova and Molchanov [773, 774] is explained by the difficulty of resolving the neighboring peaks.

In the nature of things, for glancing angles of incidence the concepts of "single" and "twofold" scattering peaks lose their original meanings. The peaks of the upper and lower arcs of the ovals correspond to a series of successive collisions, among which one or two are associated with scattering through a large angle and the rest with small angles and a low energy loss. For example, with $\psi = 12$, $\beta = 25°30'$, the peak of the lower arc (Fig. 11.42) corresponds to scattering through angles of 0°54', 3°, 18°40', 2°11', and 0°45', while the peak of the upper arc corresponds to 0°44', 1°38', 7°23', 13°12', 1°50', and 0°43'.

It is interesting to estimate the relative intensities of the two peaks. In the model under discussion this is proportional to $dP/d\beta$. Owing to the different slopes of the rising and falling parts of the $\beta(P)$ curves (Fig. 11.40) this quantity is different for the upper and lower arcs of the ovals. As a rule, the intensity on the upper arc is lower; however, near the extrema the opposite may be the case. This kind of reversal of the peaks is shown in Fig. 11.42 for $\psi = 14°$, $\beta = 40$ and 30°. The broken lines indicate the relative probabilities of single and twofold scattering for independent collisions without allowing for screening, calculated from the formulas of [773]; the continuous lines are the results of the present calculation. The "twofold" peak rising above the "single" also has been observed experimentally by Mashkova and Molchanov [773, 774].

In order to study the influence of the thermal vibrations of the lattice atoms on the phenomena in question, the Einstein model of independent harmonic oscillators is inapplicable. Here we must allow for correlation in the vibrations of neighboring atoms forming a chain. The situation is analogous to that encountered by Nelson, Thompson, and Montgomery when studying

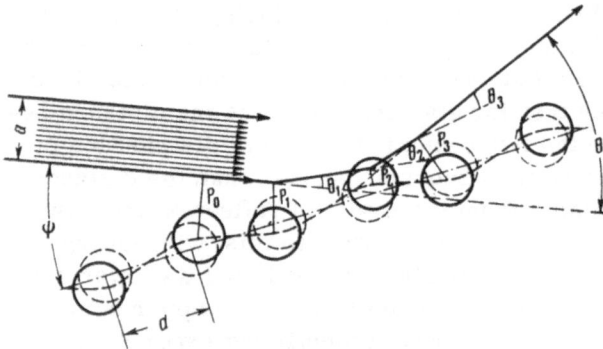

Fig. 11.43. Scattering of a particle at a chain, allowing for thermal vibrations.

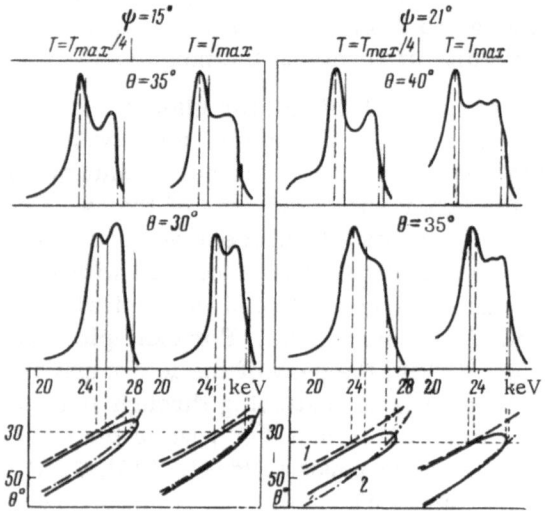

Fig. 11.44. Energy distribution of particles reflected from a chain, allowing for thermal vibrations.

the influence of thermal vibrations on focused collisions. For our own purposes we considered the chains of a monatomic face-centered lattice as one-dimensional. The well-known phonon spectrum of these contains all wavelengths above the minimum of $\lambda = 2d$ which, according to the Debye theory, corresponds to the maximum and strongest frequency W_{max}. More precisely, there are two transverse and one longitudinal vibration for each frequency. The period of these is $T > 10^{-11}$ sec. A fast ion ($v_0 \sim 10^6$–10^8 m/sec) approaches and recedes from a bent solidified chain in a time 10^{-13}–10^{-15} sec (Fig. 11.43). The amplitude of the waves increases as $\sqrt{T}$ and equals $0.1d$ near the melting point of the metal T_{max} and $0.05d$ at $T = T_{max}/4$.

Calculations of the multiple scattering of Ar^+ ions ($E_0 = 30$ keV) by a chain in the [110] direction of nickel were carried out analogously by Kivilis, Parilis, and Turaev with the help of a Firsov potential, allowing for both elastic and inelastic energy losses, for all possible collision parameters, with averaging over the phonon spectrum [785].

Typical results of such calculations are presented in Fig. 11.44. Here, $\theta = \Sigma \theta_i$ is the total scattering angle and $E = E_0 - \Sigma_i (\Delta E_{i\,el} + \Delta E_{i\,inel})$. The $E(\theta)$ relationship with specified ψ values is given for an atomic chain by the continuous curve. For comparison, the corresponding curves for single and twofold collisions are also given.

The calculated peaks for $\theta = 30, 35, 40°$ may be compared with the experimental values of Mashkova, Molchanov, and Soshka [791]. The contraction of the reflected beam due to multiple scattering and heating is shown in Fig. 11.45. The peaks representing twofold collisions fall on heating and their halfwidth increases in accordance with the calculations of Parilis [750]. The relative intensities of the peaks of twofold collisions based on chain calculations agree with experiment better than the intensities based on two-atom models.

We see from Fig. 11.44 that scattering by a chain produces a small but temperature-dependent displacement of both single (1) and twofold (2) peaks and moves them closer together.

It is possible that, as the resolving power of the instruments increases, careful measurement of the positions of the peaks on the energy scale might lead to an experimental observation of this displacement and its dependence on thermal vibrations. This would give a unique opportunity

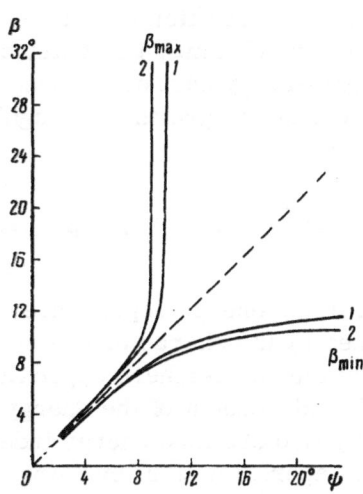

Fig. 11.45. Contraction of the boundaries of the escape angles (allowing for thermal vibrations) in the case of Ar—Ni with $E_0 = 30$ keV. Curves 1) $T = T_{max}/4$; curves 2) $T = T_{max}$.

of using fast-ion reflection for directly studying the correlation of the vibrations of the atoms forming the chains [792].

Thus, the model described in the foregoing paragraphs despite its obvious idealization, correctly embraces the essential features characterizing the reflection of fast ions from the face of a single crystal at small glancing angles and may serve as a basis for the qualitative discussion of both known experimental facts and the kind of result to be expected when studying the reflection of ions at glancing incidence.

Let us now consider the charged state of the reflected ions and its angular distribution. Experiment shows that among the reflected particles both ions and neutral atoms occur [536, 598, 698, 736]. The calculation of the charge composition of a reflected beam encounters a twofold difficulty: 1) there is no fully developed theory of multiple ionization during atomic collisions; the theory of Russek and Thomas [454] satisfactorily explains results for $Ar^+ \to Ar$, but it is not yet clear whether universal formulas can be derived from it; 2) the motion of an ion near the surface of a solid is accompanied by resonance and Auger neutralization; the Hagstrum theory [291] rather relates to the emission of Auger electrons instead of to the probability of neutralization, which for the small energies of interest to Hagstrum equalled unity.

If the first difficulty were overcome, we should be able to use the angular distribution of the charge composition to estimate the probability of Auger neutralization and determine its velocity dependence. However, the probability of ionization (only depending on the total scattering angle β) might be eliminated by keeping β constant and only varying ψ and α. This would offer the possibility of varying the component of ion velocity normal to the surface.

In actual fact, if we neglect the formation of ions with high charges, the probability of the reflection of a particle in the form of an ion (relative to the total probability of reflection with energy E in a direction β) equals

$$P(E_0, \beta) = P(E_0, \psi)\, P(E_1, \alpha)\, [1 - W(E_0, \beta)] + W(E_0, \beta)\, P(E, \alpha), \qquad (11.101)$$

where, according to [291],

$$P_0(E_0, \psi) = \exp\left(-\frac{A}{a v_0 \sin \psi}\right) \qquad (11.102)$$

is the probability that an ion moving at a velocity v_0 will not be neutralized before colliding with the surface, while $P(E_1, \alpha) = \exp(-A/av \sin \alpha)$ is the probability that a reflected ion will not be neutralized on moving away from the surface at a velocity v. The values of A and a are parameters of the moment-by-moment probability of the Auger neutralization of the ion at a distance S from the surface $R(S) = Ae^{aS}$. Then, by varying ψ and α without varying β, and measuring $P_0(E_0, \psi)$, we could find the parameters A and a. On the other hand, knowing A and a, we could derive $W(E_0, \beta)$ from experimental data.

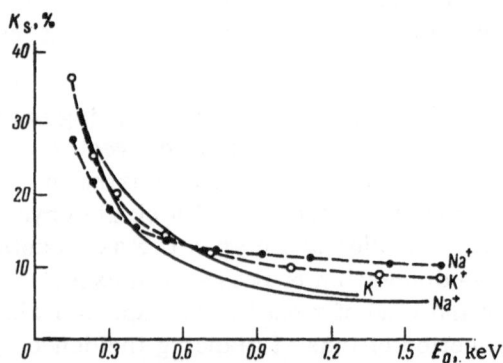

Fig. 11.46. Experimental (broken lines) and calculated (continuous lines) values of $K_s(E_0)$ for Na^+ and K^+ ions on molybdenum.

The total reflection coefficient of the ions from the surface of the solid may be obtained, if we consider the high-energy part of the spectrum and exclude diffusion, by integrating the angular-distribution function

$$\dot{K}_s(E_0, \psi) = 2\pi \int_0^\pi K(E_0, \psi, \beta) \sin\beta \, d\beta. \quad (11.103)$$

We are interested in the energy dependence $K_s(E_0, \psi)$ determined by the character of the scattering potential. In the solid-sphere approximation K_s is naturally independent of the energy of the ion. Roos [392] also obtained energy independence (to a first approximation) by using the Bohr potential [530]. This may have been due to his use of the Born approximation and other simplifying assumptions. Formulas (11.78), (11.79), (11.92), and (11.93) lead to a $K_s(E_0)$ relationship which may be put in the form

$$K_s(E_0) = K_{p_1} + K_{p_2} + K_{p_3} + \ldots, \quad (11.104)$$

where K is the integral coefficient of reflection associated with single collisions

$$K_{s_1}(E_0, \psi) = 2\pi \int_0^\pi K_1(E_0, \psi, \beta) \sin\beta \, d\beta \quad (11.105)$$

(K_{s_2}, twofold; K_{s_3}, threefold, etc.). Since each successive term is roughly an order smaller than the previous, the function $K_s(E_0)$ is mainly determined by the first term, which, in accordance with (11.78), (11.79), (11.92), and (11.93) is proportional to $1/E_0$ [if the first term equals zero, $m_2 > m_1$, $\psi > \arcsin(m_1/m_2)$, then $K_s(E_0)$ is determined by the next terms]. Figure 11.46 shows the experimental $K_s(E_0)$ curves [536] and the theoretical curves [541].

The dependence of K_s on the packing density of the face is also mainly determined by the first term $K_s \sim n$. Experiments confirm this [738].

On varying the glancing angle ψ from $\pi/2$ to 0, K_s increases at $\sim 1/\sin\psi$ until reaching unity. In the case of a single crystal there should be minima of K_s, corresponding to open channels, in this range. Graphical calculations carried out by Yurasova [701] and experimental data [692, 697] confirm this, at least in respect to the principal minimum in the [110] direction.

The foregoing semiquantitative discussion of the laws governing the reflection of ions from polycrystalline material and single-crystal solids shows that, for fairly large energies, the principal characteristics of reflection may be explained on the basis of a simple model of single and twofold paired collisions. This also partly holds for certain specific features such as the nonlinear variation in the maximum energy, the "mirror" (specular) reflection effect, and the increase in the total reflection coefficient with falling energy, which were considered as manifestations of the coupling of the atoms in the solid.

There can be no doubt that at low energies we must expect coupling of the atoms in the solid, cross sections comparable with the dimensions of the crystal cell, and possibly even a quantum character of the scattering. Special investigations will be required in order to discover the field of action and relative nature of these effects.

§4. Energy Balance When Ions Strike

a Metallic Surface

The calculation of the energy balance during the collision of ions with solid atoms, particularly the energy absorbed by the target and carried away by secondary particles, occupies a central place in the problem of the interaction between atomic particles and a solid surface. It is natural to suppose that on bombarding a metallic surface with positive ions, the energy of the primary beam will be expended in cathode sputtering, electromagnetic radiation, and the transfer of energy to the target atoms in successive elastic and inelastic collisions.

In order to set up the energy balance from experimental data and the known parameters of the colliding particles, the results of the experiments described in Chapters 4 and 6 were analyzed by Arifov et al. [381, 538].

The energy-balance equation of the bombarding ions may be written in the form

$$E_0 = E + E' + E_v + \Delta E, \tag{11.106}$$

where E_0 is the energy of the primary ion beam; E is the energy carried away by secondary scattered ions; E' is the energy carried away by primary ions neutralized at the moment of the collision and scattered in the form of neutral and excited ions; and E_v is the energy carried away by radiation. The quantity ΔE includes that part of the primary-beam energy taken up by the target in the scattering of the ions, their adsorption, or their penetration into the middle of the target, and that spent in the heating of the target and the cathode sputtering of atoms and ions.

As the result of the scattering of an ion, the target receives not the whole ion-beam energy E_0, but only a fraction of this, equal to

$$\varkappa = \frac{E_0 - E - E' - E_v}{E_0} = 1 - \frac{E}{E_0} - \frac{E'}{E_0} - \frac{E_v}{E_0}. \tag{11.107}$$

The coefficient $\varkappa$ will be called the accommodation coefficient or factor of the ion beam, in analogy with the accommodation coefficient characterizing atoms recombining at the walls of the containing vessel.

The experimental determination of the coefficient $\varkappa$ encounters the difficulty of allowing for the part of the energy carried away by primary-beam ions neutralized at the instant of collision and scattered in the form of neutral atoms. However, certain experiments on secondary phenomena carried out by bombarding tungsten, tantalum, and molybdenum surfaces with Cs^+ and Rb^+ ions [342] at low primary-beam energies indicate that E' is quite small. Electromagnetic radiation is also insignificant for low energies. Hence, expression (11.107) may be put in the form

$$\varkappa = 1 - \frac{E}{E_0}. \tag{11.108}$$

The value of $\varkappa$ thus determined will characterize the maximum value of that part of the energy transferred to the target during bombardment with positive alkali-metal ions.

The energy released as a result of the condensation of the atoms equals

$$(1 - K_s)\lambda_k,$$

where λ_k is the evaporation energy of the atom. The energy evolved on recombination of the ions with electrons equals

$$(1 - K_s)V_k,$$

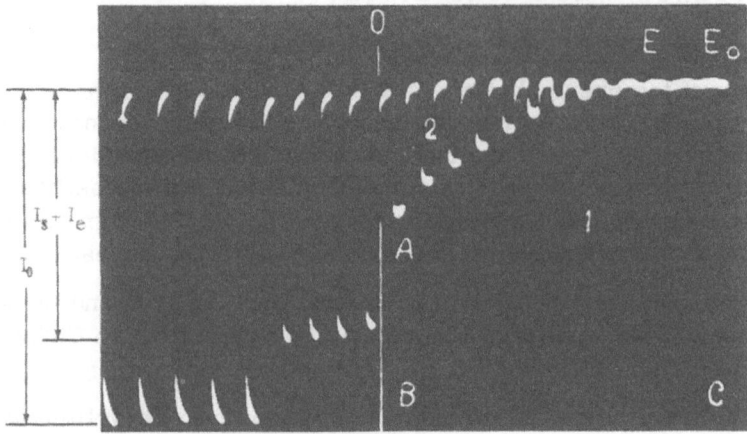

Fig. 11.47. Energy diagram of the V/A characteristics of secondary ions obtained from a pure (clean) metal on bombarding with positive ions.

where V_k is the energy of recombination of the penetrating ion in the solid. Allowing for these corrections we obtain the following expression for the proportion of the energy absorbed by the target:

$$\varkappa = 1 - \frac{E}{E_0} - \frac{E'}{E_0} - \frac{E_\nu}{E_0} + (1 - K_s) \frac{\lambda_k + V_k}{E_0}. \qquad (11.109)$$

The last term in expression (11.109) is small for energies of the order of hundreds of eV, since λ_k and V_k are no greater than a few eV. For calculating the energy balance at low energies, however, it is important that the evaporation energy and the energy of recombination of the ions in the solid should be taken into account.

If the V/A characteristic of the secondary ion current is known, then the energy E and hence $\varkappa$ may be determined by calculating the area enclosed between the "retardation" (stopping) curve and the axis of energy (Fig. 11.47).

Using the results obtained earlier by Arifov and Ayukhanov [219], we may demonstrate the possibility of calculating $\varkappa$ theoretically simply from the ion-scattering coefficient K_s and the known collision parameters of the particles.

It was shown earlier (Chapter 2) that, on bombarding clean metal surfaces, secondary ions with an almost linear energy distribution arose; the energies lay between zero and limiting values $E(\beta)$ determined by the elastic-collision relation:

$$E(\beta) = \frac{E_0}{(\mu + 1)^2} [\cos \beta \pm \sqrt{\mu^2 - \sin^2 \beta}]^2,$$

where β is the angle between the direction of the primary and secondary ions (for normal incidence of the primary beam on the target the angle β may take values from 0 to $\pi/2$), $\mu = m_1/m_2$; m_1 and m_2 are, respectively, the masses of the target atoms and bombarding ions, and E_0 is the primary-ion energy.

Let us calculate the energy E carried away by the scattered ions.

Let $n(\beta, E_0)d\omega dE$ be the number of ions scattered into a solid angle $d\omega = 2\pi \sin \beta d\beta$, having energies between E and $E + dE$. Then,

$$E = 2\pi \int_0^{\pi/2} \sin \beta \, d\beta \int_0^{E(\beta)} n(\beta, E) E \, dE. \tag{11.110}$$

Since the V/A characteristics of the secondary ions scattered at a specific angle are almost linear, particularly for low primary ion energies, the quantity

$$n(\beta, E) = -\frac{\partial N(\beta, E)}{\partial E} = \frac{n(\beta)}{E(\beta)}$$

will be independent of E [here, $N(\beta, E)$ is the number of ions scattered at an angle β and having energies greater than E]. Hence,

$$E = 2\pi \int_0^{\pi/2} \sin \beta \, d\beta \int_0^{E(\beta)} \frac{n(\beta)}{E(\beta)} E \, dE = \pi \int_0^{\pi/2} n(\beta) E(\beta) \sin \beta \, d\beta. \tag{11.111}$$

Measurements of the angular distribution of secondary ions showed that for normal incidence of the primary ion beam on the target the number of ions $n(\beta)d\omega$ scattered into a solid angle $d\omega$ at an angle β fell linearly with increasing β. The angular-distribution function $n(\beta)$ was closely approximated by a straight line:

$$n(\beta) = C\left(\frac{\pi}{2} - \beta\right).$$

The constant C may be easily expressed in terms of the number of primary ions and the ion-scattering coefficient K_s. The total number of secondary ions equals

$$n_+ = K_s n_\pi = 2\pi \int_0^{\pi/2} n(\beta) \sin \beta \, d\beta = 2\pi C \int_0^{\pi/2} \left(\frac{\pi}{2} - \beta\right) \sin \beta \, d\beta = \pi C(\pi - 2), \tag{11.112}$$

whence

$$C = \frac{K_s n_\pi}{\pi(\pi - 2)}.$$

Thus, relations (11.108), (11.111), and (11.112) completely determine the integrand in (11.110), and the accommodation coefficient $\varkappa$ takes the form

$$\varkappa = 1 - \frac{E}{n_\pi E_0} = 1 - \frac{K_s}{\pi - 2} \frac{I}{(\mu + 1)^2}, \tag{11.113}$$

where

$$I = \int_0^{\pi/2} \left(\frac{\pi}{2} - \beta\right) \sin \beta \, (\cos \beta \pm \sqrt{\mu^2 - \sin^2 \beta})^2 \, d\beta. \tag{11.114}$$

Solution of (11.114) gives

$$I = \frac{5}{9} - \frac{\pi}{6} + \frac{\pi - 2}{2} \mu^2 - \frac{\pi}{3} \mu^3 + \frac{2}{9} \mu^3 \left[2\left(2 - \frac{1}{\mu^2}\right) K_1 - \left(1 - \frac{1}{\mu^2}\right) K_2\right], \tag{11.115}$$

where K_1 and K_2 are complete elliptic integrals of the first and second kind in $1/\mu$.

For $\mu \gg 1$ (in practice, for $\mu > 1.7$), (11.114) may be more simply calculated from the approximate formula

$$\frac{2I}{\pi - 2} = \left[\mu^2 - 1.38 \left(1 - \frac{1.88}{\mu^2} - \frac{0.026}{\mu^4} \right) \mu + 0.06 \right].$$

Below we present the values of $1 - \varkappa$ calculated from formula (11.113) for values of μ corresponding to the alkali metal ions and a tantalum target; we take $K_s = 0.3$, which corresponds to the experimental value found for $E_0 = 400$ eV:

μ	 1	1.74 (Cs)	2.13 (Rb)	4.63 (K)	7.86 (Na)	25.9 (Li)	∞
$1-\varkappa$	 0	0.016	0.027	0.071	0.102	0.132	$0.5 K_s$

Table 11.2 shows the calculated values of $(1 - \varkappa)$ and also the values of $(1 - \varkappa)$ obtained by integrating the V/A curves of the secondary ion current on bombarding the target with sodium ions at various energies E_0.

We see from Table 11.2 that the difference between the calculated and measured values of $\varkappa$ is no greater than the measuring error. For low primary-beam energies, when the K_s of Na reaches 70-80%, $1 - \varkappa$ takes a very large value.

The foregoing calculations may be clearly illustrated in the form of a graph. If on the basis of the oscillograms of the V/A characteristics we plot an energy diagram for the primary ions in the form of a quadrangle OBCE (Fig. 11.47), then the energy carried away by all the secondary ions may be expressed to a first approximation by the area of the right-angled triangle OAE. Hence the maximum value of the energy transferred to the target is expressed by the area of the figure $ABCE_0E$. Then the maximum value of the accommodation coefficient of the beam energy is determined by the ratio of the area $ABCE_0E$ to the area $OBCE_0$. Thus we may calculate the accommodation coefficient of the energy of the primary ion beam from the oscillograms of the V/A characteristics of the secondary phenomena for any target temperatures and primary-ion energies.

In studying the angular distribution of the limiting energies, it was found that the secondary ions escaping in all directions carried away different amounts of energies. The limiting energies of the secondary ions E for normal incidence of the primary beam on the target lay between $E_0 \left(\frac{m_1 - m_2}{m_1 + m_2} \right)^2$ and $E_0 \left(\frac{m_1 - m_2}{m_1 + m_2} \right)$ depending on the scattering angle. If we consider that it will be near enough to take the half sum of these limiting values for the case of perpendicular beam incidence, then for the average maximum energy of the secondary ions we shall have

$$E = \frac{1}{2} E_0 \left[\left(\frac{m_1 - m_2}{m_1 + m_2} \right)^2 + \frac{(m_1 - m_2)}{(m_1 + m_2)} \right] = E_0 \frac{m_1 (m_1 - m_2)}{(m_1 + m_2)^2} .$$

It follows from the diagram that the energy loss equals

$$\delta \mathscr{E} = \Delta \mathscr{E} + \mathscr{E}_0 = \text{area } ABCE_0E = \frac{IE_0}{e} - \frac{I_s E}{2e} = \frac{IE_0}{e} - \frac{I_s E_0}{2e} \frac{m_1 (m_1 - m_2)}{(m_1 + m_2)^2} . \qquad (11.116)$$

where I_s is the secondary current of scattered ions, I is the primary ion current. Equation (11.116) enables us to calculate $\delta \mathscr{E}$ from known values of I, E_0, I_s, and E. From (11.106) and (11.116) we obtain the following for the energy $\mathscr{E}$ carried away by the secondary ions:

$$\mathscr{E} = \frac{I_s E}{2e} = \frac{I_s E_0}{2e} \frac{m_1 (m_1 - m_2)}{(m_1 + m_2)^2} .$$

Table 11.2

	Energy E_0, eV						
	200	300	400	600	800	1000	2000
K_s	0.38	0.33	0.30	0.28	0.27	0.27	0.25
$(1-\varkappa)_{calc}$	0.129	0.106	0.102	0.096	0.092	0.092	0.085
$(1-\varkappa)_{meas}$	0.13	0.100	0.090	0.080	0.080	0.080	0.080

Table 11.3

Ion	E_0, eV	K_s, %	$\beta = 0°$		$\beta = 90°$		$\beta = 0$ и $90°$	Aver. $\varkappa$
			$\eta_T = \left(\frac{m_1 - m_2}{m_1 + m_2}\right)^2$	$\varkappa$	$\eta_T = \frac{m_1 - m_2}{m_1 + m_2}$	$\varkappa$	$\eta_T = \frac{m_1(m_1 - m_2)}{(m_1 + m_2)^2}$	
Li$^+$	400	30	0.85	0.87	0.92	0.86	0.89	0.87
Na$^+$	400	30	0.60	0.91	0.77	0.88	0.69	0.89
K$^+$	400	30	0.41	0.94	0.64	0.90	0.53	0.92
Rb$^+$	400	30	0.12	0.99	0.36	0.96	0.23	0.96
Cs$^+$	400	30	0.02	0.99	0.15	0.98	0.09	0.98

Expression (11.116) may also be rewritten in the form

$$\delta\mathscr{E} = \frac{IE_0}{e}\left[1 - \frac{K_s}{2}\frac{m_1(m_1 - m_2)}{(m_1 + m_2)^2}\right] = \frac{IE_0}{e}\varkappa. \tag{11.117}$$

Expression (11.117) enables us to calculate the energy loss in terms of the current, energy, and mass of the primary ions, the mass of the target atoms, and the coefficient of secondary ion emission.

For the coefficient $\varkappa$ indicating what proportion of the primary-ion energy is transformed into heat and carried away by other incalculable processes, we obtain the following from formula (11.117):

$$\varkappa = 1 - \frac{K_s}{2}\frac{m_1(m_1 - m_2)}{(m_1 + m_2)^2}.$$

From (11.106) and (11.117) we have

$$\mathscr{E} = \frac{1}{e}IE_0(1 - \varkappa).$$

Thus we find that we may calculate the energy carried away by the secondary ions from clean metals if we know E_0, I, and $\varkappa$.

We may measure I, I_s, and $\mathscr{E}$ directly from the oscillograms. Knowing the energy of the primary ion beam and the masses of the target atoms m_1 and the bombarding ions m_2, we may calculate the values of $\delta\mathscr{E}$ and $\mathscr{E}$.

Table 11.3 gives the calculated values of $\varkappa$ for all the alkali metal ions bombarding clean tantalum at $\beta = 0$ and $90°$, and also the average value for the angles 0 and $90°$ subject to the condition $K_s = 30\%$. We see from Table 11.3 that in the case under consideration, the value of $\varkappa$ fluctuates between 0.87 and 0.99.

Usually, the efficiency of cathode sputtering is calculated from the formula

$$\eta_k = \frac{\lambda_0}{E_0}\frac{n_a}{n_\pi},\tag{11.118}$$

where λ_0 is the heat of evaporation of the target atoms, n_a is the number of sputtered target atoms; and n_π is the number of bombarding ions.

We see from (11.118) that, in order to calculate η_k, one takes the whole energy E_0 supplied by the primary ions without allowing for that taken away by the secondary ions, neutral particles, and excited secondary atoms.

It might be more correct in calculating η_k to allow for the energy carried away by the secondary ions and instead of (11.118) to use the expression

$$\eta_k = \frac{\lambda_0}{E_0\left[1 - K_s\dfrac{m_1(m_1 - m_2)}{(m_1 + m_2)}\right]}\frac{n_a}{n_\pi}$$

or

$$\eta_k = \frac{\lambda_0}{E_0 \varkappa}\frac{n_a}{n_\pi}.$$

§5.   Calculation of the Inertial Component of
Secondary Ion—Ion Emission from Metallic Targets

It was shown in Chapter 2 that ion—ion emission comprises three components: reflected ions, evaporated ions, and diffusion ions. We shall now present the calculation* for the diffusion component of the current of ion—ion emission for a particular form of primary current.

Presentation of the Problem

Let a beam of primary ions of intensity I and energy E_0 fall on a clean target at time $t = 0$. On reaching the target, some of the ions IK_s experience elastic collisions and are reflected from the target; the rest $(1 - K_s)I$ penetrate into the depths of the target, losing their energy in collisions with target atoms. The relative number of ions in a layer of thickness dx at a depth x is given by the expression $dN(x) = f(x)dx$. The function $f(x)$ is called the throwing function. Let us suppose that this function has the form

$$f(x) = Ae^{-jx}, \quad 0 < x < \infty,\tag{11.119}$$

where A is determined from the condition

$$\int_0^\infty f(x)\,dx = I(t)(1 - K_s).$$

Hence,

$$A(t) = (1 - K_s)\,jI(t),$$

in which j depends on the energy of the primary ions. The temperature dependence of j is here neglected.

---

*The calculation was carried out by A. A. Grinberg.

By introducing an effective resistance F opposing the penetration of the ions into the target, we may regard the average depth of penetration of the ions as equal to $\bar{l} = 1/j = E_0/F$, so that j is inversely proportional to the energy of the primary ions.

The choice of a function in the form of (11.119) is evidently more suitable for heavy particles, while a Gaussian distribution is better for light particles. The results of the calculation will thus primarily apply to heavy ions.

The atoms thrown into the target will diffuse to the surface and then, under the influence of surface ionization and desorption processes, will evaporate from the target surface with probabilities of W_+ and W_0, respectively.

Let us use n(x, t) to denote the concentration of atoms which have penetrated into the target and write the diffusion equation

$$\frac{\partial n(x,\,t)}{\partial t} = D\frac{\partial^2 n(x,\,t)}{\partial x^2} + f(x,\,t), \tag{11.120}$$

where D is the diffusion coefficient of the penetrating atoms. We wish to find a solution to this equation such as to satisfy the initial and boundary conditions:

$$\lim_{x \to \infty} n(x,\,t) = 0, \tag{11.121a}$$

$$D\,\frac{\partial n(x,\,t)}{\partial x} = n(x,\,t)\,a \quad \text{for} \quad x = 0, \tag{11.121b}$$

$$n(x,\,t) = 0 \quad \text{for} \quad t = 0, \tag{11.121c}$$

where $a = W_+ + W_0$.

The diffusion component of the secondary ion current is evidently determined by the relation

$$I_d(t) = W_+ n(0,\,t).$$

In order to make the problem quite specific, we still have to assign a time dependence for the primary beam I(t).

Let us first consider the case in which I(t) has the form

$$I(t) = \begin{cases} 0 & \text{for} \quad t \leqslant 0, \\ I & \text{for} \quad t > 0, \end{cases} \tag{11.122}$$

which agrees with initial condition (11.121c).

The solution of Eq. (11.120) has the form

$$I_d(0,\,t) = \frac{I(1-K_s)\,W_+}{\left(1-\frac{a}{jD}\right)a}\left[\frac{a}{jD}\,(1 - e^{j2Dt}\,\mathrm{erfc}\,\sqrt{j2Dt}) - \left(1 - e^{\frac{a^2 t}{D}}\,\mathrm{erfc}\,\sqrt{\frac{a^2 t}{D}}\right)\right]. \tag{11.123}$$

Introducing the dimensionless quantities

$$\theta = j^2 Dt, \quad \eta = \frac{a}{jD}\,,$$

we rewrite (11.123) in the form

$$I_d(\theta,\,\eta) = I(1-K_s)\frac{W_+}{a}\frac{1}{1-\eta}\{\eta\,[1 - F(\theta)] - [1 - F(\eta^2\theta)]\},$$

where

$$F(Z) = e^Z \left(1 - \frac{2}{\sqrt{\pi}} \int_0^{\sqrt{Z}} e^{-u^2} du \right).$$

Consideration of the switching off condition, i.e., the case in which I(t) has the form

$$I(t) = \begin{cases} I & \text{for} \quad t \leqslant 0, \\ 0 & \text{for} \quad t > 0, \end{cases}$$

leads to the following expression for the unknown ion current:

$$I_d(\theta, \eta) = I(1 - K_s) \frac{W_f}{a} \frac{1}{1 - \eta} [F(\eta^2 \theta) - \eta F(\theta)]. \tag{11.124}$$

For the function F(Z) with $Z \gg 1$, we have the asymptotic expansion:

$$F(Z) \simeq \frac{2}{\sqrt{\pi Z}} \left(\frac{1}{2} - \frac{1}{4Z} + \frac{3}{8Z^2} - \frac{15}{16Z^3} + \ldots \right),$$

which may be used for large t and η. If $\theta \gg 1$ and $\theta\eta \gg 1$, then

$$I_d(\theta, \eta) \approx I(1 - K_s) \frac{W_+}{a} \frac{1 + \eta}{\pi \sqrt{\pi\theta}}$$

or, transforming to dimensionless quantities

$$I_d(t) = I(1 - K_s) \frac{W_+}{a} \frac{jD + a}{aj \sqrt{\pi Dt}}. \tag{11.125}$$

For $jD \gg a$,

$$I_d(t) = I(1 - K_s) \frac{W_+}{a} \sqrt{\frac{D}{\pi t}}, \tag{11.126}$$

while for $a \gg jD$,

$$I_d(t) = I(1 - K_s) \frac{W_+}{aj} \frac{1}{\sqrt{\pi Dt}}.$$

It is supposed that j is independent of the target temperature, so that for a specific energy of the primary ions θ only depends on temperature by way of D, the generally accepted temperature dependence of the diffusion coefficient having the form

$$D = Ce^{-\frac{b}{T}}. \tag{11.127}$$

If we observe the inertial part of the current after disconnecting the primary ion beam, this should be given by formula (11.124). By plotting the time characteristics of this current for various target temperatures, we may determine the parameters of the temperature dependence of the diffusion coefficient in the following manner.

1. If in the working temperature range (T > 800°K), $jD \gg a$, i.e., $\eta \ll 1$, then instead of (11.124) we have

$$I_d(\theta, \eta) = I(1 - K_s) \frac{W_+}{a} F(\eta^2 \theta). \tag{11.128}$$

Neglecting the variation in W_+/a with temperature, we may write the following series of equations:

$$F(\eta_1^2\theta_1) = F(\eta_2^2\theta_2) = \ldots = F(\eta_k^2\theta_k), \tag{11.129}$$

from which it follows that

$$\eta_1^2\theta_1 = \eta_2^2\theta_2 = \ldots = \eta_k^2\theta_k,$$

where the index attached to η corresponds to various target temperatures $T_1, T_2, T_3, \ldots, T_k$ while the index attached to θ corresponds to the time at which the secondary current I_d reaches some specified constant value.

It follows from the equations presented in the foregoing that

$$\frac{D_1}{t_1} = \frac{D_2}{t_2} = \ldots = \frac{D_k}{t_k} \tag{11.130}$$

which, according to (11.127), gives

$$b = \left[\ln \frac{t_v + 1}{t_s}\right]\left(\frac{1}{T_v} - \frac{1}{T_{v+1}}\right)^{-1},$$

where $\nu = 1, 2, \ldots, $ k.

2. If $jD \ll a$, i.e., $\eta \gg 1$, then, according to (11.124), we have

$$I_d(\theta, \eta) = (1 - K_s)\frac{dW_+}{a} F(\theta).$$

On the same assumptions, and using the same notation as in the previous case, we obtain a series of equations

$$\theta_1 = \theta_2 = \ldots = \theta_k,$$

or

$$t_1D = t_2D_2 = \ldots = t_kD_k, \tag{11.131}$$

which gives

$$b = \left[\ln \frac{t_v}{t_{v+1}}\right]\left(\frac{1}{T_v} - \frac{1}{T_{v+1}}\right)^{-1};$$

where $\nu = 1, 2, \ldots, $ k.

The transformation from (11.128) to (11.129) was based on the temperature independence of W_+/a. However, W_+/a is none other than the ionization coefficient, which is well known to vary very slightly with temperature at high temperatures.

Since the diffusion component constitutes that part of the inertia current which is not modulated when the primary beam is interrupted at a frequency ω, it is clear that the experimentally measured diffusion current depends on the frequency ω:

$$I_d = I_s\left(\frac{2\pi}{\omega}\right). \tag{11.132}$$

If $\theta = j^2Dt$ and $\eta\theta$ is far greater than unity, the time dependence of I_d will be determined by relation (11.125), so that, in this case,

$$I_d = I(1 - K_s)\frac{W_+}{a^2}\frac{(a + jD)}{j\sqrt{\pi Dt_0}}.$$

If $jD \gg a$, then, according to (11.126),

$$I_d = \frac{j(1-K_s)}{\sqrt{\pi}} \frac{W_+}{a^2} \sqrt{\frac{D}{t_0}}, \tag{11.133}$$

while for $a \gg jD$,

$$I_d = \frac{I(1-K_s)}{j} \frac{W_+}{a} \frac{1}{\sqrt{\pi D t_0}}. \tag{11.134}$$

Since in measuring the time dependence $I_d(t)$ the time is counted from the moment of disconnecting the primary ion beam, the time dependence of the diffusion current may finally be written in the form

$$I_d = \frac{I(1-K_s)}{j} \frac{W_+}{a^2} \frac{jD+a}{\sqrt{\pi D(t_0+t)}} \tag{11.135}$$

with corresponding simplified expressions in the case of inequalities (11.122).

Let us compare these conclusions with experiment. It follows from the curves reflecting the energy dependence of the primary current that $I_d \sim E_0$. We see from formulas (11.132)-(11.135) that this can only be so when $a \gg jD$, as only in this case does formula (11.134) explain the experimentally observed linear dependence on energy. Hence, in practice, the second case is realized. According to (11.134),

$$\frac{(1-K_s)W_+}{Fa} \frac{1}{\sqrt{\pi D t_0}} = \tan \varphi, \tag{11.136}$$

where F is the "frictional force" and φ is the slope of the straight line defining the dependence of I_d / I on the energy E_0. Putting (11.136) into (11.135) we obtain

$$\frac{I_d(t)}{I} = K_d(t) = E_0 \tan \varphi \sqrt{\frac{t_0}{t+t_0}} = K_d \sqrt{\frac{t_0}{t+t_0}}, \tag{11.137}$$

where K_d is the stationary value of the coefficient $K_d(t)$. This formula may easily be verified experimentally, and a relation of this kind has already been observed qualitatively.

It follows from formula (11.134) that

$$\frac{I_d}{I} = K_d = \frac{(1-K_s)W_+}{ja} \frac{1}{\sqrt{\pi D t_0}}. \tag{11.138}$$

Since the temperature dependence of K_d is known from experiment, by measuring $K_d(t, T)$ at various temperatures we obtain the following relation from (11.138) and the generally accepted expression for D (11.127):

$$\frac{b}{T} = 2 \ln \frac{W_+}{aK_d} + \text{const.}$$

Table 11.4

Temp. T, °K	K_d %	$\frac{W}{K_d}$, %	$\frac{W_+}{aK_d}$	$\ln \frac{W_+}{aK_d}$	$1/T$
1500	13	56.5	4.34	1.468	$0.0667 \cdot 10^{-2}$
1600	10	60.5	5.05	1.8	$0.0625 \cdot 10^{-2}$
1700	5	63.0	12.6	2.534	$0.0588 \cdot 10^{-2}$

Table 11.4 shows the experimental data and the values of W_+/K_d calculated from these in the case of potassium ions striking a tungsten target.

From these data we obtain the estimate b = 16,600°, which on the energy scale corresponds to 1.43 eV. This value of b is entirely reasonable.

Regarding the value of the coefficients j, we may make some approximate assumptions. Having determined t_0 from (11.137), we may find D from formula (11.138), where all the other parameters are known except, of course, j. Thus, taking $j = 10^6$ cm^{-1}, $t_0 = 10^{-3}$ sec, $K_s = 0.3$, T = 1700°K, we obtain $D \approx 10^{-7}$ cm^2/sec.

We see from formula (11.134) that the fall in diffusion current with rising temperature takes place on account of the rise in the diffusion coefficient. The vanishing at low temperatures must be sought in the variation of W_+/a only.

APPENDIX*

In the last few years our study of the interaction of slow ions with solid surfaces has undergone considerable further development. The appearance of every fresh paper on the subject has usually tended to support the validity of the point of view developed in the monograph regarding the pair interaction of incident ions with atoms in the solid target. The anisotropy of the angular distributions of secondary particles and the structured nature of their energy distribution, which are associated with the anisotropy of the structure of single crystals, have opened wide possibilities for studying the nature of the interaction and a number of the structural characteristics of solid targets.

In this connection a number of investigations have been carried out into the value of the total scattering coefficient for various angles of incidence of the primary ions on various faces of single crystals. Thus, in a paper by Arifov, Khadzhimukhamedov, and Yunusov [793] concerning the normal incidence of primary Li^+, Na^+, K^+, Rb^+, and Cs^+ ions on the (001) face of a molybdenum single crystal, the scattering of ions at angles greater than 90° was considered. The scattering coefficient for all the ions in question fell with increasing primary ion energy. This fall took place the more sharply, the greater the mass of the incident particle. The ion energy for which the scattering coefficient reached saturation depended on the nature of the ion; this relationship presumably arose from the linear dimensions of the bombarding ions and the phenomenon of penetration. The greater the diameter of the ion, the earlier the corresponding curve passed to saturation, i.e., penetration "saturation" set in. In this series of ions the Li^+ ions occupied a special place, their diameters being much smaller than the linear dimensions of the channel normal to the (001) face.

In another paper by Arifov, Khadzhimukhamedov, and Yunusov [794], the model of hard spheres was used as theoretical basis for considering possible relationships between the number of Li^+ ions scattered at angles larger than 90° and the penetration of these ions into the (112) faces of tungsten. The calculated results agreed satisfactorily with the experimental values of the scattering coefficients for Li^+ ions striking the (112) face of a tungsten single crystal.

On the other hand, work has also continued on the energy distribution of ions scattered through various angles on bombarding the faces of a single crystal at various angles. Thus, in a paper by Arifov and Aliev [795], a Hughes—Rojansky electrostatic analyzer was used to study the energy distribution of ions scattered by the (110) and (100) faces of a molybdenum single crystal for various incident angles of the primary K^+ ions. The energy spectra had a specific structure corresponding to two groups of ions with specific energy values. Calculations showed that these groups of ions corresponded to scattering as a result of single and twofold collisions of the primary ion with individual atoms of the single crystal. In addition to this, there were

---

*Brief summary of work carried out by the author and his colleagues after the publication of the monograph in Russian.

349

certain not-very-clearly-expressed groups of ions with energies less than those of the ions which had experienced twofold collisions; the origin of these was attributed to twofold collisions with atoms lying in different directions relative to the atom undergoing the first collision.

The structured nature of the energy spectra of the scattered ions was retained on varying the energy of the primary ions (25-5000 eV) and on varying the angles of incidence (0-80°). In all cases there were also scattered ions possessing energies exceeding those of the ions which had experienced twofold collisions, and this component was explained by the possibility of multiple collisions.

In further work by Arifov and Aliev [796, 797], the structures of the energy spectra of the scattered ions were considered as functions of the energy and angle of incidence on bombarding the (100) face of a tungsten single crystal with Cs^+ ions. In this case also there was a structured nature of the scattered-ion energy spectra. The quantities η_1, η_2, and η_3 (ratios of the energies of the secondary ions experiencing one, two, and more collisions to the energy of the primary ions) were considered as functions of the primary ion energy. On reducing the primary-ion energy, the values of these ratios deviated from the calculated values based on considering the atoms of the target as free; the greater the primary energy, the more collisions the scattered ion suffered. This is a direct indication of the fact that for such primary-ion energies the target atoms cannot be regarded as free. The angular dependence of the scattered-ion intensity was also shown to suffer from anisotropy, in that, when the primary ions fell normally to the loose (110) face there was a sharp drop in scattering intensity.

Arifov and Aliev [798] further considered the characteristics of scattering for glancing angles of incidence of the primary ions on the surface of incandescent molybdenum and tungsten. For angles of incidence close to glancing, there was a preferential scattering in the specular direction. By comparing the experimental curves of the angular distribution with the calculations of Firsov [789] obtained by solving the kinetic Boltzmann equation, it was concluded that a law of interaction very similar to the Coulomb law played a major part in the scattering. For angles of incidence and reflection close to glancing, the maximum energies of the secondary ions approached the energy of the primary ions.

Individual attention was paid to the case in which the mass of the bombarding ions was greater than that of the single-crystal atoms. In this case, of course, the limiting scattering angle arises. A study of the angular relationships of the intensity and the energy spectra in this case should give some valuable and specific information regarding the reality of single and multiple pair collisions. Thus, Arifov and Aliev [799] bombarded a molybdenum single crystal with Cs^+ ions and showed that, in recording secondary ions at an angle greatly exceeding the limiting angle, the energy spectrum of the scattered ions only exhibited a group of slow ions. On observing at an angle close to the limiting value, the energy spectra contained maxima corresponding to twofold collisions. The maxima corresponding to scattering as a result of single collisions only occurred for scattering at angles not exceeding the limiting angle.

As shown in the monograph, on bombarding single crystals with ions, both anisotropy and ion—electron emission were observed. Here also, great interest is aroused in separately studying the kinetic and potential emission of the electrons on bombarding different faces of a single crystal and determining their energy distributions.

Arifov, Rakhimov, Dzhurakulov, and Karetnikova [800] described a method of using a cylindrical analyzer to study the energy-distribution curves of electrons for potential and kinetic emission. The electron energy-distribution curves were given for a polycrystalline molybdenum target bombarded with helium and argon atoms and ions. The electron energy spectrum obtained on bombarding molybdenum with helium atoms (kinetic emission) was much "softer" than that obtained by bombarding molybdenum with Ar atoms for equal velocities.

Arifov, Dzhurakulov, and Rakhimov [801] studied the effect of the adsorption of oxygen on molybdenum and the oxidation of the latter on the energy distribution of kinetic-emission electrons produced by beams of neutral He atoms in the energy range 1.5-5 keV. The adsorption of oxygen and the oxidation of the molybdenum led to considerable changes in the character of the electron energy distribution. When oxygen was adsorbed on molybdenum the energy spectrum of the electrons was enriched with slower electrons. This effect was expressed even more strongly on oxidation of the surface. The most probable energy of the electrons in the case of oxidation was 0.8 eV further to the low-energy side. Starting from the assumption that molybdenum oxide was a semiconductor, a reasonable explanation was offered for the observed electron-energy spectra.

The anisotropy of the secondary kinetic and potential emission and of the energy distribution of the electrons (associated with the anisotropy of metallic single crystals) was also studied.

Arifov, Rakhimov, and Dzhurakulov [802] also studied the coefficients of secondary electron emission and the energy distribution of the electrons on bombarding the (110) and (111) faces of molybdenum single crystals with He and Ar atoms and ions in the energy range $E_0 = 0.2$-5 keV.

On passing from the (110) to the (111) face, the coefficient of potential electron emission γ_p for Ar^+ and He^+ increased by 1.5 and 1.1 times, respectively, while the energy spectra of the electrons were enriched with faster electrons. The dependence of the coefficient of kinetic electron emission γ_k on E_0 for the (110) face was rather sharper than for the (111) face; this was mainly associated with the transparency of the crystal lattice in the direction in question. The energy spectrum of the kinetic-emission electrons from the more close-packed face was "softer" than the spectrum from the loose face.

The change in the energy-distribution curve of both the kinetic- and potential-emission electrons on passing from one face to the other did not correlate directly with the change in the work function of the faces. The difference between the kinetic and potential emission of electrons for the two faces was associated with the character of the energy spectrum of the emitted electrons, the interaction between the electrons excited in the metal and the electrons of the conduction band, the crystal structure, as well as the work function of the face.

The anisotropy of γ_k was found to diminish with decreasing mass of the bombarding atom; this was explained by the relation between the cross section of the incident atom and the diameter of the open channels in the lattice, and also by the mass ratio of the colliding particles.

Arifov, Rakhimov, and Dzhurakulov [803] made some special investigations in order to discover the part played by the work function of the surface in the potential emission of electrons. The investigations were carried out by bombarding the (110) and (111) faces of a molybdenum single crystal with He^+ and Ar^+ ions in the energy range 200-1200 eV. The coefficients of potential emission and the energy distribution of the electrons were obtained for these faces. It was a characteristic feature that the energy distribution of the electrons in the case of the He^+ ions had two maxima; on passing from the (111) to the (110) face these maxima moved in the direction of lower electron energies. In the case of Ar^+ ions the distribution curves had only one maximum, and on passing from the (111) to the (110) face in this case also there was a displacement in the direction of lower electron energies. Hence, the change in the value of the electron potential-emission coefficient with changing crystallographic direction is due to the character of the electron-energy distribution and is associated with the different transparencies of the potential barrier for different parts of the spectrum.

Arifov, Flyants, and Rakhimov [804] studied the possibility of the potential emission of electrons occurring on bombarding thin alkali films deposited on molybdenum. On introducing alkali films onto the surface of electronegative metals, the work function of the surface falls sharply, and this may satisfy the condition $V > 2\varphi$ so as to make the potential emission of electrons possible. A study of the relation between the coefficient of ion—electron emission on bombarding such surfaces with Li^+, Rb^+, and Cs^+ ions and the primary-ion energy showed that the electron emission thus engendered was in fact of the potential type. A study of the $\gamma(V_i)$ relationship for the ions Li^+, Na^+, K^+, Rb^+, and Cs^+ gave the work function of a molybdenum surface optimally covered with potassium as 1.9 eV.

The study of the phenomena taking place on bombarding solids with ions was also extended to semiconducting materials and alloys. Turmashev and Ayukhanov [805] studied the dependence of the scattering coefficient on the energy of alkali ions bombarding the (111) face of a silicon single crystal. The silicon sample, carefully cleaned by standard methods outside the vacuum, revealed traces of various contaminations on the surface right from the start on bombarding with ions. Scattering could only be observed after heat treatment in very high vacuum. It was found that the scattering of the ions at the (111) face of silicon was not very great. For Li^+ and Na^+ at primary-ion energies of over 500 eV the value never exceeded 8%. A certain degree of ion emission also occurred on bombarding with K^+, Rb^+, and Cs^+ ions (~3% for $E_0 > 500$ eV). The energy dependence of the scattering coefficients was analogous to that associated with metals. Study at high temperatures revealed a considerable quantity of evaporated ions.

In another paper [806], Arifov, Gruich, and Khamidova analyzed K^+ ions scattered from the surface of a tungsten—molybdenum alloy. The energy spectra of the scattered ions showed two maxima, the energy of which corresponded to the scattering of K^+ ions resulting from single collisions with W and Mo atoms. For low primary ion energies and a high temperature there was only one maximum, corresponding to the collision of a K^+ ion with an Mo atom, indicating that the molybdenum had come to the surface of the tungsten; this suggested a method of determining the thickness of molybdenum films on tungsten surfaces.

In some other investigations associated with determining the part played by the charged state of the particles bombarding the surface, some extremely difficult measurements of the numbers of atoms bombarding the target had to be made. Dzhurakulov, Rakhimov, and Printsev [807] developed a calorimetric method for detecting a flow of atoms accelerated to an energy of over 300 eV on the basis of microthermistors. The experiments showed that this method could be used to record flows of atoms of ~10^{-9} A/cm with an energy of 500 eV ($5 \cdot 10^{-7}$ W).

Arifov, Flyants, and Khadzhimukhamedov [808] studied the scattering of neutral Cs atoms by a tantalum surface. The total scattering coefficients obtained in these experiments agreed with the total scattering coefficients of analogous ion beams. This agreement occurred not only for high particle energies, but also in the region in which the scattering coefficient reached its maximum, i.e., for energies of the order of a few tens of electron volts.

In 1967-1968, investigations were also continued into the theory of the interaction between atomic particles and a solid surface.

Parilis, Turaev, and Kivilis [809], developing their calculations relating to scattering by a chain of atoms, calculated the scattering of ions from the face of a single crystal. The calculations were carried out on an electronic computer and succeeded in analyzing the scattered beam into a group of ions reflected from the surface chain of atoms and a group of ions channeled along the surface half channel formed by these chains. Allowance was made for the thermal vibrations of the atoms.

Parilis [810] analyzed the state of current theoretical investigations into the kinetic emission of electrons, and also the possibility of using potential emission under the influence of multicharged ions in order to identify the latter on a background of singly charged atoms [811].

Kishinevskii and Parilis [787] made a more detailed study of the elementary act of ionization by the collision of heavy atoms, forming the basis of the mechanism of kinetic electron emission. These workers showed that the Auger mechanism could be used to give a satisfactory explanation of not only the relative multiplicity of the formation of multicharged ions in atomic collisions, but also the low-energy characteristics of energy losses [812]; they were also able to calculate the cross section for the Auger ionization of atoms under the influence of multi-charged ions, constituting an analog of the Auger neutralization of ions at the surface of a metal and the potential emission of electrons [813]. This phenomenon was very efficient, and should be interesting to study experimentally [814].

REFERENCES

1. W. R. Grove, Phil. Trans. Roy. Soc., 42:87 (1852).
2. W. Hittorf, Ann. Phys., 21:90 (1884).
3. A. Berliner, Ann. Phys., 33:291 (1888).
4. M. P. Villard, J. Phys., 8:5 (1899).
5. L. Austin and H. Starke, Ann. Phys., 9:271 (1902).
6. I. Townsend, Phil. Mag., 3:557 (1902).
7. L. Holborn and L. Austin, Wiss. Abhandl. Phys. Techn. Reichsanstalt, 4:101 (1903).
8. E. Rutherford, Phil. Mag., 10:193 (1905).
9. J. J. Thomson, Proc. Cam. Phil. Soc., 13:49 (1905).
10. W. H. Logeman, Proc. Roy. Soc., 78:212 (1907).
11. G. Stark, Z. Electrochem., 14:752 (1908); 15:509 (1909).
12. V. Kohlschütter, Jahrb. Radioakt. und Electronik, 9:355 (1912).
13. N. Campbell, Phil. Mag., 29:783 (1915).
14. R. Milliken, Phys. Rev., 7:362 (1916).
15. W. L. Cheney, Phys. Rev., 10:325 (1917).
16. W. Schottky, Ann. Phys., 62:142 (1920).
17. I. Franck and E. Einsporn, Z. Phys., 2:18 (1920).
18. J. J. Thomson, Rays of Positive Electricity, Longmans, Orech., New York (1921).
19. J. Buch and G. Smith, J. Am. Inst. Electr. Engrs., 42:627 (1922).
20. G. Holst and E. Oosterhuis, Compt. Rend. Acad. Sci., 175:577 (1922).
21. H. E. Farnsworth, Phys. Rev., 20:868 (1922).
22. I. Langmuir and K. Kingdon, Phys. Rev., 21:380 (1923).
23. L. B. Taylor and I. B. Langmuir, Phys. Rev., 44:423 (1923).
24. P. L. Kapitsa, Phil. Mag., 45:989 (1923).
25. I. Langmuir and K. Kingdon, Phys. Rev., 22:148 (1923).
26. G. Holst and E. Oosterhuis, Phil. Mag., 46:1117 (1923).
27. I. Langmuir, Phys. Rev., 22:357 (1923).
28. J. A. Becker, Phys. Rev., 23:664 (1924).
29. I. B. Langmuir and K. H. Kingdon, Phys. Rev., 23:112 (1924).
30. G. Holst, Physika, 4:68 (1924).
31. H. W. Webb, Phys. Rev., 24:113 (1924).
32. A. L. Klein, Phys. Rev., 26:800 (1925).
33. P. Auger, J. Phys. Radium, 6:205 (1925).
34. H. E. Farnsworth, Phys. Rev., 25:41 (1925).
35. C. H. Kunsmann, Phys. Rev., 25:292 (1925); 27:249 (1926).
36. W. Jackson, Phys. Rev., 28:524 (1926).
37. E. Blechschmidt, Ann. Phys., 81:999 (1926).
38. A. Hippel, Ann. Phys., 80:672 (1926).
39. A. Hippel, Ann. Phys., 81:1043 (1926).
40. A. Güntherschulze, Z. Phys., 36:563 (1926).

41. H. A. Messenger, Phys. Rev., 28:962 (1926).
42. J. A. Becker, Phys. Rev., 28:341 (1926).
43. T. J. Killian, Phys. Rev., 27:578 (1926).
44. W. Jackson, Phys. Rev., 30:473 (1927).
45. L. B. Loeb, Science, 66:627 (1927).
46. F. M. Penning, Physika, 8:12 (1928).
47. M. L. Oliphant, Proc. Cam. Phil. Soc., 24:451 (1928).
48. G. E. Read, Phys. Rev., 31:629 (1928).
49. R. W. Gurney, Phys. Rev., 32:467 (1928).
50. F. M. Penning, Proc. Roy. Acad. Sci., 31:14 (1928).
51. H. J. Couliette, Phys. Rev., 32:636 (1928).
52. W. Uyterhoeven, Phys. Rev., 31:913 (1928).
53. I. A. Becker, Trans. Am. Electrochem., 55:153 (1929).
54. A. L. Hughes and V. Rojansky, Phys. Rev., 34:284 (1929).
55. M. L. Oliphant, Proc. Roy. Soc., A124:228 (1929).
56. G. G. Found, Phys. Rev., 34:1625 (1929).
57. P. I. Lukirskii, Uspekhi. Fiz. Nauk, Vol. 9 (1929).
57a. P. I. Lukirskii and N. N. Semenov, Izv. GFTI, 1:122 (1929).
58. J. H. Hughes and McMillen, Phys. Rev., 34:291 (1929).
59. R. B. Sawyer, Phys. Rev., 35:1090 (1930).
60. F. M. Penning, Proc. Amst. Acad., 33:841 (1930).
61. M. L. Oliphant, Proc. Roy. Soc., A127:373 (1930).
62. M. L. Oliphant and P. B. Moon, Proc. Roy. Soc., A127:388 (1930).
63. D. B. Medved and Y. E. Strausser, Advances in Electronics and Electron Physics, Vol. 21 (1965), p. 101.
64. W. Uyterhoeven and M. S. Harrington, Phys. Rev., 36:709 (1930).
65. H. A. Bethe, Ann. Phys., 5:325 (1930).
66. K. T. Compton and I. Langmuir, Rev. Modern Phys., 2:286 (1930).
67. L. A. Kubetskii, Author's Certificate No. 24040 [in Russian] (1930).
68. H. S. W. Massey, Proc. Cam. Phil. Soc., 26:386 (1930); 27:460 (1931).
69. G. Schneider, Ann. Phys., 11:357 (1931).
70. P. B. Moon, Proc. Cam. Phil. Soc., 27:570 (1931).
71. I. Langmuir and D. S. Villars, J. Am. Chem. Soc., 53:486 (1931).
72. M. L. Oliphant and P. B. Moon, Proc. Roy. Soc., A137:463 (1932).
73. F. Wolf, Z. Phys., 74:575 (1932).
74. I. Langmuir and C. C. Found, Phys. Rev., 39:237 (1932).
75. H. Frölich, Ann. Phys., 13:229 (1932).
76. H. Bethe, Z. Phys., 76:293 (1932).
77. R. C. Evans, Proc. Cam. Phil. Soc., 29:161 (1933).
78. R. C. Evans, Proc. Roy. Soc., A139:604 (1933).
79. C. Snoek and J. Kistemaker, Advances in Electronics and Electron Physics, Vol. 21 (1965), p. 67.
80. C. J. Brasefield, Phys. Rev., 44:1002 (1933).
81. S. Sonkin, Phys. Rev., 43:788 (1933).
82. K. T. Compton and E. S. Lamar, Phys. Rev., 44:338 (1933).
83. H. S. W. Massey and G. Smith, Proc. Roy. Soc., A142:142 (1933).
84. C. Ramsauer and R. Kollath, Ann. Phys., 16:570 (1933).
85. H. Bethe, Handbuch der Physik, Vol. 24, Pt. 1, Berlin (1933).
86. F. Bloch, Ann. Phys., 16:285 (1933).
87. F. Bloch, Z. Phys., 81:363 (1933).
88. P. I. Lukirskii, Photoeffect [in Russian], Gostekhizdat, Moscow (1933).

89. Yu. P. Maslakovets, Cathode Sputtering [in Russian], Gostekhizdat, Leningrad (1934).
90. N. D. Morgulis, Zh. Eksp. Teor. Fiz., 4:499 (1934).
91. I. A. Abroyan and N. N. Petrov, Uspekhi Fiz. Nauk, 92:105 (1967).
92. N. D. Morgulis, Zh. Eksp. Teor. Fiz., 4:684 (1934).
93. V. I. Pavlov and A. N. Dobrolyubskii, Uch. Zap. LGU, Ser. Fiz., 1:65 (1934).
94. A. Longacre, Phys. Rev., 46:407 (1934).
95. L. H. Linford, Phys. Rev., 46:325 (1934).
96. A. Rostagni, Z. Phys., 88:55 (1934).
97. R. L. Stewart, Phys. Rev., 43:483 (1934).
98. E. Parilis, Eighth Intern. Conf. on Phenomena in Ionized Gases (Invited Papers), Vienna (1967).
99. E. S. Lamar and K. T. Compton, Science, 80:541 (1934).
100. J. B. McBain, Sorption of Gases and Vapors by Solids [Russian translation], Gostekhizdat, ONTI (1934).
101. R. Seeliger and K. Sommermeyer, Z. Phys., 93:692 (1935).
102. M. L. Copley and T. E. Phipps, Phys. Rev., 48:960 (1935).
103. R. Gurney, Phys. Rev., 47:479 (1935).
104. S. A. Slutskin, Zh. Tekh. Fiz., 5:1362 (1935).
105. J. Koch, Z. Phys., 100:685 (1936).
106. K. Sommermeyer, Ann. Phys., 25:481 (1936).
107. N. D. Morgulis, M. P. Bernadiner, and A. M. Patiocha, Z. Phys. Sow. Union, 9:302 (1936).
108. L. N. Dobretsov and G. A. Morozov, Zh. Eksp. Teor. Fiz., 6:243 (1936).
109. M. Healea and E. L. Chaffee, Phys. Rev., 49:925 (1936).
110. N. D. Morgulis, Sputtering of a Metal Surface under the Impact of Positive Ions [in Ukrainian] (1936).
111. E. Rudberg, Phys. Rev., 50:127 (1936).
112. F. L. Arnot and I. C. Milligan, Proc. Roy. Soc., A156:538 (1936).
113. F. L. Arnot, Nature, 138:162 (1936).
114. V. S. Lukoshkov, Zh. Tekh. Fiz., 6:26 (1936).
115. V. I. Pavlov and S. V. Starodubtsev, Zh. Eksp. Teor. Fiz., 7:409 (1937).
116. V. I. Pavlov and S. V. Starodubtsev, Zh. Eksp. Teor. Fiz., 7:424 (1937).
117. Sh. Sh. Shekhter, Zh. Eksp. Teor. Fiz., 7:750 (1937).
118. W. Veith, Ann. Phys., 29:189 (1937).
119. J. Terzic and B. Perovic, Eighth Intern. Conf. on Phenomena in Ionized Gases, Vienna (1967), p. 56.
120. P. V. Timofeev and A. V. Afanas'eva, Zh. Eksp. Teor. Fiz., 7:2145 (1937).
121. F. W. Saris, H. Boukens, and J. Kistemaker, Eighth Intern. Conf. on Phenomena in Ionized Gases, Vienna (1967), p. 55.
122. A. Rouse, Phys. Rev., 52:1238 (1937).
123. F. L. Arnot, Proc. Roy. Soc., A158:137 (1937); 158:157 (1937).
124. H. Paetow and W. Walcher, Z. Phys., 110:69 (1938).
125. H. Bruining, Physica, 5:913 (1938).
126. J. C. Ternbull and H. E. Farnsworth, Phys. Rev., 54:509 (1938).
127. R. E. Trelor and D. H. London, Proc. Phys. Soc., 50:628 (1938).
128. F. L. Arnot and C. Beckett, Nature, 141:1011 (1938).
129. F. L. Arnot and C. Beckett, Proc. Roy. Soc., A168:103 (1938).
130. R. H. Sloane and R. Press, Nature, 141:872 (1938); Proc. Roy. Soc., A168:285 (1938).
131. L. M. Nemenov and A. S. Fedyurko, Zh. Eksp. Teor. Fiz., 9:532 (1939).
132. S. V. Izmailov, Zh. Eksp. Teor. Fiz., 9:1473 (1939).
133. N. D. Morgulis, Zh. Eksp. Teor. Fiz., 9:1484 (1939).
134. A. B. Migdal, Zh. Eksp. Teor. Fiz., 9:1163 (1939).

135. G. Hill, W. W. Buechner, J. S. Clark, and I. B. Fisk, Phys. Rev., 55:463 (1939).
136. A. E. Kadyshevich, Zh. Eksp. Teor. Fiz., 9:930 (1939).
137. M. Healea, Phys. Rev., 55:984 (1939).
138. I. Allen, Phys. Rev., 55:336 (1939).
139. N. D. Morgulis and M. P. Bernadiner, Zh. Eksp. Teor. Fiz., 9:998 (1939).
140. L. A. McColl, Phys. Rev., 56:662 (1939).
141. H. Frölich and D. E. Wooldridge, Phys. Rev., 56:562 (1939).
142. A. Ya. Vyatskin, Zh. Eksp. Teor. Fiz., 9:1049 (1939).
143. Yu. M. Kushnir, Yu. I. Milyutin, and V. P. Goncharov, Zh. Eksp. Teor. Fiz., 9:1956 (1939).
144. M. E. Gurtovoi, Zh. Eksp. Teor. Fiz., 10:483 (1940).
145. V. M. Dukel'skii and N. I. Ionov, Zh. Eksp. Teor. Fiz., 10:1248 (1940).
146. N. I. Ionov, Dokl. Akad. Nauk SSSR, 28:512 (1940).
147. M. Healea and C. Houtermans, Phys. Rev., 58:608 (1940).
148. I. Amdur and J. Pearlmann, J. Chem. Phys., 8:7 (1940).
149. P. S. Tartakovskii, Internal Photoeffect in a Dielectric [in Russian], Gostekhizdat, Moscow (1940).
150. A. T. Finkelstein, Rev. Sci. Instrum., 11:94 (1940).
151. E. M. Penning and J. H. A. Moubis, Proc. Kon. Nederl. Acad. Wet., 43:41 (1940).
152. A. E. Kadyshevich, Zh. Eksp. Teor. Fiz., 10:1384 (1940).
153. L. P. Smith and P. Z. Hantzman, J. Appl. Phys., 11:220 (1940).
154. Ya. I. Frenkel', Zh. Eksp. Teor. Fiz., 11:706 (1941).
155. G. J. Timoshenko, J. Appl. Phys., 12:69 (1941).
156. N. D. Morgulis, Zh. Eksp. Teor. Fiz., 11:300 (1941).
157. M. E. Gurtovoi, Zh. Eksp. Teor. Fiz., 11:489 (1941).
158. W. F. v.d. Weg, D. I. Bierman, and I. Kistemaker, Eighth Intern. Conf. on Phenomena in Ionized Gases, Vienna (1967), p. 54.
159. I. Amdur and J. Pearlmann, J. Chem. Phys., 9:503 (1941).
160. C. Ruthemann, Nature, 29:648 (1941); 30:145 (1941).
161. R. Dorrestein, Physica, 9:443 (1942).
162. Yu. M. Kushnir and N. I. Frumin, Zh. Tekh. Fiz., 11:4, 317 (1941).
163. P. M. Morozov, Zh. Eksp. Teor. Fiz., 11:402, 410 (1941).
164. I. Brophy, Phys. Rev., 83:534 (1941).
165. M. A. Eremeev and M. V. Zubchaninov, Zh. Eksp. Teor. Fiz., 12:358 (1942).
166. F. Fetz, Z. Phys., 119:590 (1942).
167. I. Amdur and J. Pearlmann, J. Chem. Phys., 11:157 (1943).
168. I. Chempel and O. Richardson, Proc. Roy. Soc., 182:17 (1943).
169. Ch. H. Townes, Phys. Rev., 65:319 (1944).
170. Yu. V. Gott and V. G. Telkovskii [Tel'kovskii], Eighth Intern. Conf. on Phenomena in, Ionized Gases, Vienna (1967), p. 53.
171. A. Cobas and W. E. Lamb, Phys. Rev., 65:327 (1944).
172. A. E. Kadyshevich, Zh. Eksp. Teor. Fiz., 15:600 (1945).
173. G. Ruthemann, Ann. Phys., 2:133 (1945).
174. N. D. Morgulis, Uspekhi Fiz. Nauk, 28:202 (1946).
175. S. V. Starodubtsev and U. A. Arifov, Byull. Akad. Nauk UzbSSR, 4:8 (1946).
176. U. Fano, Phys. Rev., 70:44 (1946).
177. U. A. Arifov, S. V. Starodubtsev, and G. N. Shuppe, Trudy, FTI Akad. Nauk UzbSSR, 1:82 (1947).
178. E. Formann, F. Viehbock, and H. Wotke, Eighth Intern. Conf. on Phenomena in Ionized Gases, Vienna (1967), p. 52.
179. N. D. Morgulis, Uspekhi Fiz. Nauk, 31:288 (1947).

180. L. Haworth, Phys. Rev., 48:88 (1947).

181. R. H. Sloane and H. M. Love, Nature, 159:302 (1947).

182. N. I. Ionov, Dokl. Akad. Nauk SSSR, 59:467 (1948); Zh. Eksp. Teor. Fiz., 18:174 (1948).

183. U. A. Arifov and G. N. Shuppe, Trudy FTI Akad. Nauk UzbSSR, 2:19 (1948).

184. P. Dahl and N. Sandager, Eighth Intern. Conf. on Phenomena in Ionized Gases, Vienna (1967), p. 51.

185. R. M. Chaudhri and A. W. Khan, Proc. Phys. Soc., 61: 526 (1948).

186. H. W. Berry, Phys. Rev., 74:848 (1948).

187. R. H. Sloane and C. S. Wall, Proc. Roy. Soc., 61:217 (1948).

188. L. A. Sena, Collisions of Electrons and Ions with Gas Atoms [in Russian], Gostekhizdat, Moscow (1948).

189. L. P. Smith and W. E. Parkins, Uspekhi Fiz. Nauk, 35:4, 556 (1948).

190. S. V. Starodubtsev, Zh. Eksp. Teor. Fiz., 19:215 (1949).

191. N. A. Tolstoi and P. P. Feofilov, Zh. Eksp. Teor. Fiz., 19:421 (1949).

192. U. A. Arifov, V. M. Lovtsov, and A. Kh. Ayukhanov, Dokl. UzbSSR, 10:7 (1949).

193. U. A. Arifov, V. M. Lovtsov, and A. Kh. Ayukhanov, Dokl. Akad. Nauk SSSR, 68:461 (1949).

194. V. E. Yurasova, V. I. Shulga, and D. S. Karpuzov, Eighth Intern. Conf. on Phenomena in Ionized Gases, Vienna (1967), p. 50.

195. U. A. Arifov, V. M. Lovtsov, and A. Kh. Ayukhanov, Yubil. Sb. Akad. Nauk UzbSSR, 95 (1949).

196. F. Seitz, Disc. Faraday Soc., 5:271 (1949).

197. V. M. Dukel'skii and E. Ya. Zandberg, Zh. Eksp. Teor. Fiz., 19:731 (1949).

198. G. W. C. Kaye and T. H. Laby, Experimental Physicists' Handbook [Russian translation], IL, Moscow (1949).

199. V. Kohlschütter, Jahrb. Radioakt. und Elektronik, 9:385 (1949).

200. G. Möllenstedt, Optik, 5:439 (1949).

201. G. Möllenstedt and F. Heise, Phys. Rev., 75:80 (1949).

202. F. Reichertz and H. E. Farnsworth, Phys. Rev., 75:1902 (1949).

203. E. Jahnke and F. Emde, Tables of Functions with Formulae and Curves [Russian translation], Gostekhteoretizdat, Moscow (1949), p. 129. [Bilingual (German and English) edition, Dover, New York (1945).]

204. P. T. Landsberg, Proc. Phys. Soc., A62:806 (1949).

205. V. M. Lovtsov and S. V. Starodubtsev, Trudy FTI Akad. Nauk UzbSSR, 3:45 (1950).

206. N. A. Kaptsov, Electrical Phenomena in Gases and Vacuum [in Russian], Gostekhizdat, Moscow (1950).

207. U. A. Arifov and V. M. Lovtsov, Dokl. Akad. Nauk SSSR, 75:365 (1950).

208. R. T. Bayard and D. Alpert, Rev. Sci. Instrum., 21:571 (1950).

209. H. B. Michelson, J. Appl. Phys., 21:536 (1950).

210. S. Dushman, Scientific Foundations of Vacuum Techniques, Wiley, New York (1949).

211. D. Greene, Proc. Phys. Soc., 63:11 (1950).

212. S. V. Izmailov and A. M. Furman, Zh. Eksp. Teor. Fiz., 20:799 (1950).

213. U. A. Arifov and V. M. Lovtsov, Trudy FTI Akad. Nauk UzbSSR, 3:117 (1950).

214. V. M. Lovtsov and S. V. Starodubtsev, Trudy FTI Akad. Nauk UzbSSR, 3:111 (1950).

215. V. M. Lovtsov and S. V. Starodubtsev, Trudy FTI Akad. Nauk UzbSSR, 3:57 (1950).

216. H. Myers, Proc. Roy. Soc., A215:329 (1950).

217. A. Ya. Vyatskin, Zh. Eksp. Teor. Fiz., 20:557 (1950).

218. L. M. Baroody, Phys. Rev., 78:780 (1950).

219. U. A. Arifov and A. Kh. Ayukhanov, Dokl. Akad. Nauk UzbSSR, 4:12 (1951).

220. M. E. Eremeev, Dokl. Akad. Nauk SSSR, 79:775 (1951).

221. W. Ploch, Z. Phys., 130:174 (1951).

360 REFERENCES

222. G. M. Avak'yants, Dokl. Akad. Nauk UzbSSR, 6:3 (1951).
223. G. M. Avak'yants, Dokl. Akad. Nauk UzbSSR, 9:9 (1951).
224. N. Bohr, Penetration of Atomic Particles Through Matter, Hafner, New York (1948).
225. N. Mott and H. Massey, Theory of Atomic Collisions, Oxford University Press, New Jersey (1949).
226. W. Ploch and W. Walcher, Rev. Sci. Instrum., Vol. 22, No. 12 (1952).
227. J. H. Jonker, Philips Res. Repts., 6:372 (1951).
228. M. A. Eremeev, Transactions of the Conference on Cathode Electronics [in Russian], Kiev (1951), p. 91.
229. L. N. Dobretsov and P. Ya. Uvarov, Transactions of the Conference on Cathode Electronics [in Russian], Kiev (1951), p. 43.
230. J. P. Molnar, Phys. Rev., 83:940 (1951).
231. L. D. Landau and E. M. Lifshits, Quantum Mechanics [in Russian], Gostekhizdat, Moscow (1951).
232. G. M. Avak'yants, Dokl. Akad. Nauk UzbSSR, 12:9 (1951).
233. L. N. Dobretsov, Emission Electronics [in Russian], Nauka, Moscow (1966).
234. E. A. Shpol'skii, Atomic Physics [in Russian], Gostekhizdat, Moscow (1952).
235. V. M. Chicherov, Eighth Intern. Conf. on Phenomena in Ionized Gases, Vienna (1967), p. 49.
236. L. N. Dobretov, V. N. Lepeshinskaya, and É. I. Bronshtein, Zh. Tekh. Fiz., 22:961 (1952).
237. M. E. Eremeev and V. V. Shestukhina, Zh. Tekh. Fiz., 22:1262 (1952); 22:1268 (1952); M. A. Eremeev and V. G. Yur'ev, Zh. Tekh. Fiz., 22:1290 (1952).
238. U. A. Arifov and A. Kh. Ayukhanov, Transactions of the Conference on Cathode Electronics [in Russian], Kiev (1951), p. 99.
239. V. I. Veksler, G. A. Klein, and G. N. Shuppe, Trudy FTI Akad. Nauk UzbSSR, 4:62 (1952).
240. L. N. Dobretsov and N. M. Karnaukhova, Dokl. Akad. Nauk SSSR, 85:745 (1952).
241. M. A. Eremeev and G. L. Matskevich, Zh. Tekh. Fiz., 22:1296 (1952).
242. V. I. Veksler, Author's Abstract of Dissertation [in Russian], Tashkent (1952).
243. E. H. S. Burhop, The Auger Effect and Other Radiationless Transitions, Cambridge (1952).
244. U. A. Arifov and A. A. Aliev, Eighth Intern. Conf. on Phenomena in Ionized Gases, Vienna (1967), p. 48.
245. S. Blanter, Methods of Studying Metals and Analyzing Experimental Data [Russian translation], Metallurgizdat, Moscow (1952).
246. M. A. Eremeev and L. K. Éstrinov, Zh. Tekh. Fiz., 21:1554 (1954).
247. F. Keywell, Phys. Rev., 87:160 (1952).
248. A. R. Shul'man and E. I. Myakinin, Zh. Eksp. Teor. Fiz., 22:1542 (1952).
249. J. L. H. Jonker, Philips Res. Repts., 7:1 (1952).
250. A. I. Dekker and Van der Ziel, Phys. Rev., 86:755 (1952).
251. G. Möllenstedt, Optik, 9:473 (1952).
252. H. S. W. Massey, Advances in Electronics, 4:2 (1952).
253. A. Ya. Vyatskin, Zh. Eksp. Teor. Fiz., 23:147 (1952).
254. N. P. Myers, Proc. Roy. Soc., A215:329 (1952).
255. H. C. Bourne, Doctoral Thesis, MIT (1952).
256. C. Couchet, Compt. Rend. Acad. Sci., 235:944 (1952).
257. G. R. Rick, Mass Spectroscopy [Russian translation], IL, Moscow (1953)
258. L. Francken and E. Bonnyns, Eighth Intern. Conf. on Phenomena in Ionized Gases, Vienna (1967), p. 46.
259. B. Navinšek and G. Carter, Eighth Intern. Conf. on Phenomena in Ionized Gases, Vienna (1967), p. 43.
260. U. A. Arifov, A. Kh. Ayukhanov, and S. V. Starodubtsev, Dokl. Akad. Nauk UzbSSR, 1:12 (1953).

261. V. I. Veksler and G. N. Shuppe, Zh. Tekh. Fiz., 23:1573 (1953).
262. H. D. Hagstrum, Phys. Rev., 89:244 (1953); 91:543 (1953).
263. J. J. Okano, Phys. Soc. Japan, 8:562 (1953).
264. H. D. Hagstrum, Rev. Sci. Instrum., 24:1122 (1953).
265. I. Becker and C. I. Hartmann, J. Phys. Chem., 57:153 (1953).
266. S. U. Umarov, G. M. Avak'yants, and L. G. Gurvich, Dokl. Akad. Nauk UzbSSR, 6:12 (1953); 8:23 (1953).
267. Yu. A. Dunaev and I. P. Flaks, Dokl. Akad. Nauk SSSR, 91:1 (1953).
268. V. M. Lovtsov and A. S. Smirnov, Trudy FTI Akad. Nauk UzbSSR, 5:82 (1953).
269. A. Günterschulze, Vacuum, 3:4 (1953).
270. A. R. Shul'man and E. I. Myakinin, Dokl. Akad. Nauk SSSR, 91:1075 (1953).
271. S. S. Lander, Phys. Rev., 91:1382 (1953).
272. S. H. Anson and K. Wright, Trans. J. Bull. Am. Phys. Soc., 28:7 (1953).
273. S. A. Marshall, M. A. Pomerantz, and R. Shatas, J. Franklin Inst., 256:574 (1953).
274. D. Bohm and D. Pines, Phys. Rev., 92:603 (1953).
275. D. H. Jonker, Philips Res. Repts., 8:434 (1953).
276. R. H. Sloane and R. M. Hobson, Proc. Roy. Soc., 66:663 (1953).
277. J. A. Phillips, Phys. Rev., 91:455 (1953).
278. O. Klemperer, Electron Optics, CUP, London (1953).
279. G. Philbert, Compt. Rend. Acad. Sci., 237:882 (1953).
280. M. I. Ionov, Author's Abstract of Dissertation [in Russian], LPI, Leningrad (1954).
281. U. A. Arifov, A. Kh. Ayukhanov, and S. V. Starodubtsev, Zh. Eksp. Teor. Fiz., 26:714 (1954).
282. U. A. Arifov and A. Kh. Ayukhanov, Zh. Eksp. Teor. Fiz., 27:87 (1954).
283. U. A. Arifov, Doctoral Dissertation [in Russian], Leningrad (1954).
284. U. A. Arifov, Author's Abstract of Dissertation [in Russian], LFTI AN SSSR, Leningrad (1954).
285. P. Parker, Phys. Rev., 93:1148 (1954).
286. I. P. Flaks, Author's Abstract of Dissertation [in Russian], LPI, Leningrad (1954).
287. N. D. Morgulis and N. D. Nakhodkin, Dokl. Akad. Nauk SSSR, 94:1029 (1954).
288. B. Aarset and R. W. Cloud, J. Appl. Phys., 25:1365 (1954).
289. M. L. Higatsberger et al., J. Appl. Phys., 25:883 (1954).
290. H. D. Hagstrum, Phys. Rev., 96:325 (1954).
291. H. D. Hagstrum, Phys. Rev., 96:336 (1954).
292. Lepeshinskaja [Lepeshinskaya] and E. M. Zarutskiy [Zarutskii], Eighth Intern. Conf. on Phenomena in Ionized Gases, Vienna (1967), p. 44.
293. N. V. Fedorenko, Zh. Tekh. Fiz., 24:2113 (1954).
294. A. A. Francken and L. O. Goche, Eighth Intern. Conf. on Phenomena in Ionized Gases, Vienna (1967), p. 45.
295. V. M. Gavrilyuk, Trudy Inst. Fiz. Akad. Nauk UkrSSR, No. 5, p. 87 (1954).
296. R. M. Stier, C. E. Barnett, and G. E. Evans, Phys. Rev., 96:943 (1954).
297. R. C. Bradley, Phys. Rev., 93:719 (1954).
298. G. K. Wehner, Phys. Rev., 93:633 (1954); J. Appl. Phys., 25:698 (1954).
299. E. Sternglass, Phys. Rev., 95:345 (1954).
300. N. B. Gornii and L. M. Rakhovich, Zh. Eksp. Teor. Fiz., 26:4 (1954).
301. N. B. Gornii, Zh. Eksp. Teor. Fiz., 27:171 (1954).
302. G. Tautfest and H. B. Fechter, Phys. Rev., 96:35 (1954).
303. L. Marton and L. B. Leder, Phys. Rev., 94:103 (1954).
304. L. B. Leder and L. Marton, Phys. Rev., 95:1345 (1954).
305. H. Watanabe, J. Phys. Soc. Japan, 9:920 (1954).
306. A. Ya. Vyatskin, Zh. Eksp. Teor. Fiz., 17:162 (1954).

307. L. S. Lenard, Z. Naturforsch., 9A: 727, 1019 (1954).

308. H. Fridmann, Naturwiss., 41:569 (1954).

309. L. Marton, S. A. Simpson, and T. F. McGrow, Phys. Rev., 95:634 (1954).

310. J. Sommeria, J. Phys. Radium, 15:1126 (1954).

311. M. Schwartz and P. Copeland, Phys. Rev., 96:1466 (1954).

312. R. N. Varnery, Phys. Rev., 93:1156 (1954).

313. P. A. Wolff, Phys. Rev., 95:56 (1954).

314. I. P. Flaks, Zh. Tekh. Fiz., 25:2463 (1955).

315. É. Ya. Zandberg, Zh. Tekh. Fiz., 23:1386 (1955).

316. H. C. Bourne, R. W. Cloud, and J. H. Trump, J. Appl. Phys., 26:596 (1955).

317. J. C. Mignolet, Recueil Trav. Chim., 74:685 (1955).

318. H. Fetz, J. Appl. Phys., 6:288 (1955).

319. A. Kh. Ayukhanov, Author's Abstract of Dissertation [in Russian], Tashkent (1955).

320. S. U. Umarov, G. M. Avak'yants, and L. G. Gurvich, Trudy FTI Akad. Nauk UzbSSR, 6:34 (1955).

321. E. S. Mashkova and V. A. Molchanov, Eighth Intern. Conf. on Phenomena in Ionized Gases, Vienna (1967), p. 42.

322. C. C. Thornton and L. D. Hanley, Proc. Inst. Rad. Eng., 43:2 (1955).

323. M. A. Eremeev, Trudy LPI, 181:157 (1955).

324. C. K. Wehner, Advances in Electronics and Electron Physics, 7:239 (1955).

325. F. Keywell, Phys. Rev., 97:6 (1955).

326. A. Günterschulze, Z. Phys., 141:3 (1955).

327. V. A. Vlasov, Neutrons [in Russian], Gostekhteoretizdat, Moscow (1955).

328. A. R. Shul'man and Farbshtein, Dokl. Akad. Nauk SSSR, 104:56 (1955).

329. A. R. Shul'man and S. A. Fridrikhov, Zh. Tekh. Fiz., 25:344 (1955).

330. B. L. Miller and W. C. Porter, J. Franklin Inst., 261:31 (1955).

331. B. L. Miller, J. Chem. Phys., 23:599 (1955).

332. H. Watanabe, J. Electron Microscopy, 4:24 (1955).

333. A. W. Blackstock, R. H. Ritchie, and R. D. Birkhoff, Phys. Rev., 100:1078 (1955).

334. L. Marton and H. Mendlowitz, Advances in Electronics and Electron Physics, 7:183 (1955).

335. H. Watanabe, Phys. Soc. Japan, 10:321 (1955).

336. K. T. Dolder and O. Klemperer, J. Appl. Phys., 26(12):1461 (1955).

337. L. Pincherle, Proc. Phys. Soc., B68:319 (1955).

338. S. V. Izmailov, Uch. Zap. LGPI, 1:11 (1955).

339. H. Frölich, Proc. Phys. Soc., B68:9 (1955).

340. G. Leibfried, Handbuch der Physik, Vol. 7 (1955), p. 263.

341. E. Everhart, C. Stone, and R. Carbone, Phys. Rev., 99:1287 (1955).

342. U. A. Arifov and A. Kh. Ayukhanov, Izv. Akad. Nauk SSSR, Ser. Fiz., 20:1165 (1956).

343. C. Brunnee, Vakuum-Technik, 5:96 (1956).

344. M. E. Gurtovoi and P. P. Sosenko, Nauk. Pov. KGU, Kiev, 1:11 (1956).

345. H. D. Hagstrum, Phys. Rev., 104:672 (1956).

346. H. Burrekoven, Doctoral Dissertation, Faculty of Mathematical Sciences, Cologne Univ. (1956).

347. H. D. Hagstrum, Phys. Rev., 104:309, 1516 (1956).

348. D. E. Harrison, Phys. Rev., 103:1473 (1956).

349. V. G. Tel'kovskii, Dokl. Akad. Nauk SSSR, 108:444 (1956); Izv. Akad. Nauk SSSR, Ser. Fiz., 20:1179 (1956).

350. N. K. Koshkin, I. A. Ryzhov, I. I. Shkarban, and Kalmikov, Eighth Intern. Conf. on Phenomena in Ionized Gases, Vienna (1967), p. 41.

351. Y. J. Takeishi, Phys. Soc. Japan, 11:676 (1956).

352. G. A. Harrower, Phys. Rev., 102:340 (1956).

353. R. M. Chaudhri, I. K. Mustafa, and L. T. Abdul, Nature, 177:1226 (1956); Phys. Rev., 104:1492 (1956).

354. N. D. Morgulis and V. N. Gavrilyuk, Fiz. Tverd. Tela, 1:149 (1956).

355. G. K. Wehner, Phys. Rev., 102:690 (1956).

356. J. M. Strachan and N. L. Harris, Proc. Phys. Soc., B443:1148 (1956).

357. N. D. Morgulis and V. D. Tishchenko, Izv. Akad. Nauk SSSR, Ser. Fiz., No. 10, p. 1120 (1956).

358. N. D. Morgulis and V. D. Tishchenko, Izv. Akad. Nauk SSSR, 30:54 (1956).

359. V. I. Veksler and M. B. Ben'yaminovich, Zh. Tekh. Fiz., 26:8 (1956).

360. V. I. Veksler, Dokl. Akad. Nauk UzbSSR, 9:11 (1956).

361. D. Pines, Rev. Modern Phys., 28:184 (1956).

362. A. R. Shul'man and E. I. Myakinin, Zh. Eksp. Teor. Fiz., 26:2223 (1956).

363. N. D. Morgulis and D. A. Gorodetskii, Zh. Eksp. Teor. Fiz., 30:667 (1956).

364. D. A. Gorodetskii, Izv. Akad. Nauk SSSR, Ser. Fiz., 20:1023 (1956).

365. A. R. Shul'man, Izv. Akad. Nauk SSSR, Ser. Fiz., 20:9 (1956).

365a. Yu. A. Morozov and A. R. Shul'man, Fiz. Tverd. Tela, 6:943 (1964).

366. N. B. Gornii, Zh. Eksp. Teor. Fiz., 30:160 (1956).

367. G. A. Harrower, Phys. Rev., 104:52 (1956).

368. H. Watanabe, J. Phys. Soc. Japan, 11:112 (1956).

369. W. Klein, Optik, 11:226 (1956).

370. L. B. Leder, G. Mendlovitz, and L. Marton, Phys. Rev., 101:1460 (1956).

371. E. J. Sternglass, Nature, 178:1387 (1956).

372. C. W. Jull, Proc. Phys. Soc., B69:1232 (1956).

373. L. B. Leder, Phys. Rev., 103:1721 (1956).

374. B. Canthe, Compt. Rend. Acad. Sci., 242:2634 (1956).

375. G. Haberstroh, Z. Phys., 145:20 (1956).

376. N. B. Gornii, Zh. Eksp. Teor. Fiz., 31:132 (1956).

377. H. Fridmann, Z. Naturforsch., 11A:373 (1956).

378. F. G. Bol'shov and V. K. Seleznev, Zh. Eksp. Teor. Fiz., 24:1657 (1956).

379. W. A. Depp, Phys. Rev., 104:672 (1956).

380. O. B. Firsov, Zh. Eksp. Teor. Fiz., 33:696 (1957).

381. U. A. Arifov, A. Kh. Ayukhanov, S. V. Starodubtsev, and B. M. Nosenko, Izv. Akad. Nauk UzbSSR, Ser. Fiz.-Mat. Nauk, 3:5 (1957).

382. E. S. Mironov and L. M. Nemenov, Zh. Eksp. Teor. Fiz., 32:269 (1957).

383. C. Brunnee, Z. Phys., 147:161 (1957).

384. M. B. Ben'yaminovich and V. I. Veksler, Trudy SAGU, Tashkent, 91:57 (1957).

385. R. Goutte and C. Guilland, J. Phys. Radium, 18:202 (1957).

386. I. M. Mitropan and V. S. Gumenyuk, Zh. Eksp. Teor. Fiz., 32:214 (1957).

387. K. N. Kuznetsov, Trudy NII MRTP SSSR, 2:36 (1957).

388. H. D. Hagstrum, J. Appl. Phys., 28:323 (1957).

389. O. Roos, Z. Phys., 147:210 (1957).

390. E. J. Sternglass, Phys. Rev., 108:1 (1957).

391. W. Kohl, Technology of Materials for Electrical Vacuum Apparatus [Russian translation], Gosenergoizdat, Moscow (1957).

392. O. Roos, Z. Phys., 147:184 (1957).

393. U. A. Arifov, A. Kh. Ayukhanov, and S. V. Starodubtsev, Zh. Eksp. Teor. Fiz., 33:845 (1957).

394. M. M. Bredov and I. M. Okuneva, Dokl. Akad. Nauk SSSR, 113:795 (1957).

395. A. M. Furman, Zh. Eksp. Teor. Fiz., 32:1591 (1957).

396. L. I. Varnerin and J. H. Carmichael, J. Appl. Phys., 28:913 (1957).

397. A. I. Kondrashev and N. N. Petrov, Eighth Intern. Conf. on Phenomena in Ionizing Gases, Vienna (1967), p. 40.

398. V. V. Belyakova and M. A. Mittsev, Zh. Tekh. Fiz., 27:803 (1957).

399. G. K. Wehner, Phys. Rev., 108:1, 35 (1957).

400. E. B. Henschke, J. Appl. Phys., 28:411 (1957); 28:4 (1957).

401. E. B. Henschke, Phys. Rev., 106:4 (1957); 101:1286 (1956).

402. R. E. Stebbings, Proc. Roy. Soc., A241:270 (1957).

403. R. S. Silsbee, J. Appl. Phys., 28:11 (1957).

404. H. Fridmann, Fortschr. Phys., 5:51 (1957).

405. E. J. Sternglass and J. E. Holliday, J. Appl. Phys., 28:1189 (1957).

406. V. I. Milyutin and A. M. Kabanov, Uspekhi Fiz. Nauk, 61:673 (1957).

407. N. B. Gornii and A. Yu. Raisakas, Zh. Eksp. Teor. Fiz., 33:572 (1957).

408. R. H. Rinchie, Phys. Rev., 106:884 (1957).

409. H. Kanter, Ann. Phys., 20:144 (1957).

410. L. N. Dobretsov and T. L. Matskevich, Zh. Tekh. Fiz., 27:734 (1957).

411. L. Sosnovskii, Izv. Akad. Nauk SSSR, 21:70 (1957).

412. L. Bess, Phys. Rev., 105:1469 (1957).

413. E. Fuls, P. Jones, F. Ziemba, and E. Everhart, Phys. Rev., 107:704 (1957).

414. D. A. Gorodetskii, Zh. Eksp. Teor. Fiz., 34:7 (1958).

415. D. A. Gorodetskii, Radiotekh. i Elektronika, 3:345 (1958).

416. I. M. Bronshtein, Izv. Akad. Nauk SSSR, Ser. Fiz., 22:441 (1958).

417. I. M. Bronshtein and V. V. Roshin, Zh. Tekh. Fiz., 28:2200, 2476 (1958).

418. U. A. Arifov, A. Kh. Ayukhanov, S. V. Starodubtsev, and Kh. Khadzhimukhamedov, Izv. Akad. Nauk UzbSSR, Ser. Fiz.-Mat. Nauk, 5:15 (1958).

419. U. A. Arifov and R. Rakhimov, Izv. Akad. Nauk UzbSSR, Ser. Fiz.-Mat. Nauk, 5:5 (1958).

420. U. A. Arifov and R. Rakhimov, Izv. Akad. Nauk UzbSSR, Ser. Fiz.-Mat. Nauk, 6:49 (1958).

421. U. A. Arifov and R. Rakhimov, Dokl. Akad. Nauk UzbSSR, 12:15 (1958).

422. U. A. Arifov, A. Kh. Ayukhanov, and S. V. Starodubtsev, Izv. Akad. Nauk UzbSSR, Ser. Fiz.-Mat. Nauk, 2:107 (1958).

423. A. Riddoch and I. H. Leck, Proc. Phys. Soc., 72:467 (1958).

424. P. M. Watters, Phys. Rev., 111:1053 (1958).

425. G. Slodzian, Compt. Rend. Acad. Sci., 246:3631 (1958).

426. R. Goutte and C. Guilland, J. Phys. Radium, 19:911 (1958).

427. F. Pradal and R. Simon, Compt. Rend. Acad. Sci., 247:438 (1958).

428. S. V. Izmailov, Zh. Tekh. Fiz., 28:2209 (1958).

429. Ya. A. Tashkhanova, U. A. Arifov, and A. Kh. Ayukhanov, Eighth Intern. Conf. on Phenomena in Ionized Gases, Vienna (1967), p. 39.

430. Yu. N. Talanin, Author's Abstract of Dissertation [in Russian], Leningrad (1958).

431. H. Massey and E. Burhop, Electrons and Ionic Impact Phenomena, Oxford University Press (1952).

432. H. W. Berry, J. Appl. Phys., 29:1219 (1958).

433. Y. Takeishi, J. Phys. Soc. Japan, 13:766 (1958).

434. S. V. Izmailov, Uch. Zap. LGPI, 166:309 (1958).

435. O. B. Firsov, Zh. Eksp. Teor. Fiz., 34:447 (1958).

436. L. D. Landau and E. M. Lifshits, Mechanics [in Russian], Fizmatgiz, Moscow (1958).

437. J. H. Carmichael and E. A. Trendelenberg, J. Appl. Phys., 29:1570 (1958).

438. I. A. Abroyan, Eighth Intern. Conf. on Phenomena in Ionized Gases, Vienna (1967), p. 38.

439. M. Koedam, Physica, 24:8 (1958).

440. R. C. Honig, J. Appl. Phys., 29:549 (1958).

441. V. E. Yurasova, Zh. Tekh. Fiz., 28:1966 (1958).

442. E. Langsberg, Phys. Rev., 111:1120 (1958).

443. G. K. Wehner, Phys. Rev., 112:1120 (1958).
444. D. T. Goldman and A. Simon, Phys. Rev., 111:91 (1958).
445. C. D. O'Brian, A. Lindner, and W. J. Moore, J. Chem. Phys., 29:3 (1958).
446. R. H. Fowler and H. E. Farnsworth, Phys. Rev., 111:103 (1958).
447. F. Pradal, R. Saporte, and C. R. Aradse, Compt. Rend. Acad. Sci., 246:2880 (1958).
448. H. Brunning, Physics and Application of Secondary-Electron Emission [Russian transla-
      tion], Sovetskoe Radio, Moscow (1958).
449. N. B. Gornii, Zh. Eksp. Teor. Fiz., 35:281 (1958).
450. N. B. Gornii, Izv. Akad. Nauk SSSR, Ser. Fiz., 22:475 (1958).
451. A. Ya. Vyatskin, Zh. Eksp. Teor. Fiz., 28:2217 (1958).
452. A. Ya. Vyatskin, Zh. Eksp. Teor. Fiz., 28:2455 (1958).
453. C. J. Powell, J. L. Robins, and J. B. Swan, Phys. Rev., 110:657 (1958).
454. A. Russek and M. Thomas, Phys. Rev., 109:2015 (1958).
455. D. E. Moe and O. H. Petsh, Phys. Rev., 110:1358 (1958).
456. T. P. Matsui, Visnik Kiev Univ., Ser. Fiz.-Ta Khim., No. 1, pp. 115-116 (1958).
457. U. A. Arifov, A. Kh. Ayukhanov, S. V. Starodubtsev, and Kh. Khadzhimukhamedov, Dokl.
      Akad. Nauk SSSR, 124:60 (1959).
458. U. A. Arifov and Kh. Khadzhimukhamedov, Izv. Akad. Nauk UzSSR, Ser. Fiz.-Mat. Nauk,
      2:47 (1959).
459. R. C. Bradley, J. Appl. Phys., 30:1 (1959).
460. P. Cousinie, N. Colombie, R. Fert, and R. Simon, Compt. Rend. Acad. Sci., 219:387
      (1959).
461. A. I. Akishin and S. S. Vasil'ev, Fiz. Tverd. Tela, 1:833 (1959).
462. J. B. Hasted, J. Appl. Phys., 30:22 (1959).
463. V. I. Veksler, Izv. Akad. Nauk UzSSR, Ser. Fiz.-Mat. Nauk, 4:34 (1959).
464. S. V. Izmailov, Fiz. Tverd Tela, 1:1546 (1959).
465. S. V. Izmailov, Fiz. Tverd. Tela, 1:1557 (1959).
466. O. B. Firsov, Zh. Eksp. Teor. Fiz., 36:1517 (1959).
467. R. C. Abbott and H. W. Berry, J. Appl. Phys., 30:871 (1959).
468. V. N. Lepeshinskaya and V. N. Belogurov, Fiz. Tverd. Tela, 1:1806 (1959).
469. A. I. Krokhina and G. V. Spivak, Izv. Akad. Nauk SSSR, Ser. Fiz., 23:741 (1959).
470. M. V. Gomoyunova, Fiz. Tverd. Tela, 1:329 (1959).
471. G. N. Shuppe, Electron Emission of Metal Crystals [in Russian], Izd. SAGU, Tashkent
      (1959).
472. N. I. Guseva, Fiz. Tverd. Tela, 1:10 (1959).
473. G. K. Wehner, Phys. Rev., 114:5 (1959).
474. V. E. Yurasova, G. V. Spivak, and F. F. Kushnir, Izv. Akad. Nauk SSSR, Ser. Fiz.,
      23:744 (1959).
475. V. E. Yurasova, N. V. Pleshivtsev, and I. V. Orfanov, Zh. Eksp. Teor. Fiz., 37:966
      (1959).
476. M. W. Thompson, Phil. Mag., 4:139 (1959).
477. N. Loegreid and G. R. Wehner, J. Appl. Phys., 30:3 (1959).
478. G. Leibfried, J. Appl. Phys., 30:1388 (1959).
479. Characteristic Energy Losses of Electrons in Solids [Russian translation], IL, Moscow
      (1959).
480. I. M. Bronshtein and R. B. Segal', Fiz. Tverd. Tela, 1:1489 (1959).
481. I. M. Bronshtein and R. B. Segal', Fiz. Tverd. Tela, 1:1500 (1959).
482. I. M. Bronshtein and A. M. Shuchinskii, Zap. Leningr. Gorn. Inst., 27:98 (1959).
483. H. Frank, Z. Naturforsch., 14a:247 (1959).
484. A. N. Kabanov and V. I. Milyutin, Radiotekhnika i Elektronika, 4:109 (1959).
485. N. B. Gornii, Zh. Eksp. Teor. Fiz., 37:340 (1959).

486. I. M. Bronshtein and R. B. Segal', Fiz. Tverd. Tela, 1:1246 (1959).

487. V. G. Bol'shov and V. V. Zarubin, Fiz. Tverd. Tela, 1:462 (1959).

488. G. A. Ganichev and T. T. Utkin, 1:648 (1959).

489. I. A. Abroyan, Fiz. Tverd. Tela, 1(12):1854 (1959).

490. C. J. Powell and J. B. Swan, Phys. Rev., 115:869 (1959).

491. A. von Engel, Ionized Gases. 2nd ed., Oxford University Press, New York (1965).

492. N. V. Fedorenko, Uspekhi Fiz. Nauk, 68:481 (1959).

493. U. A. Arifov and R. Rakhimov, Izv. Akad. Nauk SSSR, Ser. Fiz., 24:657 (1960).

494. U. A. Arifov, A. Kh. Ayukhanov, and D. D. Gruich, Izv. Akad. Nauk SSSR, Ser. Fiz.,
      24:710 (1960).

495. U. A. Arifov, Kh. Kh. Khadzhimukhamedov, Izv. Akad. Nauk SSSR, Ser. Fiz., 24:705
      (1960).

496. U. A. Arifov and D. A. Tashkhanova, Izv. Akad. Nauk SSSR, Ser. Fiz., 24:664 (1960).

497. U. A. Arifov and D. A. Tashkhanova, Izv. Akad. Nauk UzbSSR, Ser. Fiz.-Mat. Nauk,
      2:61 (1960).

498. V. I. Veksler, Zh. Eksp. Teor. Fiz., 38:324 (1960).

499. F. Catoni and R. Gilson, Eighth Intern. Conf. on Phenomena in Ionized Gases, Vienna
      (1967), p. 37.

500. N. N. Petrov, Fiz. Tverd. Tela, 2:940 (1960).

501. V. N. Vapnik, L. G. Gurvich, and N. V. Zinov'ev, Izv. Akad. Nauk SSSR, Ser. Fiz.,
      24:685 (1960).

502. Ya. M. Fogel', R. P. Slabospitskii, and A. B. Rastrepin, Zh. Tekh. Fiz., 30:63 (1960).

503. Ya. M. Fogel, R. P. Slabospitskii, and I. M. Karnaukhov, Zh. Tekh. Fiz., 30:824 (1960).

504. N. N. Petrov, Fiz. Tverd. Tela, 2:949 (1960).

505. N. N. Petrov, Fiz. Tverd. Tela, 2:1300 (1960).

506. N. N. Petrov, Izv. Akad. Nauk SSSR, Ser. Fiz., 24:6 (1960).

507. N. N. Petrov, NTI, Byull. LPI, Leningrad, 3:63-72 (1960).

508. H. Stanton, J. Appl. Phys., 31:678 (1960).

509. R. Castainy, B. Jourffrey, and C. Slodzian, Compt. Rend. Acad. Sci., 251:1010 (1960).

510. S. Thomas and E. B. Pattinson, Eighth Intern. Conf. on Phenomena in Ionized Gases,
      Vienna (1967), p. 36.

511. E. A. Trendelenberg and J. H. Carmichael, Advances in Vacuum Science and Tech-
      nology, 2:657 (1960).

512. A. Kh. Ayukhanov and G. Iskhakov, Izv. Akad. Nauk SSSR, Ser. Fiz., 24:715 (1960).

513. L. N. Dobretsov, Atomic Physics [in Russian], Moscow (1960).

514. H. D. Hagstrum, Phys. Rev., 119:940 (1960).

515. W. J. Moore, Am. Scientist, 48:109 (1960).

516. P. J. Rol, J. M. Fluit, and J. Kistemaker, Physica, 26:1000 (1960); 26:1009 (1960).

517. G. K. Wehner, J. Appl. Phys., 31:8, 1392 (1960).

518. M. Koedam and A. Hoogensoorn, Physica, 26:351 (1960).

519. R. V. Stuart and G. K. Wehner, Phys. Rev. Letters, 4:403 (1960).

520. R. S. Pease, Rend. S. J. F.—Gorso B158 (1960).

521. D. E. Harrison, J. Chem. Phys., 32:1336 (1960).

522. N. V. Pleshivtsev, Zh. Eksp. Teor. Fiz., 37:10, 878 (1960).

523. F. Gronlund and M. J. Moore, J. Chem. Phys., 32:150 (1960).

524. G. Leibfried, J. Appl. Phys., 31:117 (1960).

525. J. B. Gibson, A. N. Goland, M. Milgram, and C. H. Vineyard, Phys. Rev., 120:1299
      (1960).

526. R. van Yan and R. S. Nelson, Eighth Intern. Conf. on Phenomena in Ionized Gases,
      Vienna (1967), p. 35.

527. I. L. Robins and I. B. Swan, Proc. Phys. Soc., 76:857 (1960).

528. L. P. Levine and H. W. Berry, Phys. Rev., 118:1 (1960).

529. V. I. Gaponov, Electronics, Vol. 1 [in Russian], Fizmatgiz, Moscow, p. 34.

530. N. Bohr, Passage of Atomic Particles Through Matter [Russian translation], IL, Moscow (1960).

531. G. M. Batanov, Fiz. Tverd. Tela, 2:2048 (1960).

532. H. D. Hagstrum, J. Phys. Chem. Solids, 14:33 (1960).

533. N. N. Petrov, Nauchno-Tekhn. Byull. Leningr. Politekh. Inst., 3:46 (1960).

534. J. Leroy and K. Prelec, Rapp. CEA, No. 1445, p. 17 (1960).

535. G. M. Batanov, Nauchno.-Tekhn. Byull. Leningr. Politekh. Inst., No. 9, pp. 92-100, 101-107 (1960).

536. U. A. Arifov, Interaction of Atomic Particles with a Metal Surface [in Russian], Tashkent (1961).

537. M. Gurtovii and B. Krulikivs"kii, Visnik. Kiev Univ., Ser. Astronom., Fiz. Ta Khim., 3(2): 33-35 (1960, 1961).

538. U. A. Arifov, Transactions of the Tashkent Conference on the Peaceful Uses of Atomic Energy, Vol. 1 [in Russian], Tashkent (1961), p. 210.

539. É. S. Parilis and L. M. Kishinevskii, Fiz. Tverd. Tela, 3:1219 (1961).

540. L. G. Gurvich, Dokl. Akad. Nauk UzbSSR, 12:45 (1961).

541. U. A. Arifov, Kh. Khadzhimukhamedov, E. S. Parilis, and L. M. Kishinevskii, Izv. Akad. Nauk UzbSSR, Ser. Fiz.-Mat. Nauk, 6:50 (1961).

542. U. A. Arifov, Kh. Kh. Khadzhimukhamedov, and A. P. Sokolov, Izv. Akad. Nauk UzbSSR, Ser. Fiz.-Mat. Nauk, 5:56 (1961).

543. U. A. Arifov, Kh. Kh. Khadzhimukhamedov, and A. P. Sokolov, Izv. Akad. Nauk UzbSSR, Ser. Fiz.-Mat. Nauk, 5:62 (1961).

544. U. A. Arifov and Kh. Kh. Khadzhimukhamedov, Izv. Akad. Nauk UzbSSR, Ser. Fiz.-Mat. Nauk, 5:65 (1961).

545. U. A. Arifov and Kh. Kh. Khadzhimukhamedov, Izv. Akad. Nauk UzbSSR, Ser. Fiz.-Mat. Nauk, 6:47 (1961).

545a. U. A. Arifov and A. Kh. Ayukhanov, Izv. Akad. Nauk UzbSSR, Ser. Fiz.-Mat. Nauk, 6:40 (1961).

546. U. A. Arifov, Kh. Kh. Khadzhimukhamedov, and A. P. Sokolov, Izv. Akad. Nauk UzbSSR, Ser. Fiz.-Mat. Nauk, 6:40 (1961).

547. U. A. Arifov, Kh. Kh. Khadzhimukhamedov, A. P. Sokolov, and M. Karimova, Izv. Akad. Nauk UzbSSR, Ser. Fiz.-Mat. Nauk, 6:44 (1961).

548. É. S. Parilis, Izv. Akad. Nauk UzbSSR, Ser. Fiz.-Mat. Nauk, 6:78 (1961).

549. U. A. Arifov, D. D. Gruich, and Kh. Mirrakhimova, Dokl. Akad. Nauk UzbSSR, 10:14 (1961).

550. U. A. Arifov, N. N. Flyants, and A. Kh. Ayukhanov, Dokl. Akad. Nauk UzbSSR, 10:10 (1961).

551. U. A. Arifov, R. R. Rakhimov, and Kh. Dzhurakulov, Dokl. Akad. Nauk UzbSSR, 9:7 (1961).

552. N. N. Petrov, Fiz. Tverd. Tela, 3:1 (1961).

553. U. A. Arifov, A. Kh. Ayukhanov, and A. A. Aliev, Izv. Akad. Nauk UzbSSR, Ser. Fiz.-Mat. Nauk, 6:57 (1961).

554. J. Vineyard, Uspekhi Fiz. Nauk, 4:435 (1961).

555. M. Koedam, Philips Res. Reports, 16:101, 166 (1961).

556. W. Laegreid and G. K. Wehner, J. Appl. Phys., 32:365 (1961).

557. V. A. Molchanov and V. G. Tel'kovskii, Dokl. Akad. Nauk SSSR, 136:4 (1961).

558. S. P. Wolsky and E. J. Zdanuk, Phys. Rev., 121:8 (1961).

559. O. C. Yonts, C. E. Norman, and D. E. Harrison, J. Appl. Phys., 31:447 (1961).

560. P. L. Hines and R. Wallor, J. Appl. Phys., 32:2 (1961).

561. C. H. Weijsenfeld, A. Hoogendoorn, and M. Koedam, Physica, 27:763 (1961).

562. G. K. Wehner and D. Rosenberg, J. Appl. Phys., 32:5 (1961).

563. I. M. Bronshtein and B. S. Fraiman, Fiz. Tverd. Tela, 3:1638, 3220 (1961).

564. D. McKlown, Rev. Sci. Instrum., 32:133 (1961).

565. I. I. Dushkov, V. A. Molchanov, V. G. Tel'kovskii, and V. M. Chicherov, Zh. Tekh. Fiz., 31:8 (1961).

566. R. C. Bradley, A. Arking, and D. S. Beers, J. Chem. Phys., 33:764 (1961).

567. B. V. Panin, Zh. Eksp. Teor. Fiz., 41:3 (1961).

568. K. Kopizky and H. E. Stier, Z. Naturforsch., 16a:1257 (1961).

569. R. S. Nelson and M. W. Thompson, Proc. Roy. Soc., A259:458 (1961).

570. B. M. Nosenko and N. A. Strukov, Izv. Akad. Nauk SSSR, Ser. Fiz., 25:314 (1961).

571. J. A. Davies and G. A. Sims, Canad. J. Chem., 39:601 (1961).

572. A. N. Kabanov, Yu. N. Kushnir, and G. V. Fetisov, Izv. Akad. Nauk SSSR, Ser. Fiz., 25:748 (1961).

573. J. L. Robins, Proc. Phys. Soc., 78:1177 (1961).

574. C. J. Powell, J. Nucl. Energy, 2:57 (1961).

575. L. Marton, J. A. Simpson, H. A. Fawbe, and N. Swanson, Phys. Rev., 127:182 (1961).

576. J. L. Robins and P. E. Best, Proc. Phys. Soc., 75:110 (1961).

577. I. A. Abroyan, Fiz. Tverd. Tela, 3:588 (1961).

578. G. M. Batanov et al., Fiz. Tverd. Tela, 2:2048 (1960); 3:558 (1961).

579. H. D. Hagstrum, Phys. Rev., 122:83; 123:758 (1961).

580. D. Onderdelinden and I. Kistemaker, Eighth Intern. Conf. on Phenomena in Ionized Gases, Vienna (1967), p. 33.

581. S. V. Izmailov, Fiz. Tverd. Tela, 3:2804 (1961); 4:2554 (1962).

582. A. Abrahamson, R. Hatcher, and J. H. Vineyard, Phys. Rev., 121:159 (1961).

583. N. N. Petrov, Nauchno-Tekhn. Inform. Byull. LPI, 1:65 (1961).

584. T. Sugiura and T. Hayakawa, Bull. Chem. Soc. Japan, 34:58 (1961).

585. H. Fetz, A. Diener, and E. Meyer, Z. Angew. Phys., 13:292 (1961).

586. T. Sugiura, Bull. Chem. Soc. Japan, 34:1475 (1961).

587. S. Kronenberg, K. Nilson, and M. Basso, Phys. Rev., 124:1309 (1961).

588. J. S. Colligon, Vacuum, 11:272 (1961).

589. P. H. Diderichs and G. Leibfried, Z. Phys., 170:320 (1962).

590. V. E. Yurasova, V. M. Buchanov, and N. N. Rimskii, Eighth Intern. Conf. on Phenomena in Ionized Gases, Vienna (1967), p. 32.

591. R. V. Stuart and G. K. Wehner, J. Appl. Phys., 33:2345 (1962).

592. H. G. Scott, J. Appl. Phys., 33:6 (1962).

593. S. P. Wolsky, Phys. Rev., 108:5 (1962).

594. M. I. Guseva, Radiotekhnika i Elektronika, No. 9 (1962).

595. D. B. Medved and H. Poppa, J. Appl. Phys., 33:1759 (1962).

596. J. Kramar and R. Ceskosl, Casop. Fys., A12(5-6):539 (1962).

597. R. C. Bradley and E. Ruedel, J. Appl. Phys., 33:6, 80 (1962).

598. B. V. Panin, Zh. Eksp. Teor. Fiz., 42:313 (1962).

599. M. W. Thompson, R. S. Nelson, and B. W. Farmery, Properties of Reactor Materials and Effects of Radiation Damage, Bull. Hersvorth, London (1962), p. 98.

600. C. H. Weijsenfeld, Phys. Rev. Letters, 2(6):295 (1962).

601. V. E. Yurasova and V. M. Bukhanov, Kristallografiya, 7:2 (1962).

602. V. E. Endzhevets, V. A. Molchanov, V. G. Tel'kovskii, and M. A. Faruk, Zh. Tekh. Fiz., 32:8 (1962).

603. V. A. Molchanov and V. G. Tel'kovskii, Izv. Akad. Nauk SSSR, Ser. Fiz., 26:11 (1962).

604. M. Kaminsky, Phys. Rev., 126:1267 (1962).

605. M. W. Thompson, Phil. Mag., 7:2015 (1962).

606. G. M. Batanov, Fiz. Tverd. Tela, 4:1778 (1962); Radioelektronika, 5:852 (1963).
607. U. A. Arifov, R. R. Rakhimov, M. Abdullaeva, and S. Gaipov, Izv. Akad. Nauk SSSR, Ser. Fiz., 26:11 (1962).
608. J. L. Robins and P. E. Best, Proc. Phys. Soc., 79:110, 119 (1962).
609. I. M. Bronshtein and V. S. Kovalenko, Fiz. Tverd. Tela, 4:2047 (1962).
610. J. Geiger, Z. Naturforsch., 17a:696 (1962).
611. H. Beorch, J. Geiger, H. Hellwig, and H. L. Michel, Phys. Rev., 169:1252 (1962).
612. V. E. Krohn, J. Appl. Phys., 33:3523 (1962).
613. Ya. P. Zingermann and V. A. Ishchuk, Fiz. Tverd. Tela, 1:8, 2212 (1962).
614. V. I. Veksler, Zh. Eksp. Teor. Fiz., 42:325 (1962).
615. V. I. Veksler, Fiz. Tverd. Tela, 4:419 (1962).
616. P. E. Best, Proc. Phys. Soc., 79:139 (1962).
617. L. M. Kishinevskii, Izv. Akad. Nauk SSSR, 26:1410 (1962).
618. L. M. Kishinevskii and É. S. Parilis, Izv. Akad. Nauk SSSR, 26:1409 (1962).
619. S. Ghosh and S. Khare, Phys. Rev., 125:1254 (1962).
620. É. S. Parilis, Dokl. Akad. Nauk UzbSSR, 12:8 (1962).
621. É. S. Parilis, Radiotekhnika i Elektronika, 7:1979 (1962).
622. E. S. Mashkova and V. A. Molchanov, Dokl. Akad. Nauk SSSR, 146:585 (1962).
623. L. G. Filippenko, Zh. Tekh. Fiz., 32:356 (1962).
624. U. A. Arifov, A. A. Aliev, and A. Kh. Ayukhanov, Izv. Akad. Nauk SSSR, Ser. Fiz., 26:11, 1440 (1962).
625. L. Large and W. Whitelock, Proc. Phys. Soc., 79:148 (1962).
626. J. Takeishi, J. Phys. Soc. Japan, 17(2):326 (1962).
627. J. Sugiura, Bull. Chem. Soc. Japan, 35(2):326 (1962).
628. H. J. Klein, Dtsch. Nationalbibliogr., BN7:553 (1962).
629. N. N. Petrov, Izv. Akad. Nauk SSSR, Ser. Fiz., 26:11, 1327 (1962).
630. K. H. Krebs, Ann. Phys., 10:213 (1962).
631. P. J. Hauman, Vech. Centke, Nat. Vech. Scient., No. 61, p. 357 (1962).
632. L. P. Moroz and A. Kh. Ayukhanov, Izv. Akad. Nauk SSSR, 26:1322 (1962).
633. V. A. Shustrov, V. I. Poltoratskii, A. Kh. Ayukhanov, Izv. Akad. Nauk UzbSSR, Ser. Fiz.-Mat. Nauk, No. 2, p. 65 (1962).
634. A. I. Akishin, Ion Bombardment in Vacuum, Gostekhizdat, Moscow (1963).
635. A. I. Akishin and V. S. Zaiulin, Pribory i Tekh. Eksperim., 1:152 (1963).
636. R. V. Stuart and G. K. Wehner, in: Progress in Vacuum Science and Technology (ed.: A. S. D. Barrett), Pergamon, New York (1959).
637. V. I. Veksler, Radiotekhnika i Elektronika, 8(1):145 (1963); 5:2737 (1963).
638. M. B. Ben'yaminovich and V. I. Veksler, Izv. Akad. Nauk UzbSSR, Ser. Fiz.-Mat. Nauk, 3:29 (1963).
639. U. A. Arifov, Interaction of Atomic Particles with the Surface of a Metal, Washington (1963).
640. M. W. Thompson, Phys. Rev. Letters, 6:24 (1963).
641. M. Kaminsky, Bull. Am. Phys. Soc., 8:338 (1963); 8:428 (1963).
642. N. G. Nakhodkin and P. V. Mel'nik, Fiz. Tverd. Tela, 6:2441 (1963).
643. L. P. Moroz and A. Kh. Ayukhanov, Radiotekhnika i Elektronika, 8:322 (1963).
644. U. A. Arifov and A. Kh. Kasymov, Radiotekhnika i Elektronika, 8:138 (1963).
645. Kh. Dzhurakulov, R. R. Rakhimov, and U. A. Arifov, Radiotekhnika i Elektronika, 8:299 (1963).
646. D. A. Tashkhanova, R. R. Rakhimov, and U. A. Arifov, Radiotekhnika i Electronika, 8:2, 294 (1963).
647. U. A. Arifov and D. A. Tashkhanova, Dokl. Akad. Nauk UzbSSR, 9:13 (1963).
648. U. A. Arifov, D. A. Tashkhanova, and R. R. Rakhimov, Dokl. Akad. Nauk UzbSSR, 10:5 (1963).

649. E. S. Mashkova, V. A. Molchanov, and D. D. Odintsov, Fiz. Tverd. Tela, 5:3426 (1963).

650. C. D. Magnuson and C. E. Carlston, Phys. Rev., 129:2403 (1963).

651. C. D. Magnuson and C. E. Carlston, Phys. Rev., 129:2409 (1963).

652. U. A. Arifov, R. R. Rakhimov, and Kh. Dzhurakulov, Radiotekhnika i Elektronika, 8:299 (1963).

653. U. A. Arifov and A. Kh. Kasymov, Radiotekhnika i Elektronika, 8:138 (1963).

654. U. A. Arifov and A. Kh. Kasymov, Izv. Akad. Nauk UzbSSR, Ser. Fiz.-Mat. Nauk, 4:93 (1963).

655. U. A. Arifov and A. Kh. Kasymov, Dokl. Akad. Nauk SSSR, 158:82 (1964).

656. U. A. Arifov, R. R. Rakhimov, and Kh. Dzhurakulov, Radiotekhnika i Elektronika, 8:299 (1963).

657. L. P. Moroz and A. Kh. Ayukhanov, Radiotekhnika i Elektronika, 8:322 (1963).

658. V. I. Veksler, Zh. Eksp. Teor. Fiz., 44:14 (1963).

659. V. I. Veksler, Trudy Tashkent Gos. Univ., No. 221, p. 128 (1963).

660. Ya. M. Fogel' et al., Radioelektronika, 8:684 (1963).

661. D. B. Medved, P. Mahadevan, and J. K. Layton, Phys. Rev., 129:2086 (1963).

662. É. S. Parilis and N. Yu. Turaev, Summaries of Contributions to the Eleventh Conference on the Physical Basis of Cathode Electronics [in Russian], Kiev (1963), p. 79.

663. U. A. Arifov and R. R. Rakhimov, Izv. Akad. Nauk SSSR, Ser. Fiz., 34:657 (1960).

664. P. M. Propst, Phys. Rev., 129:7 (1963).

665. U. A. Arifov, A. A. Aliev, and A. Kh. Ayukhanov, Izv. Akad. Nauk UzbSSR, Ser. Fiz.-Mat. Nauk, No. 4, p. 86 (1963).

666. A. B. Laponsky, J. Appl. Phys., 34:1568 (1963).

667. P. Mahadevan and J. K. Layton, Phys. Rev., 129:79 (1963); J. Appl. Phys., 34:2810 (1963).

668. L. N. Large, Proc. Phys. Soc., 81:1101 (1963); 81:175 (1963).

669. A. A. Dorozhkin and N. N. Petrov, Zh. Tekh. Fiz., 33:350 (1963).

670. É. S. Parilis, Candidate's Dissertation [in Russian], Tashkent (1963).

671. F. M. Propst and E. Lüscher, Rev. Sci. Instrum., 34:574 (1963); Phys. Rev., 132:1037 (1963).

672. H. D. Hagstrum, Ann. N. Y. Acad. Sci., 101:674 (1963).

673. S. N. Ghosh and S. P. Khare, Phys. Rev., 129:1638 (1963).

674. E. S. Mashkova, V. A. Molchanov, and D. D. Odintsov, Dokl. Akad. Nauk SSSR, 151:1074 (1963).

675. V. Walther and M. Hintenberger, Z. Naturforsch., 18a:843 (1963).

676. M. S. Gorodezsky and A. M. Bergoldt, J. Phys. Radium, 24:374 (1963).

677. K. F. Halt and G. F. Weston, J. Sci. Instrum., 40:573 (1963).

678. N. N. Flyants, U. A. Arifov, and A. Kh. Ayukhanov, Radiotekhnika i Elektronika, 8:34 (1963).

679. D. B. Medved, J. Appl. Phys., 34:3142 (1963).

680. F. Propst, Doctoral Dissertation, University of Illinois (1963).

681. H. D. Hagstrum and J. Takeischi, Transactions of Intern. Conf. on Phenomena in Ionized Gases, Vol. 2, Paris (1963).

682. U. A. Arifov, A. Kh. Ayukhanov, and S. V. Starodubtsev, Radiotekhnika i Elektronika, 8:669 (1963).

683. V. A. Shustrov, R. M. Khasanov, and A. Kh. Ayukhanov, Izv. Akad. Nauk UzbSSR, Ser. Fiz.-Mat. Nauk, No. 1, p. 31 (1964).

684. V. A. Shustrov and A. Kh. Ayukhanov, Dokl. Akad. Nauk UzbSSR, 10:22 (1964).

685. N. V. Pleshivstev, Pribory i Tekh. Eksperim., No. 5 (1964).

686. H. Oechsner, Eighth Intern. Conf. on Phenomena in Ionized Gases, Vienna (1967), p. 31.

687. H. A. James and S. C. James, J. Appl. Phys., 34:3 (1964).

688. M. B. Ben'yaminovich and V. I. Veksler, Zh. Tekh. Fiz., 34:361 (1964).

689. V. A. Kvilidze, E. S. Mashkova, and V. A. Molchanov, Izv. Akad. Nauk SSSR, Ser. Fiz., 28:9 (1964).

690. V. N. Lepeshinskaya and E. M. Zarutskii, Izv. Akad. Nauk SSSR, Ser. Fiz., 28:1390 (1964).

691. U. A. Arifov, A. A. Aliev, and A. Kh. Ayukhanov, Izv. Akad. Nauk UzbSSR, Ser. Fiz.-Mat. Nauk, 4:20 (1964).

692. V. E. Yurasova, V. A. Brzhezinskii, and G. M. Ivanov, Zh. Eksp. Teor. Fiz., 47:473 (1964).

693. H. Zscheile, Phys. Status Solidi, 6:K87 (1964).

694. I. M. Bronshtein and S. S. Denisov, Fiz. Tverd. Tela, 6:2644 (1964).

694a. I. M. Bronshtein and Ya. M. Shuchinskii, Radiotekhnika i Elektronika, 9:904 (1964).

695. U. A. Arifov and A. Kh. Kasymov, Dokl. Akad. Nauk UzbSSR, 8:15.(1964).

696. D. D. Gruich, N. Rakhimbaeva, G. Ikramov, and T. Arifov, Izv. Akad. Nauk UzbSSR, Ser. Fiz.-Mat., 1:53 (1964).

697. J. M. Fluit, J. Kistemaker, and E. Snoek, Physica, 30:870 (1964).

698. S. Datz and C. Snoek, Phys. Rev., 134:A347 (1964).

699. É. S. Parilis and N. Yu. Turaev, Dokl. Akad. Nauk UzbSSR, Ser. Fiz., 12:16 (1964).

700. W. L. Gay and D. E. Harrison, Phys. Rev., 135:A1780 (1964).

701. V. E. Yurasova, Izv. Akad. Nauk SSSR, Ser. Fiz., 28:1470 (1964).

702. U. A. Arifov, D. D. Gruich, and L. Yu. Chastukhina, Izv. Akad. Nauk SSSR, Ser. Fiz., 28:1402 (1964).

703. K. D. Schuy, J. Franzen, and H. Hintenberger, Z. Naturforsch., 19a:153 (1964).

704. U. A. Arifov, A. Kh. Ayukhanov, V. A. Shustrov, R. M. Khasanov, and V. I. Poltoratskii, Dokl. Akad. Nauk SSSR, 155:306 (1964).

705. J. H. Fluit and C. Snoek, Physica, 30:345 (1964).

706. U. A. Arifov, N. N. Flyants, and R. R. Rakhimov, Dokl. Akad. Nauk UzbSSR, 10:15 (1964).

707. U. A. Arifov, N. N. Flyants, and R. R. Rakhimov, Dokl. Akad. Nauk UzbSSR, 10:18 (1964).

708. V. A. Molchanov and V. Soshka, Dokl. Akad. Nauk SSSR, 155:170 (1964).

709. V. I. Veksler, Fiz. Tverd. Tela, 6:2228 (1964).

710. U. A. Arifov and D. D. Gruich, Dokl. Akad. Nauk UzbSSR, No. 7, p. 18 (1964).

711. U. A. Arifov and D. D. Gruich, Dokl. Akad. Nauk UzbSSR, No. 11, p. 20 (1964).

712. G. Baynon, Mass Spectrography and Its Use in Chemistry [Russian translation], Mir, Moscow (1964).

713. L. Lehmann and M. Robinson, Phys. Rev., 134:A37 (1964).

714. Yu. V. Martynenko, Fiz. Tverd. Tela, 6:2003 (1964); 6:3529 (1964).

715. A. I. Erofeev, Inzh. Zh., 4:36 (1964).

716. R. R. Rakhimov and Kh. Dzhurakulov, Radiotekhnika i Elektronika, 9:333 (1964).

717. U.A. Arifov, A. A. Aliev, and A. Kh. Ayukhanov, Dokl. Akad. Nauk UzbSSR, No. 9, p. 22 (1964).

718. E. S. Chambers, Phys. Rev., 133:A1202 (1964).

719. F. M. Devienne, J. C. Roustan, and J. Souquet, Compt. Rend. Acad. Sci., 258:140 (1964); 260:4701 (1965).

720. J. H. Krebs, Ann. Phys. (DDR), 13(3/4):97–100 (1964).

721. E. S. Mashkova and V. A. Molchanov, Zh. Tekh. Fiz., 34:2081 (1964).

722. N. Daly and R. Powell, Proc. Phys. Soc., 84:595 (1964).

723. E. S. Mashkova and V. A. Molchanov, Fiz. Tverd. Tela, 6(11):3486 (1964).

724. E. S. Mashkova and V. A. Molchanov, Fiz. Tverd. Tela, 6:3704 (1964).

725. L. P. Moroz and A. Kh. Ayukhanov, Izv. Akad. Nauk SSSR, Ser. Fiz., 28:1395 (1964).

726. U. A. Arifov, S. Gaipov, M. Ikramova, and R. R. Rakhimov, Dokl. Akad. Nauk UzbSSR, No. 11, p. 19 (1965).

727. V. A. Shustrov, Candidate's Dissertation [in Russian], Physicotechnical Institute, AN UzbSSR (1965).

728.   V. A. Shustrov, R. M. Khasanov, and A. Kh. Ayukhanov, Radiotekhnika i Elektronika, 10:3 (1965).

729.   E. J. Zdanuk and S. P. Wolsky, J. Appl. Phys., 36:5 (1965).

730.   M. Kaminsky, Atomic and Ionic Impact Phenomena on Metal Surfaces, Academic Press, New York (1964).

731.   R. S. Nelson, Phil. Mag., 10:110, 291 (1965).

732.   B. Navinsek, J. Appl. Phys., 36:5 (1965).

733.   L. P. Moroz, Candidate's Dissertation [in Russian], Tashkent (1965).

734.   D. A. Tashkhanova, Candidate's Dissertation [in Russian], Tashkent (1965).

735.   U. A. Arifov and A. Kh. Kasymov, Izv. Akad. Nauk UzbSSR, Ser. Fiz.-Mat. Nauk, 2:23 (1965).

736.   E. S. Mashkova, V. A. Molchanov, and V. Soshka, Dokl. Akad. Nauk SSSR, 161:813 (1965).

737.   E. S. Mashkova and V. A. Molchanov, Fiz. Tverd. Tela, 7:1872 (1965).

738.   U. A. Arifov, Kh. Kh. Khadzhimukhamedov, and A. I. Yunusov, Dokl. Akad. Nauk UzbSSR, 1:20 (1965).

739.   U. Fano and W. Lichten, Phys. Rev. Letters, 14:627 (1965).

740.   D. D. Gruich, Candidate's Dissertation [in Russian], Tashkent (1965).

741.   H. D. Hagstrum and Y. Takeishi, Phys. Rev., 137:A304 (1965); 137:A641 (1965).

742.   H. D. Hagstrum and P. Pretzer, Proc. Seventh Intern. Conf. on Phenomena in Ionized Gases, Belgrade (1965).

743.   H. D. Hagstrum, Y. Takeishi, and P. Pretzer, Phys. Rev., 139:A526 (1965).

744.   C. Carlson, G. Magnuson, P. Mahadevan, and D. Harrison, Phys. Rev., 139:A729 (1965).

745.   D. Harrison, C. Carlson, and G. Magnuson, Phys. Rev., 139:A737 (1965).

746.   B. L. Schram, J. Berboom, W. Kleine, and J. Kistemaker, Proc. Seventh Intern. Conf. on Phenomena in Ionized Gases, Belgrade (1965).

747.   É. S. Parilis and N. Yu. Turaev, Dokl. Akad. Nauk SSSR, 161:84 (1965).

748.   L. L. Myasnikov, L. D. Raigorodskii, and B. A. Finagin, Zh. Tekh. Fiz., 35:542 (1965).

749.   E. S. Mashkova, V. A. Molchanov, É. S. Parilis, and N. Yu. Turaev, Phys. Letters, 18:7 (1965).

750.   E. (É. S.) Parilis, Proc. Seventh Intern. Conf. on Phenomena in Ionized Gases, Belgrade (1965).

750a. Proc. Seventh Intern. Conf. on Phenomena in Ionized Gases, Belgrade (1965).

751.   I. A. Abroyan, V. N. Lavrov, and A. I. Titov, Fiz. Tverd. Tela, 7:3159 (1965).

752.   V. I. Veksler, Zh. Eksp. Teor. Fiz., 49:90 (1965).

753.   P. Dahl and J. Magyar, Phys. Rev., 140:A1420 (1965).

754.   A. Aliev, Candidate's Dissertation [in Russian], Tashkent (1965).

755.   U. A. Arifov, R. R. Rakhimov, S. Gaipov, and N. Karetnikova, Dokl. Akad. Nauk UzbSSR, No. 12, p. 16 (1965).

756.   A. A. Dorozhkin and N. N. Petrov, Fiz. Tverd. Tela, 7:118 (1965).

757.   E. S. Mashkova and V. A. Molchanov, Zh. Tekh. Fiz., 35:575 (1965).

758.   M. Yar and R. H. Chaudri, Nature, 205:997 (1965).

759.   A. I. Kondrashov and N. P. Petrov, Fiz. Tverd. Tela, 7:1559 (1965).

760.   F. E. Jamerson, C. B. Leffert, and D. B. Rees, J. Appl. Phys., 36:355 (1965).

761.   R. J. Euring, Phys. Rev., 139:A1840 (1965).

762.   H. Zscheile, Phys. Stat. Solidi, 11:159 (1965).

763.   H. J. Klein, Z. Phys., 188:78 (1965).

764.   B. Fagot, N. Colombie, and C. Fert, Compt. Rend. Acad. Sci., 261:2855 (1965).

765.   P. Mahadevan, G. D. Magnuson, J. K. Layton, and C. Carlson, Phys. Rev., 140:A1407 (1965).

766.   Z. Jurela and B. Perovic, Eighth Intern. Conf. on Phenomena in Ionized Gases, Vienna (1967), p. 30.

767. U. A. Arifov, D. D. Gruich, and Kh. M. Hamidova, Eighth Intern. Conf. on Phenomena in Ionized Gases, Vienna (1967), p. 29.

768. F. M. Devienne, Compt. Rend. Acad. Sci., 260:5739 (1965).

769. U. A. Arifov, Kh. Kh. Khadzhimukhamedov, and A. Kh. Ayukhanov, Izv. Akad. Nauk UzbSSR, Ser. Fiz.-Mat. Nauk, 1:57 (1966).

770. U. A. Arifov, Kh. Kh. Khadzhimukhamedov, A. I. Yunusov, and A. A. Aliev, Izv. Akad. Nauk SSSR, Ser. Fiz., 30(12):1995 (1966).

771. Yu. V. Marynenko, Fiz. Tverd. Tela, 8:637 (1966).

772. A. F. Tulinov, Dokl. Akad. Nauk SSSR, 162:546 (1966).

773. E. S. Mashkova and V. A. Molchanov, Fiz. Tverd. Tela, 8:1517 (1966).

774. E. S. Mashkova and V. A. Molchanov, Dokl. Akad. Nauk SSSR, 172:813 (1967).

775. U. A. Arifov, Kh. Dzhurakulov, and R. R. Rakhimov, Izv. Akad. Nauk UzbSSR, Ser. Fiz.-Mat. Nauk, 6:33 (1966).

776. U. A. Arifov, A. A. Aliev, and A. Kh. Ayukhanov, Izv. Akad. Nauk UzbSSR, Ser. Fiz.-Mat. Nauk, No. 2, p. 47 (1966).

777. U. A. Arifov, A. A. Aliev, and A. Kh. Ayukhanov, Izv. Akad. Nauk UzbSSR, No. 3, p. 42 (1966).

778. A. A. Aliev and U. A. Arifov, Dokl. Akad. Nauk SSSR, 172:65 (1967).

779. E. S. Mashkova, V. A. Molchanov, É. S. Parilis, and N. Yu. Turaev, Dokl. Akad. Nauk SSSR, 166:330 (1966).

780. É. S. Parilis and N. Yu. Turaev, Izv. Akad. Nauk SSSR, Ser. Fiz., 30:1983 (1966).

781. C. H. Weijsenfeld, Yield, Energy, and Angular Distributions of Sputtered Atoms, Oeldrop (1966).

782. U. A. Arifov, S. Gaipov, M. Ikramova, and R. R. Rakhimov, Izv. Akad. Nauk SSSR, Ser. Fiz., 30:896 (1966).

783. É. Mukhamadiev and R. R. Rakhimov, Izv. Akad. Nauk SSSR, Ser. Fiz., 30:892 (1966).

784. A. Kh. Ayukhanov and M. K. Abdullaeva, Izv. Akad. Nauk SSSR, Ser. Fiz., 30(12):2000 (1966).

785. V. M. Kivilis, É. S. Parilis, and N. Yu. Turaev, Dokl. Akad. Nauk SSSR, 173:805 (1967).

786. V. I. Veksler, Doctoral Dissertation [in Russian], Leningrad (1966).

787. L. M. Kishinevskii and É. S. Parilis, Dokl. Akad. Nauk UzbSSR, No. 9, p. 10 (1966).

788. É. S. Parilis, Transactions of the Third International Congress of Crystallographers [in Russian], Moscow (July, 1966).

789. O. B. Firsov, Dokl. Akad. Nauk SSSR, 169:1311 (1966).

790. I. N. Evdokimov, V. A. Molchanov, D. D. Odintsov, and V. M. Chicherov, Fiz. Tverd. Tela, 8:2939 (1966).

791. E. S. Mashkova, V. A. Molchanov, and V. Soshka, Phys. Stat. Solidi, 19:425 (1967).

792. É. S. Parilis, N. Yu. Turaev, and V. M. Kivilis, Proc. Eighth Intern. Conf. on Phenomena in Ionized Gases, Vienna (1967).

793. U. A. Arifov, Kh. Kh. Khadzhimukhamedov, and A. I. Yunusov, Dokl. Akad. Nauk UzbSSR, 3:15 (1967).

794. U. A. Arifov, Kh. Kh. Khadzhimukhamedov, and A. I. Yunusov, Izv. Akad. Nauk UzbSSR, Ser. Fiz.-Mat. Nauk, 4:62 (1967).

795. U. A. Arifov and A. A. Aliev, Zh. Eksp. Teor. Fiz., 54:354 (1968).

796. U. A. Arifov and A. A. Aliev, Dokl. Akad. Nauk UzbSSR, 10:37 (1967).

797. U. A. Arifov and A. A. Aliev, Dokl. Akad. Nauk SSSR, 183:60 (1968).

798. U. A. Arifov and A. A. Aliev, Dokl. Akad. Nauk SSSR, 180:312 (1968).

799. U. A. Arifov and A. A. Aliev, Dokl. Akad. Nauk SSSR, 184:4 (1968).

800. U. A. Arifov, R. R. Rakhimov, Kh. Dzhurakulov, and N. A. Karetnikova, Izv. Akad. Nauk UzbSSR, Ser. Fiz.-Mat. Nauk, 3:40 (1967).

801. U. A. Arifov, Kh. Dzhurakulov, and R. R. Rakhimov, Dokl. Akad. Nauk UzbSSR, 6:16 (1967).

374   REFERENCES

802.   U. A. Arifov, R. R. Rakhimov, and Kh. Dzhurakulov, Fiz. Tverd. Tela, 10:1166 (1968).
803.   U. A. Arifov, R. R. Rakhimov, and Kh. Dzhurakulov, Dokl. Akad. Nauk UzbSSR, 3:16
       (1968).
804.   U. A. Arifov, N. N. Flyants, and R. R. Rakhimov, Dokl. Akad. Nauk UzbSSR, 8:15 (1967).
805.   É. Turmashev and A. Kh. Ayukhanov, Izv. Akad. Nauk UzbSSR, Ser. Fiz.-Mat. Nauk,
       1:29 (1968).
806.   U. A. Arifov, D. D. Gruich, and Kh. Kh. Khamidova, Dokl. Akad. Nauk SSSR, 180:175
       (1968).
807.   Kh. Dzhurakulov, R. R. Rakhimov, and N. Printseva, Izv. Akad. Nauk UzbSSR, Ser. Fiz.-
       Mat. Nauk, 5:77 (1968).
808.   U. A. Arifov, N. N. Flyants, and Kh. Kh. Khadzhimukhamedov, Izv. Akad. Nauk UzbSSR,
       Ser. Fiz.-Mat. Nauk, 5:69 (1968).
809.   É. S. Parilis, N. Yu. Turaev, and V. M. Kivilis, Summaries of the Thirteenth All-Union
       Conference on Emission Electronics [in Russian], Moscow (1968).
810.   É. S. Parilis, A Survey of Phenomena in Ionized Gases, Invited Papers of the Eighth
       International Conference, Vienna (1968), p. 309.
811.   É. S. Parilis, Preprint of the United Institute of Nuclear Research [in Russian], Dubna,
       R7-3355 (1967).
812.   L. M. Kishinevskii and É. S. Parilis, Zh. Tekh. Fiz., 38:760 (1968).
813.   L. M. Kishinevskii and É. S. Parilis, Fifth International Conference on the Physics of
       Electron and Atomic Collisions [in Russian], Leningrad (1967), p. 100.
814.   L. M. Kishinevskii and É. S. Parilis, Zh. Eksp. Teor. Fiz., 55:1932 (1968).